PLANT IDENTIFICATION TERMINOLOGY
AN ILLUSTRATED GLOSSARY

PLANT IDENTIFICATION TERMINOLOGY

AN ILLUSTRATED GLOSSARY

JAMES G. HARRIS
MELINDA WOOLF HARRIS

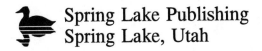

Spring Lake Publishing
Spring Lake, Utah

Publisher's Cataloging in Publication Data

Harris, James G., 1954-
Harris, Melinda Woolf, 1953-
Plant identification terminology: an illustrated glossary / James G. Harris & Melinda Woolf Harris
x, 198 p.: illus.; 26 cm.
ISBN 0-9640221-5-X (alk. paper)
1. Botany—Dictionaries. 2. Botany—Terminology. I. Title.
QK9.H37 1994 580.3 H242 94-65026

PREFACE

This is a book of phytography, the descriptive terminology of plants. It was written to fill the need we perceived for a comprehensive, illustrated guide to the terminology of systematic botany.

A formidable task facing the student of plant taxonomy is gaining a working knowledge of the vast terminology required to use a typical plant identification key. Most keys are provided with a glossary, but, because of the technical nature of many botanical terms, these glossaries are often of limited value. Either the user may find a verbal description inadequate to convey the essence of a complex botanical term, or the definition may include two or three additional terms that also must be defined to make sense of the original definition. The experience of keying out even one plant specimen may become so tedious and frustrating that the student quickly loses all enthusiasm for plant identification.

Often all that is required to quickly convey the meaning of a botanical term is a simple illustration. In this volume we have attempted to assemble a glossary that includes most of the terms a student would encounter in a typical plant identification key, and we have provided line drawings for all terms that we feel might be made clearer by an illustration.

For simplicity of use, we have attempted, whenever possible, to place illustrations on the same page (or on the facing page) as the term definition. Naturally, this has meant much duplication of some illustrations.

For example, the terms "receptacle," "calyx," "corolla," "androecium" and "gynoecium" all could be illustrated with the same drawing placed in a single location in the text. Instead, we have placed copies of the drawing throughout the text near each appropriate definition. We believe that this approach will make the book more convenient and useful.

The book is divided into two parts. Part One is the essential core of the book. It is an alphabetical glossary of more than twenty-four hundred terms commonly used in plant description and identification. Part Two is designed primarily for the student. Here we have grouped related terms together to facilitate study and comparison.

As we have examined botanical keys and descriptions over the years, we have noticed that the same term is often interpreted quite differently by various authors.

For example, some botanists use the word "scorpioid" to describe a one-sided cymose inflorescence coiled like the tail of a scorpion, while others use it to describe an inflorescence with a zigzag rachis. Historically, "scorpioid" appears in botanical literature in both connotations for at least the last one hundred fifty years.

While current usage seems to favor the zigzag interpretation of "scorpioid," it is difficult to argue with those who choose to apply the term to coiled inflorescences. The original Greek word means, literally, scorpion-like, and a coiled inflorescence is certainly more evocative of a scorpion's tail than is a zigzag inflorescence.

Plant systematics, perhaps more than any other branch of botany, includes a strong historical element. It is not uncommon, for example, to hear a taxonomist trace his or her professional roots back to Asa Gray or another prominent botanist of the last century. Perhaps the divergent current usages of some botanical terms have their origins in separate professional clans and lineages, having been passed from teacher to student over several generations.

We have not attempted to resolve these conflicts in interpretation; that is not the purpose of this volume. Instead, we have tried to include diverse usages of terms so that the student of botany can make use of the book no matter the interpretation employed by the author of the identification key or description in use.

We hope that this volume will prove useful to professional botanists and students of botany alike.

James G. Harris
Melinda Woolf Harris

CONTENTS

PART ONE

ILLUSTRATED GLOSSARY

ILLUSTRATED GLOSSARY

A- (prefix). Meaning without or lacking.

Abaxial. The side away from the axis. Figure 1. (compare **adaxial**)

Aberrant. Different from the usual; atypical; abnormal.

Abortion. The failure of a structure or organ to develop.

Abortive. Not fully or properly developed; rudimentary. Figure 2.

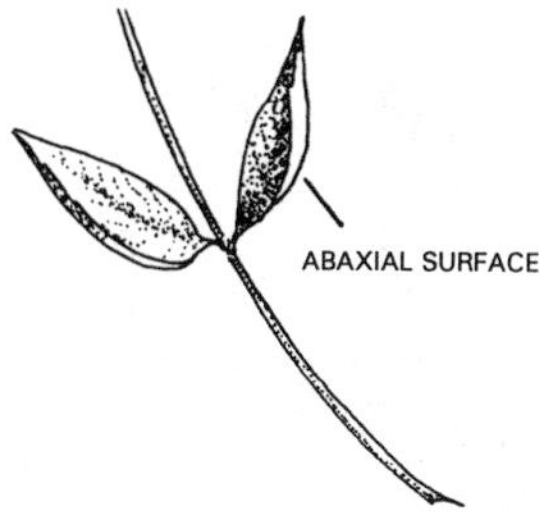

Figure 1

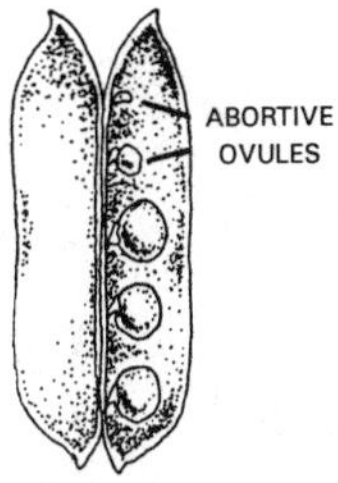

Figure 2

Abrupt. Terminating suddenly. Figure 3. (compare **truncate**)

Abruptly pinnate. Pinnate without an odd leaflet at the tip. Figure 4. (same as **even pinnate**)

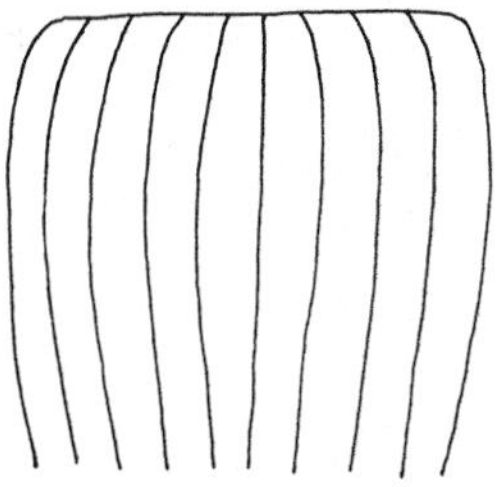

Figure 3

Figure 4

Abscission. The falling away of a leaf or other organ caused by the breakdown of thin-walled cells at the base of the structure.

Acarpous. Without carpels; lacking a gynoecium. Figure 5.

Acaulescent. Without a stem, or the stem so short that the leaves are apparently all basal, as in the dandelion. Note: the peduncle should not be confused with the stem. Figure 6.

Accessory bud. An extra bud in a leaf axil. Figure 7.

Accessory fruit. A fleshy fruit developing from a succulent receptacle rather than the pistil. The ripened ovaries are small achenes on the surface of the receptacle, as in the strawberry. Figure 8.

Figure 5

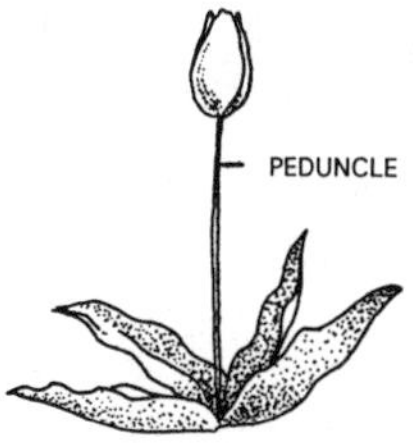

Figure 6

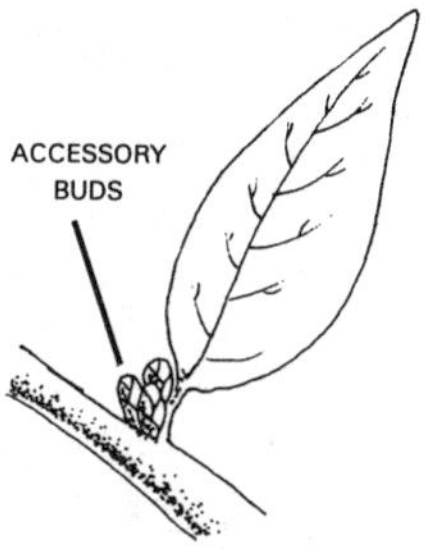

Figure 7

Figure 8

Accrescent. Becoming larger with age, as a calyx which continues to enlarge after anthesis. Figure 9.

Accumbent cotyledons. Cotyledons lying against the radicle along one edge. Figure 10. (compare **incumbent cotyledons**)

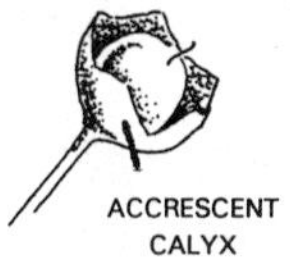

Figure 9

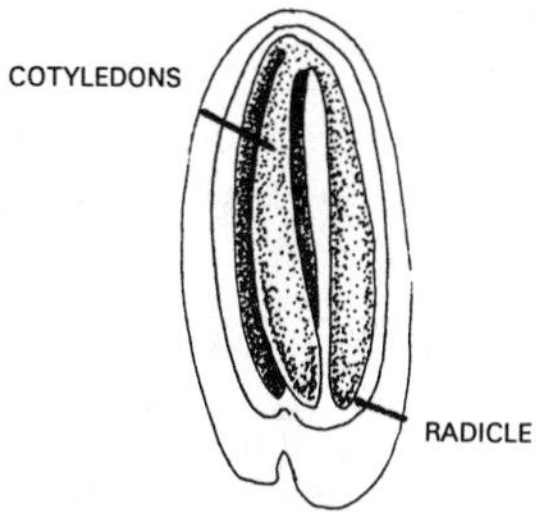

Figure 10

Acerose. Needle-shaped, as the leaves of pine or spruce. Figure 11.

Achene. A small, dry, indehiscent fruit with a single locule and a single seed (ovule), and with the seed attached to the ovary wall at a single point, as in the sunflower. Figure 12.

Achlamydeous. Lacking a perianth.

Achlorophyllous. Without chlorophyll, as in plants or plant structures which are not green.

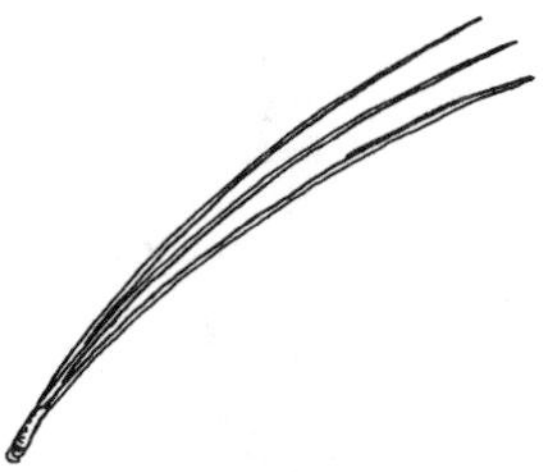

Figure 11

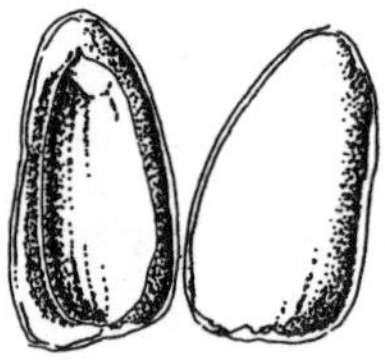

Figure 12

Acicular. Needle-shaped. (see **acerose**)

Aciculate. Marked as with pinpricks or needle scratches; needle-shaped.

Acidophilous. Acid loving, as in a plant which prefers acidic soils.

Acorn. The hard, dry, indehiscent fruit of oaks, with a single, large seed and a cup-like base. Figure 13.

Acotyledonous. Without cotyledons.

Acrid. A sharp, bitter, or biting taste.

Figure 13

Acrogen. A non-flowering plant which grows only at the apex, as in a fern.

Acrogenic. See **acrogenous**.

Acrogenous. Growing from the apex.

Acropetal. Near the tip rather than the base; produced sequentially from the base to the apex, as the flowers in an indeterminate inflorescence. Figure 14.

Acroscopic. Facing the tip or apex. Figure 15.

Figure 14

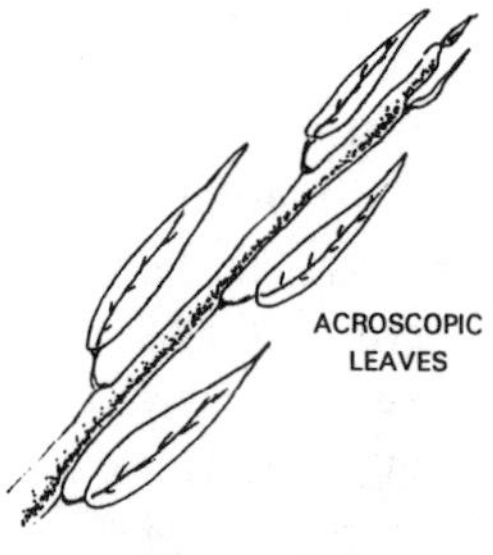

Figure 15

Actinomorphic. Radially symmetrical, so that a line drawn through the middle of the structure along any plane will produce a mirror image on either side. Figure 16. (compare **zygomorphic**, and see **regular**)

Actinomorphous. See **actinomorphic**.

Aculeate. Prickly; covered with prickles. Figure 17.

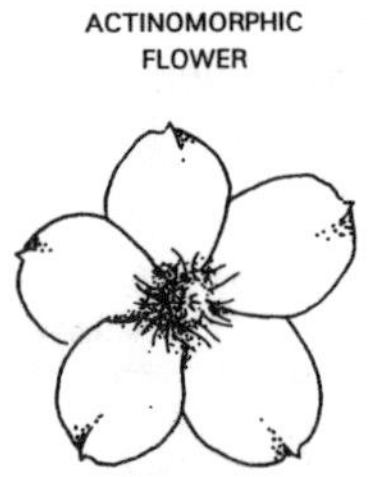

Figure 16

Figure 17

Aculeolate. Minutely prickly; covered with tiny prickles. Figure 18.

Acumen. Apex.

Acuminate. Gradually tapering to a sharp point and forming concave sides along the tip. Figure 19.

Figure 18

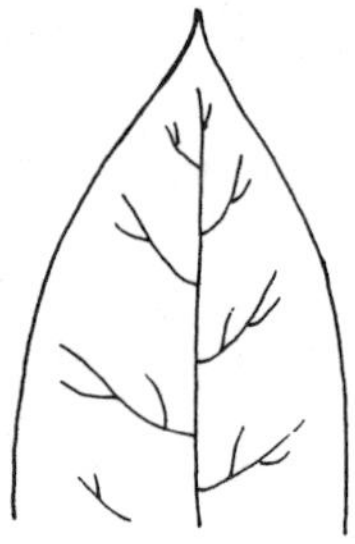

Figure 19

Acute. Tapering to a pointed apex with more or less straight sides. Figure 20.

Ad- (prefix). Meaning to or toward.

Adaxial. The side toward the axis. Figure 21. (compare **abaxial**)

Figure 20

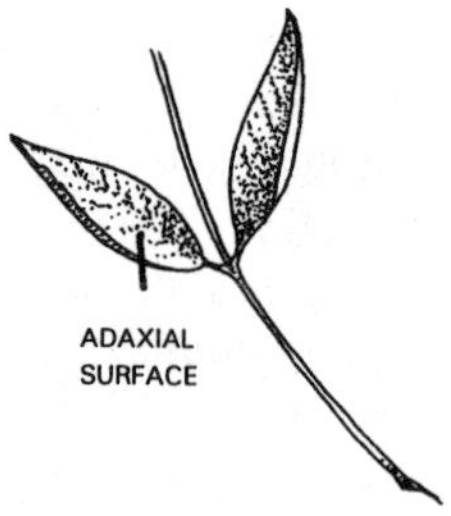

Figure 21

Adherent. Sticking together of unlike parts, as the anthers to the style. The attachment is not as firm or solid as **adnate**.

Adnate. Fusion of unlike parts, as the stamens to the corolla. Figure 22. (compare **connate**)

Adpressed. Lying close to another organ, but not fused to it.

Adscendent. See **ascending**.

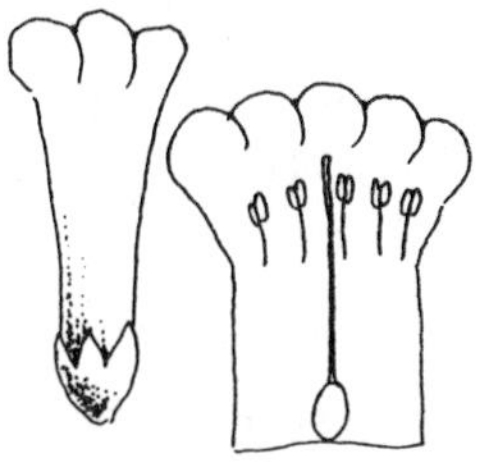

Figure 22

Adsurgent. See **ascending**.

Adventitious. Structures or organs developing in an unusual position, as roots originating on the stem. Figure 23.

Adventive. Not native; introduced and beginning to spread in the new region.

Aequilateral. Equal-sided, as opposed to oblique (in leaves). Figure 24.

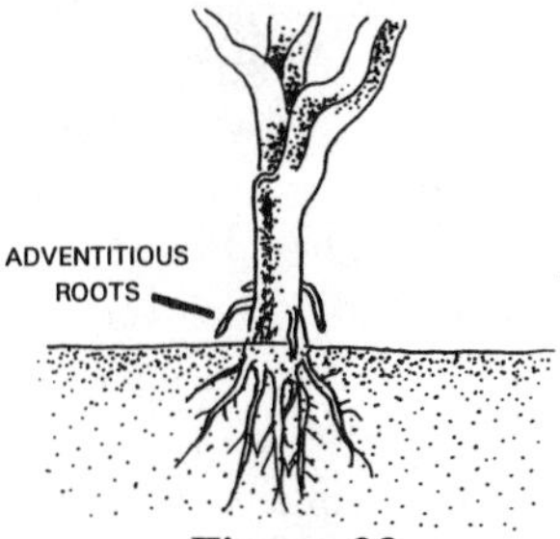

Figure 23

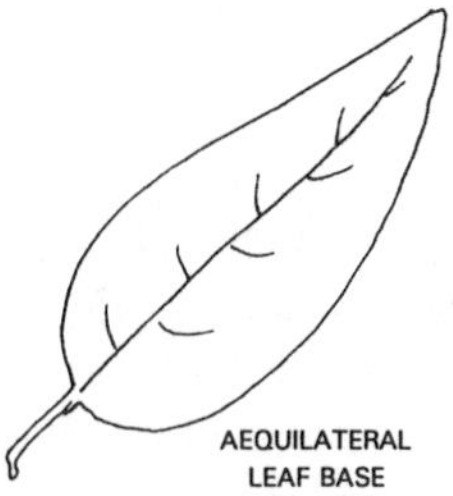

Figure 24

Aerial. Occurring above ground or water.

Aestival. Flowering or appearing in the summer.

Aestivation. The arrangement of floral parts in a bud.

Agamospecies. A species which usually produces seeds asexually, by agamospermy,

Agamospermy. Formation of seed without fertilization.

Agglomerate. Crowded into a dense cluster. Figure 25.

Aggregate. Densely clustered. Figure 25.

Aggregate fruit. Usually applied to a cluster or group of small fleshy fruits originating from a number of separate pistils in a single flower, as in the clustered drupelets of the raspberry. Figure 26.

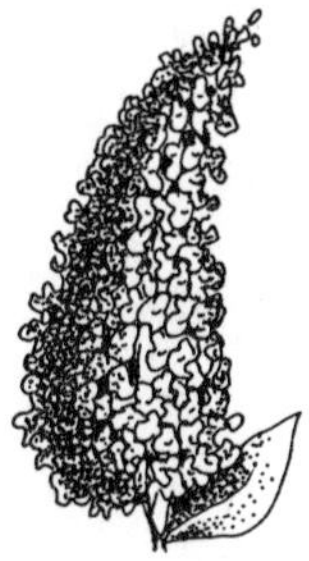

Figure 25

Figure 26

Akene. See **achene**.

Ala (pl. **alae**). A winglike extension or process; one of the two lateral petals of a papilionaceous corolla. Figure 27.

Alate. Winged. Figure 28.

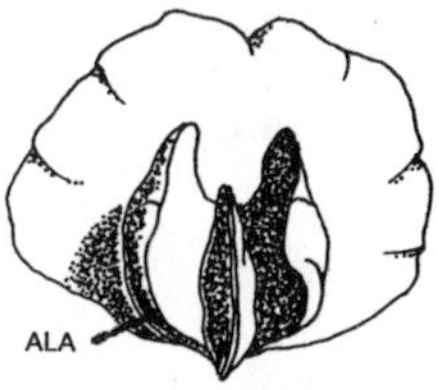

Figure 27

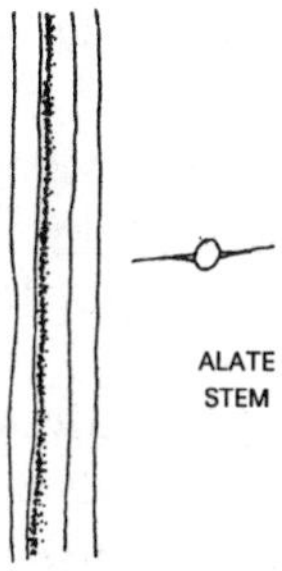

Figure 28

Albumen. The nutritive tissue in a seed. Figure 29.

Alkaline. Material that is basic rather than acidic; having a Ph greater than 7.0.

Alliaceous. Having the smell or taste of garlic.

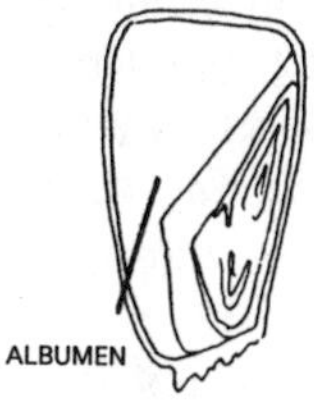

Figure 29

Allogamy. Cross-pollination.

Allopatric. Occupying different geographic regions. (compare **sympatric**)

Alluvial. Of or pertaining to alluvium (i.e. organic or inorganic materials deposited by running water).

Alpine. Of or pertaining to areas above timberline; growing above timberline.

Alternate. Borne singly at each node, as leaves on a stem. Figure 30; borne between rather than over other organs, as stamens between the petals. Figure 31. (compare **opposite**)

Figure 30

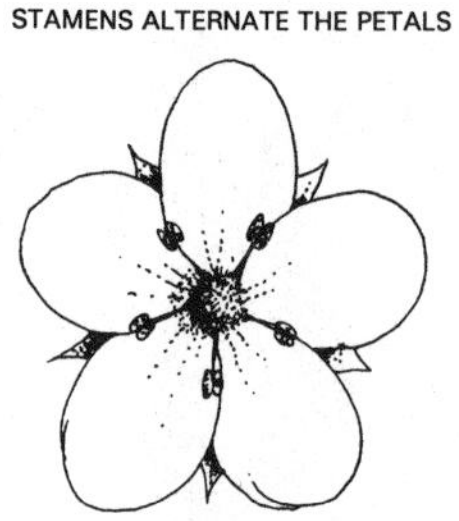

Figure 31

Alveola (pl. **alveolae**, **alveolas**). Pits arranged in a honeycomb-like pattern. Figure 32.

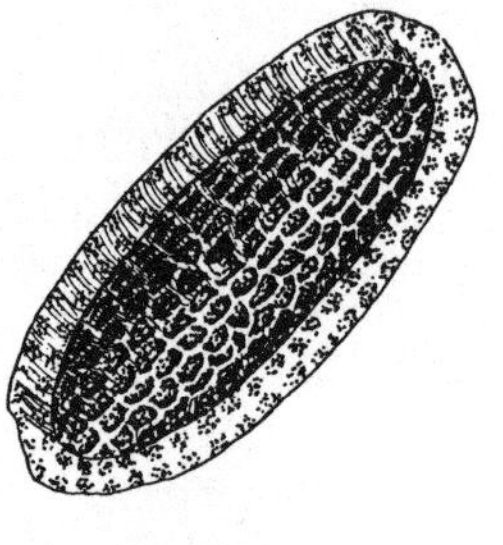

Figure 32

Alveolar. See **alveolate**.

Alveolate. Honeycombed, with pits separated by thin, ridged partitions. Figure 32.

Alveole. See **alveola**.

Alveolus (pl. **alveoli**). See **alveola**.

Ament. See **catkin**.

Amentaceous. Catkin-like, or catkin-bearing.

Amethystine. Amethyst-colored; purplish.

Amorphous. Without any definite form; shapeless; lacking symmetry.

Amphibious. Living both in water and on land.

Amphitropous ovule. An ovule which is half-inverted and straight, with the hilum lateral. Figure 33.

Amplexicaul. Clasping the stem, as the base or stipules of some leaves. Figure 34.

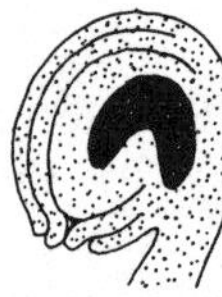

Figure 33

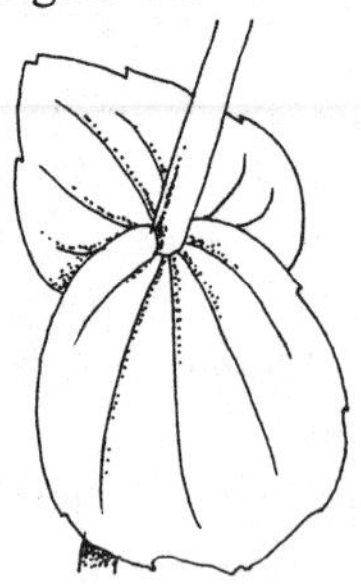

Figure 34

Ampliate. Enlarged or expanded. Figure 35.

Ampullaceous. Swelling out like a bottle or bladder. Figure 36.

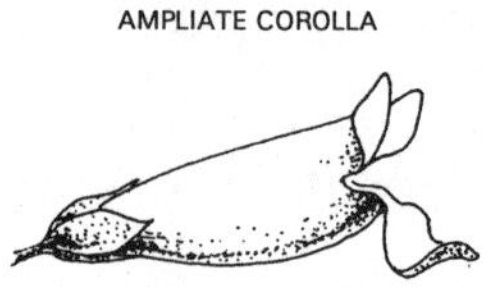

Figure 35

Figure 36

Anandrous. Without stamens; lacking an androecium. Figure 37.

Ananthous. Without flowers.

Anastomosing. Rejoining after branching and forming an intertwining network, as in some leaf veins. Figure 38.

Figure 37

Figure 38

Anatropous ovule. An ovule which is inverted and straight with the micropyle situated next to the funiculus. Figure 39.

Ancipital. Two-edged, as the winged stem of *Sisyrinchium*. Figure 40.

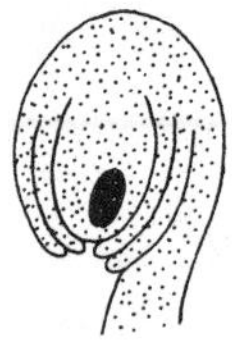

Figure 39

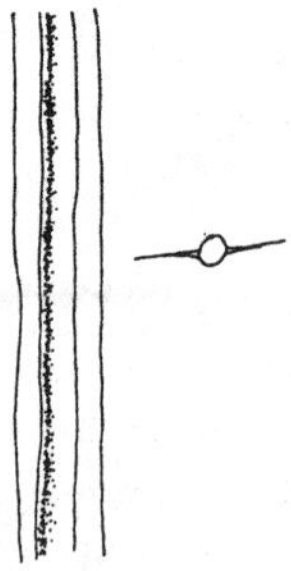

Figure 40

Androclinium. See **clinandrium**.

Andro-dioecious. Having staminate and perfect flowers on separate plants.

Androecium. All of the stamens in a flower, collectively.

Androgynophore. Stalk supporting the androe

cium and gynoecium in some flowers. Figure 41.

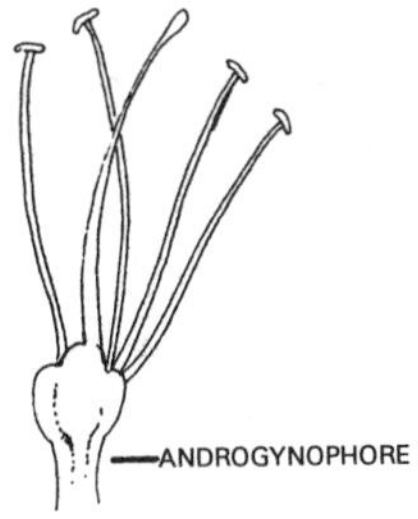

Figure 41

Androgynous. With both staminate and pistilate flowers, the staminate flowers borne above the pistilate, as in the inflorescence of some *Carex* species. (compare **gynaecandrous**)

Andro-monoecious. See **andro-polygamous.**

Androphore. A stalk supporting a group of stamens.

Andro-polygamous. Having staminate and perfect flowers on the same plant.

Androspore. A male spore of *Isoetes.*

Anemophilous. Wind pollinated; producing wind-borne pollen.

Angiosperm. A plant producing flowers and bearing ovules (seeds) in an ovary (fruit).

Angulate. Angled. Figure 42.

Angustiseptate. Of a fruit flattened at right angles to the septum; the septum crosses the narrowest diameter. Figure 43.

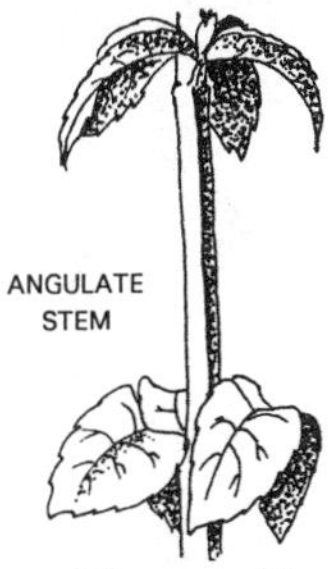

Figure 42

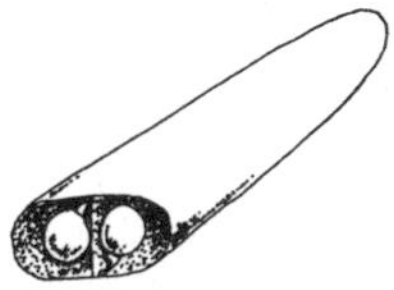

Figure 43

Anisomerous. With a different number of parts (usually less) than the other floral whorls, as in a flower with five sepals and petals, but only two stamens. Figure 44.

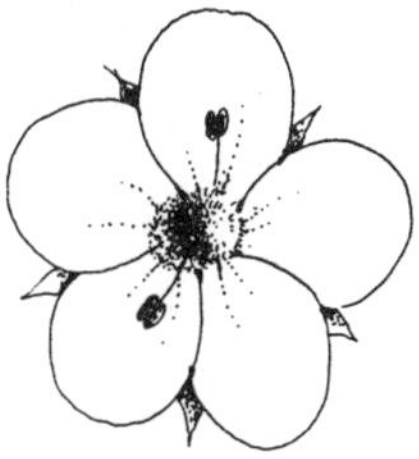

Figure 44

Annotinal. Appearing annually.

Annual. A plant which germinates from seed, flowers, sets seed, and dies in the same year.

Annular. In the form of a ring. Figure 45.

Annulate. In the form of a ring. Figure 45; with rings or ringlike markings.

Annulus. A row of specialized, thick-walled cells along one side of a fern sporangium which aids in the dispersal of spores. Figure 46; a ring-shaped structure.

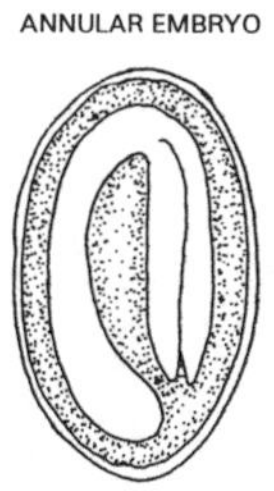

Figure 45

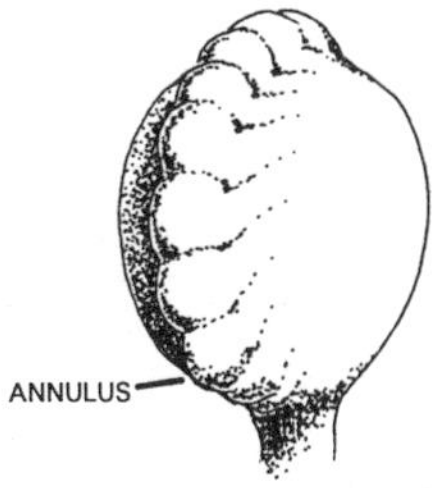

Figure 46

Anomocytic stomate. A stomate lacking differentiated subsidiary cells.

Antarctic. Distributed in those regions of the earth lying between the Antarctic Circle and the South Pole.

Antepetalous. Directly in front of (opposite) the petals. Figure 47.

Anterior. In the front; on the side away from the axis, as the lower lip of a bilabiate corolla. Figure 48. (compare **posterior**)

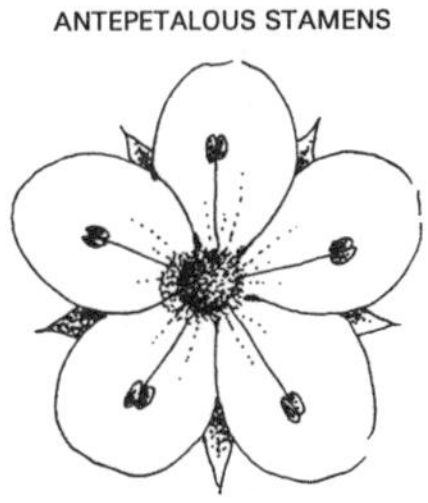

Figure 47

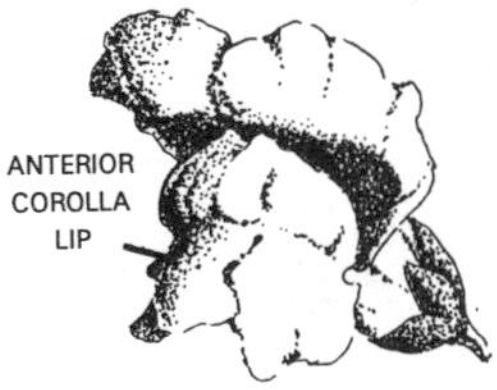

Figure 48

Antesepalous. Directly in front of (opposite) the sepals. Figure 49.

Anther. The expanded, apical, pollen bearing portion of the stamen. Figure 50.

Anther sac. One of the pollen bearing chambers of the anther. Figure 51.

Antherid. See **antheridium.**

Antheridium (pl. **antheridia).** The male reproductive structure in moss and fern gametophytes.

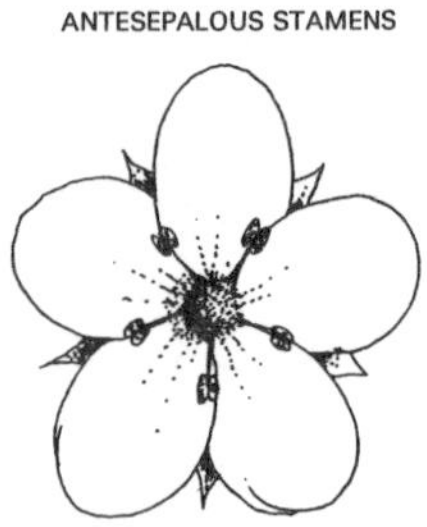

Figure 49

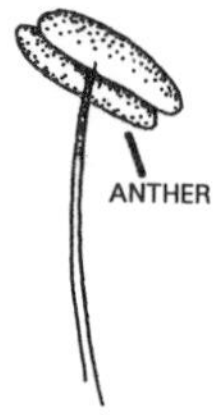

Figure 50

Antheriferous. Bearing anthers.

Antherozoid. Male sexual cells.

Anthesis. The flowering period, when the flower is fully expanded and functioning.

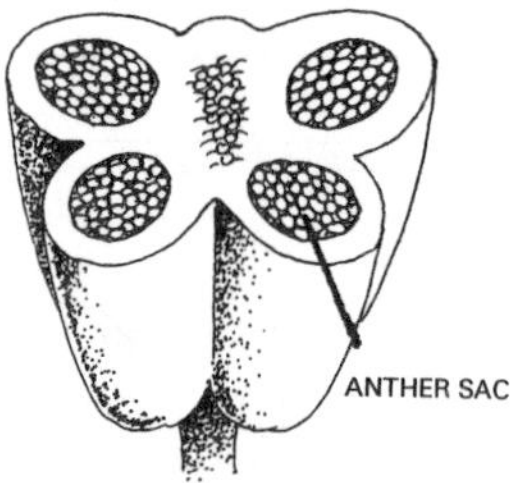

Figure 51

Anthocarp. A fruit with some portion of the flower besides the pericarp persisting, as in a pome with the fleshy perianth tube surrounding the pericarp. Figure 52.

Anthocyanic. Containing anthocyanin pigments.

Anthocyanin. Water-soluble pigments (blue, purple, or red).

Anthophore. An elongated stalk (stipe) bearing the corolla, stamens, and pistil above the receptacle and calyx. Figure 53.

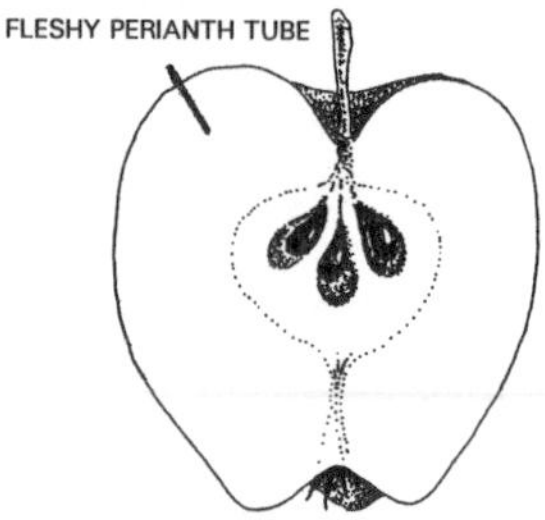

Figure 52

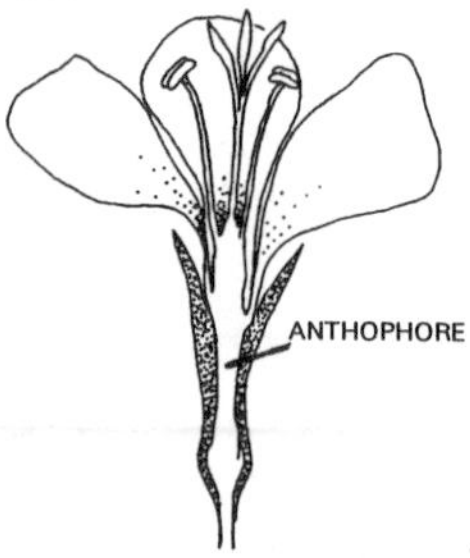

Figure 53

Anthotaxy. The arrangement of flowers on the flowering axis; inflorescence.

Anthoxanthin. Water-soluble pigments (yellow, orange, or red).

Antipetalous. See **Antepetalous.**

Antisepalous. See **Antesepalous.**

Antrorse (adv. **antrorsely**). Directed forward or upward. Figure 54. (compare **retrorse**)

Aperturate. With one or more openings or apertures. In pollen grains, these apertures may be only thin spots rather than actual perforations. Figure 55.

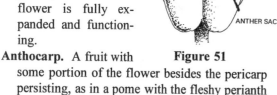

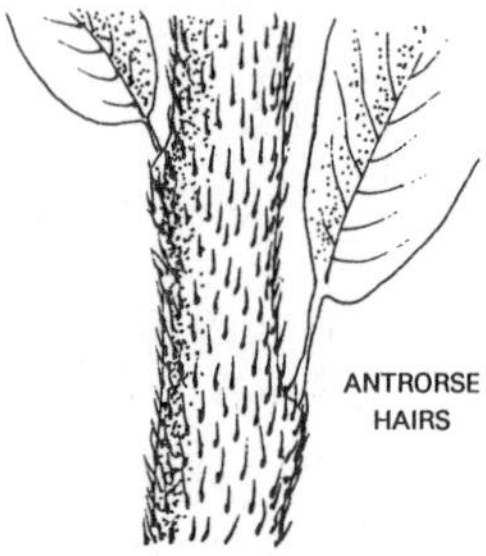

Figure 54

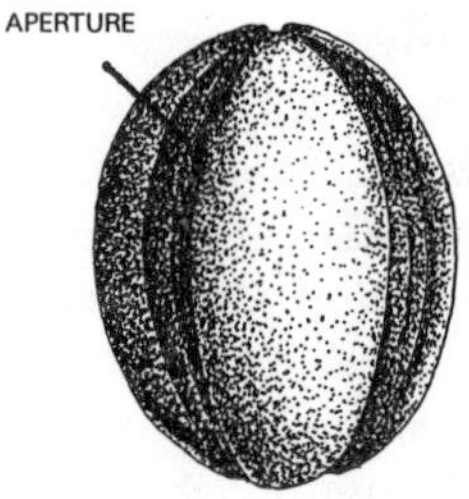

Figure 55

Apetalous. Without petals.

Apex (pl. **apices**). The tip; the point farthest from the point of attachment.

Aphyllopodic. Having the lowermost leaves reduced to small scales. Figure 56. (compare **phyllopodic**)

Aphyllous. Without leaves.

Apical. Located at the apex or tip.

Apiculate. Ending abruptly in a small, slender point. Figure 57.

Apiculation. See **apiculus.**

Apiculus. A small, slender point. Figure 57.

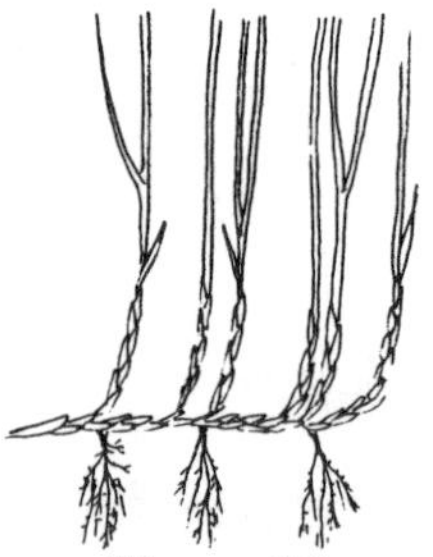

Figure 56

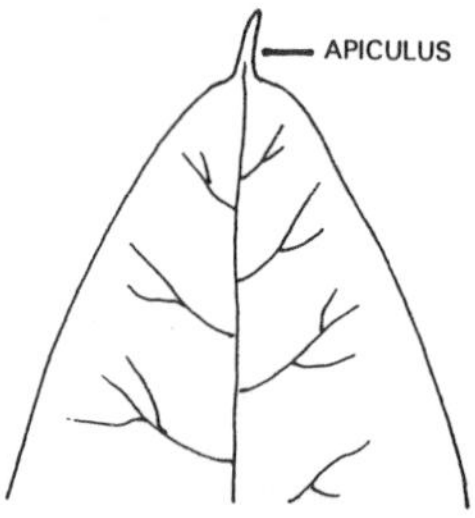

Figure 57

Apocarp. An apocarpous gynoecium.

Apocarpous. A flower with carpels forming separate pistils, as in a buttercup. Figure 58. (compare **syncarpous**)

Apogamous. See **apomictic.**

Apogamy. See **apomixis.**

Figure 58

Apomictic. Of or pertaining to apomixis.

Apomixis. Defined broadly as any form of asexual reproduction and narrowly, and more commonly, as seed production without fertilization (agamospermy).

Apopetalous. Having separate petals. Figure 59. (same as **polypetalous**; compare **sympetalous** and **gamopetalous**)

Apophysis. That portion of a cone scale which is exposed when the cone is closed; a projection or protuberance. Figure 60.

Figure 59

Figure 60

Apospory. Development of gametophytes from somatic cells.

Apostemonous. With separate stamens.

Appendage. A secondary part attached to a main structure.

Appendiculate. Bearing appendages.

Applanate. Flattened. Figure 61.

Appressed. Pressed close or flat against another organ.

Approximate. Borne close together, but not fused.

Figure 61

Aquatic. Growing in water.

Arachnoid. Bearing long, cobwebby, entangled hairs. Figure 62.

Arboreal. See **arborescent**.

Arboreous. See **arborescent**.

Arborescent. Treelike.

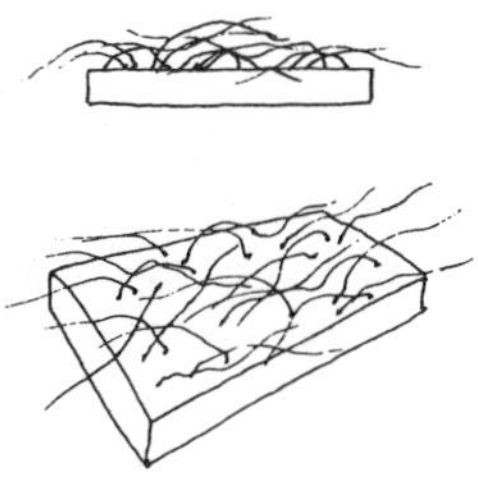

Figure 62

Archegone. See **archegonium**.

Archegonium (pl. **archegonia**). The female reproductive structure in moss and fern gametophytes.

Arctic. Distributed in those regions of the earth lying between the Arctic Circle and the North Pole.

Arcuate. Curved into an arch, like a bow.

Arenicolous. Growing in sand.

Areola (pl. **areolae**, **areolas**). A small, well-defined area on a surface, as the area between the veinlets of a leaf or the region of a cactus bearing the flowers and/or spines. Figure 63.

Areole. See **areola**.

Areolate. Marked with areolae. Figure 63.

Argenteous. Silvery.

Argillaceous. Clayey; of or pertaining to plants growing on clay soils.

Arhizous. Without roots.

Aril. An appendage growing at or near the hilum of a seed; fleshy thickening of the seed coat, as in *Taxus*. Figure 64.

Arillate. Possessing an aril. Figure 64.

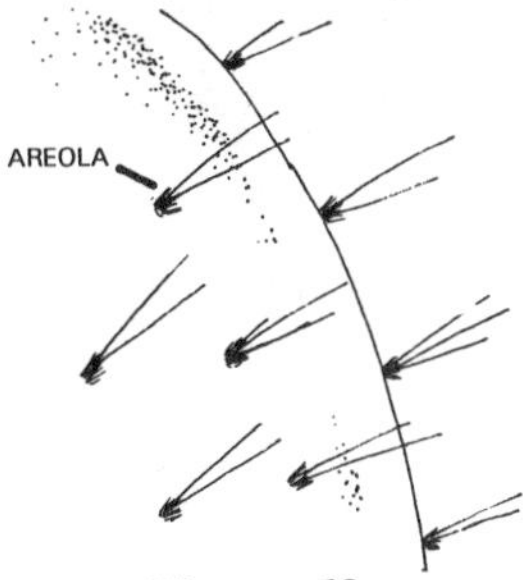

Figure 63

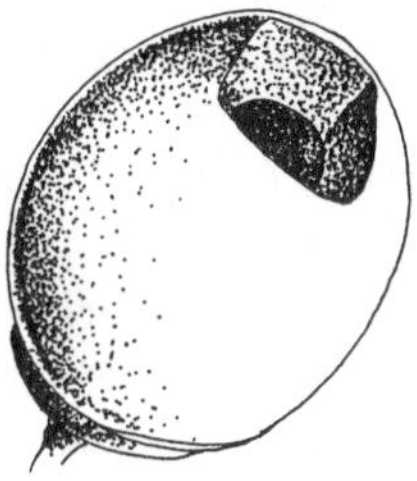

Figure 64

Arilliform. Arillike.

Arillode. A false aril.

Arista (pl. **aristae**). An awn or bristle. Figure 65.

Aristate. Bearing an awn or bristle at the tip. Figure 65.

Aristiform. Awnlike.

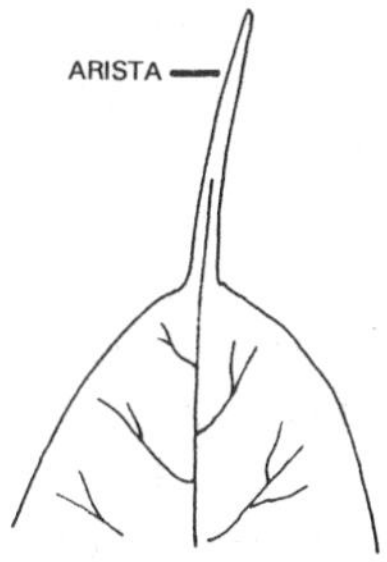

Figure 65

Aristulate. Bearing a minute awn or bristle at the tip. Figure 66.

Armature. Thorns, spines, barbs, or prickles.

Arm-cell. Cells with incomplete septae extending inward, as in the leaf mesophyll cells of some members of the grass family.

Armed. Bearing thorns, spines, barbs, or prickles.

Article. Section of a fruit separated from others by a constricted joint. Figure 67.

Articulate. Jointed. Figure 67; separating at maturity along a well-defined line of dehiscence.

Figure 66

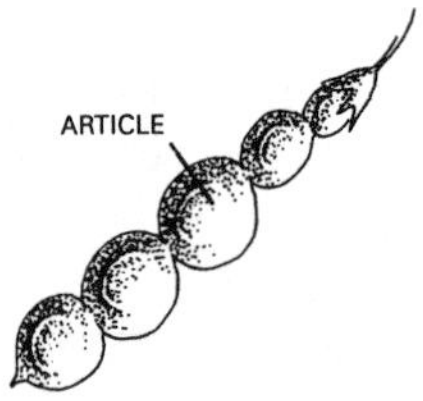

Figure 67

Articulation. A joint or point of attachment.

Ascendent. See **ascending**.

Ascending. Growing obliquely upward, usually curved. Figure 68.

Asepalous. Without sepals.

Asexual. Reproducing without sexual union.

Asperity. A tiny projection or hairlike prickle of an epidermal cell.

Figure 68

Asperous. Rough to the touch.

Assurgent. See **ascending**.

Astemonous. Without stamens.

Astringent. Constricting or contracting.

Astylocarpellous. Lacking a style and a stipe.

Astylocarpepodic. Without a style, but with a stipe.

Astylous. Without a style.

Asymmetric. Not divisible into equal halves, as in some leaves. Figure 69; irregular in shape.

Atomate. Bearing sessile or subsessile glands.

Atomiferous. See **atomate**.

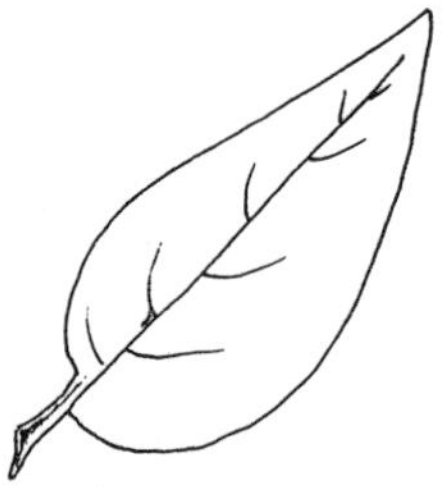

Figure 69

Atro- (prefix). Dark or blackish.

Atropous ovule. See **orthotropous ovule**.

Atropurpurea. Dark purple, often almost blackish.

Attenuate. Tapering gradually to a narrow tip or base. Figure 70.

Atypical. Not typical.

Auricle. A small, ear-shaped appendage. Figure 71.

Auriculate. With auricles. Figure 71.

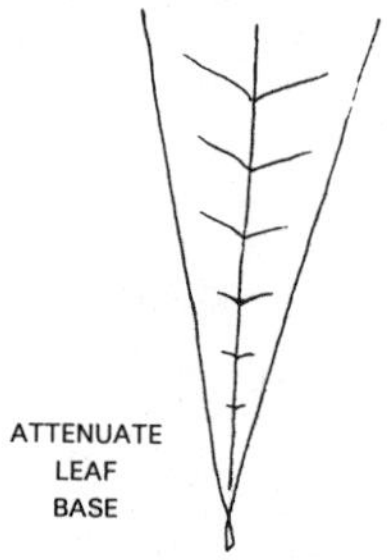

Figure 70

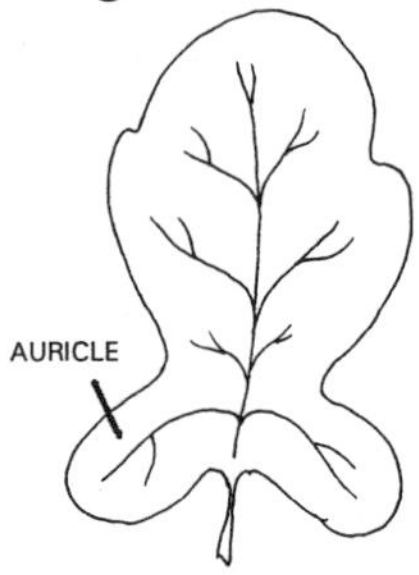

Figure 71

Auriculate-clasping. Earlike lobes at the base of a leaf, encircling the stem. Figure 72.

Austral. Southern. (compare **boreal**)

Autogamy. Self-pollination.

Autophytic. See **autotrophic**.

Autotrophic. Producing their own nutritive substances; containing chlorophyll and, therefore, green; photosynthetic.

Autumnal. Flowering or appearing in the autumn.

Awl-shaped. Short, narrowly triangular, and sharply pointed like an awl. Figure 73.

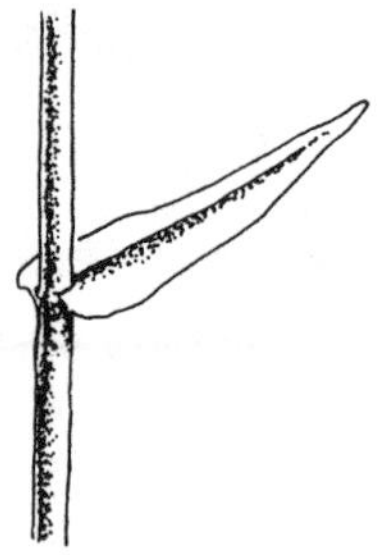

Figure 72

Figure 73

Awn. A narrow, bristlelike appendage, usually at the tip or dorsal surface. Figure 74.

Awned. Possessing an awn. Figure 74.

Axial. See **axile**.

Axil. The point of the upper angle formed between the axis of a stem and any part (usually a leaf) arising from it. Figure 75.

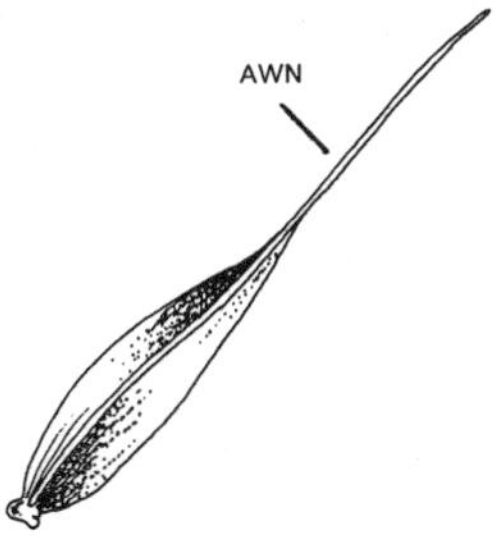

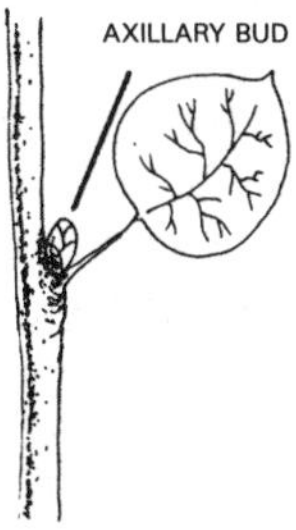

Figure 74 **Figure 75**

Axile. Positioned on the axis; pertaining to the axis.

Axile placentation. Ovules attached to the central axis of an ovary with two or more locules. Figure 76.

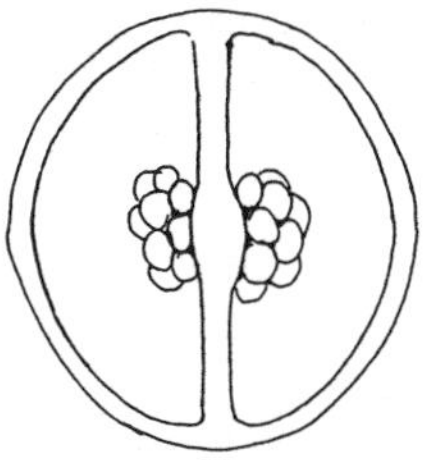

Figure 76

Axillary. Positioned in or arising in an axil. Figure 75.

Axis (pl. **axes**). The longitudinal, central supporting structure or line around which various organs are borne, as a stem bearing leaves.

Baccate. Berrylike and soft.

Balsam. A fragrant, sticky exudate from any of various tree species, especially those of the genus *Commiphora.*

Balsamiferous. Producing balsam; balsam-like.

Banded. Striped.

Banner. The upper and usually largest petal of a papilionaceous flower, as in peas and sweet peas. Figure 77.

Barbate. Bearded or tufted with long, stiff hairs. Figure 78.

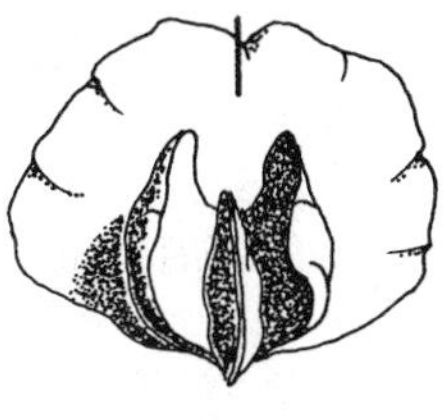

Figure 77 **Figure 78**

Barbed. With short, rigid, reflexed points, like the barb of a fishhook. Figure 79.

Barbellate. With short, stiff hairs or barbs. Figure 80.

Figure 79 **Figure 80**

Barbellulate. With very tiny short, stiff hairs or barbs. Figure 81.

Bark. The outermost layers of a woody stem including all of the living and nonliving tissues external to the cambium. Figure 82.

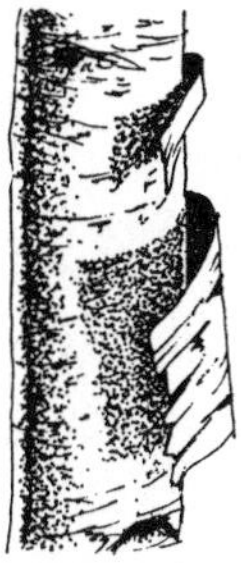

Figure 81 **Figure 82**

Basal. Positioned at or arising from the base, as leaves arising from the base of the stem. Figure 83.

Basal placentation. Ovules positioned at the base of a single-loculed ovary. Figure 84.

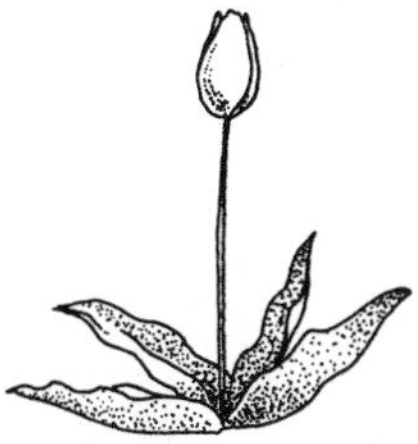

Figure 83 **Figure 84**

Basifixed. Attached by the base. Figure 85. (compare **versatile** and **dorsifixed**)

Basipetal. Near the base rather than the tip; produced sequentially from the apex toward the base, as the flowers in a determinate inflorescence. Figure 86.

Figure 85

Figure 86

Basiscopic. Facing toward the base. Figure 87.

Bast. The fibrous inner bark of some trees; phloem.

Beak. A narrow or prolonged tip, as on some fruits and seeds. Figure 88.

Beaked. Bearing a beak. Figure 88.

Figure 87

Figure 88

Bearded. Bearing one or more tufts of long hairs. Figure 89.

Berry. A fleshy fruit developing from a single pistil, with several or many seeds, as the tomato. Sometimes applied to any fruit which is fleshy or pulpy throughout, i.e. lacking a pit or core. Figure 90.

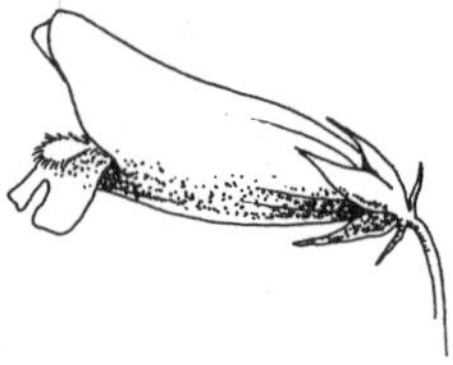

Figure 89

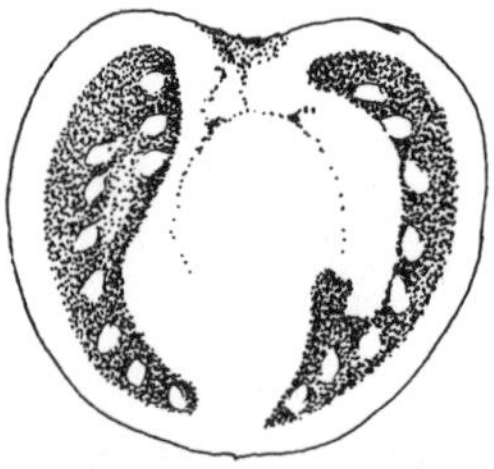

Figure 90

Betacyanin. Betalain pigment varying from blue to red.

Betalain. Water-soluble, nitrogen-containing pigments.

Betaxanthin. Betalain pigment varying from yellow to red.

Bi- (prefix). Meaning two or twice.

Bicarpellate. With two carpels. Figure 91.

Bicolored. Of two distinct colors.

Biconcave. Concave on both sides. Figure 92.

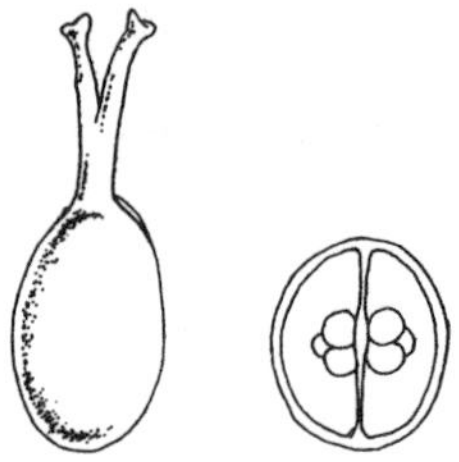

Figure 91

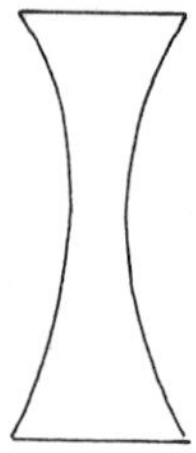

Figure 92

Biconvex. Convex on both sides. Figure 93.

Bicrenate. Doubly crenate, as when the teeth of a crenate leaf are also crenate. Figure 94.

Figure 93

Figure 94

Bidentate. With two teeth. Figure 95.

Biduous. Lasting two days.

Biennial. A plant which lives two years, usually forming a basal rosette of leaves the first year and flowers and fruits the second year.

Figure 95

Bifacial. With the opposite surfaces different in color or texture, as in some leaves.

Bifarious. In two vertical rows. Figure 96.

Biferous. Appearing twice annually.

Bifid. Deeply two-cleft or two-lobed, usually from

the tip. Figure 97.

Biflorous. Flowering in the spring and again in the autumn.

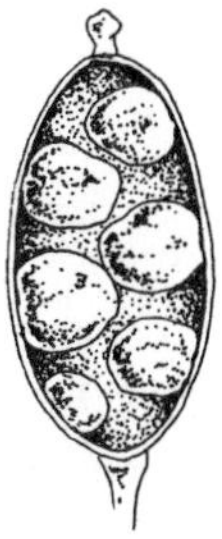

Figure 96

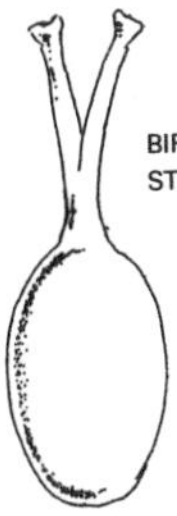

Figure 97

Bifoliate. With two leaves or two leaflets.

Bifurcate. Two-forked; divided into two branches. Figure 98.

Bigeminate. Twice divided into equal pairs.

Bijugate. See **bigeminate**.

Bilabiate. Two-lipped, as in many irregular flowers. Figure 99.

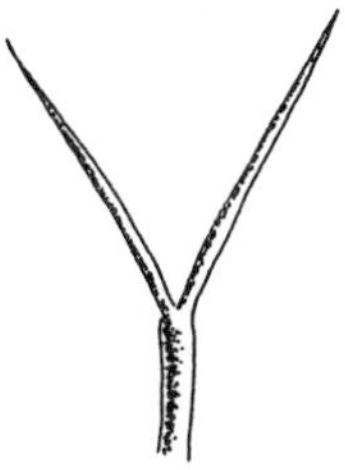

Figure 98

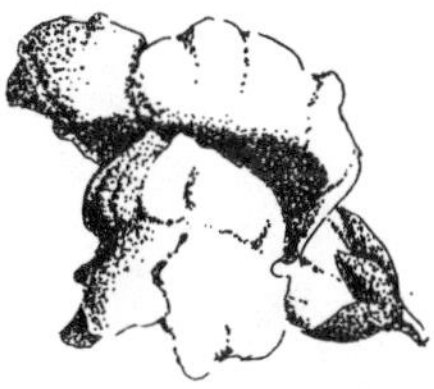

Figure 99

Bilateral. Arranged on two sides, as leaves on a stem. Figure 100.

Bilobed. Divided into two lobes. Figure 101.

Figure 100

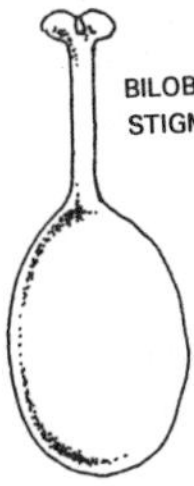

Figure 101

Bilocellate. Divided into two locelli or secondary locules, as when a main locule of an ovary is partitioned into two cavities. Figure 102.

Bilocular. With two locules, as in some ovaries. Figure 103.

Biloculate. See **bilocular**.

Bimestrial. Lasting two months; occurring every two months.

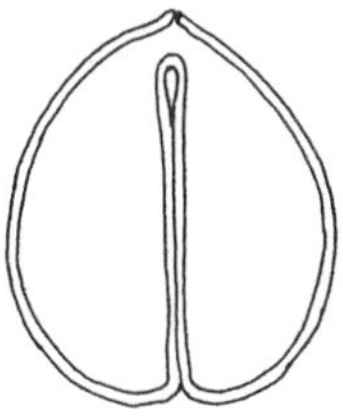

Figure 102

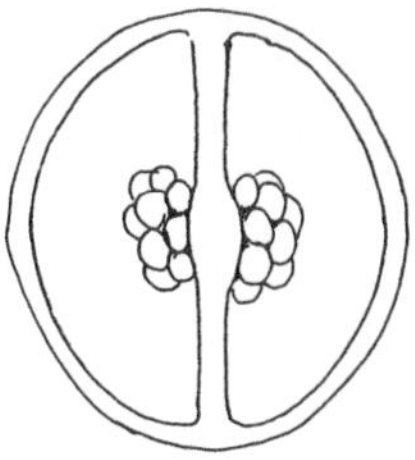

Figure 103

Binate. Borne in pairs. Figure 104.

Bipalmate. Twice palmate; with the divisions again palmately divided.

Bipartite. Divided almost to the base into two divisions. Figure 105.

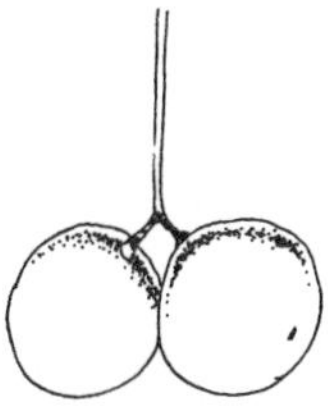

Figure 104

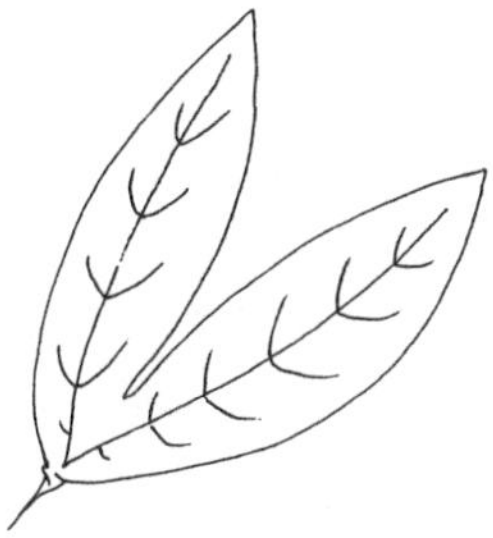

Figure 105

Bipetalous. With two petals.

Bipinnate. Twice pinnate; with the divisions again pinnately divided. Figure 106.

Bipinnatifid. Twice pinnately cleft. Figure 107.

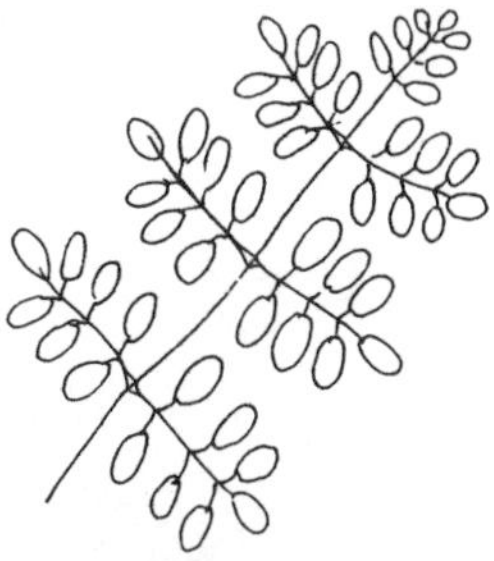

Figure 106

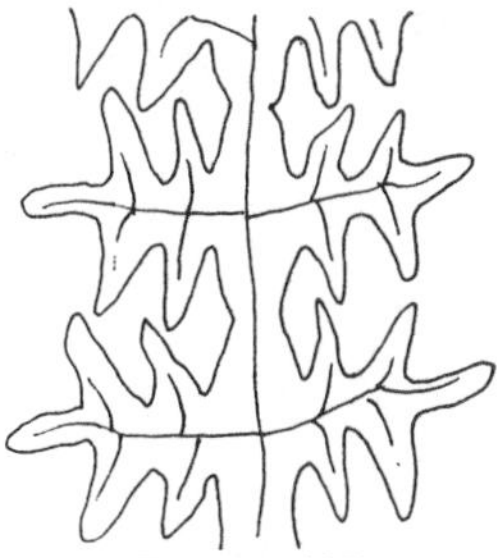

Figure 107

Bis- (prefix). See **bi-**.

Bisected. Split into two parts. Figure 105.

Biserial. See **biseriate**.

Biseriate. Arranged in two rows or series. Figure

108.

Biserrate. Doubly serrate, as when the teeth of a serrate leaf are also serrate. Figure 109.

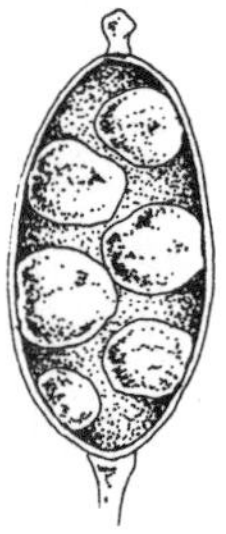

Figure 108

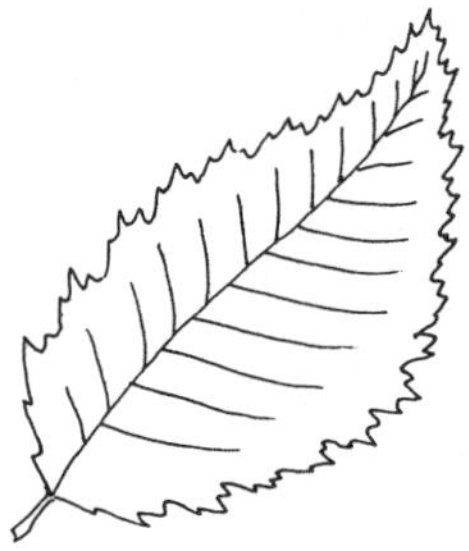

Figure 109

Bisexual. A flower with both male and female reproductive organs (stamens and pistils). Figure 110. (same as **perfect**)

Biternate. Doubly ternate; with the ternate divisions again ternately divided. Figure 111.

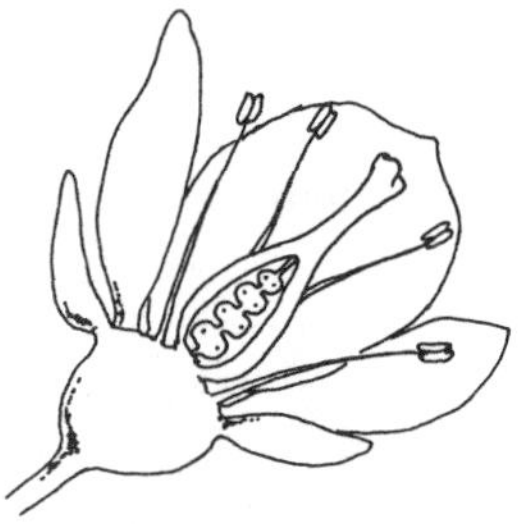

Figure 110

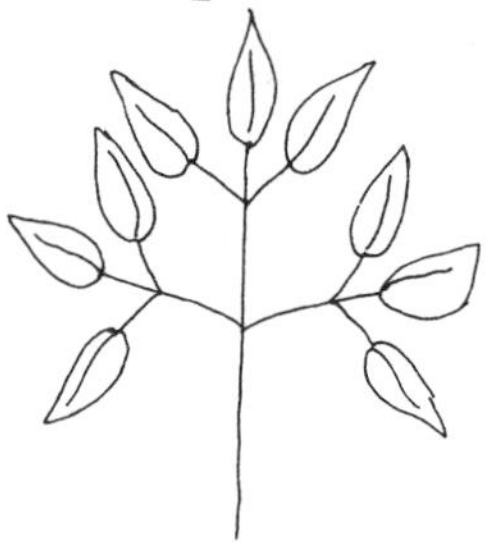

Figure 111

Biturbinate. Top-shaped, but with the widest part some distance from one end. Figure 112.

Bladder. A structure which is thin-walled and inflated. Figure 113.

Bladderlike. Thin-walled and inflated. Figure 113.

Bladdery. See **bladderlike.**

Figure 112

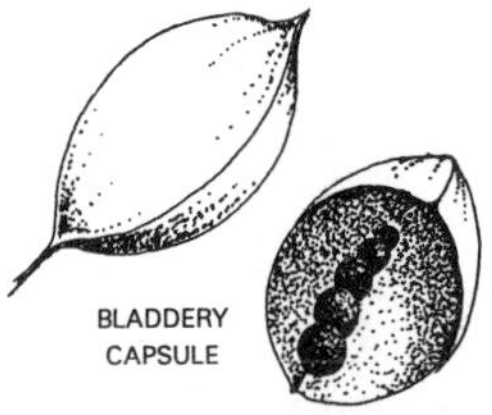

Figure 113

Blade. The broad part of a leaf or petal. Figures 114 and 115.

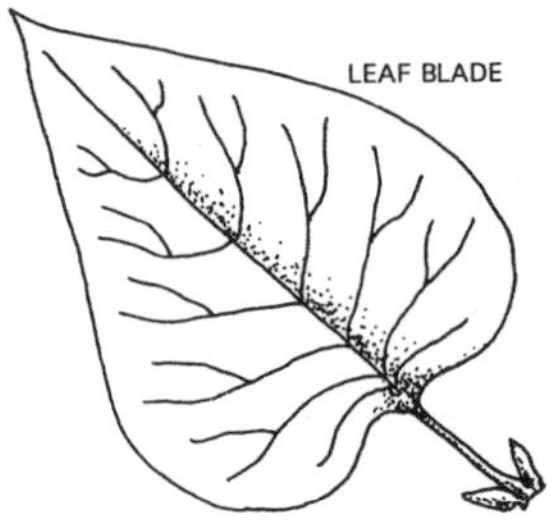

Figure 114

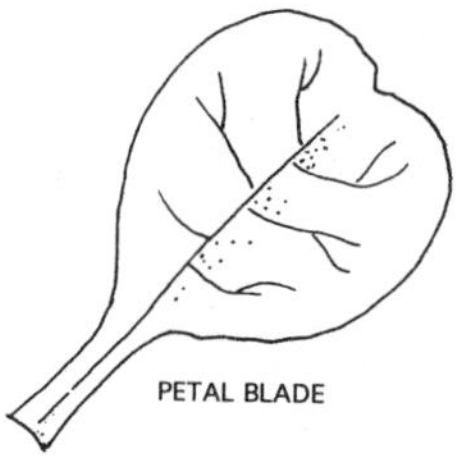

Figure 115

Bloom. A whitish, waxy, powdery coating on a surface; the flower.

Blotched. Marked with irregular spots or blots.

Bole. The trunk of a tree.

Bordered. With the edge of a different color than the main body of the organ or structure.

Boreal. Northern. (compare **austral**)

Boss. A protuberance or projection from a surface or organ. Figure 116.

Botuliform. Sausage-shaped. Figure 117.

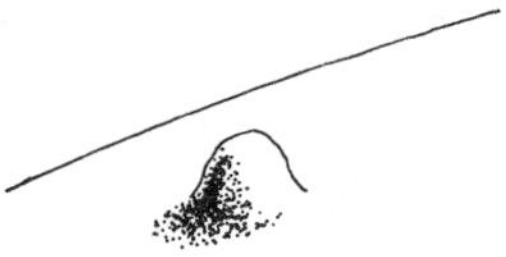

Figure 116

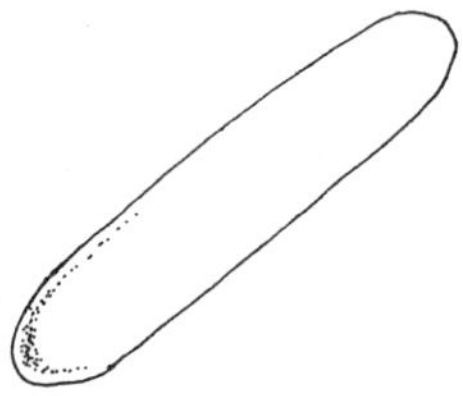

Figure 117

Brackish. Somewhat saline.

Bract. A reduced leaf or leaflike structure at the base of a flower or inflorescence. Figure 118; in conifers, one of the main structures arising from the cone axis.

Bracteal. Of or pertaining to bracts; bracteate.

Bracteate. With bracts.

Bracteiform. Bracklike.

Bracteolate. With bracteoles.

Bracteole. A small bract borne on a petiole. Figure 119.

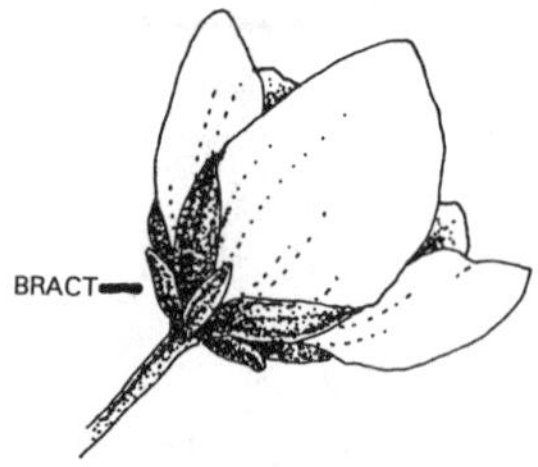

Figure 118

Bracteose. With many bracts or with conspicuous bracts.

Bractlet. See **bracteole**.

Branchlet. A small branch. Figure 120.

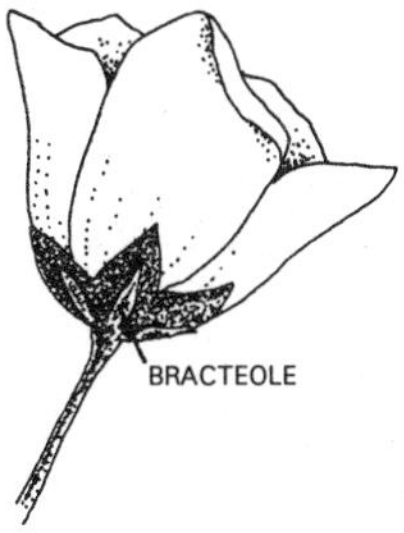

Figure 119

Figure 120

Bridge. A band of tissue connecting the corolla scales, as in *Cuscuta*. Figure 121.

Bristle. A short, stiff hair or hairlike structure. Figure 122.

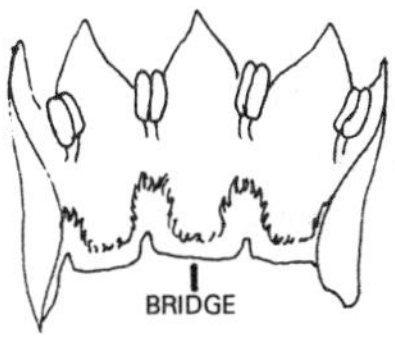

Figure 121

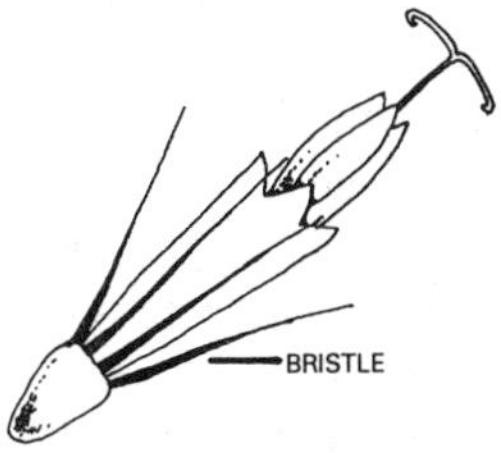

Figure 122

Bud. An undeveloped shoot or flower. Figure 123.

Bud scales. Modified leaves covering a bud. Figure 123.

Bulb. An underground bud with thickened fleshy scales, as in the onion. Figure 124.

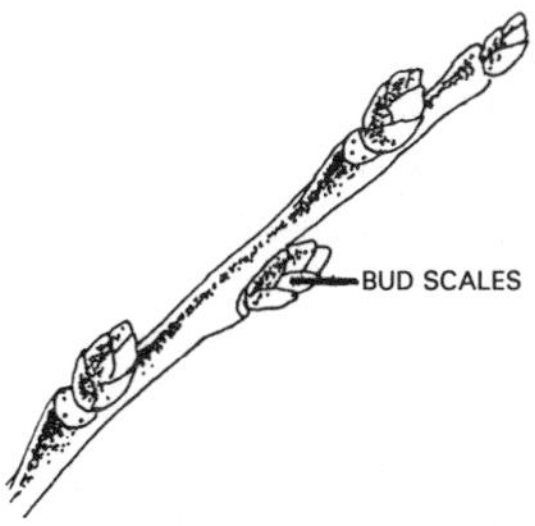

Figure 123

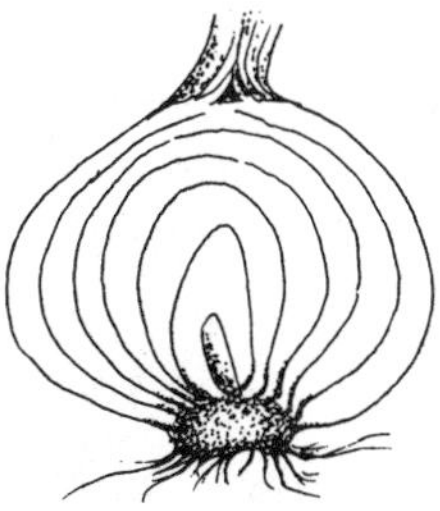

Figure 124

Bulbel. A small bulb arising from the base of a larger bulb. Figure 125.

Bulbiferous. Producing bulbs.

Bulbil. See **bulblet**.

Bulblet. A small bulb; a bulblike structure borne above ground, usually in a leaf axil. Figure 126.

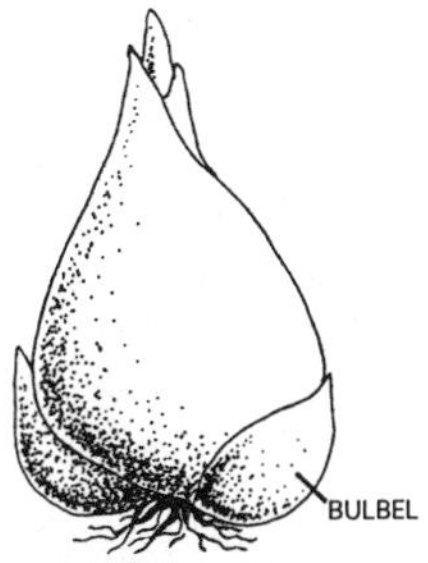

Figure 125

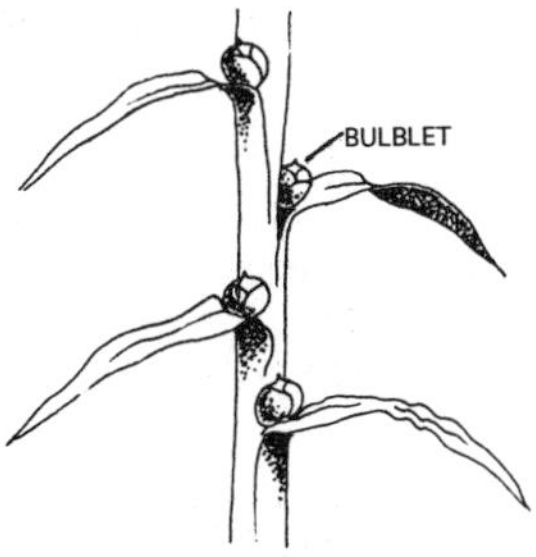

Figure 126

Bulbose. Bulblike.

Bullate. With rounded, blistery projections covering the surface. Figure 127.

Bulliform cells. Large, thin-walled epidermal cells of the intercostal zone of the leaf blade in some members of the grass family. Figure 128.

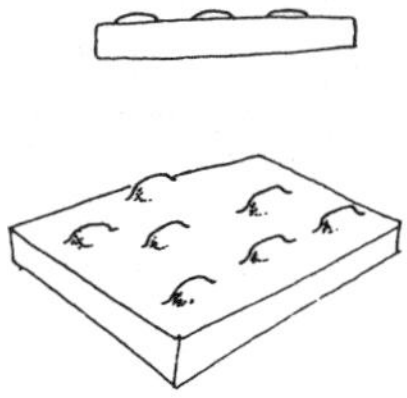

Figure 127

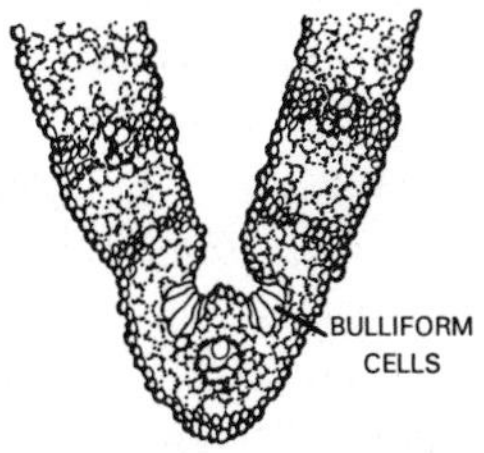

Figure 128

Bundle scar. Scar left on a twig by the vascular bundles when a leaf falls. Figure 129.

Bur (or **burr**). A structure armed with often hooked or barbed spines or appendages. Figure 130.

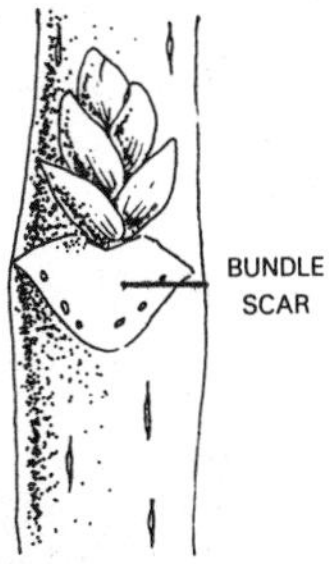

Figure 129

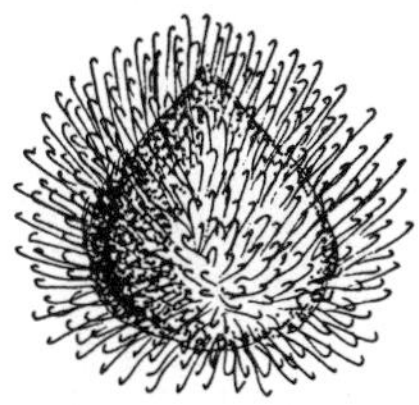

Figure 130

Bush. See **shrub**.

Buttressed. With props or supports, as in the flared trunks of some trees.

Caducous. Falling off very early compared to similar structures in other plants.

Caespitose. Growing in dense tufts. Figure 131.

Calcar. A spur or spurlike appendage. Figure 132.

Calcarate. With a calcar; spurred. Figure 132.

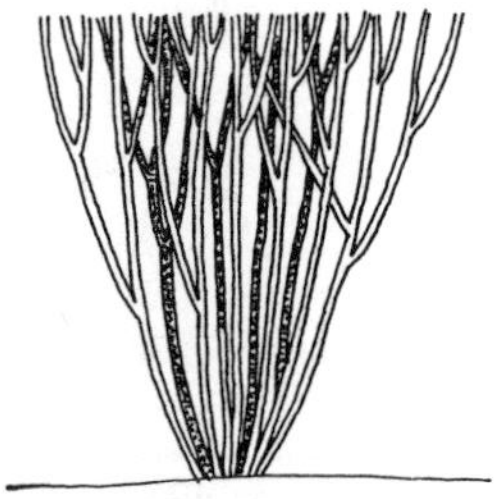

Figure 131

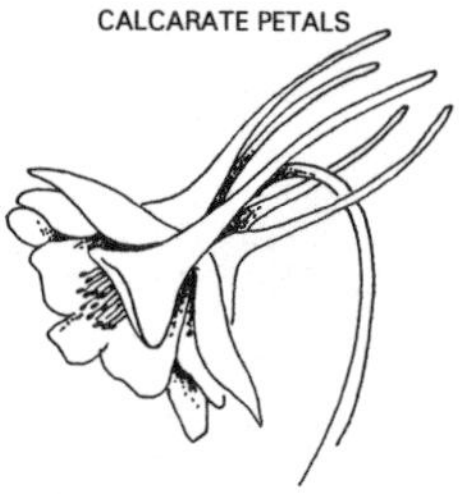

Figure 132

Calceolate. Shoe-shaped or slipper-shaped, as the labellum of some orchids. Figure 133.

Calcicole. A plant growing on calcareous soil.

Calcifuge. A plant which avoids calcareous soil.

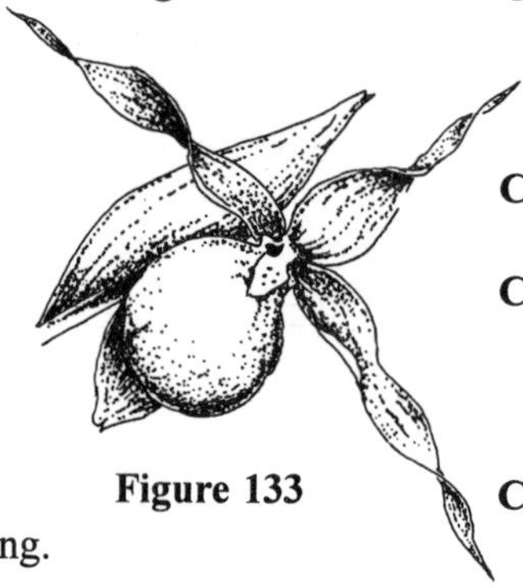

Figure 133

Calciphilous. Lime-loving.

Callose. See **callous**.

Callosity. A hardened or thickened area.

Callous. Hardened or thickened; having a callus.

Callus. A hard thickening or protuberance; the thickened basal extension of the lemma in many grasses. Figure 134.

Calyculate. With small bracts around the calyx, as if possessing an outer calyx; with small bracts around the base of the involucre. Figure 135.

Figure 134

Figure 135

Calyptra. A hood or lid. Figure 136.

Calyx (pl. **calyces, calyxes**). The outer perianth whorl; collective term for all of the sepals of a flower. Figure 137.

Calyx limb. See **calyx lobe**.

Figure 136

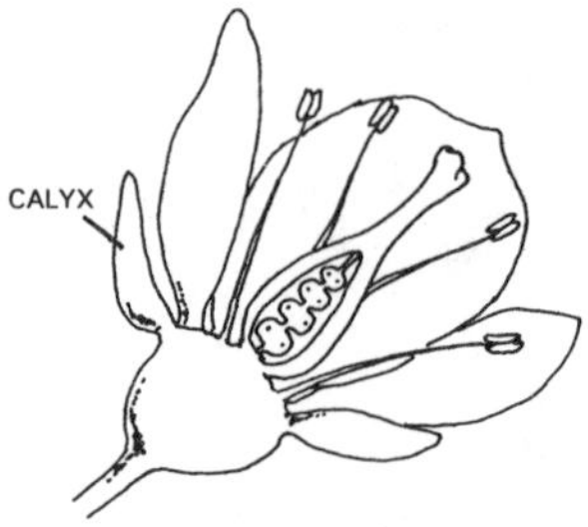

Figure 137

Calyx lobe. One of the free portions of a calyx of united sepals. Figure 138.

Calyx tooth. See **calyx lobe**.

Calyx tube. The tube-like united portion of a calyx of united sepals. Figure 138.

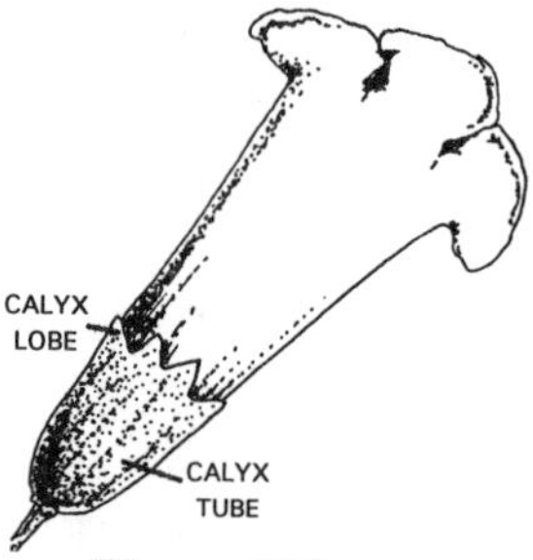

Figure 138

Cambium. A tissue composed of cells capable of active cell division, producing xylem to the inside of the plant and phloem to the outside; a lateral meristem. Figure 139.

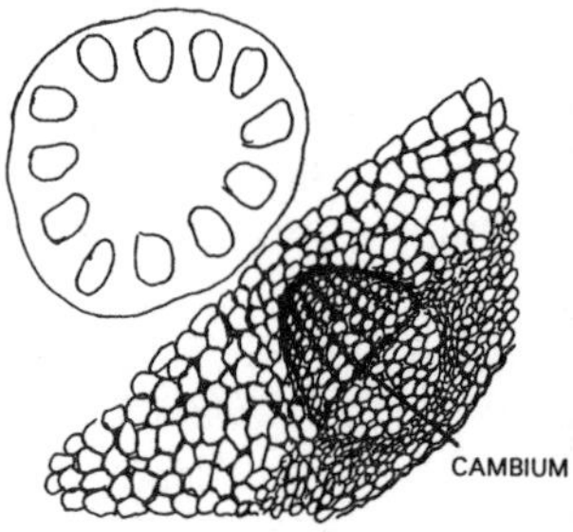

Figure 139

Campanulate. Bell-shaped. Figure 140.

Campylotropous ovule. An ovule which is curved so that the micropyle is positioned near the funiculus and the chalaza. Figure 141.

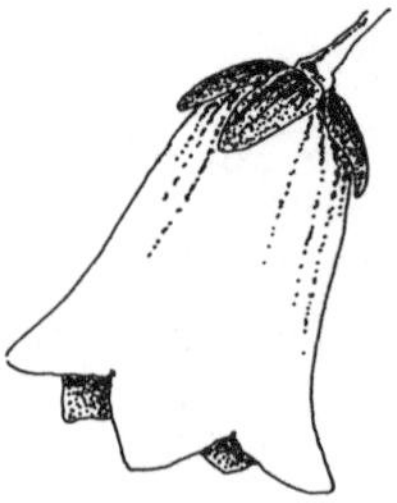

Figure 140

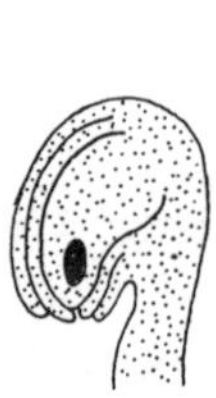

Figure 141

Canaliculate. With longitudinal channels or grooves. Figure 142.

Cancellate. Latticed with a fine, regular, reticulate pattern. Figure 143.

Figure 142

Figure 143

Canescent. Gray or white in color due to a covering of short, fine gray or white hairs. Figure 144.

Capillary. Hair-like; very slender and fine. Figure 145.

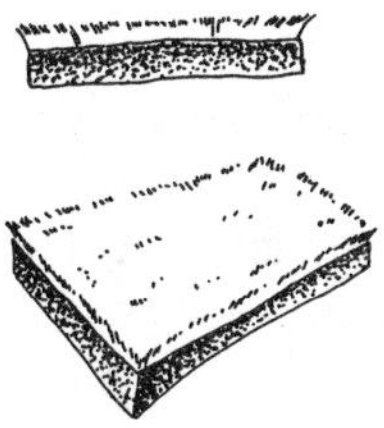

Figure 144

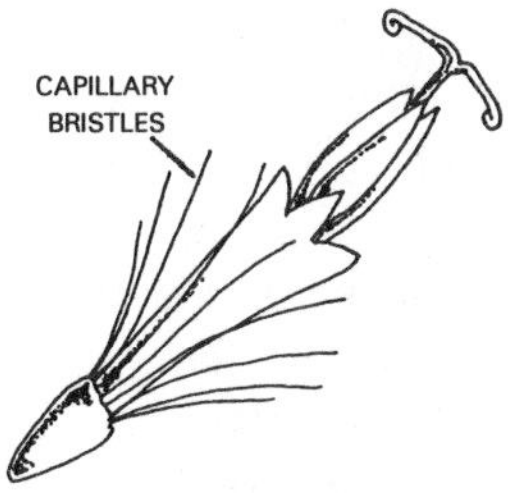

Figure 145

Capitate. Head-like, or in a head-shaped cluster, as the flowers in the Compositae (Asteraceae). Figures 146 and 147.

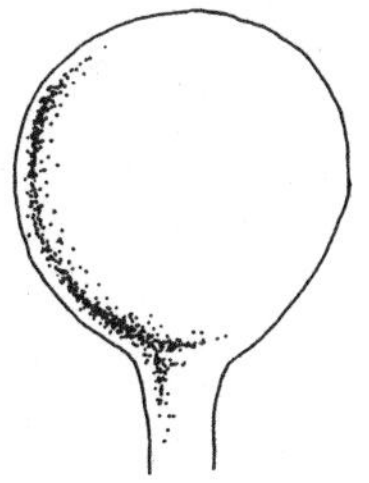

Figure 146

Figure 147

Capitellate. With small head-like structures, or with parts in very small head-shaped clusters. Figure 148.

Capitulum. A small flower head. Figure 148.

Capsular. Of or pertaining to a capsule; capsule-like.

Capsule. A dry, dehiscent fruit composed of more than one carpel. Figure 149.

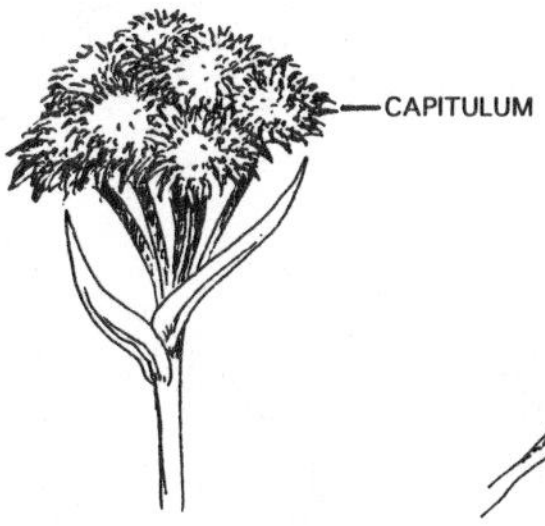

Figure 148

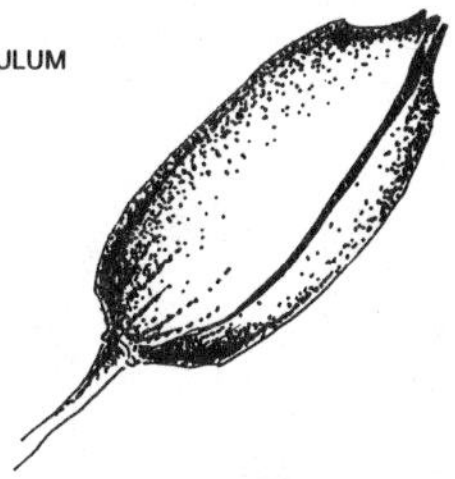

Figure 149

Carina. A keel or ridge. Figure 150.

Carinal. See **carinate**.

Carinate. Keeled with one or more longitudinal ridges. Figure 151.

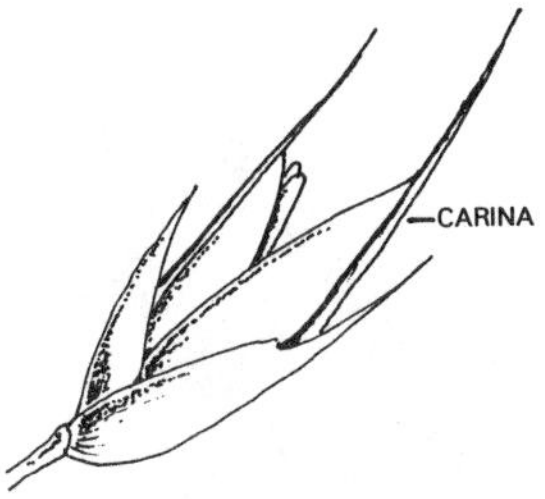

Figure 150

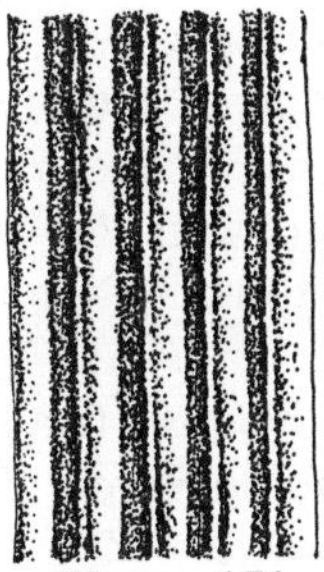

Figure 151

Cariopsis. See **caryopsis**.

Carneous. Flesh-colored.

Carnose. With a fleshy texture.

Carotene. Yellow, orange, or red fat-soluble pigments.

Carotenoid pigment. A carotene or xanthophyll pigment.

Carpel. A simple pistil formed from one modified leaf, or that part of a compound pistil formed from one modified leaf; megasporophyll. Figure 152.

Carpellate. With carpels.

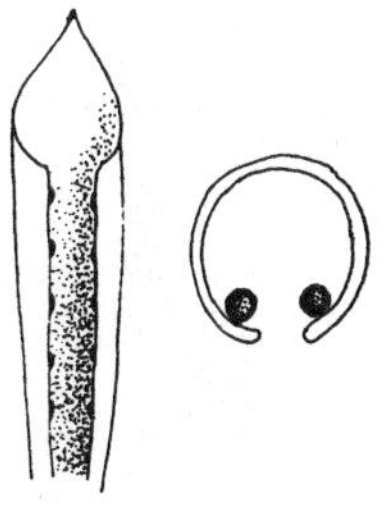

Figure 152

Carpophore. A slender prolongation of the receptacle between the carpels as a central axis, as in the fruits of some members of the Umbelliferae

(Apiaceae) and the Geraniaceae. Figure 153.

Carpopodium. A stipe supporting an ovary.

Cartilaginous. Tough and firm but elastic and flexible, like cartilage.

Caruncle. A protuberance or appendage near the hilum of a seed. Figure 154.

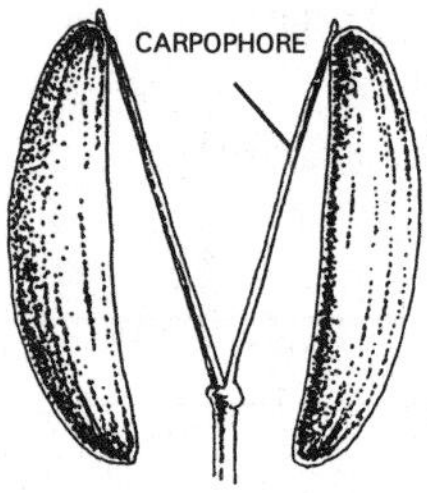

Figure 153

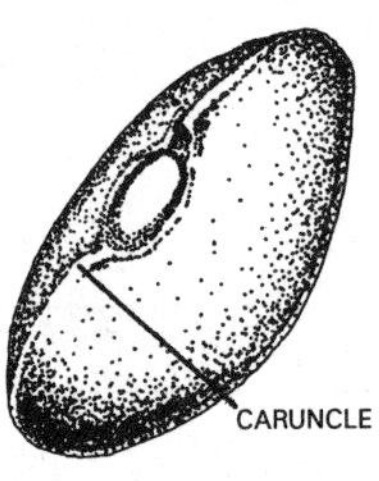

Figure 154

Caruncular. Of or pertaining to a caruncle.

Carunculate. See **caruncular**.

Caryopsis. A dry, one-seeded, indehiscent fruit with the seed coat fused to the pericarp, as in the fruits of the grass family; a grain. Figure 155.

Castaneous. Chestnut-colored; dark reddish-brown in color.

Figure 155

Catkin. An inflorescence consisting of a dense spike or raceme of apetalous, unisexual flowers as in Salicaceae and Betulaceae; an ament. Figure 156.

Caudate. With a taillike appendage. Figure 157.

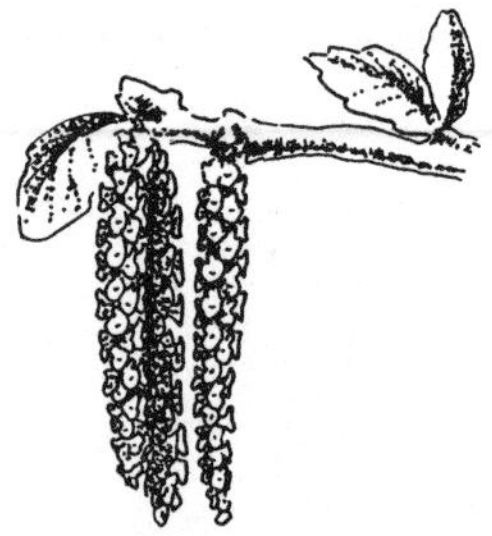

Figure 156

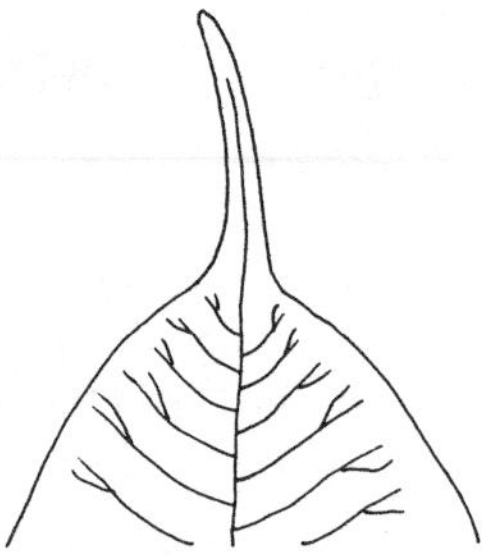

Figure 157

Caudex (pl. **caudices**, **caudexes**). The persistent and often woody base of a herbaceous perennial. Figure 158.

Caulescent. With an obvious leafy stem rising above the ground. Figure 159. (Compare **acaulescent**)

Caulicle. A small stem; a rudimentary stem.

Cauliflorous. Bearing flowers on the stem or trunk.

Cauline. Of, on, or pertaining to the stem, as leaves arising from the stem above ground level. Figure 159.

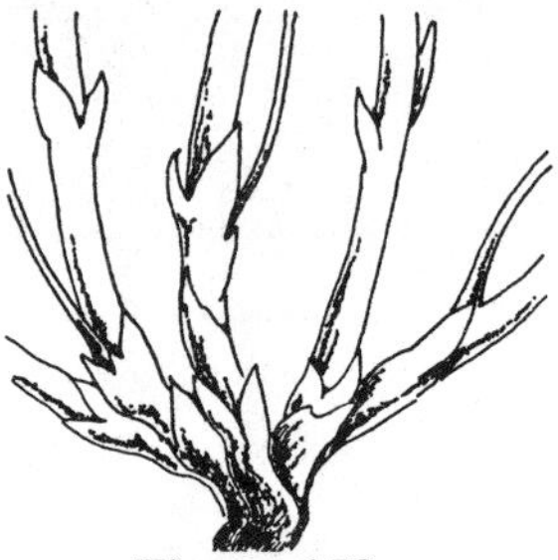

Figure 158

Figure 159

Caulis. The main stem of a herbaceous plant.

Caulocarpic. With the stem living for several years.

Cauloid. Stem-like.

Cell. As used in plant identification, a hollow cavity or compartment within a structure, as the cavity of the anther containing pollen or the cavity of the ovary containing ovules; a locule. Figure 160.

Cellular. Made up of small cavities or compartments.

Centrifugal inflorescence. A flower cluster developing from the center outward, as in a cyme. Figure 161.

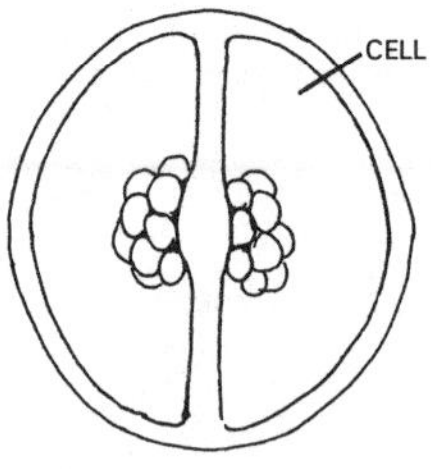

Figure 160

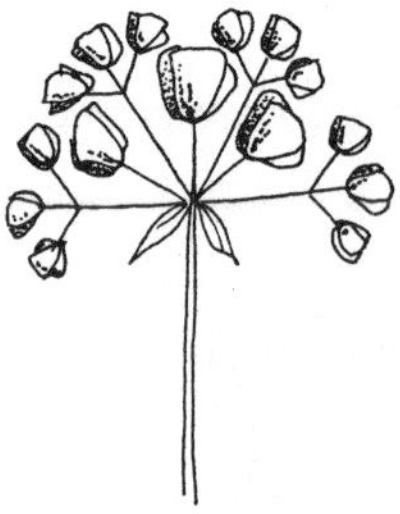

Figure 161

Centripetal inflorescence. A flower cluster developing from the edge toward the center, as in a corymb. Figure 162.

Ceraceous. Waxy in texture or appearance.

Cernuous. Drooping or nodding, as a flower.

Figure 163.

Cerogenous. Wax producing.

Cespitose. See **caespitose**.

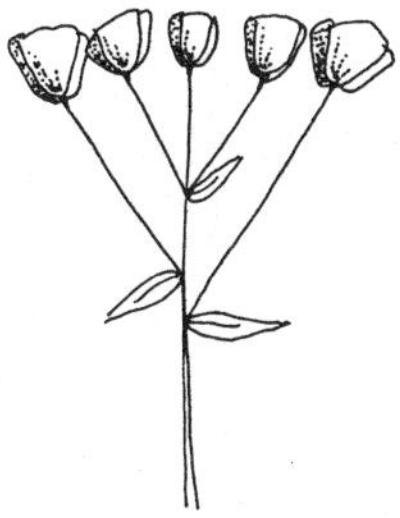

Figure 162

Figure 163

Chaff. Thin dry scales or bracts, as the bracts on the receptacle of the heads of the Compositae (Asteraceae). Figure 164.

Chaffy. With chaff; chafflike.

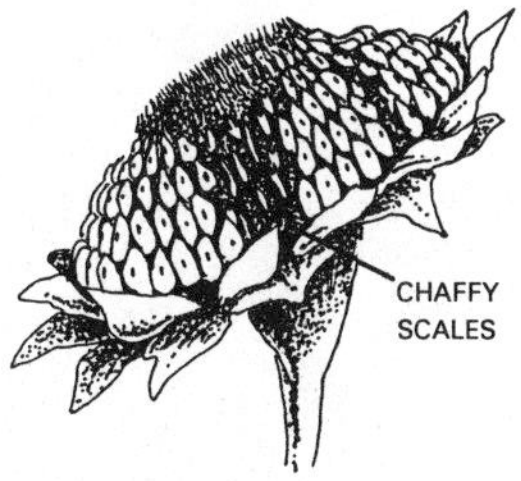

Figure 164

Chalaza. The part of an ovule or seed where the integuments are connected to the nucellus, at the opposite end from the micropyle. Figure 165.

Chamaephyte. A plant which produces resting buds just above the ground.

Chambered. With hollow spaces. Figure 166.

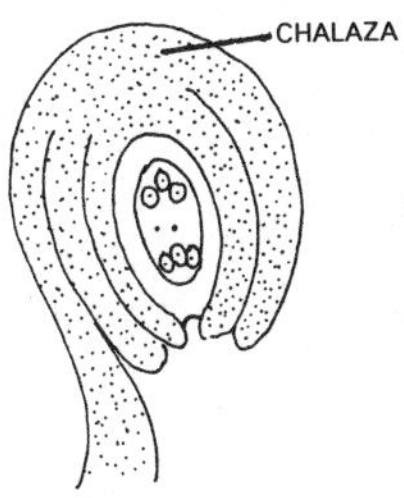

Figure 165

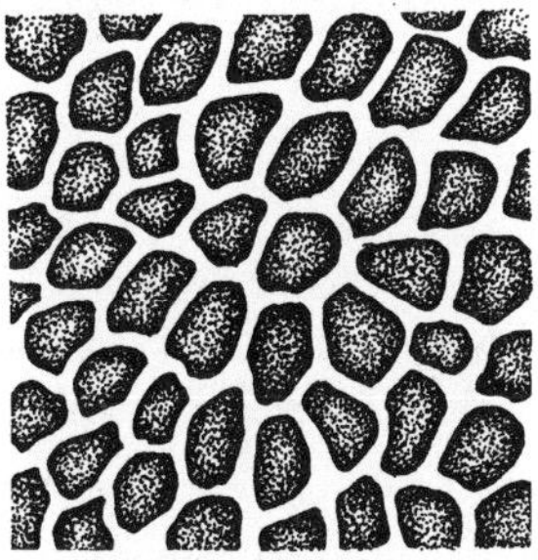

Figure 166

Channeled. With one or more deep longitudinal grooves. Figure 167.

Chaparral. A vegetation type consisting of dense thickets of evergreen shrubs.

Chartaceous. With a papery texture, usually not green.

Chasmogamous. Applied to flowers which open before fertilization and are usually cross-pollinated. (compare **cleistogamous**)

Chlamydeous. With, or pertaining to, a floral whorl.

Chlorophyll. The green pigment of plants associated with photosynthesis.

Chlorophyllous. Of or containing chlorophyll; green.

Figure 167

Choripetalous. See **apopetalous** or **polypetalous**.

Ciliate. With a marginal fringe of hairs. Figure 168.

Ciliolate. With a marginal fringe of minute hairs. Figure 169.

Cilium (pl. **cilia**). A small hair or hairlike process, usually along the margin of a structure. Figure 168.

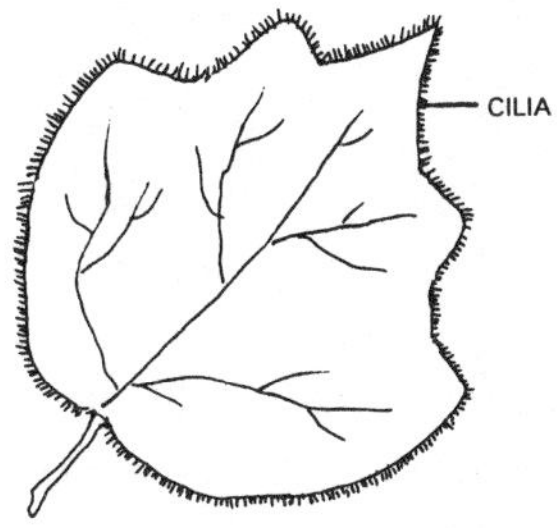

Figure 168

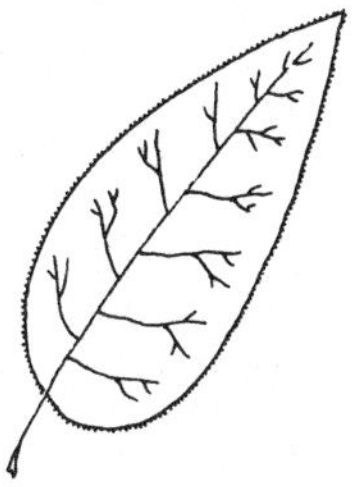

Figure 169

Cincinnus. A dense helicoid cyme with the pedicels short on the developed side. Figure 170.

Cinereous. Ash-colored; grayish due to a covering of short hairs.

Circinate. Coiled from the tip downward, as in the young leaves of a fern. Figure 171.

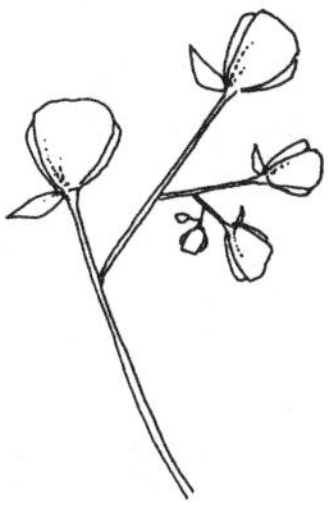

Figure 170

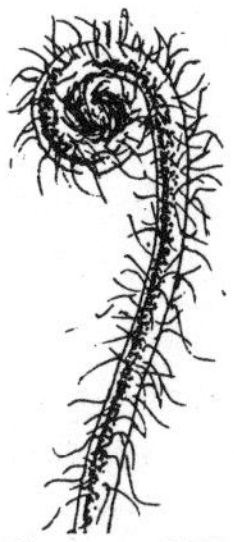

Figure 171

Circum- (prefix). Meaning around, as around an object or structure.

Circumscissile. Dehiscing along a transverse circular line, so that the top separates like a lid. Figure 172.

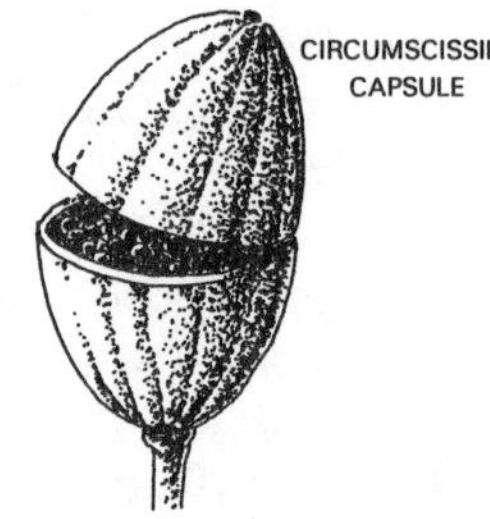

Figure 172

Cirrate. With cirri. Figure 173.

Cirrhiferous. See **cirriferous.**

Cirrhose. See **cirrose.**

Cirriferous. Bearing a tendril. Figure 173.

Cirrose. With cirri. Figure 173; resembling a cirrus. Figure 174.

Cirrus (pl. **cirri**). A tendril. Figure 173.

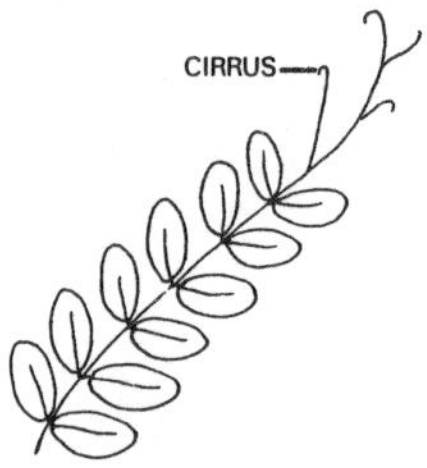

Figure 173

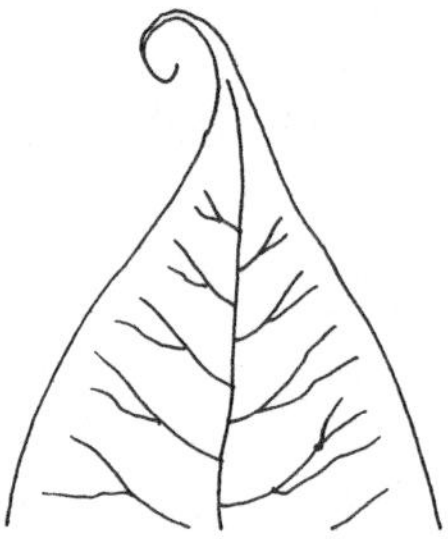

Figure 174

Cladode. See **cladophyll.**

Cladophyll. A stem with the form and function of a leaf. Figures 175 and 176. (same as **phyllo-**

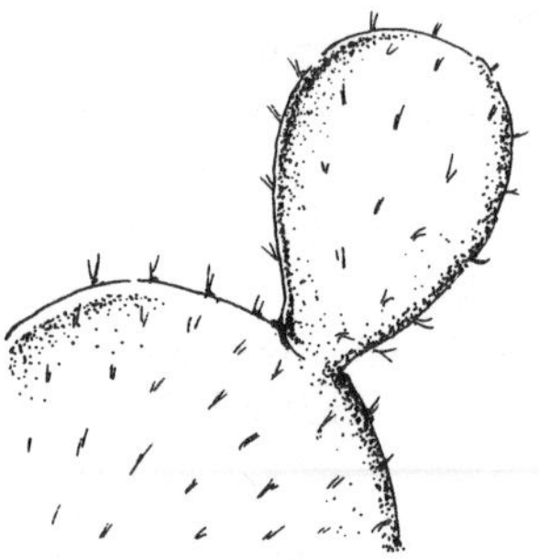

Figure 175

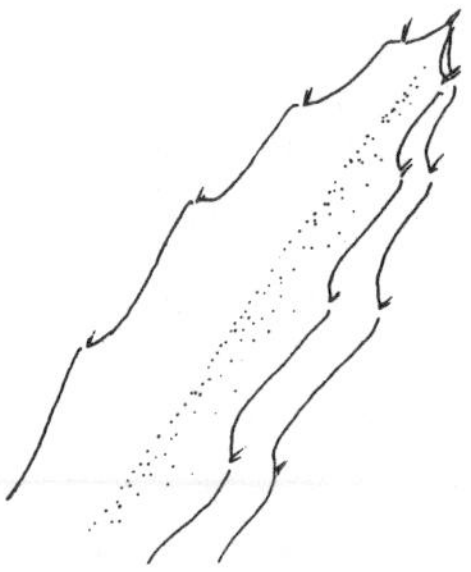

Figure 176

Cladoptosic. Dropping the leaves, branches, and stems at one time, as in *Taxodium*.

Clambering. Weakly climbing on other plants or surrounding objects. Figure 177.

Clammy-pubescent. With sticky glandular hairs.

Clasping. Wholly or partly surrounding the stem. Figure 178.

Clathrate. Lattice-like in appearance. Figure 179.

Clavate. Club-shaped, gradually widening toward the apex. Figures 180 and 181.

Clavellate. Diminutive of clavate.

Claw. The narrowed base of some petals and sepals. Figure 182.

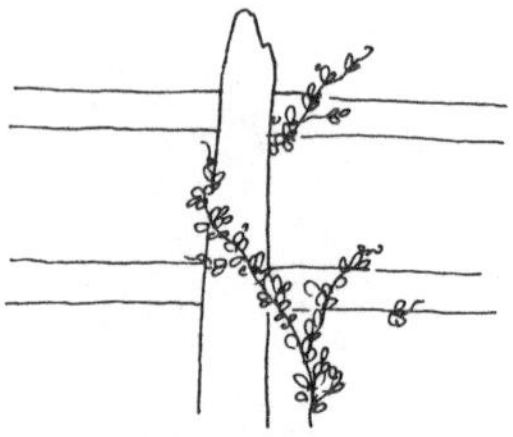

Figure 177

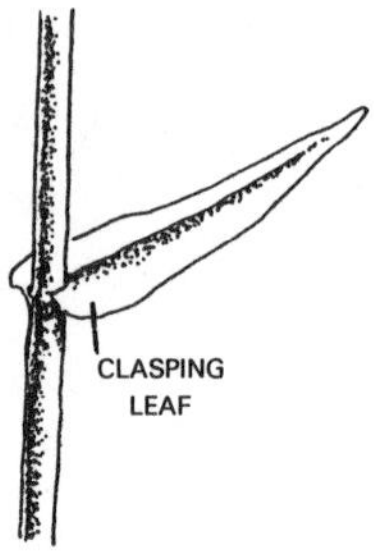

Figure 178

Figure 179

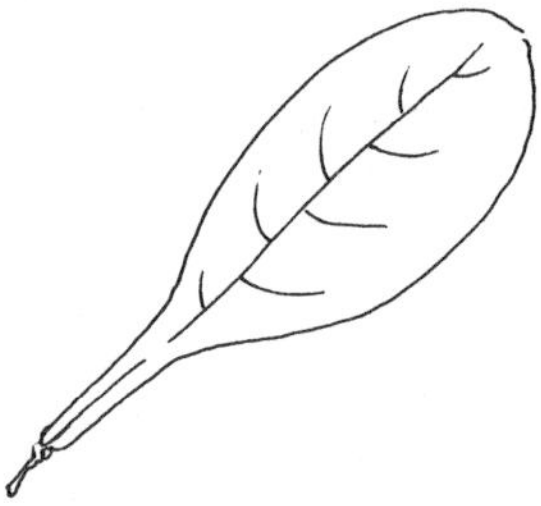

Figure 180

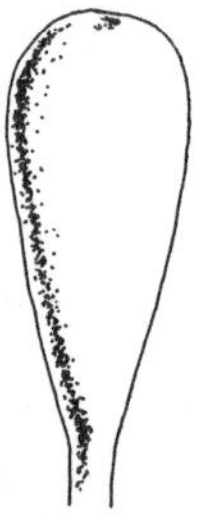

Figure 181

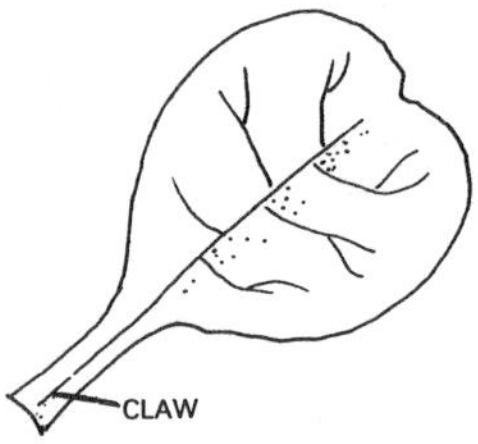

Figure 182

Cleft. Cut or split about half-way to the middle or base. Figures 183 and 184.

Figure 183

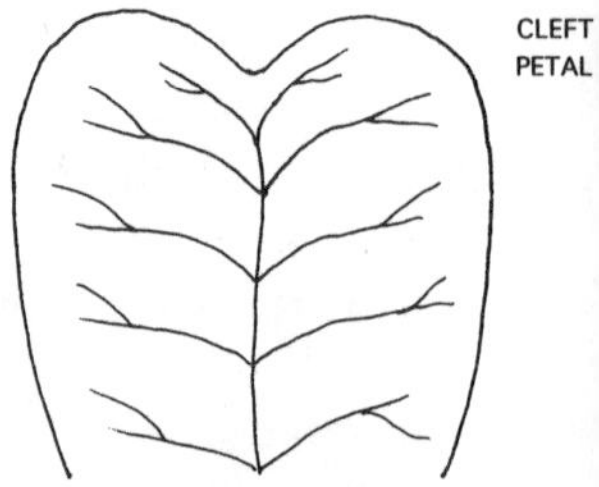

Figure 184

Cleistogamous. Flowers which self-fertilize

without opening. (compare **chasmogamous**)

Cleistogene. A plant which bears cleistogamous flowers.

Climbing. Growing more or less erect by leaning or twining on another structure for support. Figure 185.

Clinandrium. The portion of an orchid column in which the anther is concealed. Figure 186.

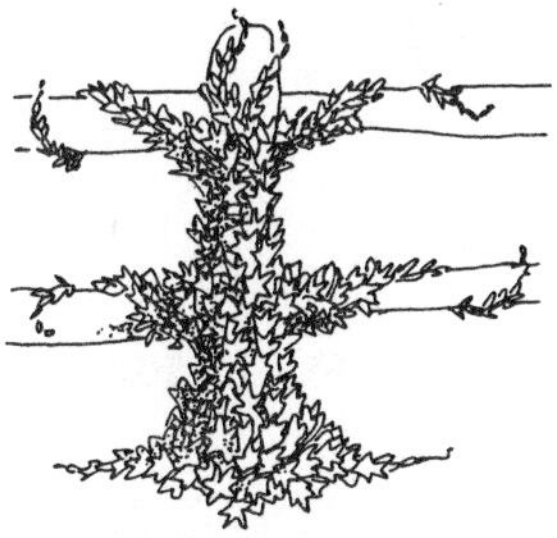

Figure 185

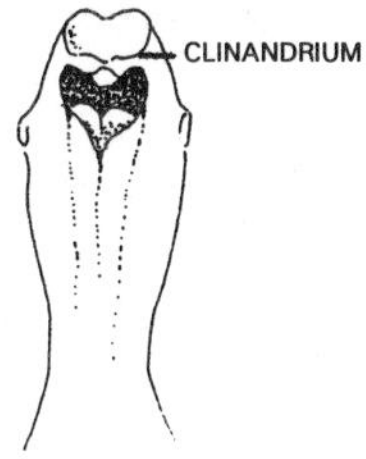

Figure 186

Clone. A group of individuals originating from a single parent plant by vegetative reproduction.

Clouded. Blended with patches of another color.

Coalescent. United together to form a single unit. Figure 187.

Coat. The covering of a seed; the outer covering of an organ or structure. Figure 188.

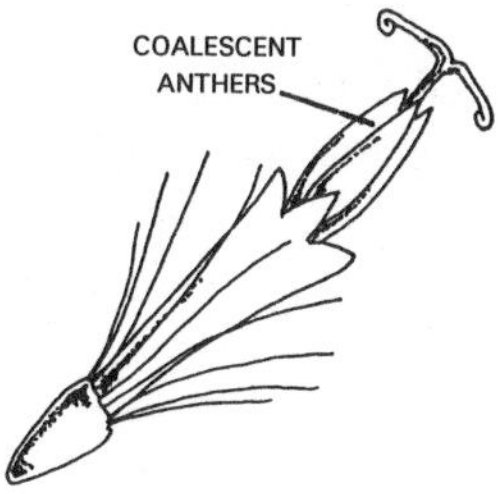

Figure 187

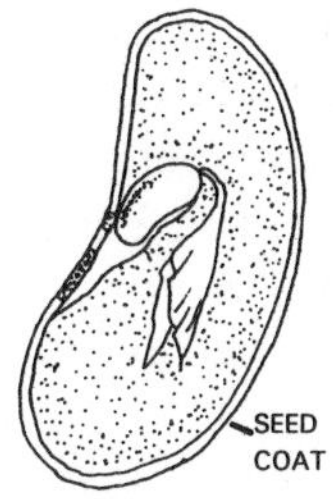

Figure 188

Coccus (pl. **cocci**). One of the segments (carpels) of a schizocarp. Figure 189.

Cochleate. Shaped like the coiled shell of a snail. Figure 190.

Coerulean. Blue or bluish.

Coetaneous. With the leaves and flowers developing at the same time.

Coherent. Sticking together of like parts. The attachment is not as firm or solid as **connate**.

Cohesion. See **coherent**.

Coleoptile. The sheath protecting the stem tip in monocotyledons. Figure 191.

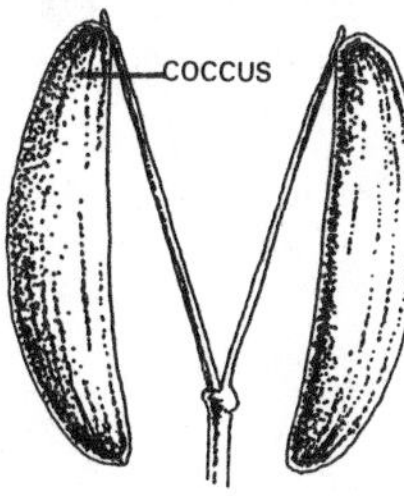

Figure 189

Figure 190

Coleorhiza. The sheath which surrounds and is penetrated by the radicle in some seeds. Figure 191.

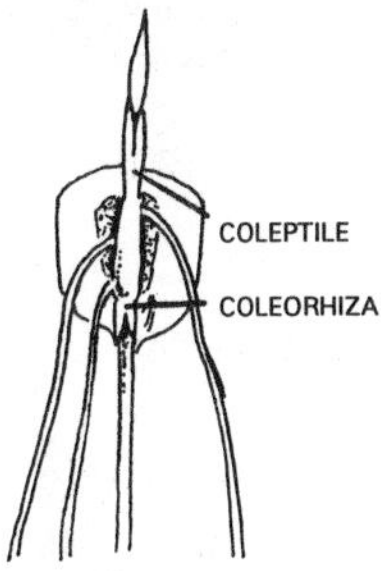

Figure 191

Collar. The area on the outside of a grass leaf at the juncture of the blade and sheath. Figure 192.

Collateral. Situated side by side.

Colonial. Forming colonies; usually refers to groups of plants connected to one another by underground organs.

Column. A structure formed by the union of staminal filaments, as in many Malvaceae; the united filaments and style in the Orchidaceae. Figures 193 and 194.

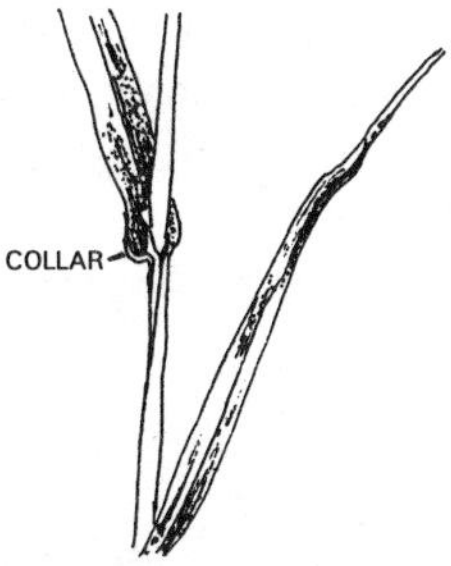

Figure 192

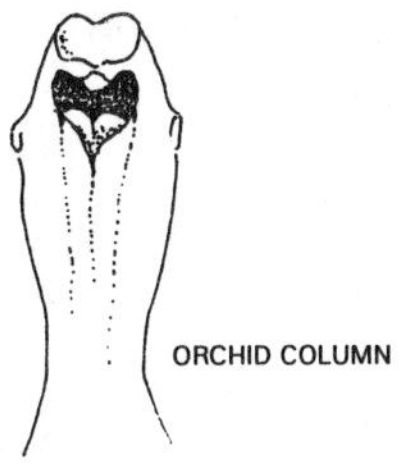

Figure 193

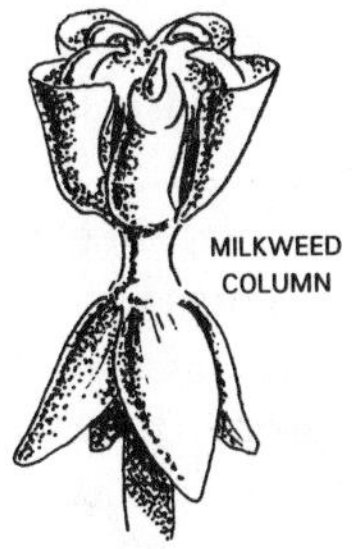

Figure 194

Columnar. Shaped like a column.

Coma. A tuft of hairs, especially on the tip of a seed. Figure 195.

Commissural. Of or pertaining to a commissure.

Commissure. The face by which two carpels join one another, as in the Umbelliferae (Apiaceae). Figure 196.

Comose. With a tuft of hairs or coma. Figure 195.

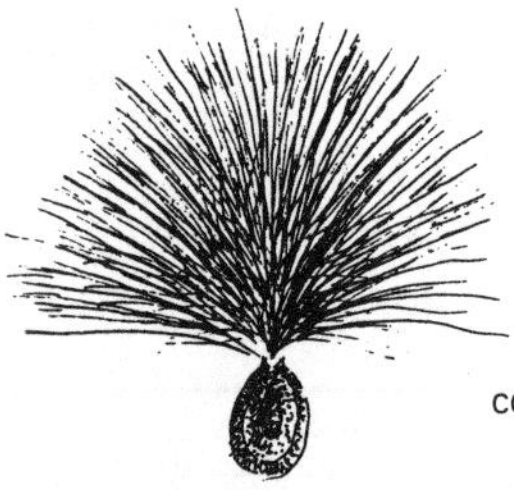

Figure 195

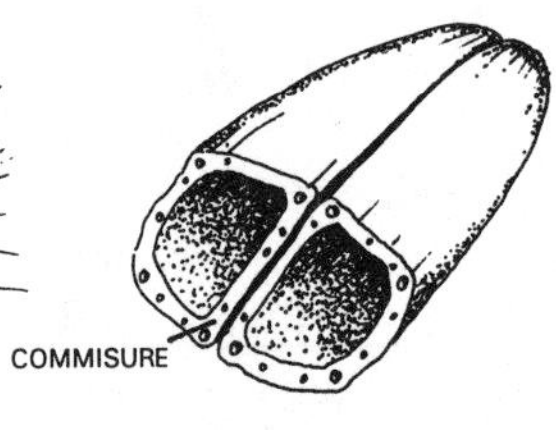

Figure 196

Complanate. Flattened. Figure 197.

Complete. With all of the parts typically belonging to it, as a flower with sepals, petals, stamens, and pistils. Figure 198.

Complicate. Folded together. Figure 199.

Figure 197

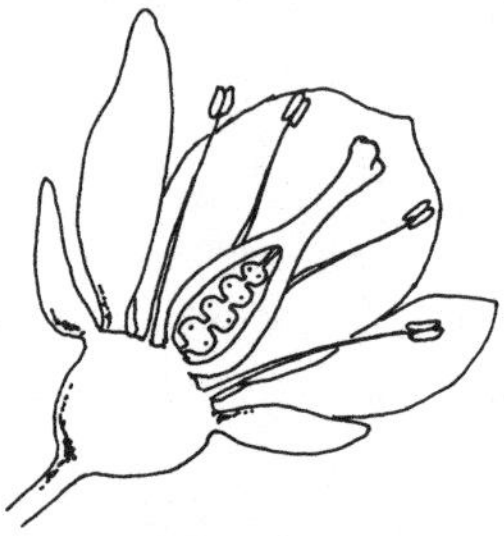

Figure 198

Figure 199

Compound. With two or more like parts in one organ.

Compound leaf. A leaf separated into two or more distinct leaflets. Figure 200.

Compound ovary. An ovary of two or more carpels. Figure 201.

Figure 200

Compressed. Flattened. Figure 197.

Concave. Hollowed out or curved inward. Figure 202.

Concavo-concave. Concave on both sides. Figure 203.

Concavo-convex. Concave on one side and convex on the other. Figure 204.

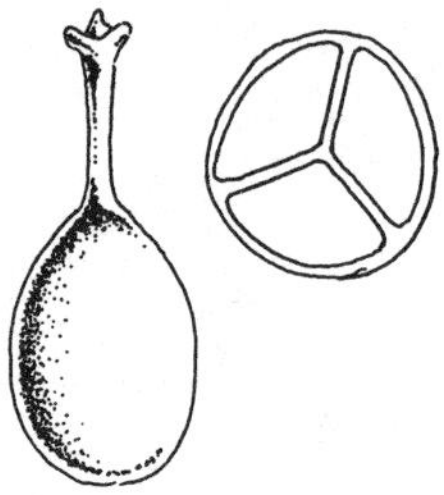

Figure 201

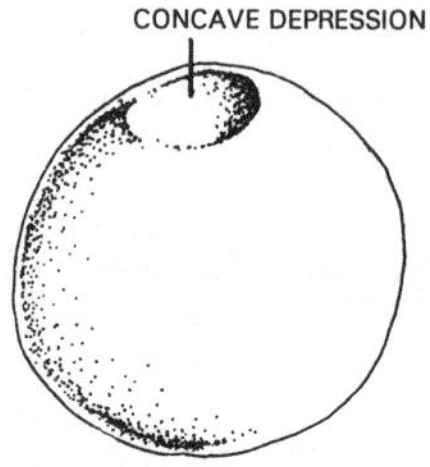

Figure 202

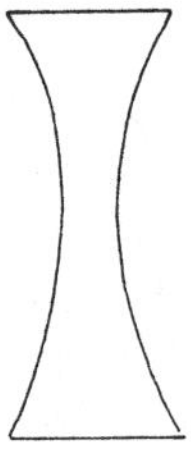

Figure 203

Figure 204

Concolored (adj. **concolorous**). With all parts of uniform color.

Conduplicate. Folded together lengthwise with the upper surface within, as the leaves of many grasses. Figure 205.

Cone. A dense cluster of sporophylls on an axis; a strobilus. Figure 206.

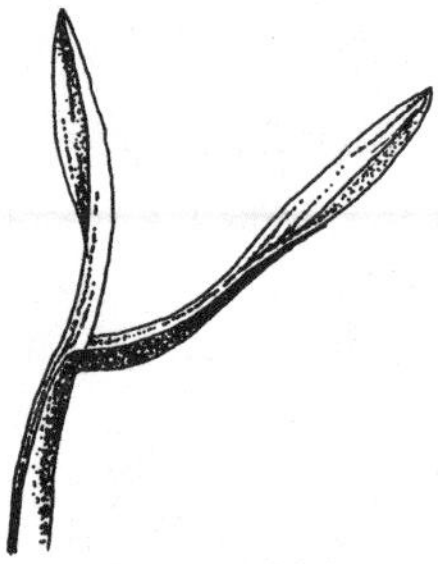

Figure 205

Figure 206

Confluent. Running together or blending of one part into another. Figure 207.

Congested. Densely crowded. Figure 208.

Conglomerate. Densely clustered. Figure 208.

Conic. Cone-shaped, with the point of attachment

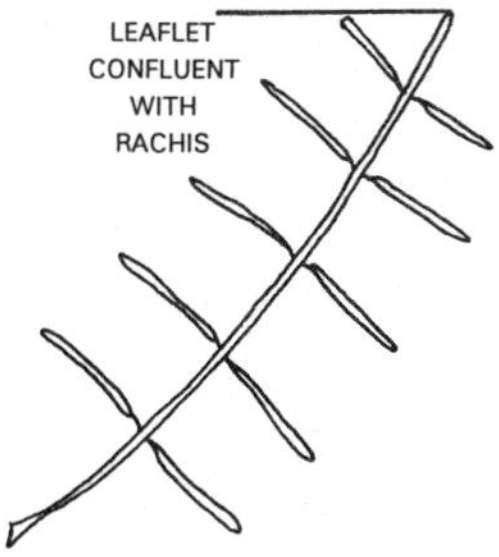

Figure 207

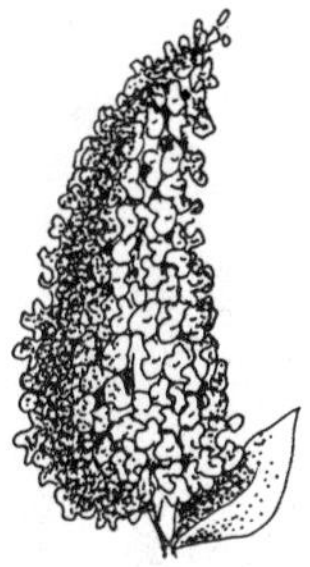
Figure 208

at the broad end. Figure 209.

Conical. See **conic**.

Coniferous. Bearing cones or strobili.

Conjugate. Coupled; in a single pair, as in a compound leaf with only two leaflets. Figure 210.

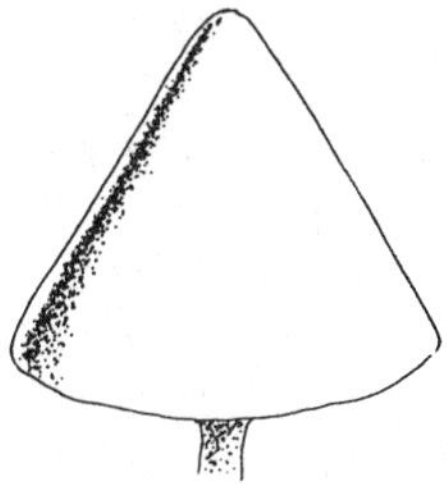
Figure 209

Connate. Fusion of like parts, as the fusion of staminal filaments into a tube. Figure 211. (compare **adnate**)

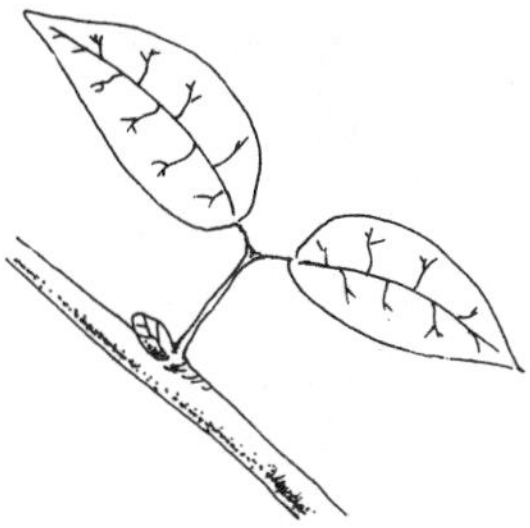
Figure 210

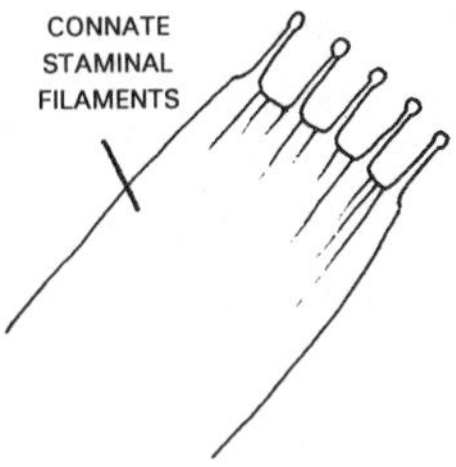

Figure 211

Connate-perfoliate. With the bases of opposite leaves fused around the stem. Figure 212.

Connective. The portion of the stamen connecting the two pollen sacs of an anther. Figure 213.

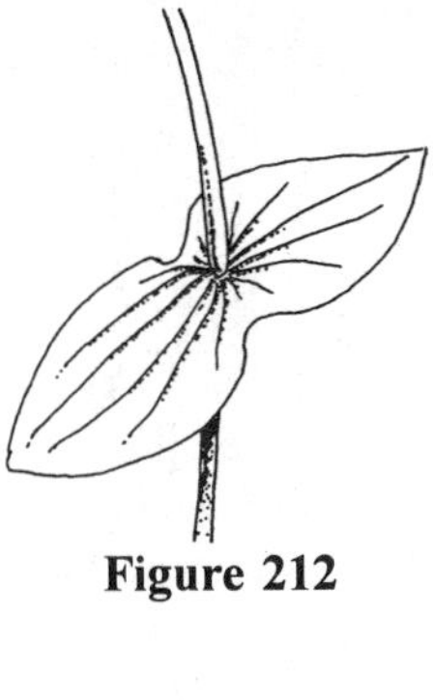
Figure 212

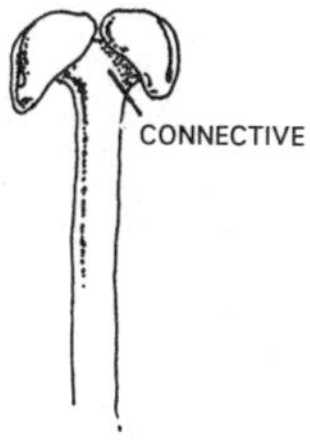

Figure 213

Connivent. Converging, but not actually fused or united. Figure 214.

Consimilar. Similar to one another.

Conspecific. Of the same species.

Constricted. Drawn together or narrowed. Figure 215.

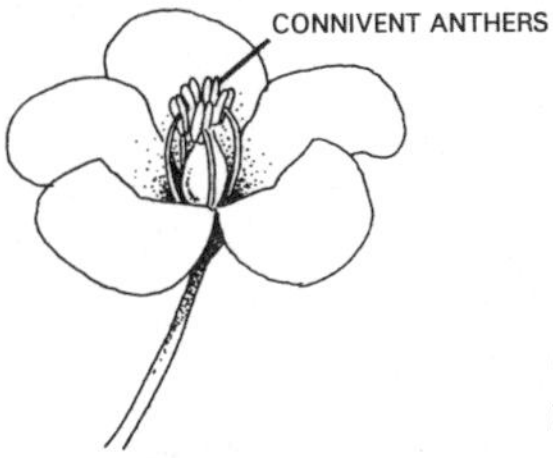

Figure 214

Figure 215

Contiguous. Adjoining; touching.

Continuous. Not jointed; not separating at maturity along a well-defined line of dehiscence.

Contorted. Twisted or bent. Figure 216.

Contracted. Narrowed; narrow, thick, and dense, as an inflorescence with crowded, short or appressed branches. Figure 217.

Figure 216

Figure 217

Convex. Rounded and curved outward on the surface. Figure 218.

Convolute. Rolled up longitudinally. Figure 219; with parts in an overlapping arrangement like shingles on a roof, as petals arranged as to be partially covered by one adjacent petal and partially overlapping the other adjacent petal. Figure 220.

Figure 218

Copious. Large in number or quantity; abundant.

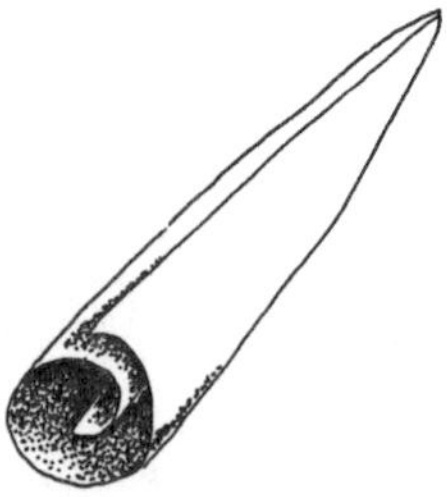

Figure 219

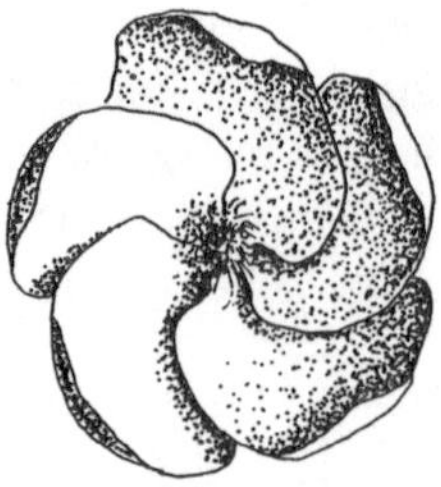

Figure 220

Coppice. A thicket of bushes or small trees; sprouts arising from a stump.

Copse. A thicket of bushes or small trees; woods.

Coralloid. Corallike. Figure 221.

Cordate. Heart-shaped, with the notch at the base. Figure 222.

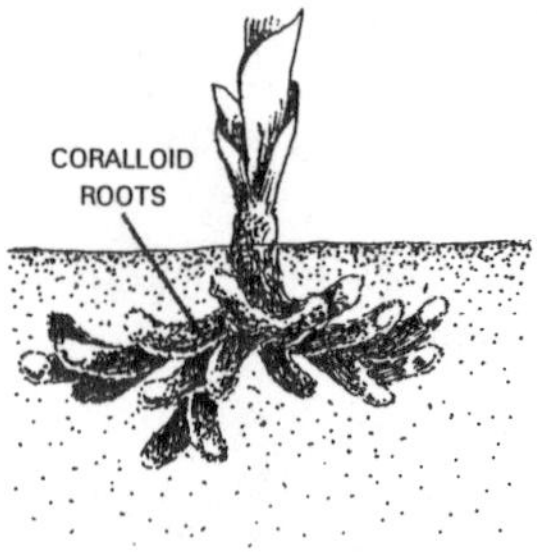

Figure 221

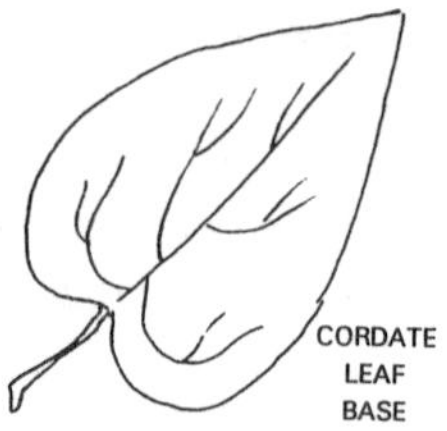

Figure 222

Cordiform. See **cordate.**

Coreaceous. See **coriaceous.**

Coriaceous. With a leathery texture.

Corm. A short, solid, vertical underground stem with thin papery leaves. Figure 223. (compare **bulb)**

Cormel. A small corm arising at the base of a larger corm.

Cormous. With corms.

Corneous. Horny.

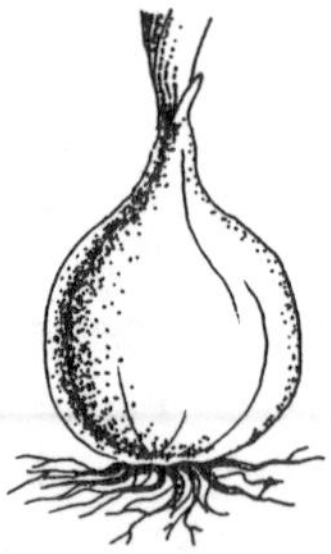

Figure 223

Corniculate. With small horn-like protuberances. Figure 224.

Cornute. Horned. Figure 224.

Corolla. The collective name for all of the petals of a flower; the inner perianth whorl. Figure 225.

Corolla lobe. One of the free portions of a corolla of united petals. Figure 226.

Corolla tube. The hollow, cylindric portion of a corolla of united petals. Figure 226.

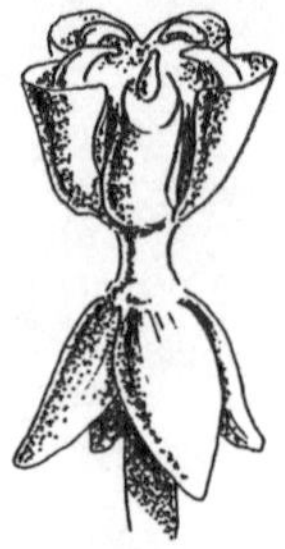

Figure 224

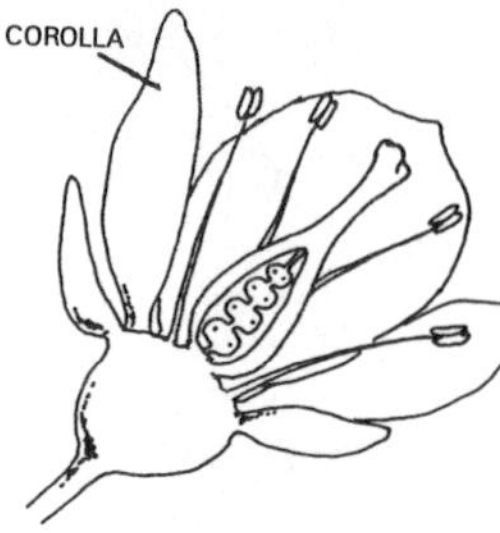

Figure 225

Corona. Petal-like or crown-like structures between the petals and stamens in some flowers; a crown. Figures 227 and 228.

Coroniform. Crown-shaped. Figures 227 and 228.

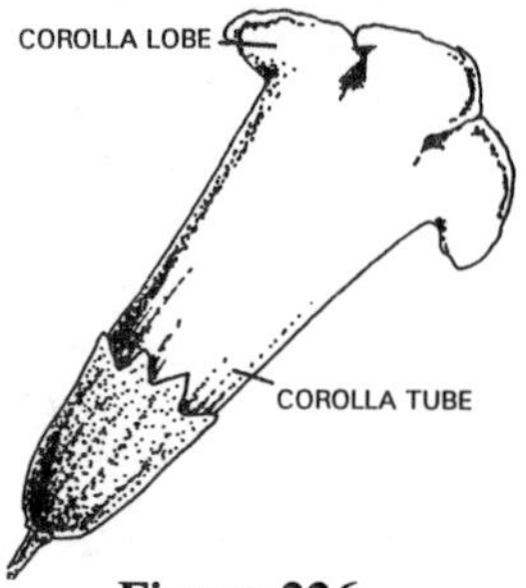

Figure 226

Corrugated. Wrinkled or folded into alternating furrows and ridges. Figures 229 and 230.

Corrugation. A wrinkle, fold, furrow, or ridge. Figures 229 and 230.

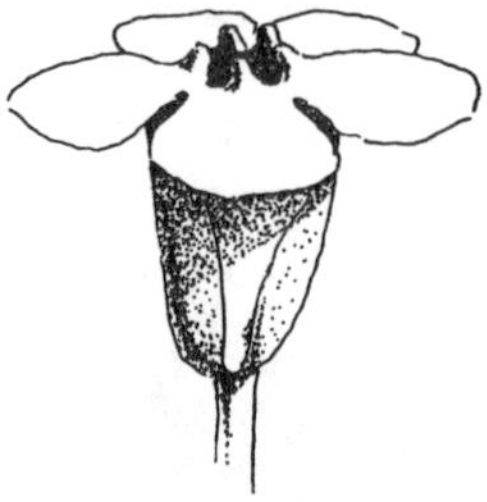

Figure 227

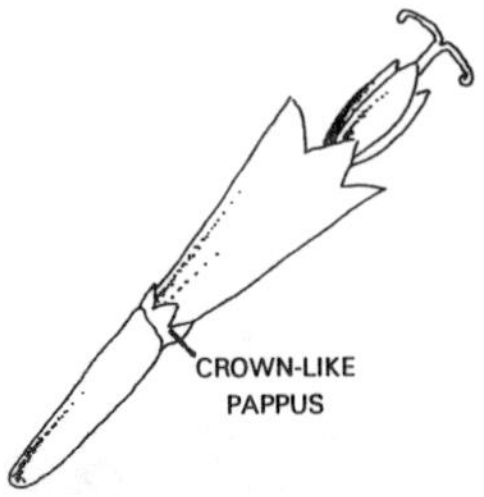

Figure 228

Figure 229

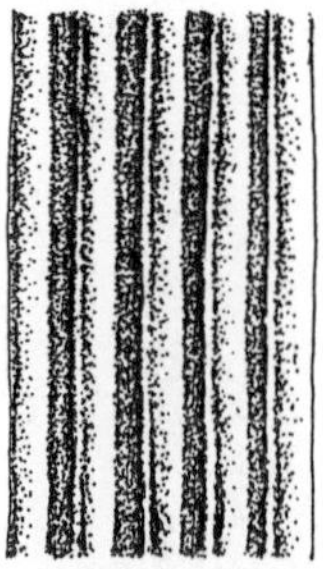

Figure 230

Cortex. Bark or rind; root tissue between the

epidermis and the stele. Figure 231.

Cortical. Of or pertaining to the cortex.

Corymb. A flat-topped or round-topped inflorescence, racemose, but with the lower pedicels longer than the upper. Figures 232 and 233.

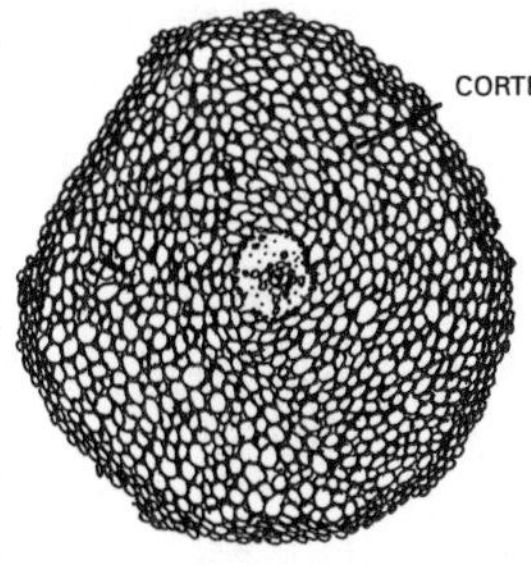

Figure 231

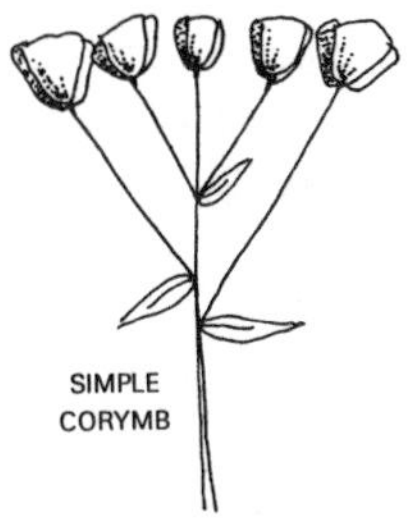

Figure 232

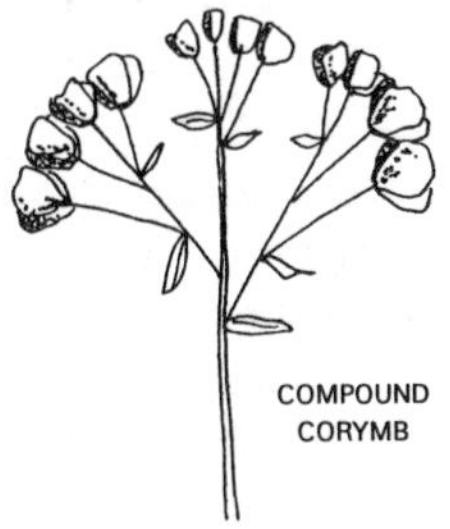

Figure 233

Corymbiform. An inflorescence with the general appearance, but not necessarily the structure, of a true corymb.

Corymbose. Having flowers in corymbs. The term is sometimes used in the same sense as **corymbiform**.

Costa (pl. **costae**). A rib or prominent midvein. Figure 234.

Costate. Ribbed. Figure 234.

Costular. Pertaining to the ribs or veins.

Cotyledon. A primary leaf of the embryo; a seed leaf. Figures 235 and 236.

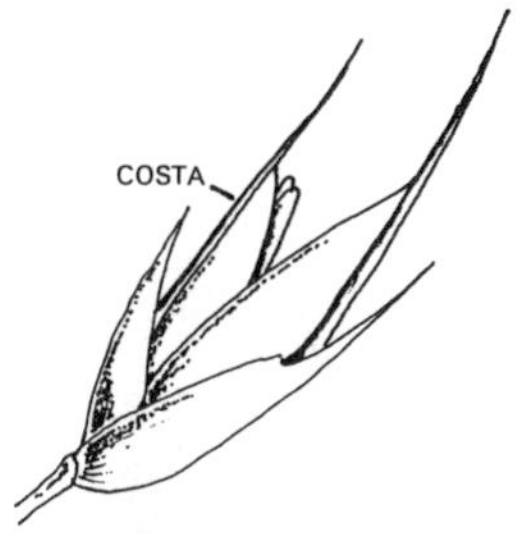

Figure 234

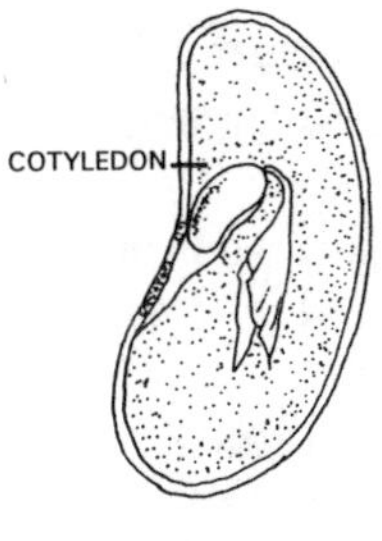

Figure 235

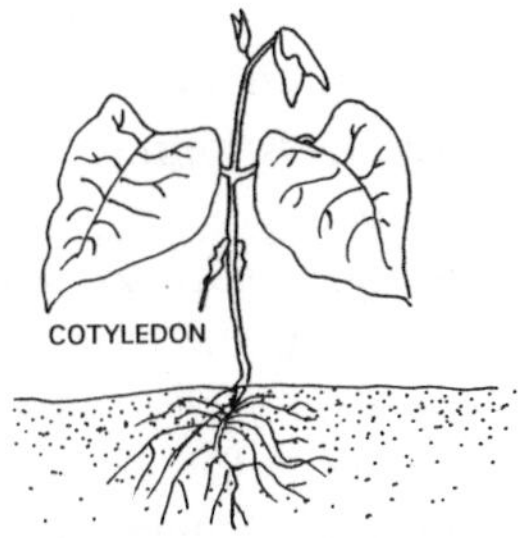

Figure 236

Cotyliform. Cup-shaped. Figure 237.

Crateriform. Bowl-shaped. Figure 237.

Creeping. Growing along the surface of the ground, or just beneath the surface, and producing roots, usually at the nodes. Figure 238.

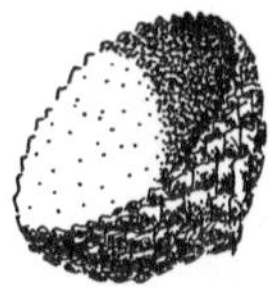

Figure 237

Cremocarp. See **schizocarp**.

Crenate. With rounded teeth along the margin. Figure 239.

Crenation. A rounded projection or tooth along the margin of a leaf. Figure 239.

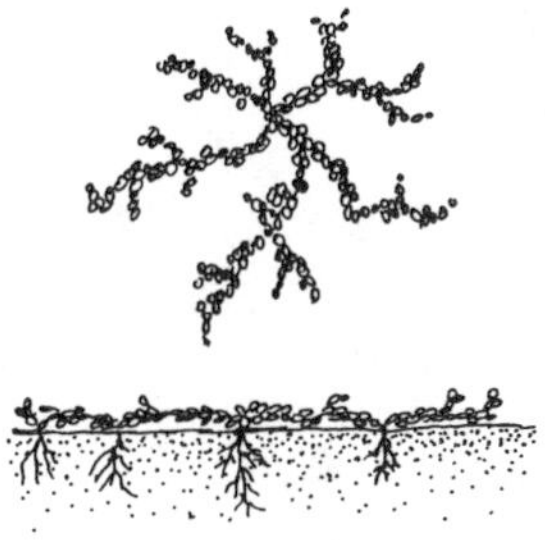

Figure 238

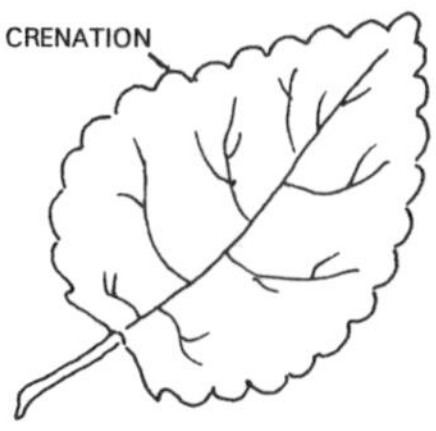

Figure 239

Crenature. See **crenation**.

Crenulate. With very small rounded teeth along the margin. Figure 240.

Crenulation. A very small rounded tooth along a margin; a minute crenation. Figure 240.

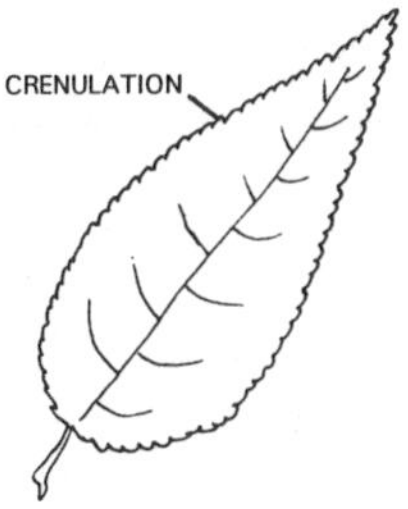

Figure 240

Creosote. An oily liquid with a strong, penetrating odor.

Crest. An elevated ridge or rib on a surface. Figure 241.

Crested. With a crest, usually on the back or at the summit. Figures 241 and 242.

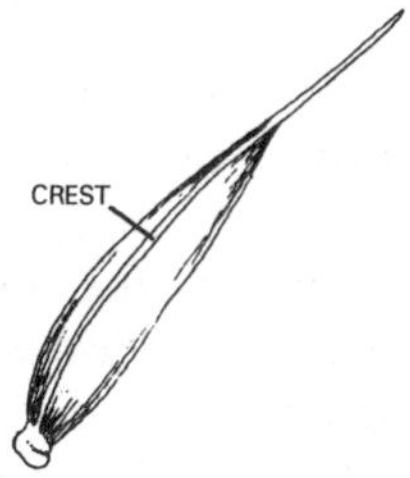

Figure 241

Cribriform. Sievelike. Figure 243.
Cribrose. See **cribriform**.
Cribrous. See **cribriform**.

Figure 242

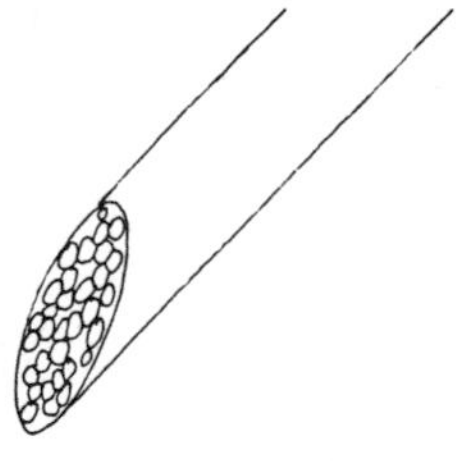

Figure 243

Crinite. With tufts of long, soft hairs. Figure 244.
Crinkled. Flattened and somewhat twisted, kinked, or curled. Figure 245.

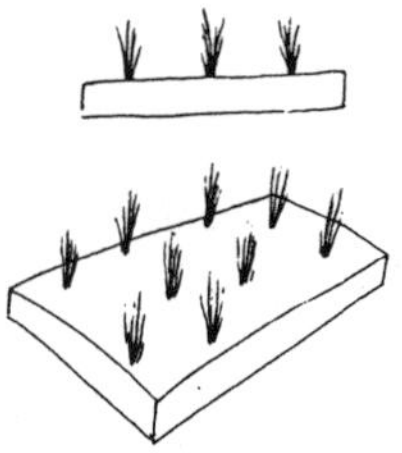

Figure 244

Figure 245

Crispate. See **crisped**.
Crisped. Curled, wavy or crinkled. Figure 246.
Cristate. With a terminal tuft or crest. Figure 247.

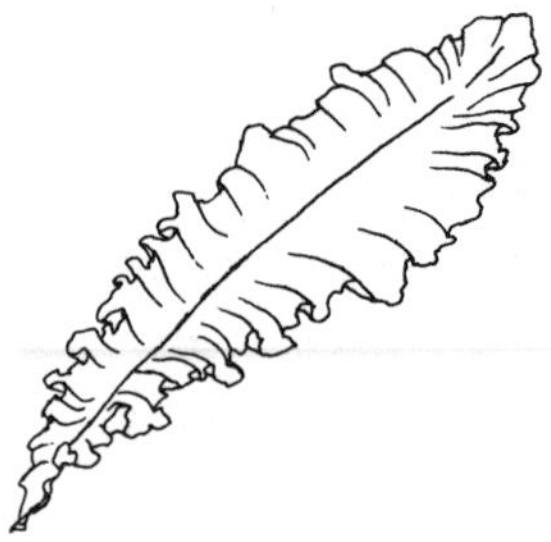

Figure 246

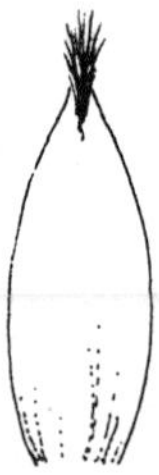

Figure 247

Crown. The persistent base of a herbaceous perennial. Figure 248; the top part of a tree; a corona. Figure 249.
Cruciate. See **cruciform**.
Cruciform. Cross-shaped. Figure 250.
Crustaceous. Dry and brittle.
Crustose. Hard and brittle.

Figure 248

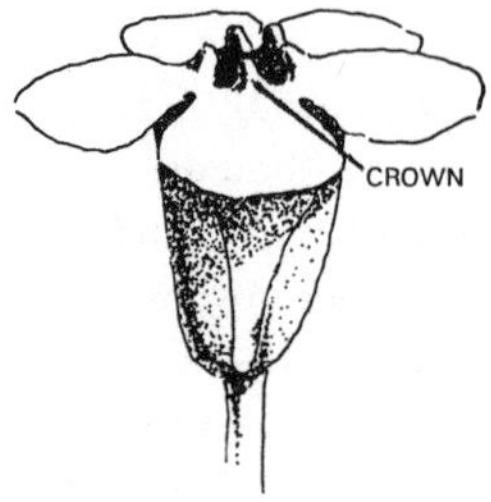

Figure 249

Cryptogam. A plant that does not produce seeds. (compare **phanerogam**)
Cucullate. Hooded or hood-shaped. Figure 251.
Cucullus. A hood. Figure 251; a seed covering external to the seed coat. Figure 252.
Cucumiform. Cucumber-shaped; cylindrical with rounded ends.
Culm. A hollow or pithy stalk or stem, as in the grasses, sedges, and rushes. Figure 253.

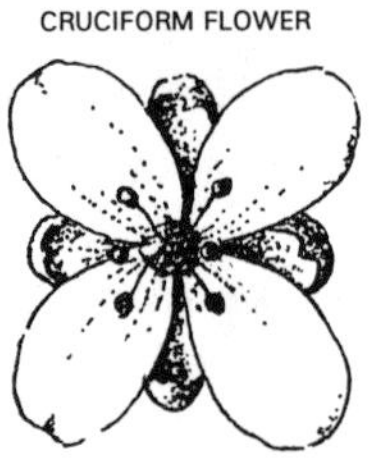

Figure 250

Figure 251

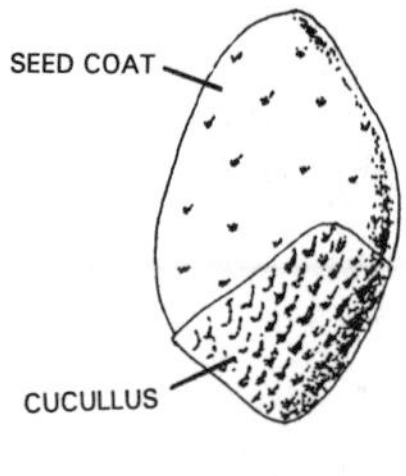

Figure 252

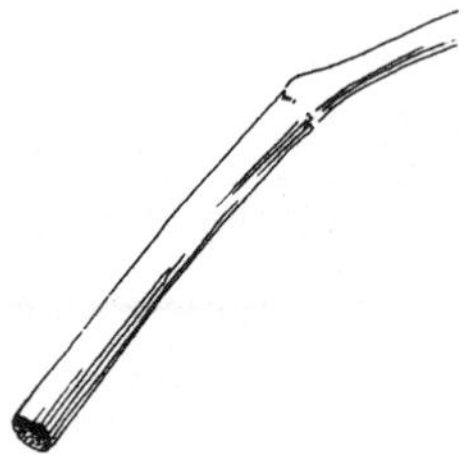

Figure 253

Cultivar. A form of plant originating under cultivation.
Cuneate. Wedge-shaped, triangular and tapering to a point at the base. Figure 254.
Cuneiform. See **cuneate**.
Cupulate. Cup-shaped; with a small cup-like structure. Figure 255.

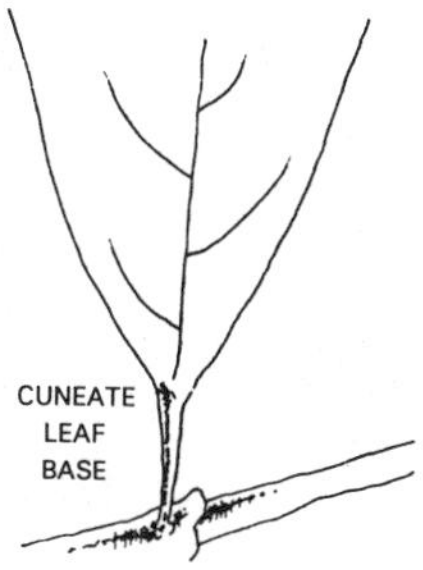

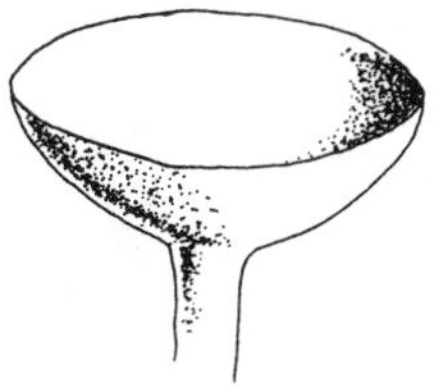

Figure 254 **Figure 255**

Cupule. A cup-shaped involucre, as in an acorn. Figures 256 and 257.

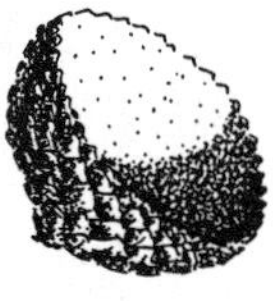

Figure 256 **Figure 257**

Cusp. A short, sharp, abrupt point, usually at the tip of a leaf or other organ. Figure 258.

Cuspidate. Tipped with a short, sharp, abrupt point (cusp). Figure 258.

Cuticle. The waxy layer on the surface of a leaf or stem.

Cyathiform. With the form of a cyathium; cup-shaped.

Cyathium (pl. **cyathia**). The inflorescence in the genus *Euphorbia*, consisting of a cup-like involucre containing a single pistil and male flowers with a single stamen. Figure 259.

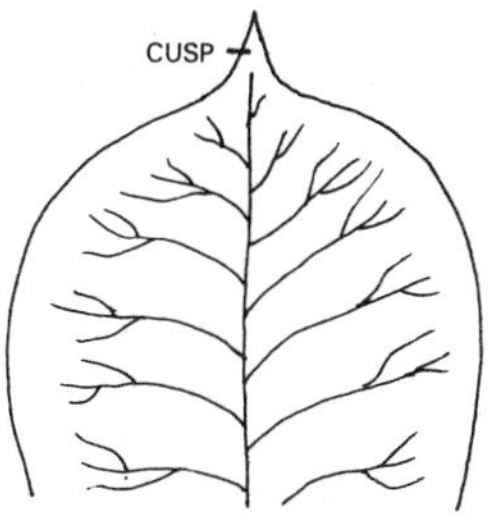

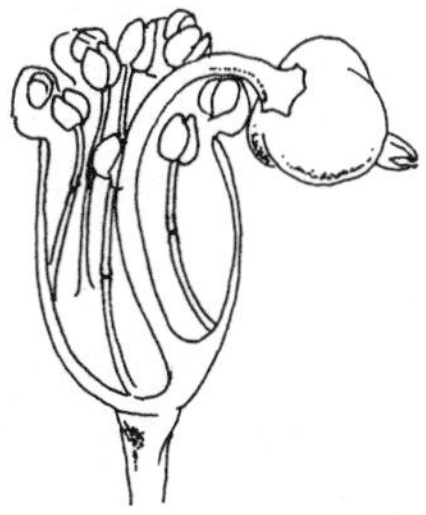

Figure 258 **Figure 259**

Cyclic. Occurring in apparent cycles or whorls, as the parts of a typical flower. Figure 260.

Cylindric. Cylinder-shaped; elongate and round in cross section. Figure 261.

Cylindrical. See **cylindric**.

Cylindroid. Shaped like a cylinder, but elliptic in cross-section. Figure 262.

Cymbiform. Boat-shaped. Figure 263.

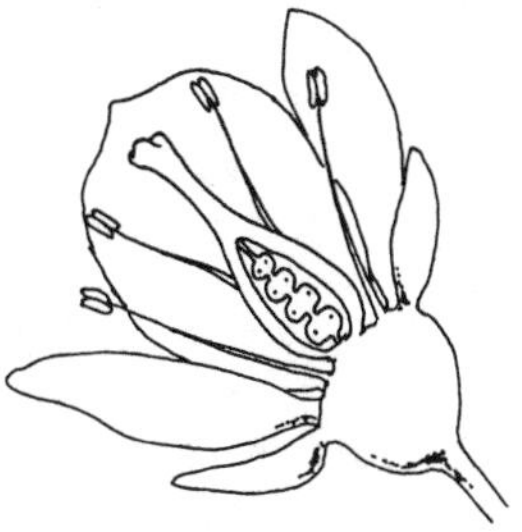

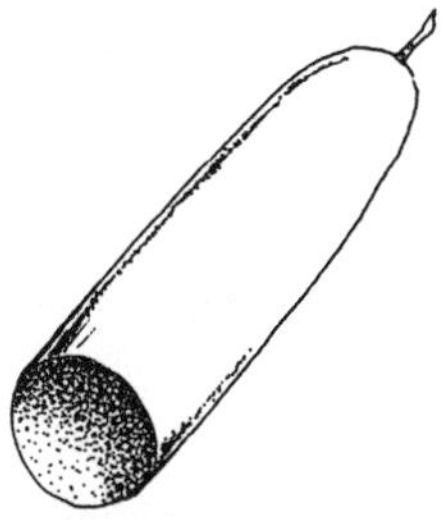

Figure 260 **Figure 261**

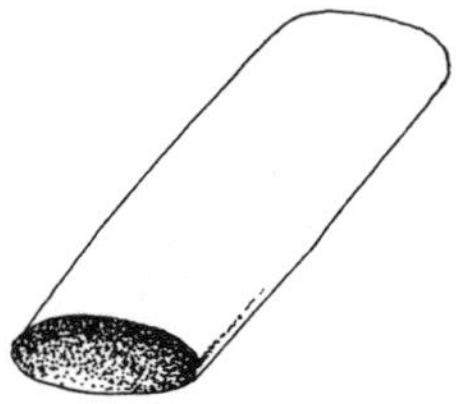

Figure 262 **Figure 263**

Cyme. A flat-topped or round-topped determinate inflorescence, paniculate, in which the terminal flower blooms first. Figure 264.

Cymose. With flowers in a cyme. Figure 264.

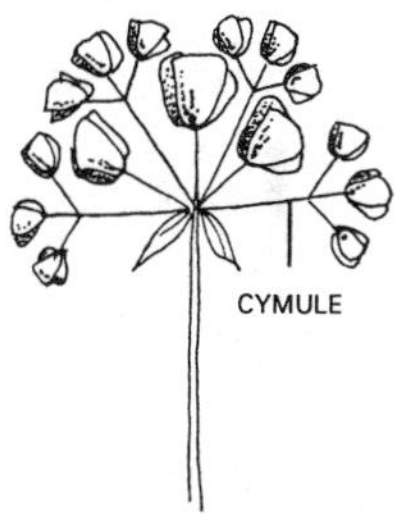

Figure 264

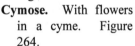

Cymule. A small cyme or a small section of a compound cyme. Figure 264.

Deciduous. Falling off, as leaves from a tree; not evergreen; not persistent.

Declinate. See **declined**.

Declined. Curved downward. Figure 265.

Decompound. More than once-compound, the leaflets again

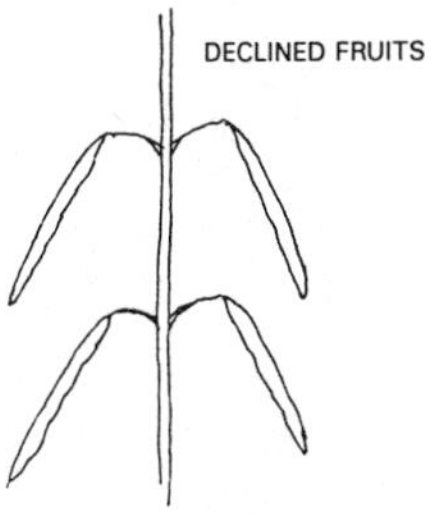

Figure 265

divided. Figure 266.

Decumbent. Reclining on the ground but with the tip ascending. Figure 267.

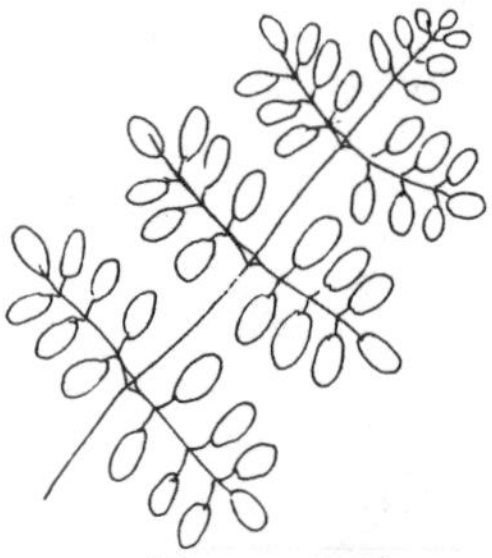

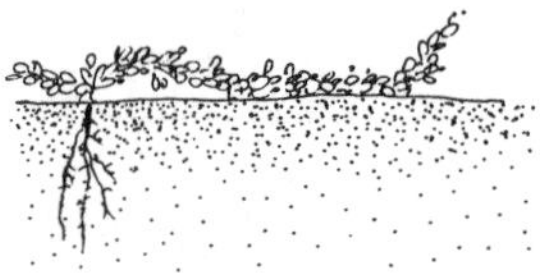

Figure 266 **Figure 267**

Decurrent. Extending downward from the point of insertion, as a leaf base that extends down along the stem. Figure 268.

Decussate. Arranged along the stem in pairs, with each pair at right angles to the pair above or below. Figure 269.

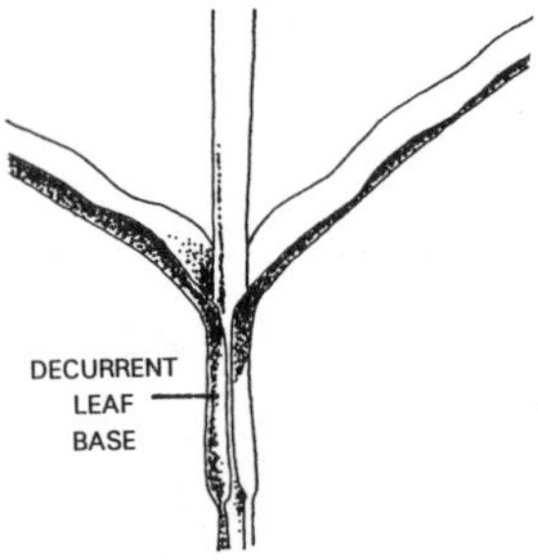

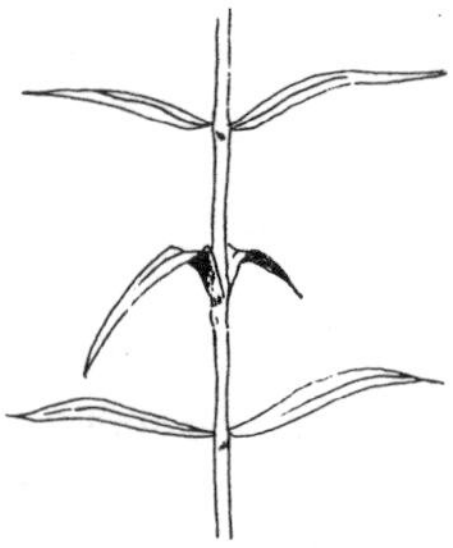

Figure 268 **Figure 269**

Deflexed. Bent abruptly downward. Figure 270.

Dehiscence. The opening at maturity of fruits and anthers.

Dehiscent. Opening at maturity or when ripe to release the contents, as a fruit or an anther. Figure 271.

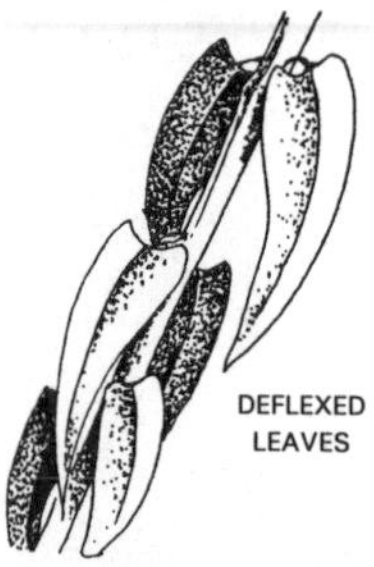

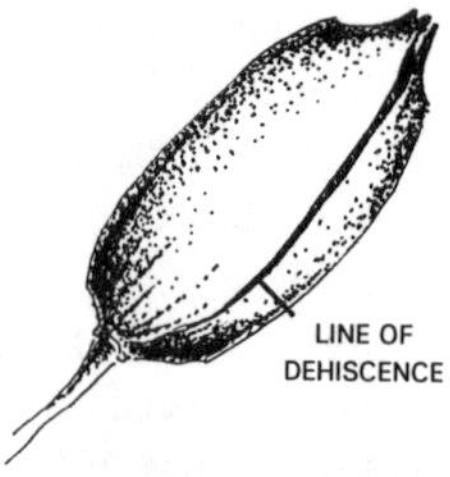

Figure 270 **Figure 271**

Deliquescent. An irregular pattern of branching without a well defined central axis from bottom to top. Figure 272.

Deltoid. With the shape of the Greek letter delta; shaped like an equilateral triangle. Figure 273.

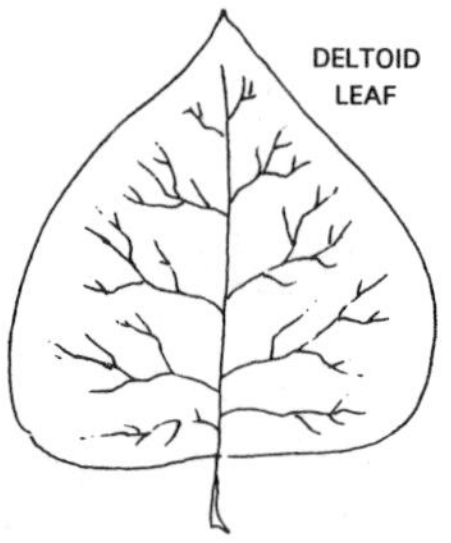

Figure 272 **Figure 273**

Dendriform. With a treelike form.

Dendritic. With a branching pattern similar to that in a tree, as in some hairs in the Cruciferae (Brassicaceae). Figure 274.

Dendroid. See **dendriform**.

Dentate. Toothed along the margin, the teeth directed outward rather than forward. Figure 275.

Dentation. The state of being dentate; a tooth along a margin, as in a leaf. Figure 275.

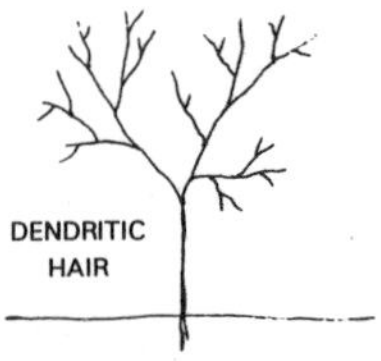

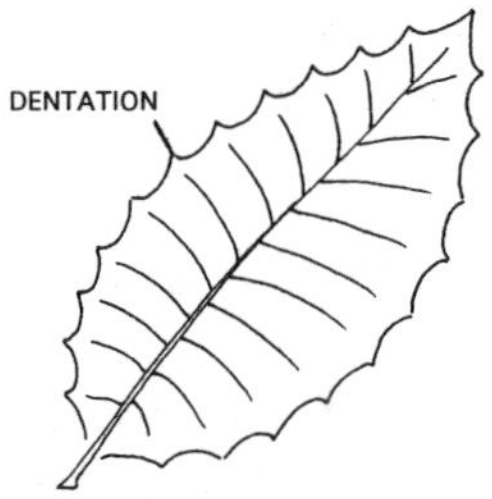

Figure 274 **Figure 275**

Denticle. A small tooth or toothlike projection. Figure 276.

Denticulate. Dentate with very small teeth. Figure 276.

Denticulation. The state of being denticulate; a denticle. Figure 276.

Dentiform. Tooth-shaped.

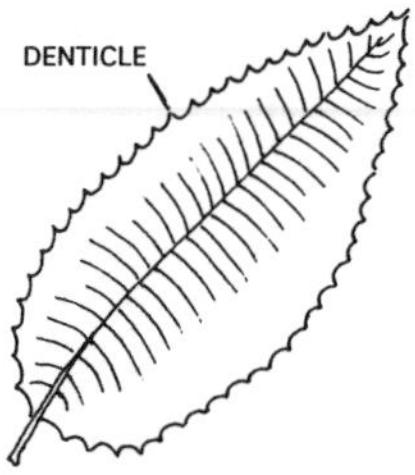

Figure 276

Denudate. Stripped bare; denuded.

Depauperate. Stunted or poorly developed, usually due to adverse environmental conditions.

Depressed. Flattened down from above. Figure 277.

Dermal. Of or pertaining to the epidermis.

Descending. Directed downward at a moderate angle. Figure 278.

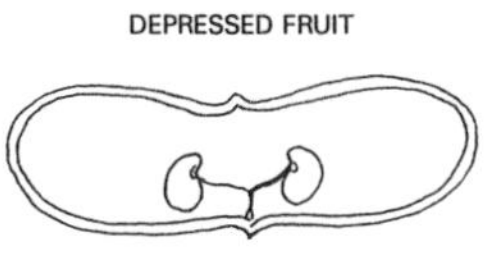

Figure 277

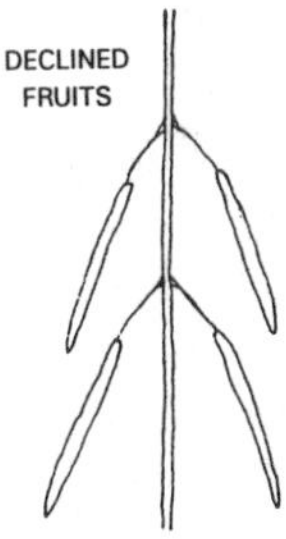

Figure 278

Determinate. Describes an inflorescence in which the terminal flower blooms first, halting further elongation of the main axis. Figure 279.

Dextrorse. Turned to the right or spirally arranged to the right, as in the leaves on some stems. Figure 280. (compare **sinistrorse**)

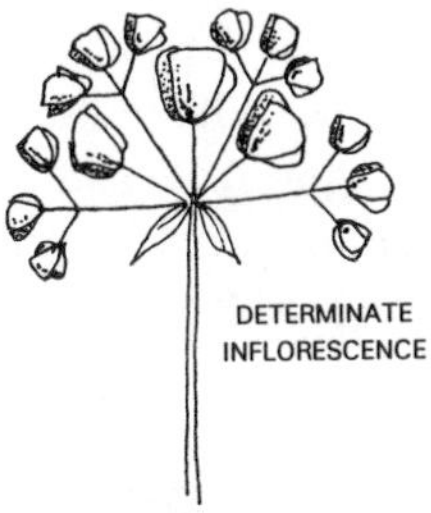

Figure 279

Figure 280

Di- (prefix). Meaning two or twice.

Diadelphous. Stamens united into two often unequal sets by their filaments. Figures 281 and 282.

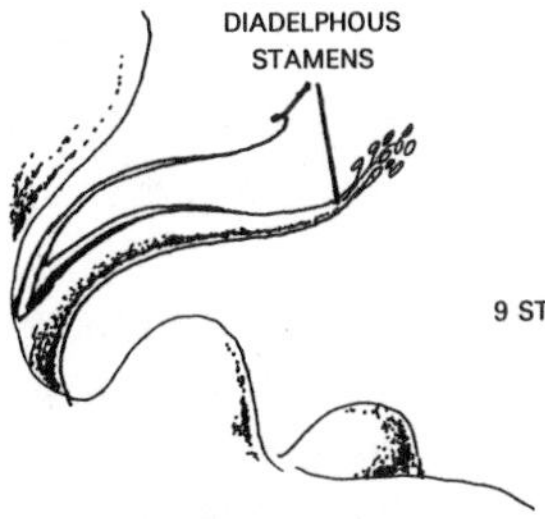

Figure 281

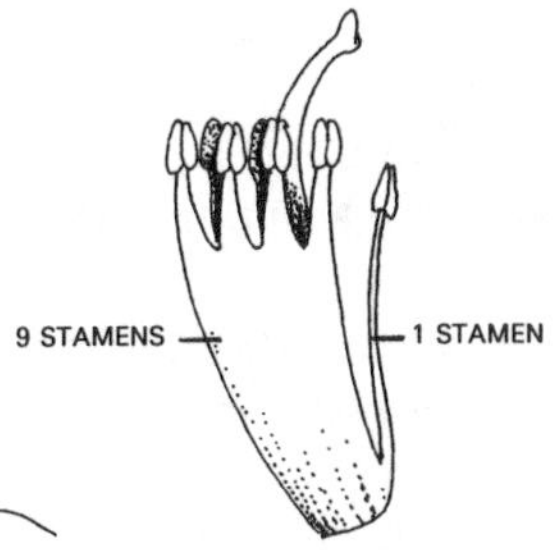

Figure 282

Diandrous. With two stamens.

Diaphanous. Translucent.

Dicarpellate. See **bicarpellate**.

Dichasium. A cymose inflorescence in which each axis produces two opposite or subopposite lateral axes. Figure 283.

Dichlamydeous. With two types of perianth whorls, i.e., calyx and corolla. Figure 284.

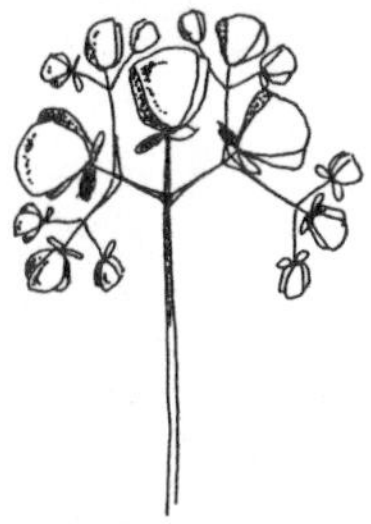

Figure 283

Figure 284

Dichogamic. See **dichogamous**.

Dichogamous. With the pistils and stamens maturing at different times to prevent self-fertilization. (compare **homogamous**)

Dichogamy. Having dichogamous flowers.

Dichotomous. Branched or forked into two more or less equal divisions. Figure 285.

Diclinous. With the stamens and pistils in separate flowers; imperfect. Figure 286.

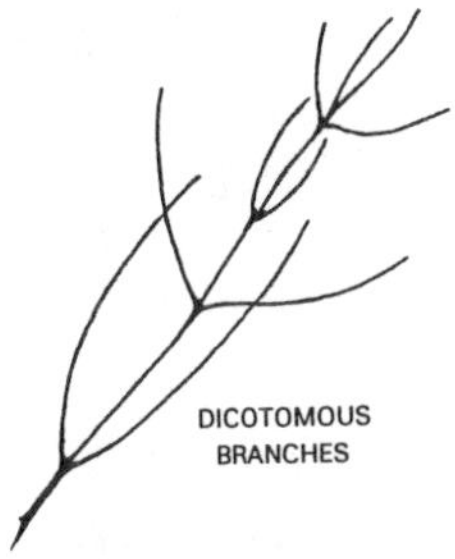

Figure 285

Dicotyledonous. With two cotyledons. Figure 287.

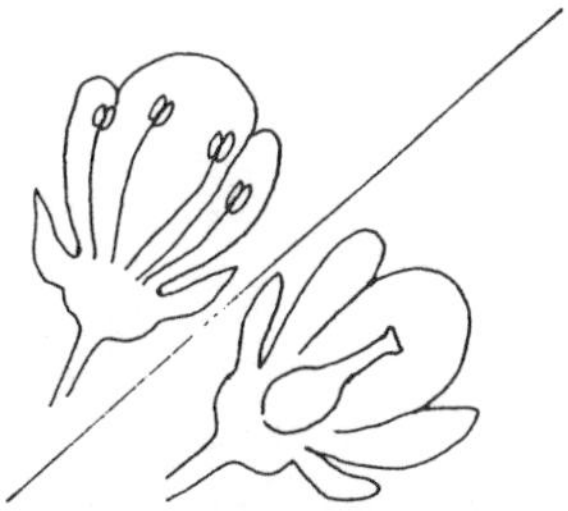

Figure 286

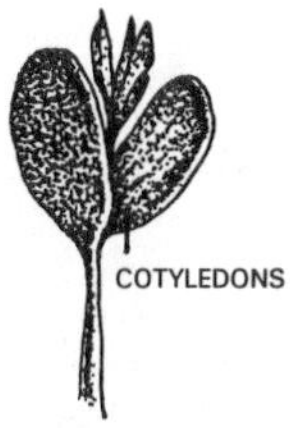

Figure 287

Dicyclic. With two whorls.

Didymous. Developing or occurring in pairs; twin. Figure 288.

Didynamous. With two pairs of stamens of

unequal length; occurring in pairs. Figure 289.

Diecious. See **dioecious**.

Diffuse. Widely or loosely spreading. Figure 290.

Digitate. Lobed, veined, or divided from a common point, like the fingers of a hand. Figure 291. (same as **palmate**)

Digitation. A digitlike lobe or division. Figure 291.

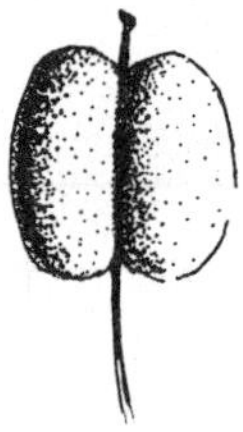

Figure 288

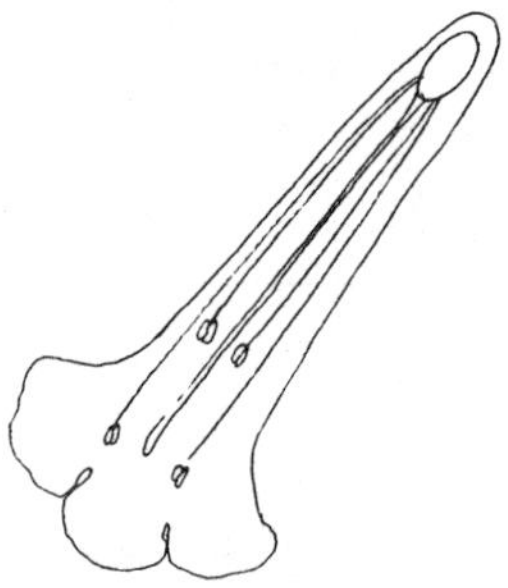

Figure 289

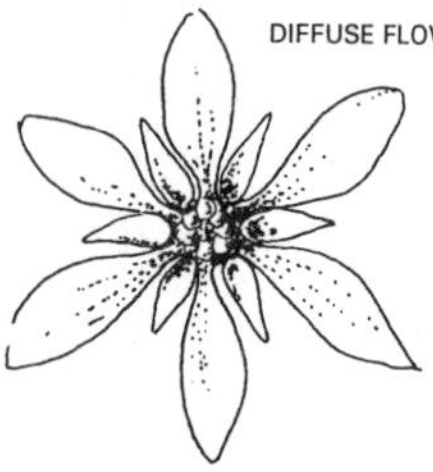

Figure 290

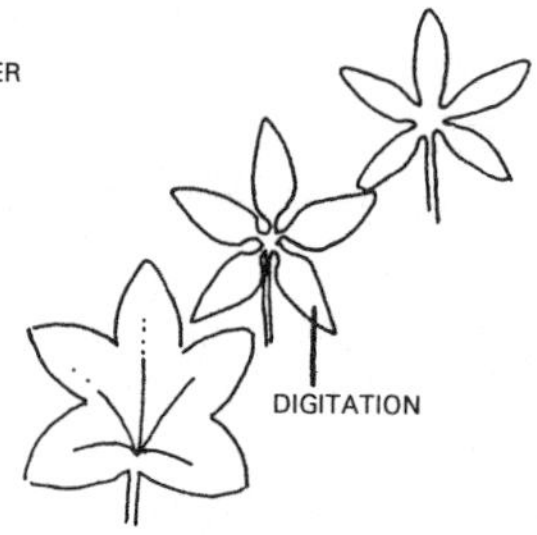

Figure 291

Digitiform. Fingerlike.

Digynous. With two pistils. Figure 292.

Dilated. Flattened or expanded. Figure 293.

Figure 292

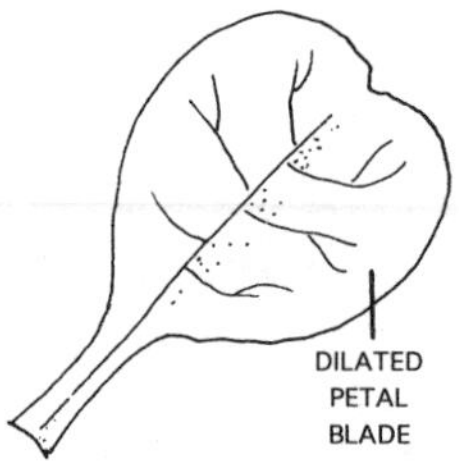

Figure 293

Dimerous. With parts arranged in sets or multiples of two.

Dimidiate. Divided into halves.

Dimorphic. With two forms.

Dimorphous. See **dimorphic**.

Dioecious. Flowers imperfect, the staminate and pistillate flowers borne on different plants. (compare **monoecious**)

Dioicous. See **dioecious**.

Dipetalous. See **bipetalous**.

Diploid. With two full sets of chromosomes in each cell.

Diplostemonous. With two series of stamens, the outer series opposite the sepals and the inner series opposite the petals; with twice as many stamens as petals. Figure 294.

Disarticulating. Separating at maturity at a joint. Figure 295.

Figure 294

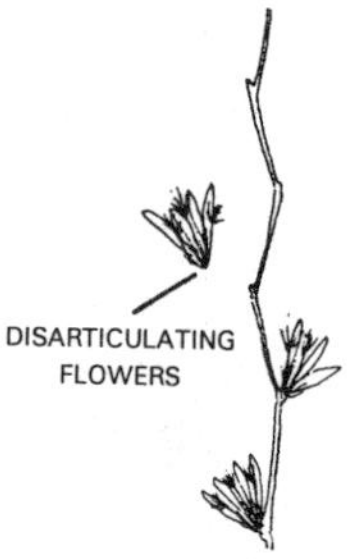

Figure 295

Disc. See **disk**.

Disciform. In the form of a disk. Figure 296.

Discoid. Resembling a disk. Figures 296 and 297; with disk flowers, as in an involucrate head of the Compositae (Asteraceae) which lacks ray flowers.

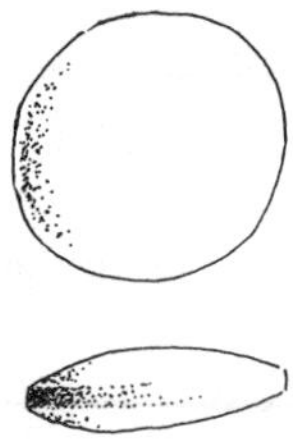

Figure 296

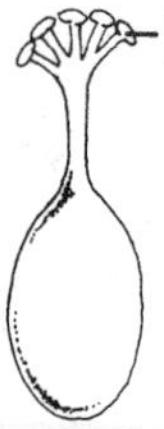

Figure 297

Disjunct. Occurring in widely separated geographic areas.

Disk. An enlargement or outgrowth of the receptacle around the base of the ovary; in the Compositae (Asteraceae) the central portion of the involucrate head bearing tubular or disk flowers. Figure 298.

Disk flower. A regular flower of the Compositae (Asteraceae). Figure 299.

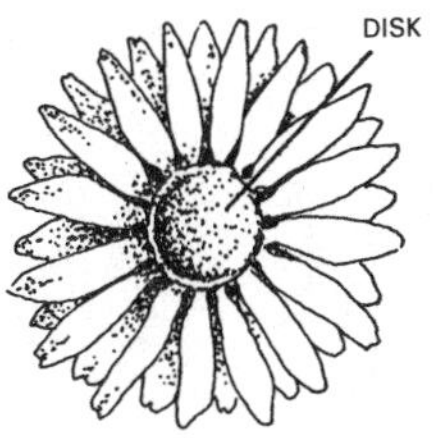

Figure 298

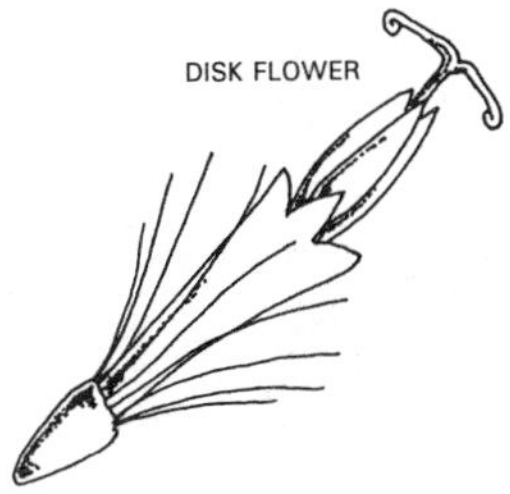

Figure 299

Dissected. Deeply divided into many narrow segments. Figure 300.

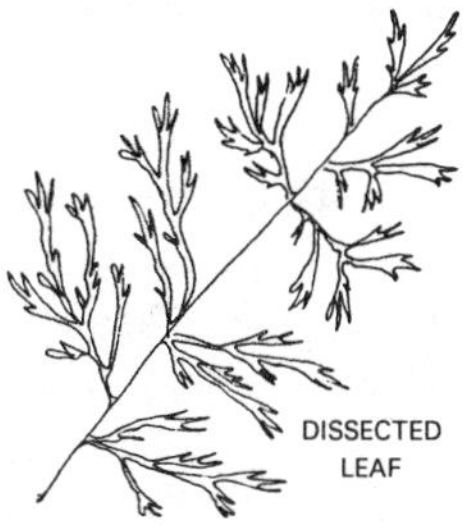

Figure 300

Dissepiment. See **septum**.

Distal. Toward the tip, or the end of the organ opposite the end of attachment. (compare **proximal**)

Distended. Stretched or swollen.

Distichous. In two vertical ranks or rows on opposite sides of an axis; two-ranked. Figure 301.

Distinct. Separate; not attached to like parts. Figure 302. (compare **connate**)

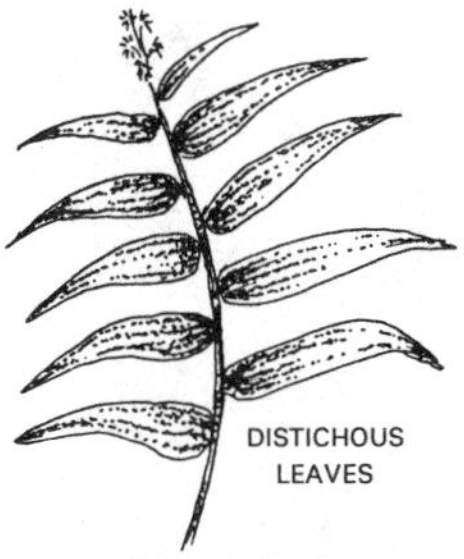

Figure 301

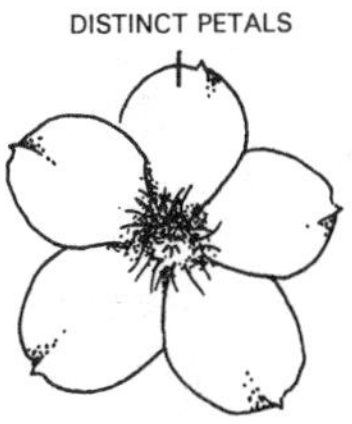

Figure 302

Diurnal. Occurring or opening in the daytime.

Divaricate. Widely diverging or spreading apart. Figure 303.

Divergent. Diverging or spreading. Figure 304.

Divided. Cut or lobed to the base or to the midrib. Figure 305.

Dolabriform. Ax-shaped or cleaver-shaped; pick-shaped; attached at some point other than the base, usually near the middle. Figure 306.

Dorsal. Pertaining to the back or outward surface of an organ in relation to the axis, as in the lower surface of a leaf; abaxial. Figure 307. (compare **ventral**)

Dorsiferous. Borne on the back. Figure 308.

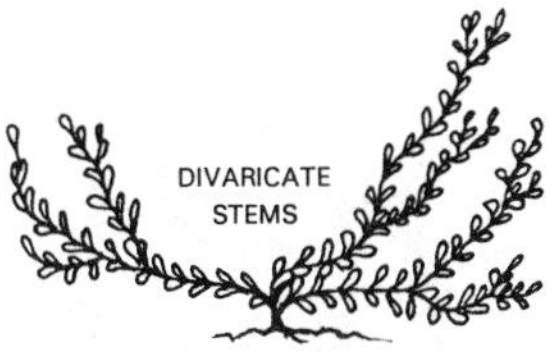

Figure 303

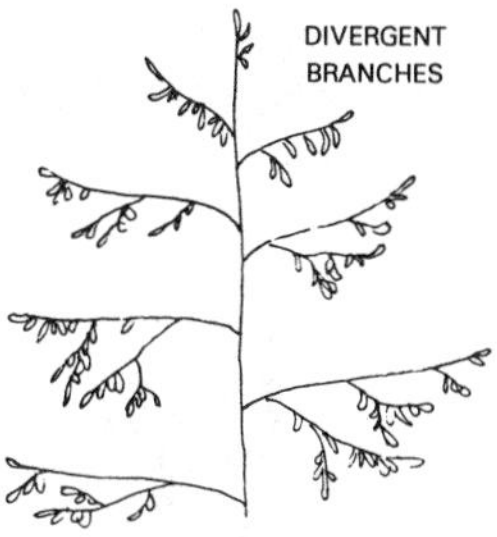

Figure 304

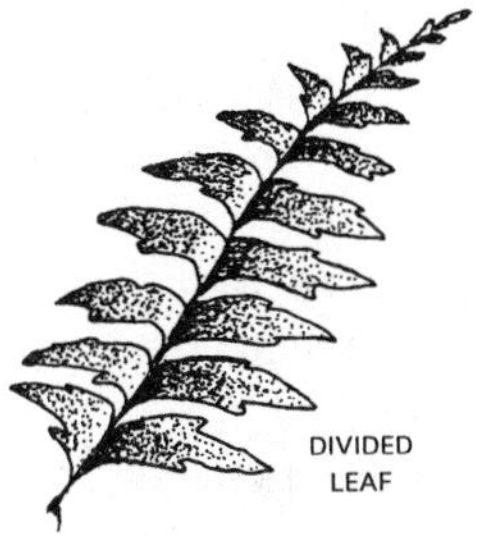

Figure 305

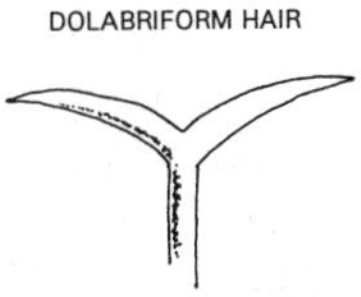

Figure 306

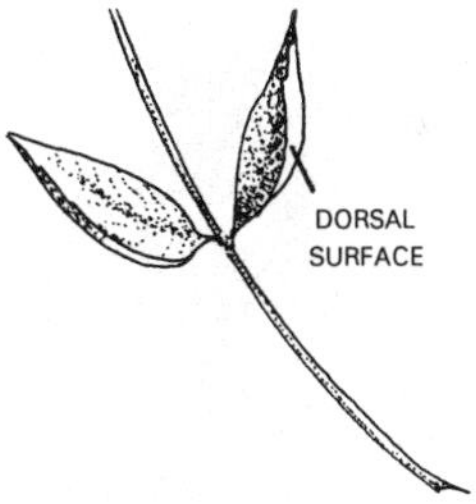

Figure 307

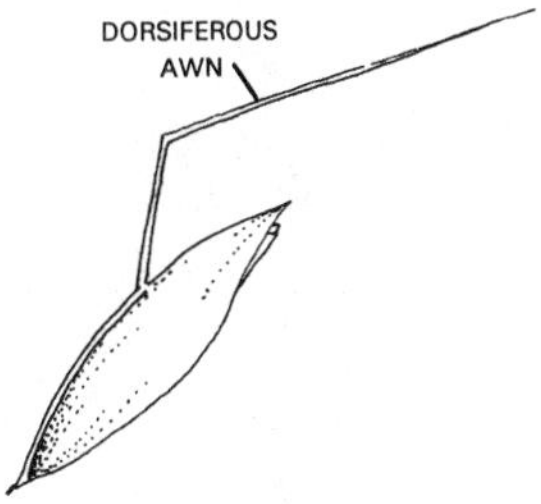

Figure 308

Dorsifixed. Attached at the back. Figure 309. (compare **versatile** and **basifixed**)

Dorsiventral. Having an upper and a lower surface; flattened from top to bottom. Figure 310.

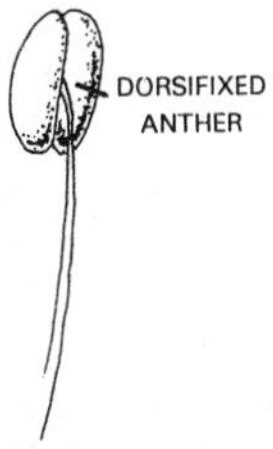

Figure 309

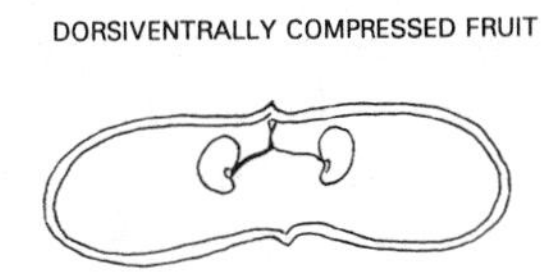

Figure 310

Dorsoventral. See **dorsiventral**.

Double. Having a larger number of petals than usual.

Double-serrate. See **biserrate**.

Downy. Covered with soft, fine hairs. Figure 311.

Drooping. Bending or hanging down. Figure 312.

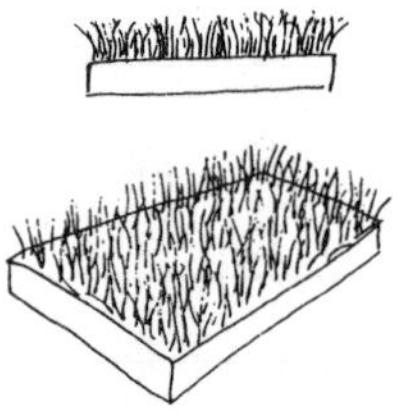

Figure 311

Figure 312

Drupaceous. Bearing drupes; resembling a drupe or consisting of drupes.

Drupe. A fleshy, indehiscent fruit with a stony endocarp surrounding a usually single seed, as in a peach or cherry. Figure 313.

Drupelet. A small drupe, as in the individual segments of a raspberry fruit. Figure 314.

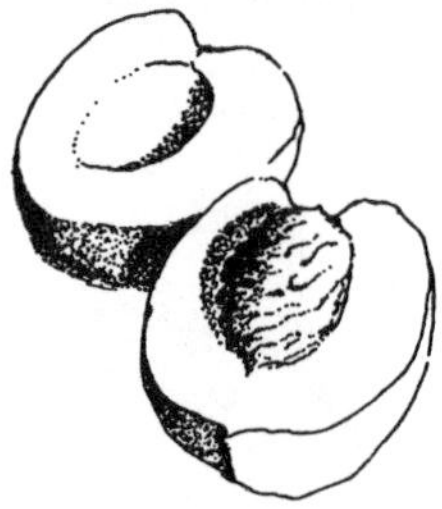

Figure 313

Figure 314

Dyad. A group of two. Figure 315.

E- (prefix). Meaning without, or from, or away from.

Eared. See **auriculate**.

Ebracteate. Without bracts.

Eccentric. Off-center; not positioned directly on the central axis. Figure 316.

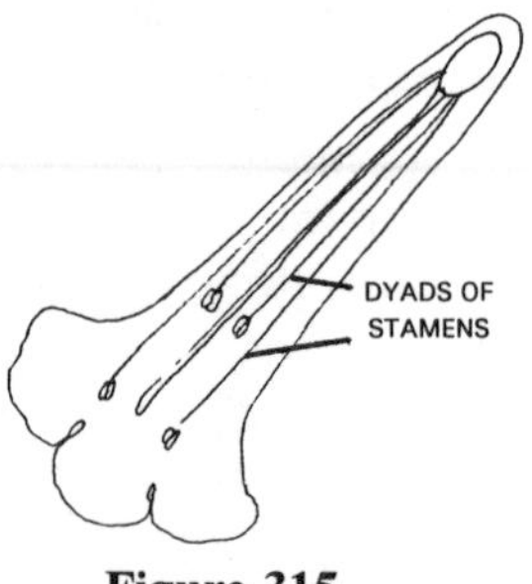

Figure 315

Echinate. With prickles or spines. Figure 317.

Echinulate. With very small prickles or spines. Figure 318.

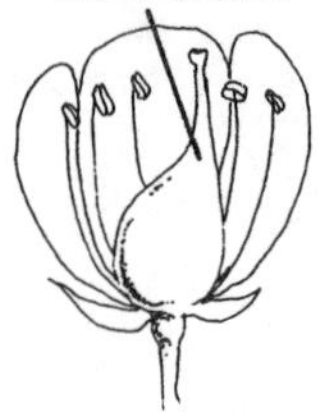

Figure 316

Figure 317

Ecostate. Without a midrib.

Ecotype. Those individuals adapted to a specific environment.

Ectocarp. See **exocarp**.

Edaphic. Due to, or pertaining to, the soil.

Edentate. Without teeth. Figure 319.

Figure 318

Eglandular. Without glands.

Elaminate. Without a blade.

Ellipsoid. A solid body elliptic in long section and circular in cross section. Figure 320.

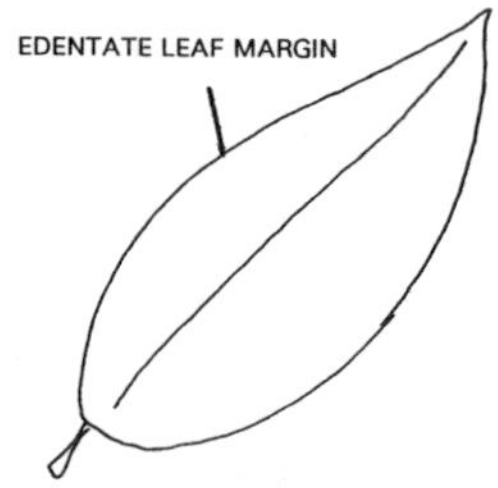

Figure 319

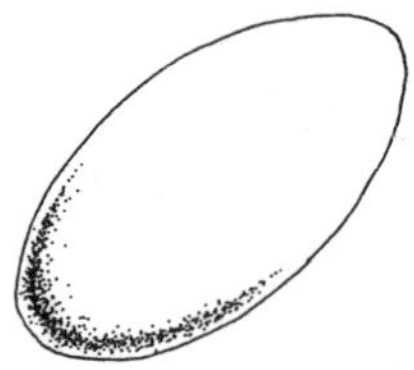

Figure 320

Elliptic. In the shape of an ellipse, or a narrow oval; broadest at the middle and narrower at the two equal ends. Figure 321.

Elliptical. See **elliptic**.

Elongate. Drawn out; lengthened.

Figure 321

Emarginate. With a notch at the apex. Figure 322.

Embryo. The young plant within a seed. Figure 323.

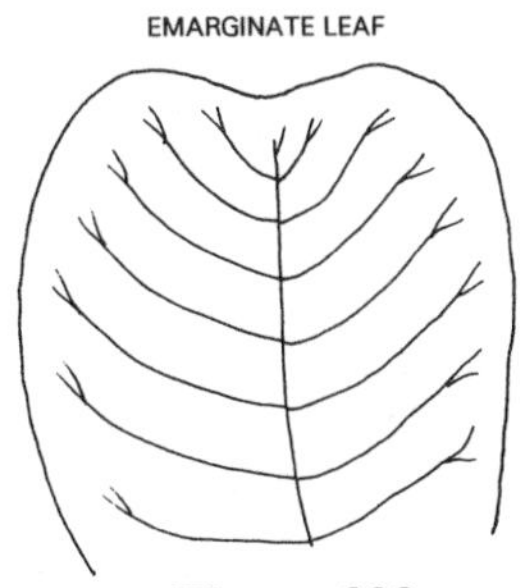

Figure 322

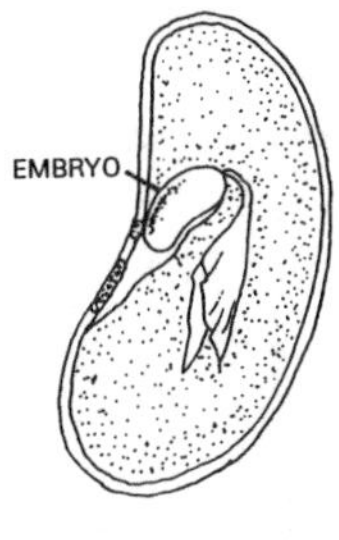

Figure 323

Embryo sac. The megagametophyte within the ovule of a flowering plant. Figure 324.

Emergent. See **emersed**.

Emersed. Rising from, or standing out of, water.

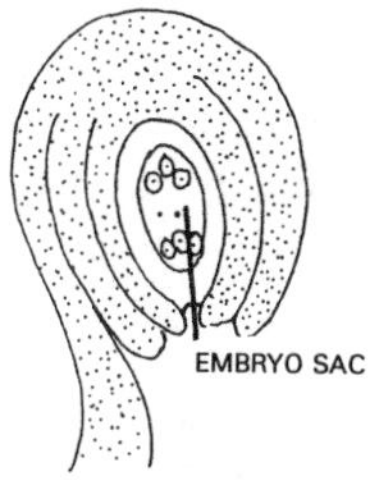

Figure 324

Enation. A projection or outgrowth from the surface of an organ or structure. Figure 325.

Endemic. Peculiar to a specific geographic area or edaphic type.

Endocarp. The inner layer of the pericarp of a fruit. Figure 326. (compare **mesocarp** and **exocarp**)

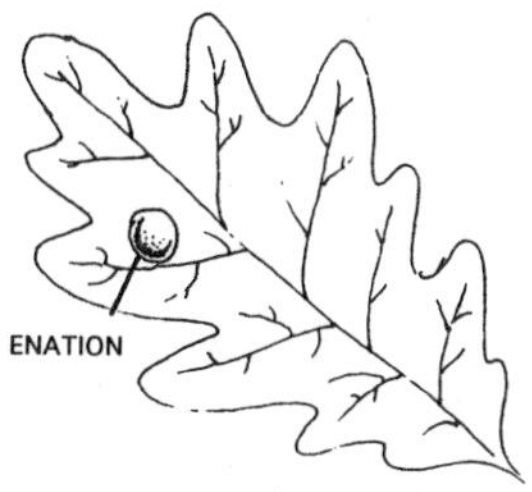

Figure 325

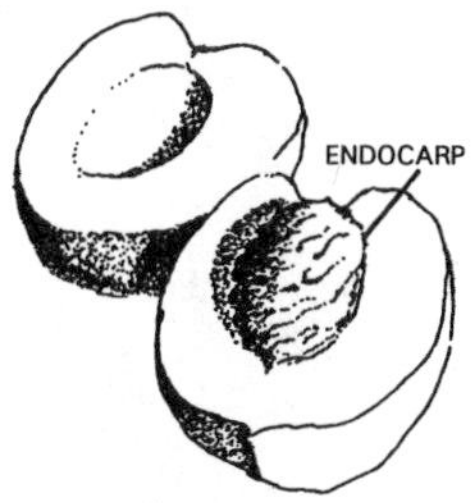

Figure 326

Endogenous. Growing from, or originating from, within.

Endosperm. The nutritive tissue surrounding the embryo of a seed derived from the fusion of a sperm cell with the polar nuclei of the embryo sac. Figure 327.

Ensiform. Sword-shaped, as an *Iris* leaf. Figure 328.

Entire. Not toothed, notched, or divided, as the continuous margins of some leaves. Figure 329.

Entomophagous. Insect-eating; insectivorous.

Entomophilous. Insect pollinated.

Epappose. Without a pappus. Figure 330.

Figure 327

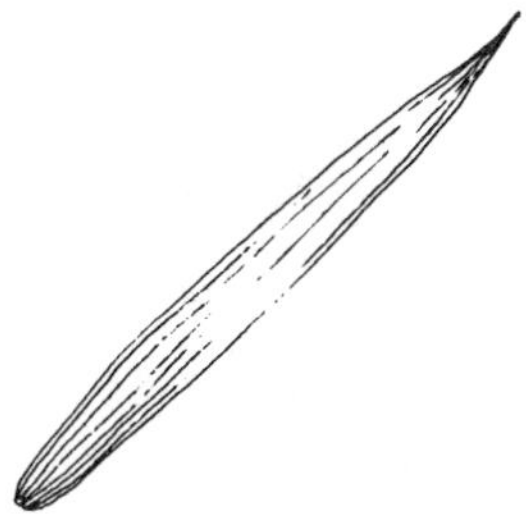

Figure 328

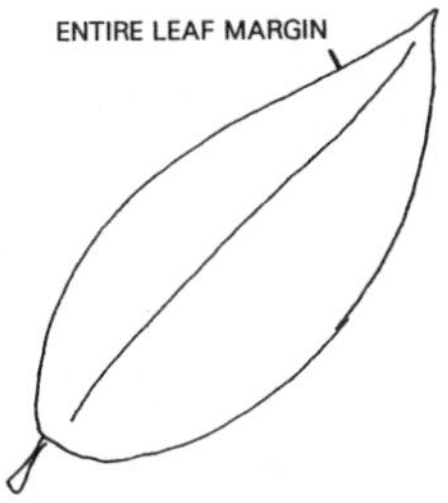

Figure 329

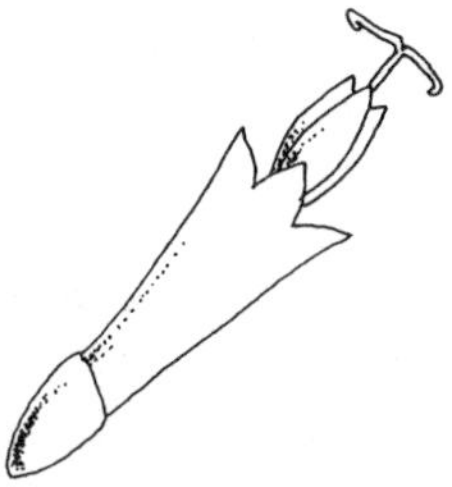

Figure 330

Epetiolate. Without a petiole, as in a sessile leaf. Figure 331.

Epetiolulate. Without a petiolule, as in a sessile leaflet. Figure 332.

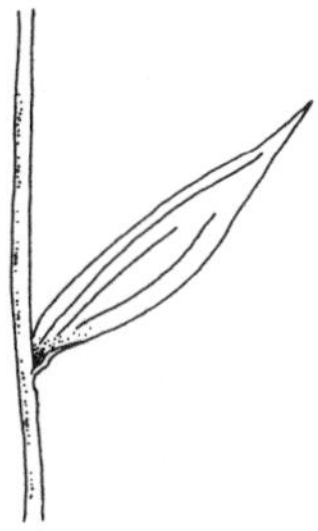

Figure 331

Figure 332

Ephemeral. Lasting a very short time.

Epi- (prefix). Meaning upon.

Epiblast. A small flap of tissue on the embryo of some members of the grass family.

Epicalyx. An involucre which resembles an outer calyx, as in *Malva*. Figure 333.

Epicarp. See **exocarp**.

Epicotyl. That portion of the embryonic stem above the cotyledons. Figure 334.

Epidermis. The outermost cellular layer of a

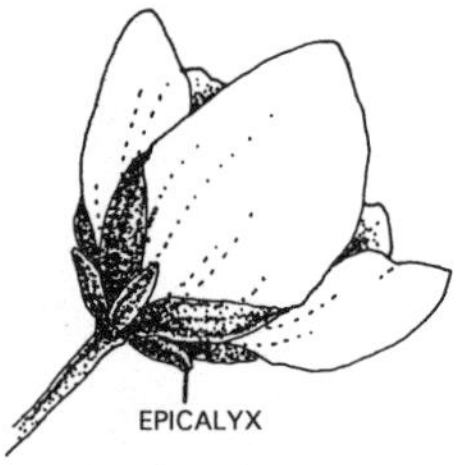

Figure 333

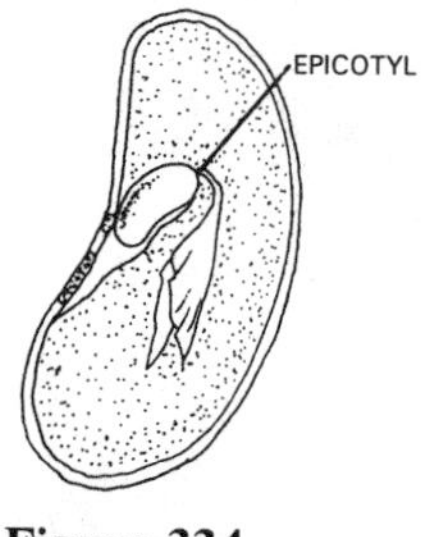

Figure 334

nonwoody plant organ. Figure 335.

Epigaeous. See **epigeous**.

Epigenous. Growing on the surface, as a fungus growing on the surface of a leaf.

Epigeous. Growing near the ground; said of a seedling which raises its cotyledons above the ground. Figure 336.

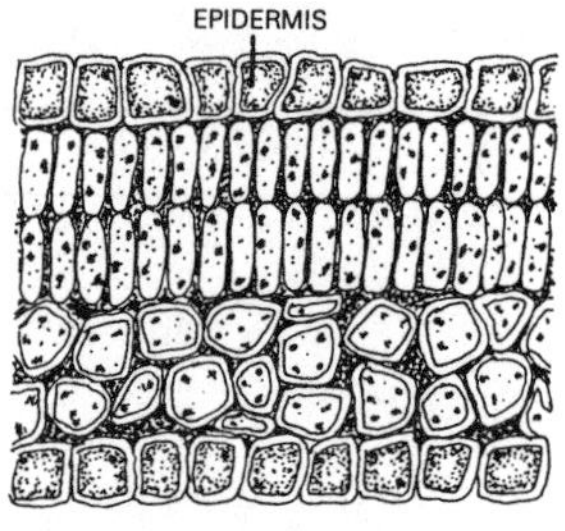

Figure 335

Epigynous. With stamens, petals, and sepals attached to the top of the ovary, the ovary inferior to the other floral parts. Figure 337.

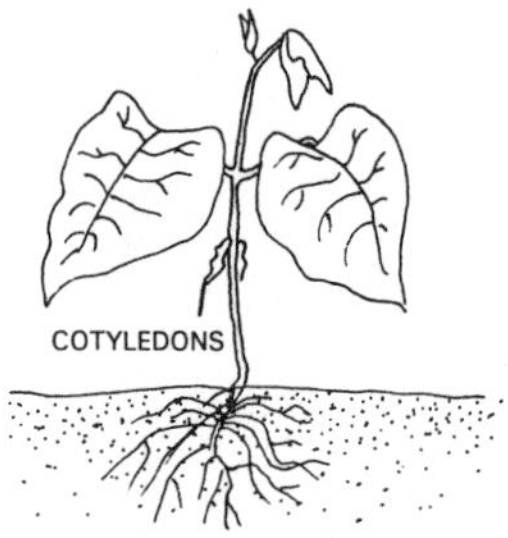

Figure 336

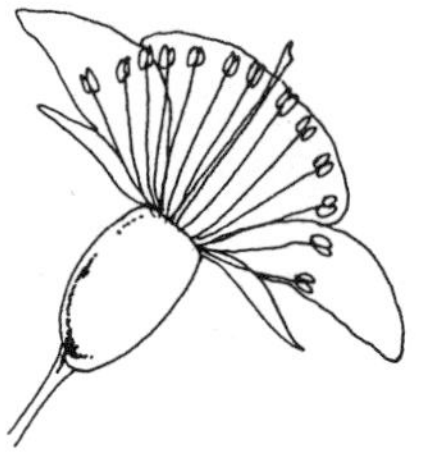

Figure 337

Epigyny. An epigynous condition.

Epipetalous. Attached to the petals. Figure 338.

Epipetric. Growing on a rock.

Epiphyte. A plant which grows upon another plant but does not draw food or water from it. (compare **parasite**)

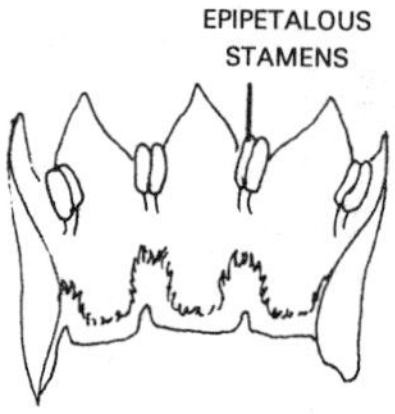

Figure 338

Episepalous. Attached to the sepals.

Equilateral. With sides of equal shape and length. Figure 339.

Equinoctial. With flowers that open regularly at a particular hour of the day.

Equisetoid. Resembling *Equisetum*.

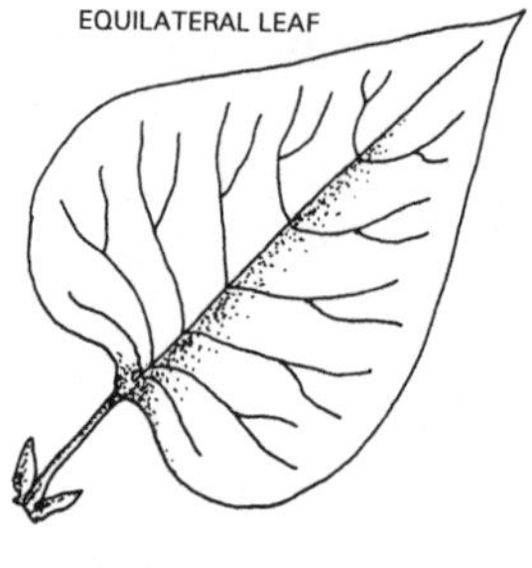

Figure 339

Equitant. Overlapping or straddling in two ranks, as the leaves of *Iris*. Figure 340.

Eramous. With unbranched stems. Figure 341.

Figure 340

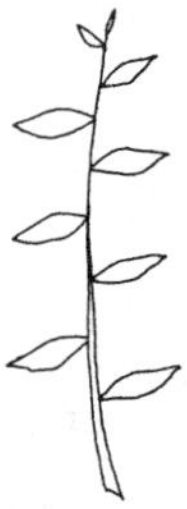

Figure 341

Erect. Vertical, not declining or spreading. Figure 342.

Erose. With the margin irregularly toothed, as if gnawed. Figure 343.

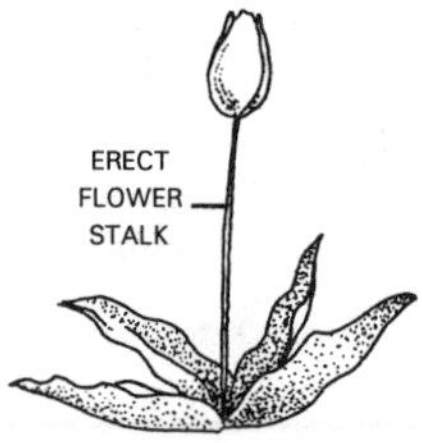

Figure 342

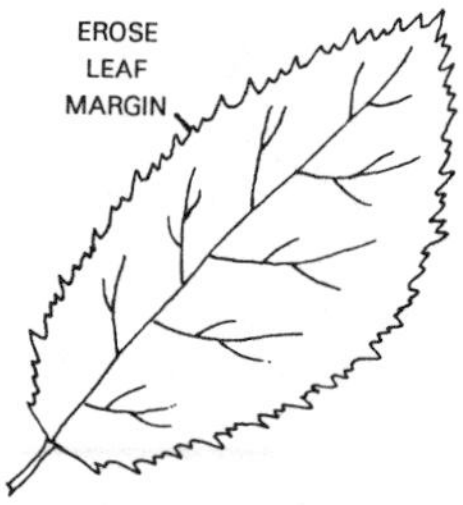

Figure 343

Erosulate. More or less erose.

Estipellate. Without stipels. Figure 344.

Estipulate. Without stipules. Figure 344.

Estival. See **aestival**.

Estivation. See **aestivation**.

Etiolated. White due to a lack of chlorophyll.

Eu- (prefix). Meaning true or real.

Evanescent. Fleeting; remaining only a very short time.

Even-pinnate. Pinnately compound with a terminal pair of leaflets or a tendril rather than a single terminal leaflet, so that there is an even number of leaflets. Figures 345 and 346.

Figure 344

Evergreen. Having green leaves through the winter; not deciduous.

Figure 345

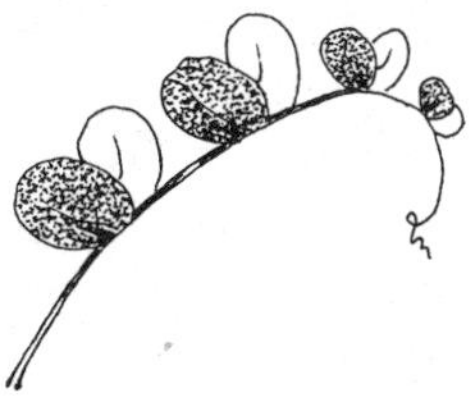

Figure 346

Ex- (prefix). Same as **e-**.

Exalbuminous. Without albumen.

Excavated. Hollowed out or concave, as the surface of some seeds. Figure 347.

Excrescence. An irregular growth or protuberance. Figure 348.

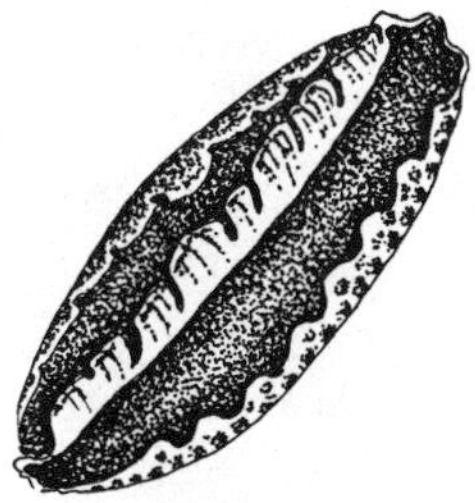

Figure 347

Figure 348

Excurrent. Extending beyond the apex, as the midrib in some leaves. Figure 349; extending beyond what is typical, as in a leaf base which extends down the stem. Figure 350; with a prolonged main axis from which lateral branches arise. Figure 351.

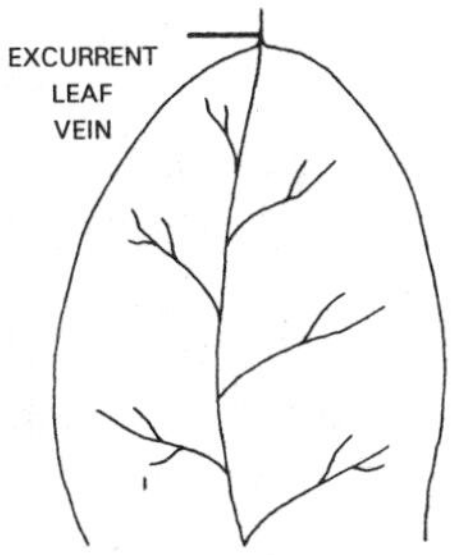

Figure 349

Excurved. Curving outward, away from the axis. Figures 352 and 353.

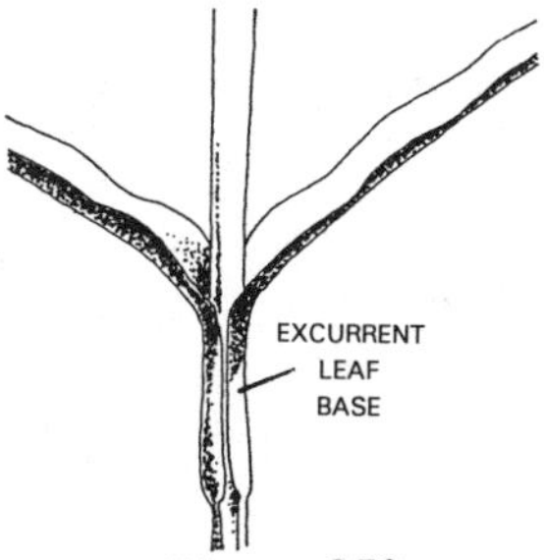

Figure 350

Figure 351

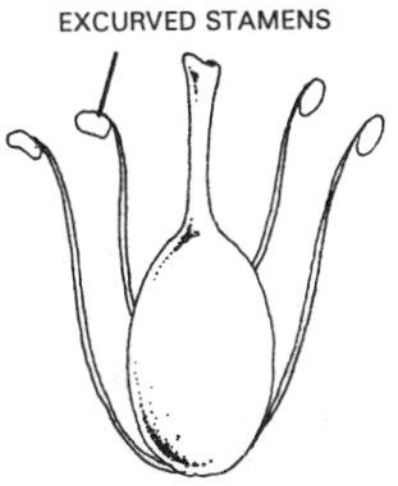

Figure 352

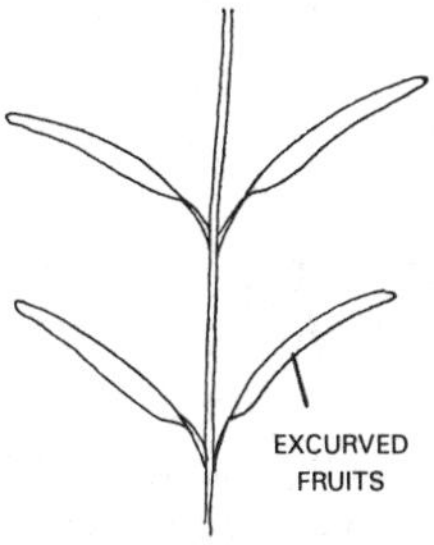

Figure 353

Exfoliate. To peel off in flakes or layers, as the bark of some trees. Figure 354.

Exine. The outer layer of the two-layered wall of a pollen grain. Figure 355.

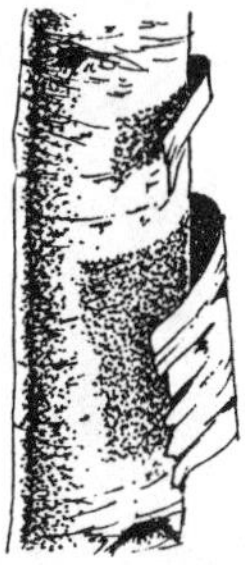

Figure 354

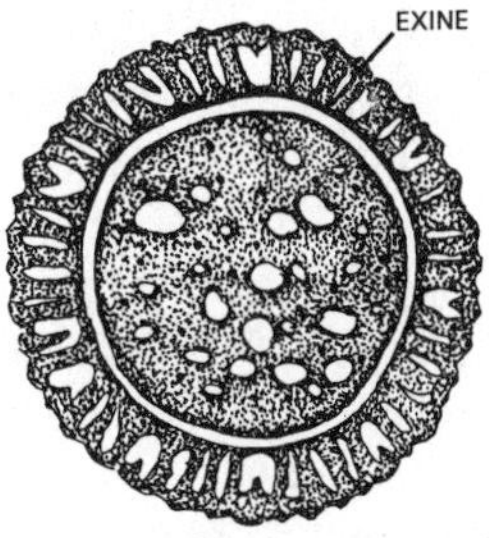

Figure 355

Exocarp. The outer layer of the pericarp of a fruit. Figure 356. (compare **mesocarp** and **endocarp**)

Exogenous. Growing from, or originating from, without.

Exomorphic. Pertaining to the external form.

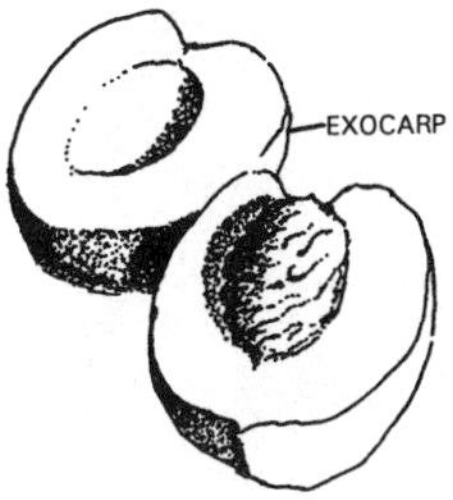

Figure 356

Exotic. Not native; introduced from elsewhere, but not completely naturalized.

Explanate. Spread out flat. Figure 357.

Exserted. Projecting beyond the surrounding parts, as stamens protruding from a corolla; not included. Figure 358.

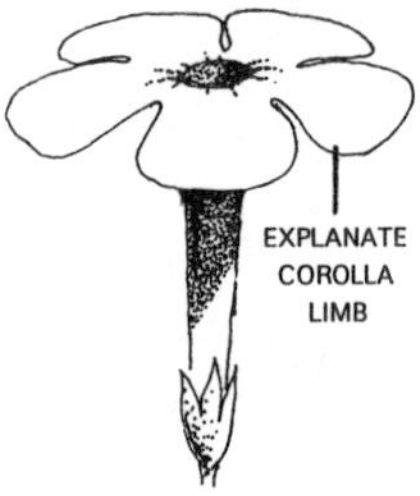

Figure 357

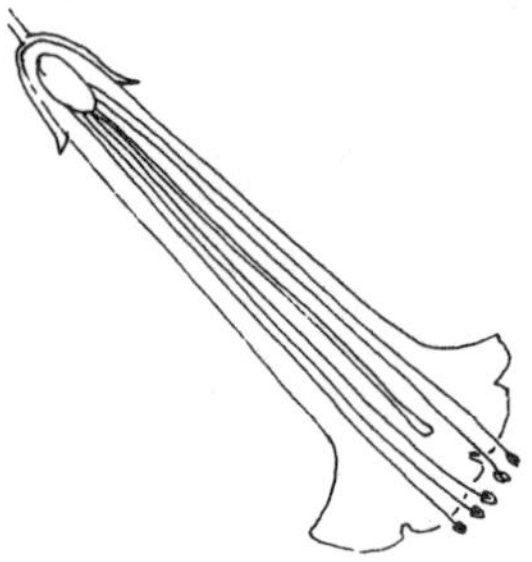

Figure 358

Exstipellate. See **estipellate**.

Exstipulate. See **estipulate**.

Extra- (prefix). Meaning outside or beyond.

Extra-axillary. Outside of but close to the axil.

Extrafloral. Outside the flower.

Extrastaminal. Outside of the stamens.

Extrorse. Turned outward, away from the axis. Figure 359. (compare **introrse**)

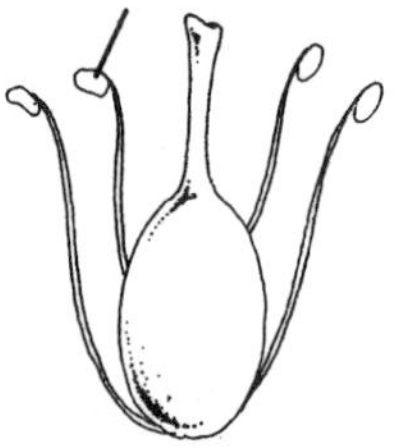

Figure 359

Exudate. A substance exuded or excreted from a plant.

Faboid. Bean-like.

Faceted. With many plane surfaces, like a cut gem, as in some seeds. Figure 360.

Facial. Pertaining to or on the face, rather than the sides or edges.

Falcate. Sickle-shaped; hooked; shaped like the beak of a falcon. Figure 361.

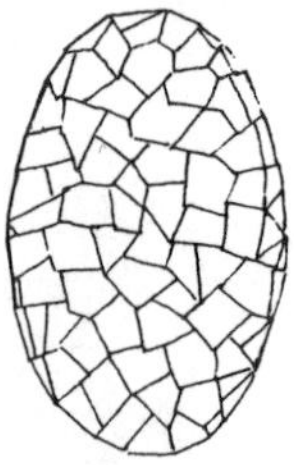

Figure 360

Figure 361

Falciform. See **falcate**.

Falls. The sepals of an *Iris*. Figure 362.

Farinaceous. Mealy in texture; starchy. Figure 363.

Farinose. Covered with a mealy, powdery substance. Figure 363.

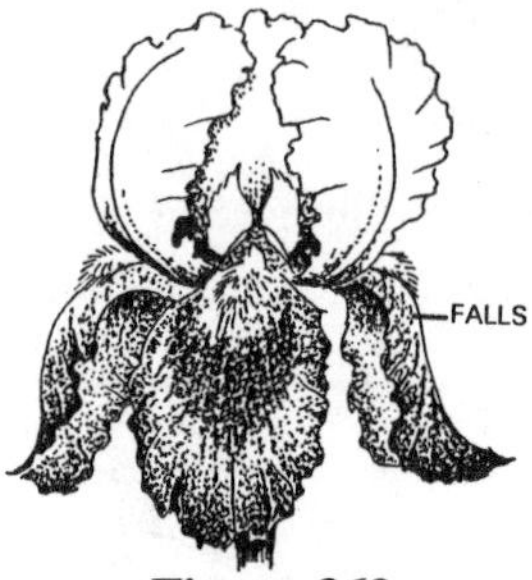

Figure 362

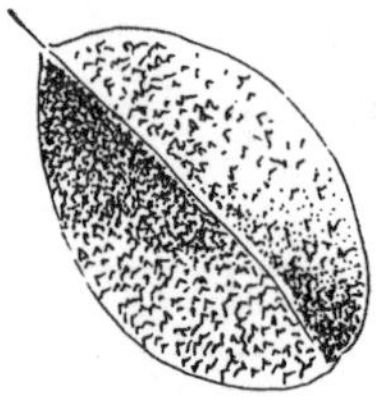

Figure 363

Fasciated. Compressed into a bundle or band; grown closely together; with the stems malformed and flattened as if several separate stems had been fused together. Figure 364.

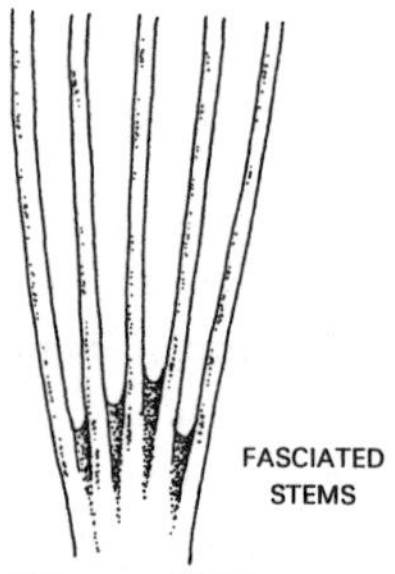

Figure 364

Fascicle. A tight bundle or cluster. Figure 365.

Fasciculate. Arranged in fascicles. Figure 365.

Fastigiate. Clustered, parallel, and erect, giving a broom-like appearance. Figure 366.

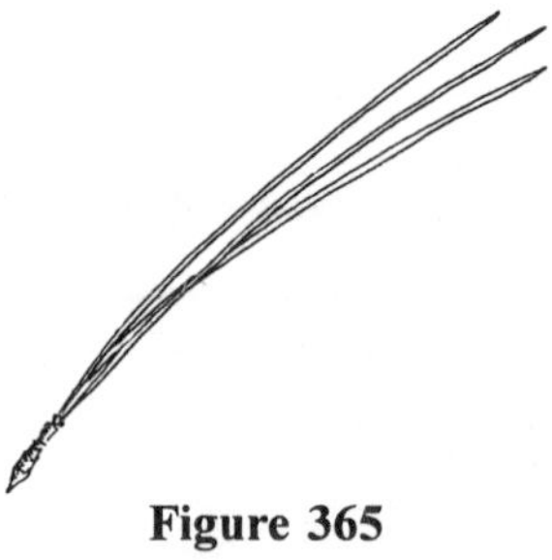

Figure 365

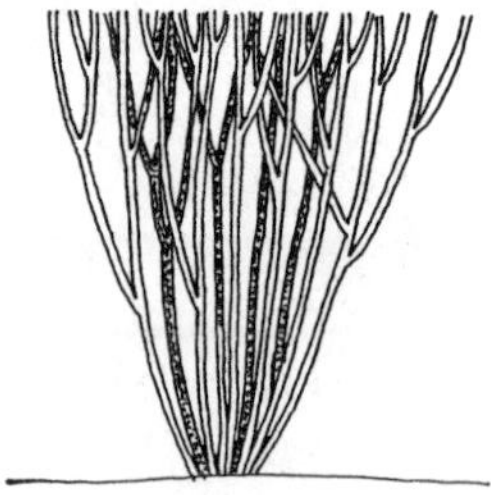

Figure 366

Faveolate. Honeycombed or pitted; alveolate. Figure 367.

Favose. See **faveolate**.

Fenestrate. With window-like perforations, openings, or translucent areas. Figure 368.

Ferruginous. Rust-colored.

Fertile. Capable of bearing seeds; capable of

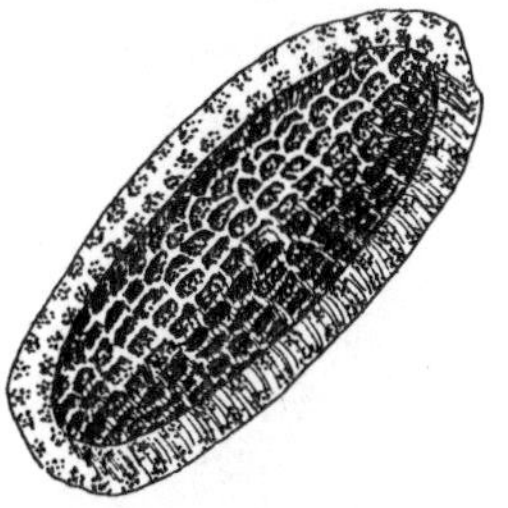

Figure 367

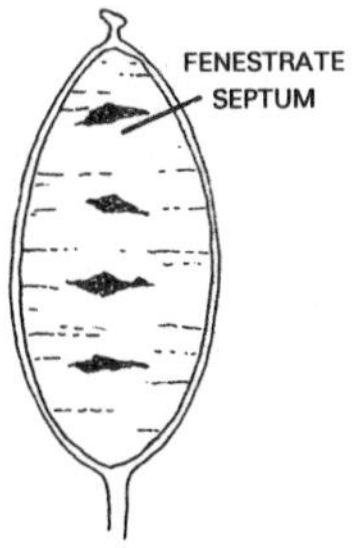

Figure 368

bearing pollen.

Festucoid. Resembling the grass *Festuca*; a member of the Festucoidea group of grasses.

Fetid. With an offensive odor; stinking.

Fibril. A delicate fiber or hair. Figure 369.

Fibrillate. See **fibrillose.**

Fibrillose. Bearing fibrils. Figure 369.

Fibrous. Containing or resembling fibers.

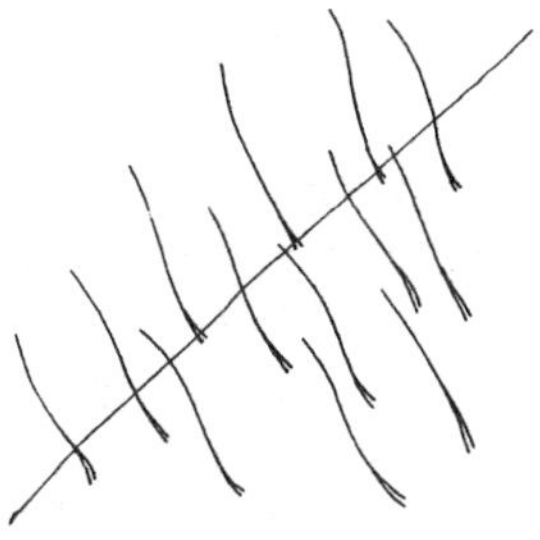

Figure 369

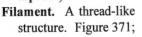

Fibrous roots. A root system with all of the branches of approximately equal thickness, as in the grasses and other monocots. Figure 370. (compare **taproot**)

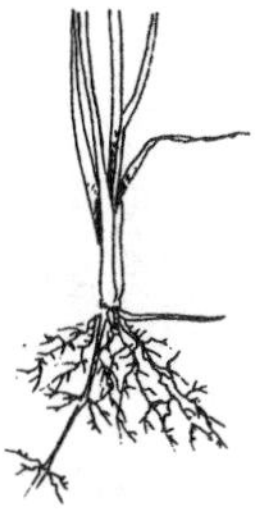

Figure 370

Filament. A thread-like structure. Figure 371; the stalk of the stamen which supports the anther. Figure 372.

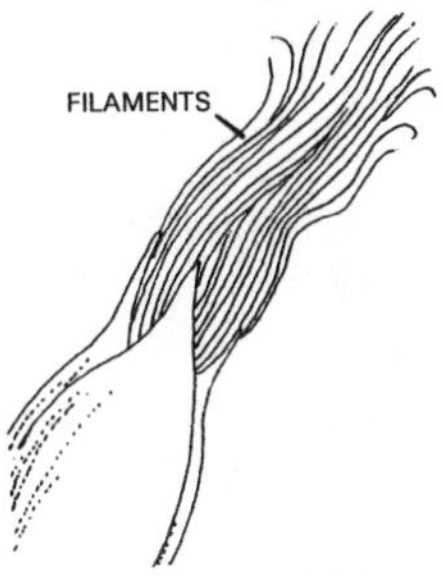

Figure 371

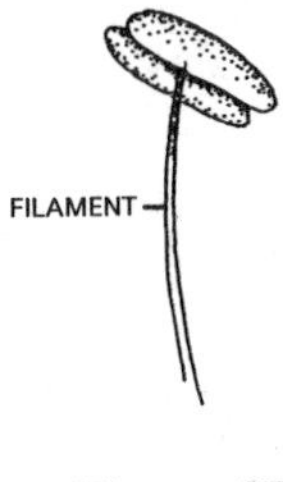

Figure 372

Filamentose. See **filamentous.**

Filamentous. Bearing or resembling filaments. Figure 371.

Filantherous. Of a stamen with a distinct anther and filament. Figure 372.

Filiferous. Bearing filaments or filament-like growths. Figure 371.

Filiform. Threadlike; filamentous. Figure 371.

Fimbriate. Fringed, usually with hairs or hairlike structures (fimbrillae) along the margin. Figure 373.

Fimbriation. A fringe. Figure 373.

Fimbrilla (pl. **fimbrillae**). A single unit of marginal fringe. Figure 373.

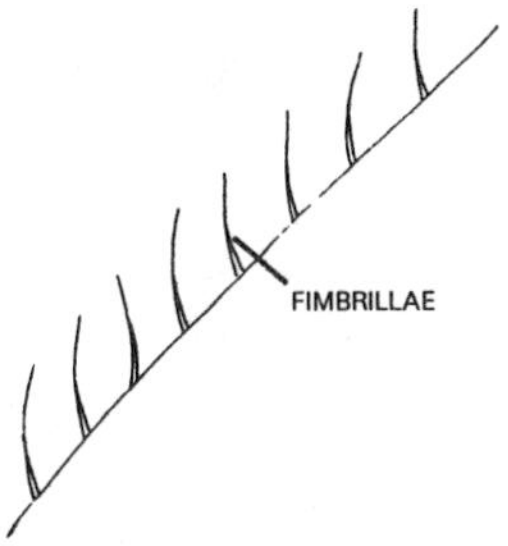

Figure 373

Fimbrillate. Fringed with very fine hairs. Figure 374.

Fimbriolate. See **fimbrillate.**

Fistular. See **fistulose.**

Fistulose. Hollow and cylindrical; tubular. Figure 375.

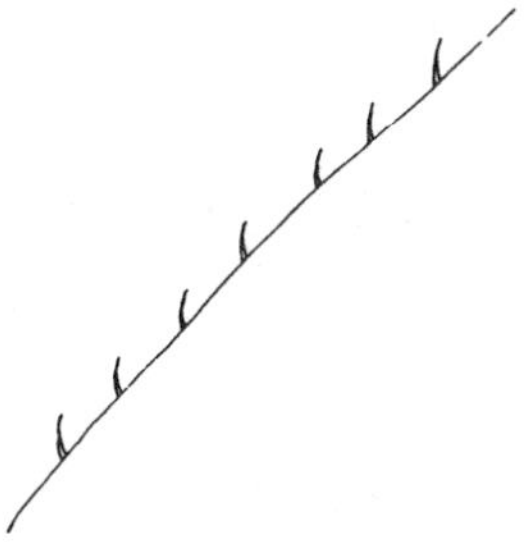

Figure 374

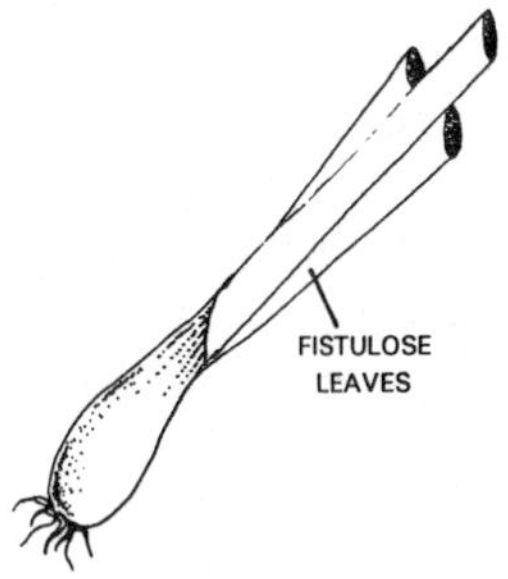

Figure 375

Fistulous. See **fistulose.**

Flabellate. Fan-shaped. Figure 376.

Flabelliform. See **flabellate.**

Flaccid. Limp or flabby; not rigid.

Flagellate. With long, slender runners. Figure 377.

Figure 376

Flagelliform. Elongate and slender; whiplike. Figure 377.

Flange. A projecting rim or edge. Figure 378.

Fleshy. Thick and pulpy; succulent. Figure 379.

Flexuose. With curves or bends; sinuous; somewhat zigzagged. Figure 380.

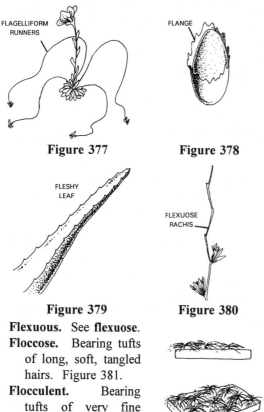

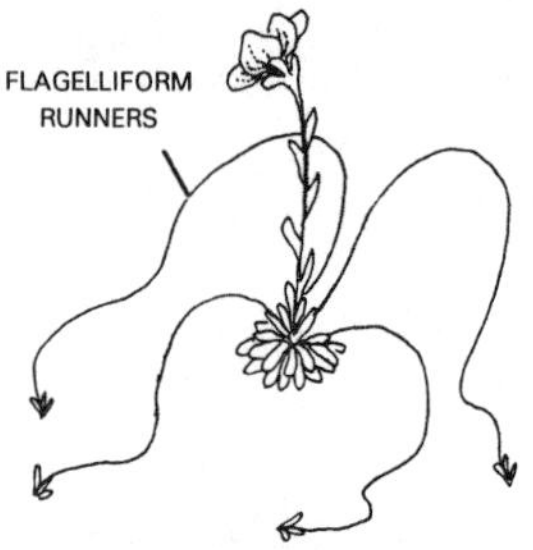

Figure 377

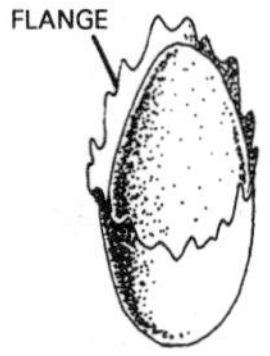

Figure 378

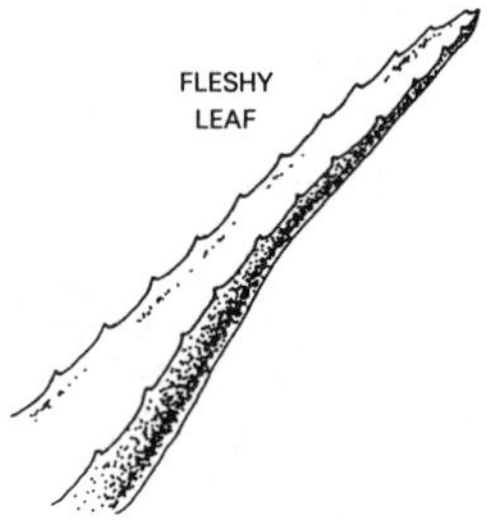

Figure 379

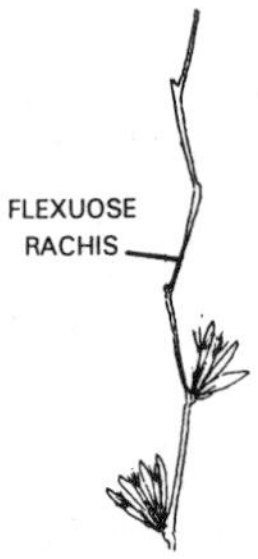

Figure 380

Flexuous. See **flexuose.**

Floccose. Bearing tufts of long, soft, tangled hairs. Figure 381.

Flocculent. Bearing tufts of very fine woolly hairs; floccose. Figure 381.

Figure 381

Floral envelope. A collective term for the calyx and corolla. Figure 382; the calyx in a flower lacking a corolla. (same as **perianth**)

Floral tube. An elongated tubular portion of a perianth. Figure 383.

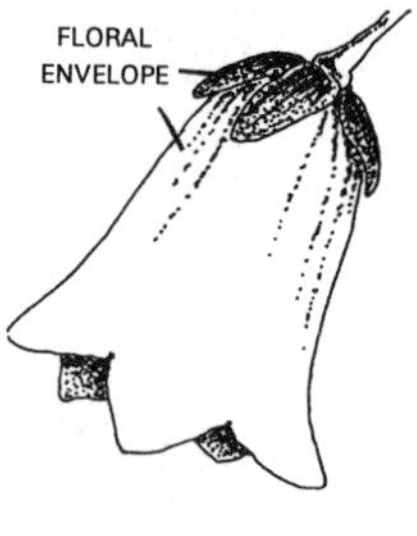

Figure 382

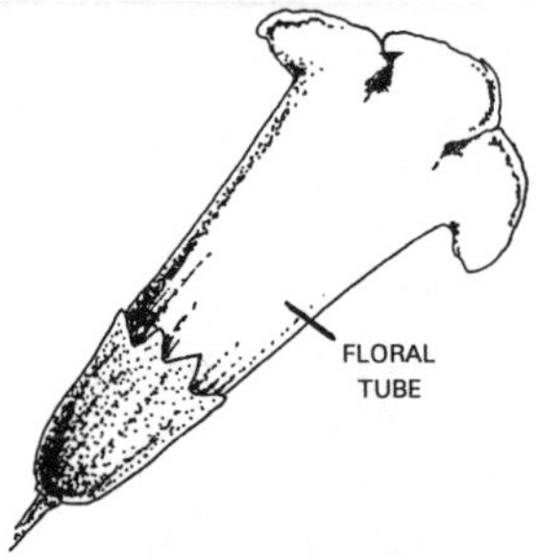

Figure 383

Floret. A small flower; an individual flower within a dense cluster, as a grass flower in a spikelet, or a flower of the Compositae (Asteraceae) in an involucrate head. Figures 384 and 385.

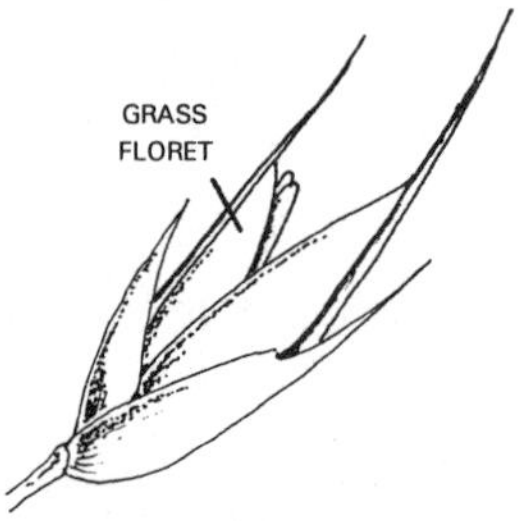

Figure 384

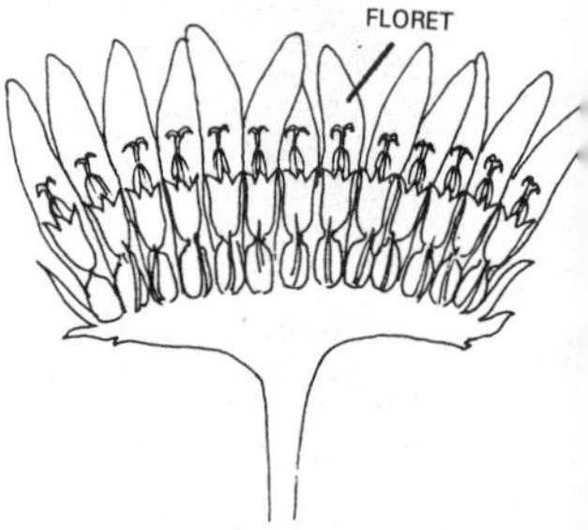

Figure 385

Floricane. The second-year flowering and fruiting cane (shoot) of *Rubus*. (compare **primocane**)

Floriferous. Flower-bearing.

Flower. The reproductive portion of the plant, consisting of stamens, pistils, or both, and usually including a perianth of sepals or both sepals and petals. Figure 386.

Fluted. With furrows or grooves. Figure 387.

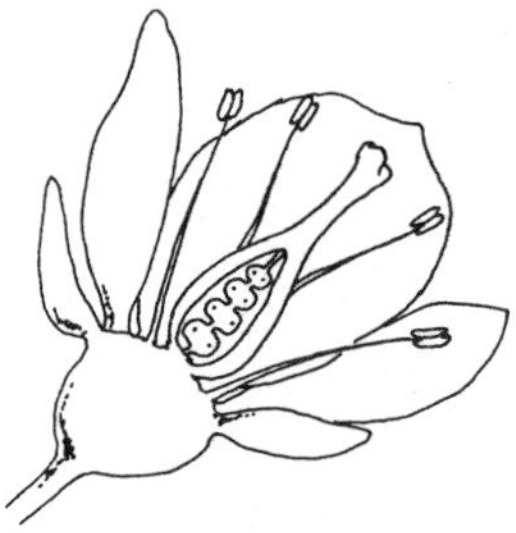

Figure 386

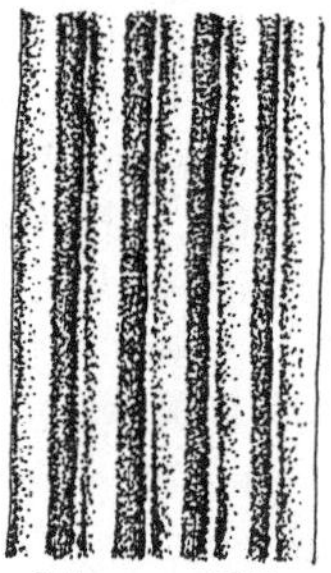

Figure 387

Foliaceous. Leaflike in color and texture; bearing leaves; of or pertaining to leaves.

Foliage. The leaves of a plant, collectively.

Foliar. Pertaining to leaves; leaflike.

Foliate. Having leaves; leaflike.

Foliated. Leaf-shaped.

Foliation. The act of producing leaves; the arrangement of leaves within a bud; foliage.

Foliature. A leaf cluster; foliage.

Foliolate. Pertaining to or having leaflets; usually used in compounds, such as **bifoliolate** or **trifoliolate**.

Foliose. Leafy.

Follicle. A dry, dehiscent fruit composed of a single carpel and opening along a single side, as a milkweed pod. Figure 388.

Follicular. Of or pertaining to a follicle. Figure 388.

Forb. A non-grasslike herbaceous plant.

Forked. Divided into two or more essentially equal branches. Figure 389.

Figure 388

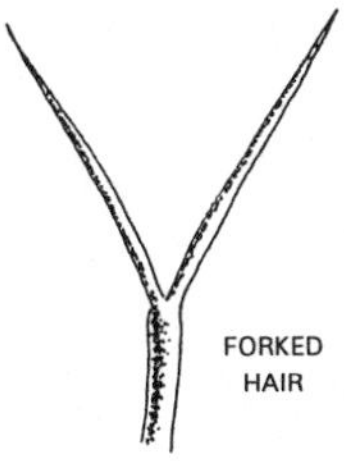

Figure 389

Fornix (pl. **fornices**). One of a set of small crests or scales in the throat of a corolla, as in many of the Boraginaceae. Figure 390.

Fovea (pl. **foveae**). A small pit or depression. Figure 391.

Foveate. With foveae; pitted. Figure 391.

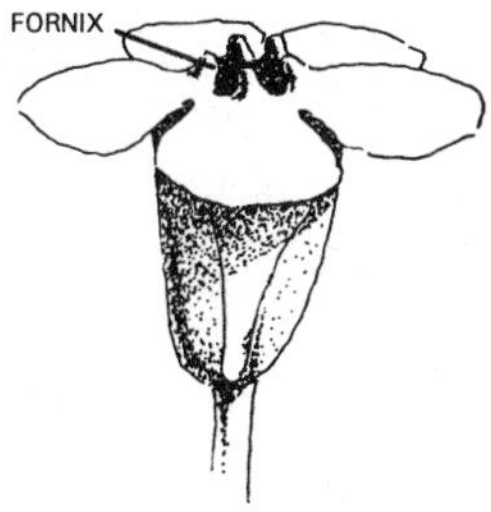

Figure 390

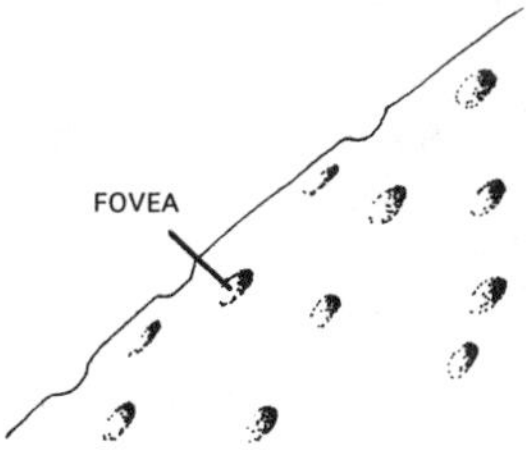

Figure 391

Foveola (pl. **foveolae**). A little fovea; a very small pit or depression. Figure 392.

Foveolate. With foveolae; minutely pitted. Figure 392.

Free. Not attached to other organs. Figure 393.

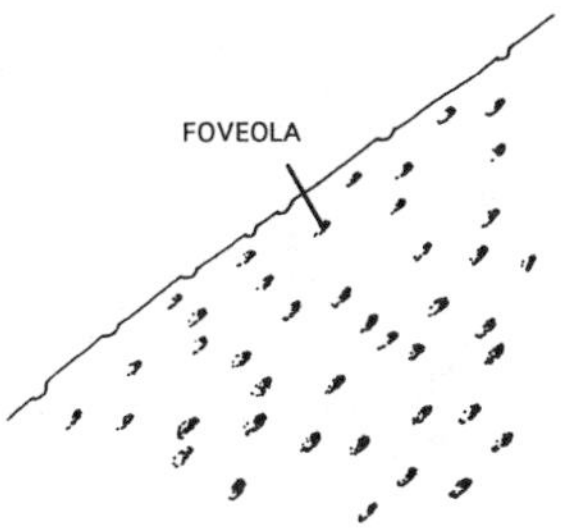

Figure 392

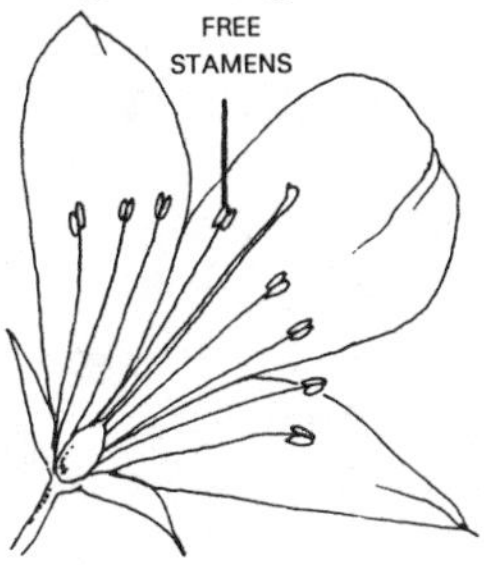

Figure 393

Free-central placentation. Ovules attached to a free-standing column in the center of a unilocular ovary. Figure 394.

Fringed. With hairs or bristles along the margin. Figure 395.

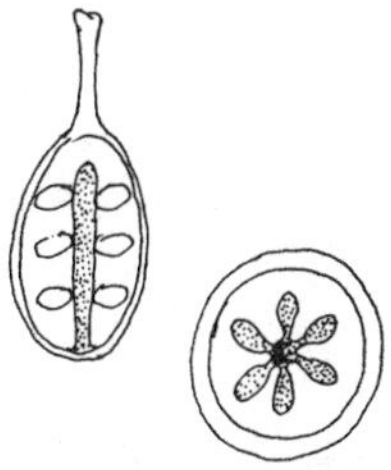

Figure 394

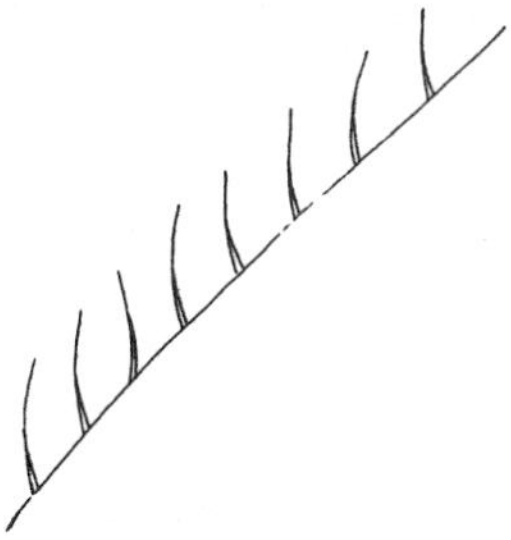

Figure 395

Frond. A large, divided leaf; a fern or palm leaf. Figure 396.

Fructiferous. Fruit-bearing.

Fructification. The fruiting process of a plant; the fruit of a plant.

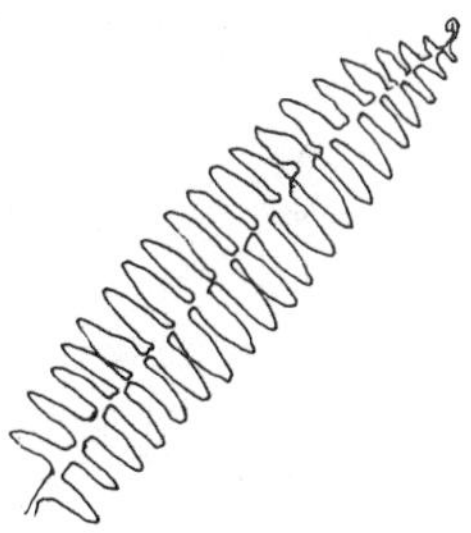

Figure 396

Fruit. A ripened ovary and any other structures which are attached and ripen with it.

Frutescent. Shrubby or shrublike.

Fruticose. See **frutescent**.

Fruticulose. Somewhat shrubby; small and shrubby.

Fugacious. Falling or withering early; ephemeral. (compare **caducous**)

Fulvous. Tawny; dull yellowish-brown or yellowish-gray.

Funicle. See **funiculus**.

Funiculate. With a funiculus. Figure 397.

Funiculus (pl. **funiculi**). The stalk connecting the ovule to the placenta; the stalk of a seed. Figure 397.

Funnelform. Gradually widening from base to apex; funnel-shaped. Figure 398.

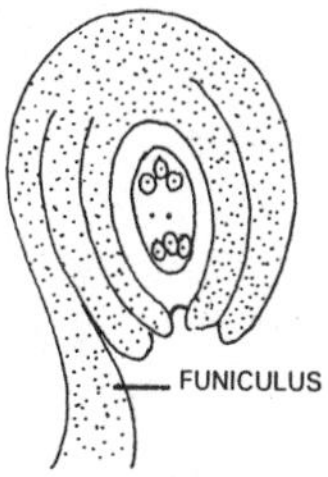

Figure 397

Furcate. Forked; often used in compounds, such as **bifurcate** or **trifurcate**.

Furfuraceous. Scurfy; branlike; flaky. Figure 399.

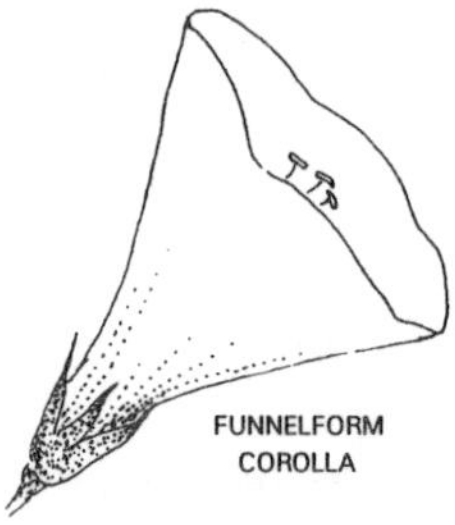

Figure 398

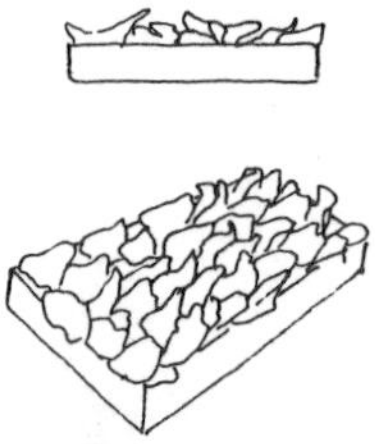

Figure 399

Fuscous. Dark grayish-brown; dusky.

Fusiform. Spindle-shaped; broadest near the middle and tapering toward both ends. Figure 400.

Galea. The helmet-shaped or hoodlike upper lip of some two-lipped corollas. Figure 401.

Galeate. With a galea; galea-like. Figure 401.

Galeiform. Helmet-shaped; galea-like. Figure 401.

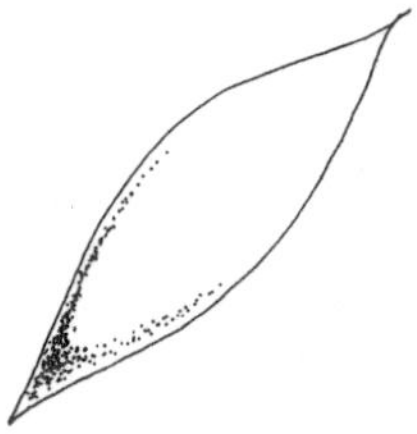

Figure 400

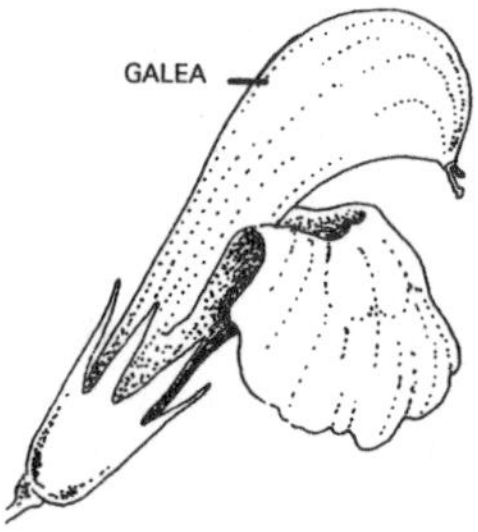

Figure 401

Gametophyte. The haploid (1n), gamete-producing generation of the plant reproductive cycle, the reduced and inconspicuous portion of the life cycle in the vascular plants. (compare **sporophyte**)

Gamo- (prefix). Meaning union of like parts.

Gamopetalous. With the petals united, at least partially. Figure 402.

Gamophyllous. With the leaves united, usually by the margins.

Gamosepalous. With the sepals united. Figure 402.

Geitonogamy. Pollination between flowers of the same plant.

Gelatinous. Jellylike in texture.

Geminate. In equal pairs like twins. Figure 403.

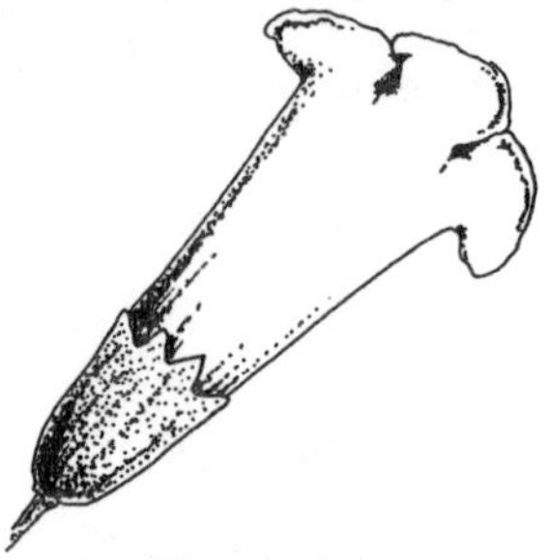

Figure 402

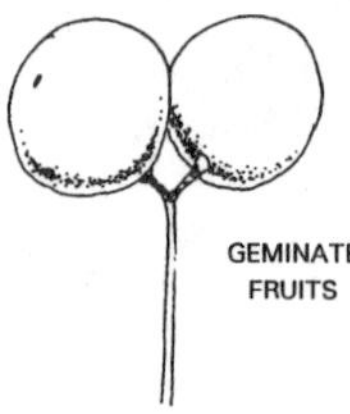

Figure 403

Gemma (pl. **gemmae**). A bud or budlike structure, or cluster of cells which separate from the parent plant and propagate offspring plants.

Gemmate. With gemmae; reproducing by gemmae.

Gemmation. The process of reproduction by gemmae.

Gemmule. See **gemma**.

Geniculate. With abrupt kneelike bends and joints. Figure 404.

Genome. A complete chromosome set.

Gibbosity. A swelling or protuberance; the state of being gibbous. Figure 405.

Gibbous. Swollen or enlarged on one side; ventricose. Figure 405.

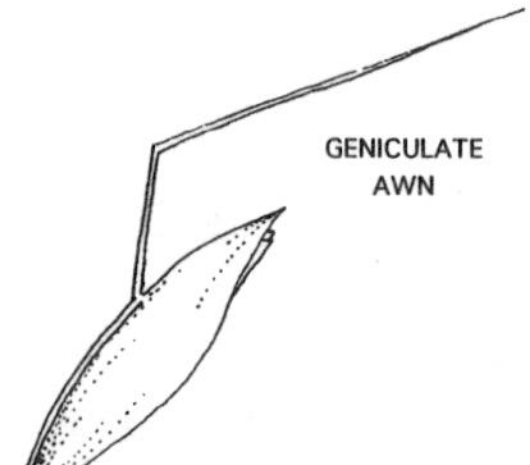

Figure 404

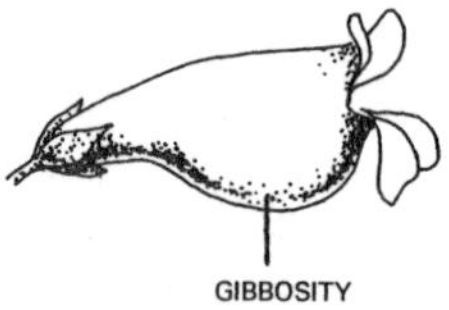

Figure 405

Glabrate. Becoming glabrous; almost glabrous.

Glabrescent. See **glabrate**.

Glabrous. Smooth; hairless.

Gladiate. Sword-shaped. Figure 406.

Gland. An appendage,

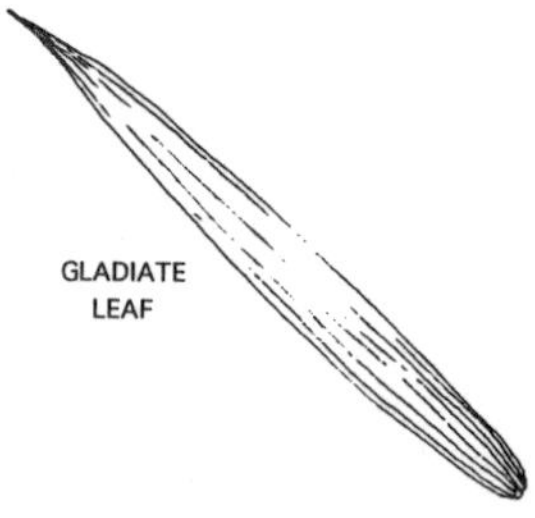

Figure 406

protuberance, or other structure which secretes sticky or oily sub-stances. Figure 407.

Glandular. Bearing glands. Figure 407.

Glanduliferous. See **glandular**.

Glaucescent. Somewhat glaucous; becoming glaucous.

Glaucous. Covered with a whitish or bluish waxy coating (bloom), as on the surface of a plum.

Globose. Globe-shaped; spherical. Figure 408.

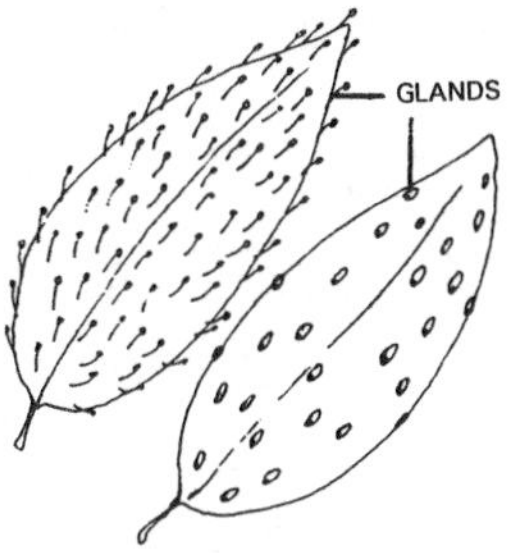

Figure 407

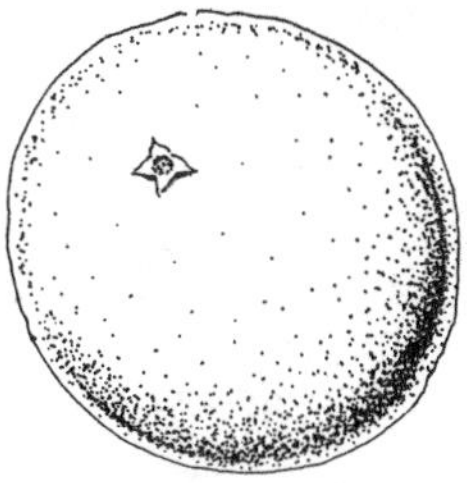

Figure 408

Globular. See **globose**.

Glochid (pl. **glochidia**). A barbed hair or bristle, as the fine hairs in *Opuntia*. Figures 409 and 410.

Glochidiate. Barbed at the tip. Figure 409; bearing glochids. Figure 410.

Figure 409

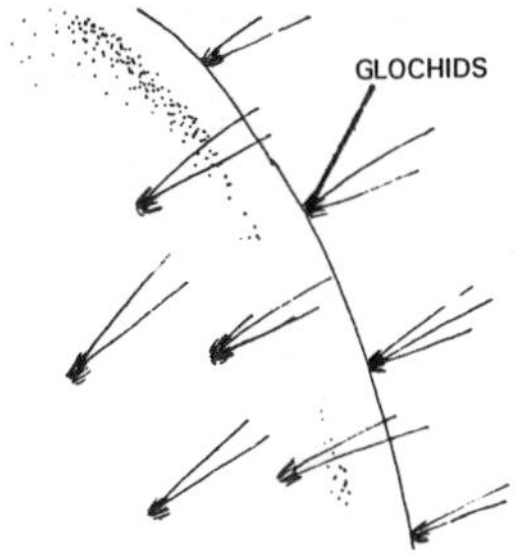

Figure 410

Glomerate. Densely clustered. Figure 411.

Glomerulate. Arranged in very small, dense clusters. Figure 412.

Glomerule. A dense cluster; a dense, head-like cyme.

Glumaceous. Resembling a glume; bearing glumes.

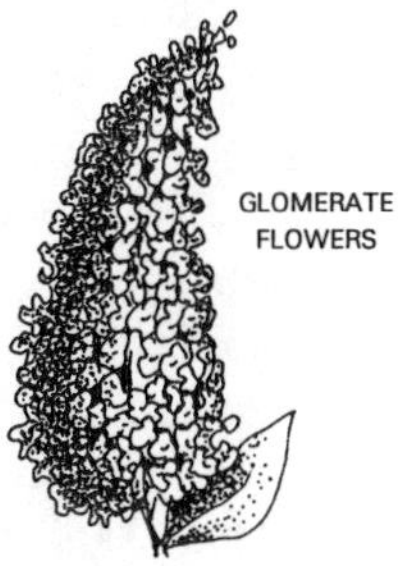

Figure 411

Glume. One of the paired bracts at the base of a grass spikelet; a chaffy bract in the grasses or sedges. Figure 413.

Glutinous. Gluey; sticky; gummy; covered with a sticky exudation.

Graduate. Divided or marked at regular intervals; with parts of progressively different lengths, as in some Compositae (Asteraceae) in which the outer involucral bracts are shorter than the inner. Figure 414.

Grain. A seedlike structure, as on the fruit of some *Rumex* species; a caryopsis. Figure 415.

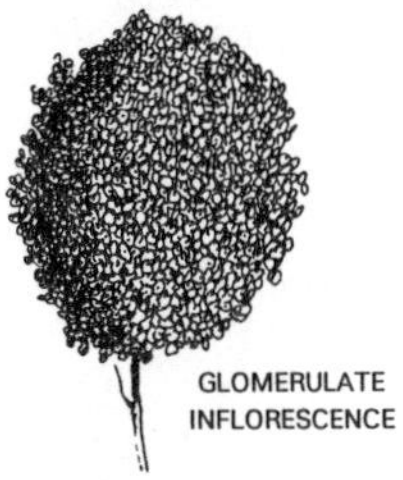

Figure 412

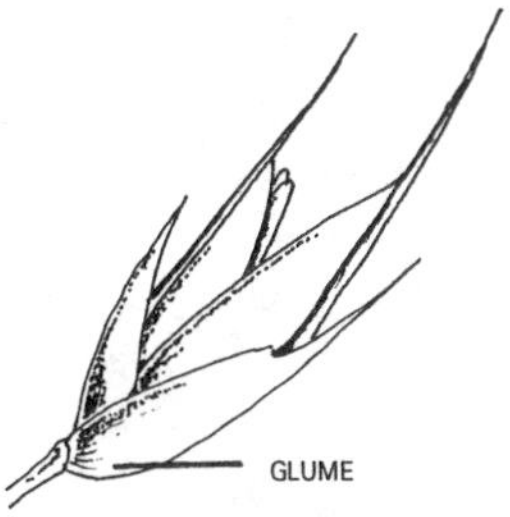

Figure 413

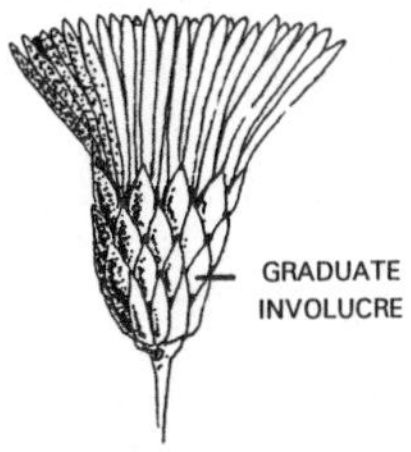

Figure 414

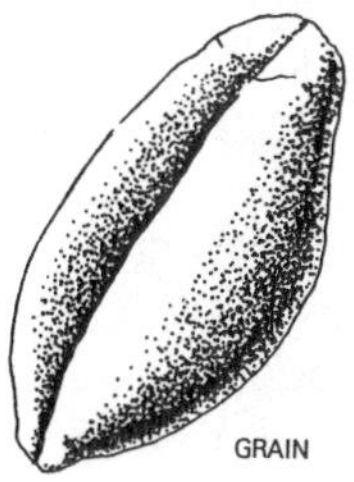

Figure 415

Granular. With small granules or grains.

Granulate. See **granular**.

Granule. A small grain.

Granuliferous. See **granular**.

Grenadine. Bright red; the color of pomegranate juice.

Gymnosperm. Plants producing seeds which are not borne in an ovary (fruit), the seeds usually borne in cones.

Gynaecandrous. An inflorescence with the pistilate flowers borne above the staminate, as in some *Carex* species. (compare **androgynous**)

Gynaeceum. See **gynoecium**.

Gynaecium. See **gynoecium**.

Gynandrial. See **gynandrous**.

Gynandrium. A column bearing stamens and

pistils. Figures 416 and 417.

Gynandrous. With the stamens adnate to the pistil. Figure 417.

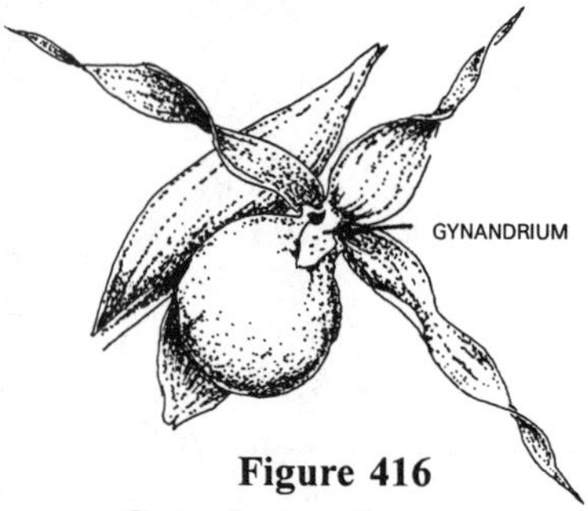

Figure 416

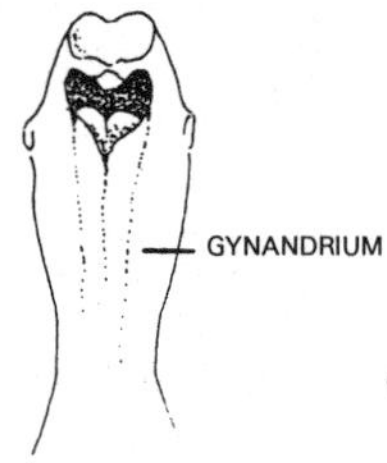

Figure 417

Gynecium. See **gynoecium**.

Gynobase. An elongation or enlargement of the receptacle, as in the flowers of the Boraginaceae. Figure 418.

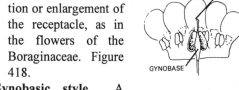

Gynobasic style. A style which is attached to the gynobase as well as to the carpels. Figure 418.

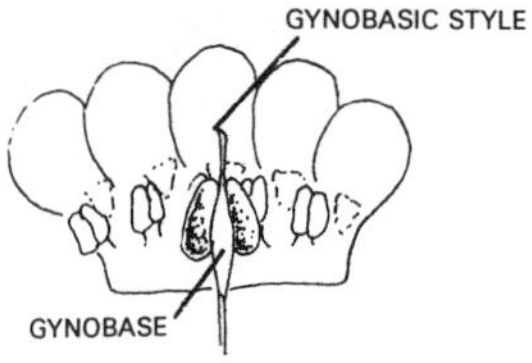

Figure 418

Gyno-dioecious. Having pistillate and perfect flowers on separate plants.

Gynoecium. All of the carpels or pistils of a flower, collectively. Figure 419.

Gynophore. An elongated stalk bearing the pistil in some flowers. Figure 420.

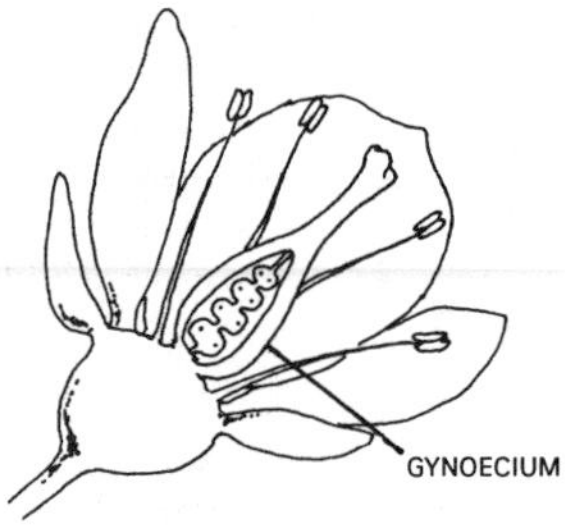

Figure 419

Figure 420

Gynostegium. A structure formed from the fusion of the anthers with the stigmatic region of the gynoecium, as in the Asclepiadaceae. Figure 421.

Gynostemial. See **gynandrous**.

Habit. The general appearance, characteristic form, or mode of growth of a plant.

Habitat. The environmental circumstances or kind of place where a plant grows.

Halberd-shaped. See **hastate**.

Half-inferior. Attached below the lower half, as a flower with a hypanthium that is fused to the lower half of the ovary, giving the appearance that the other floral whorls are arising from about the middle of the ovary. Figure 422.

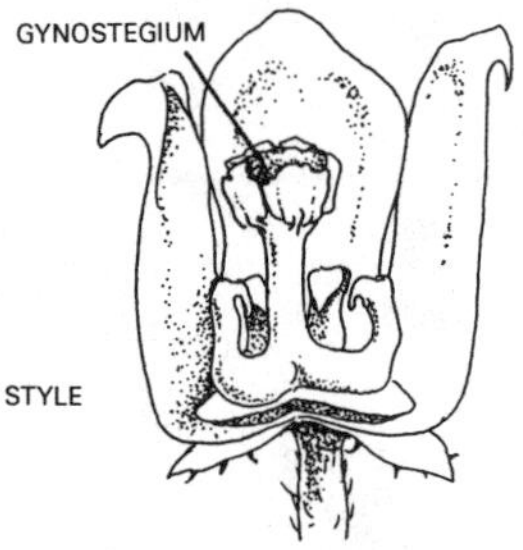

Figure 421

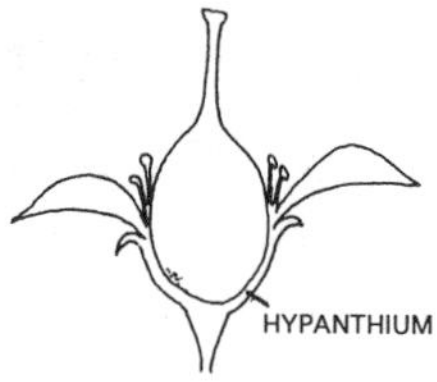

Figure 422

Halophyte. A plant which grows in salty soil.

Hamate. Hook-shaped; hooked at the tip. Figure 423.

Hapaxanthic. Flowering only once.

Haploid. With a single full set of chromosomes in each cell.

Haplostemonous. With one series of stamens; with as many stamens as petals. Figure 424.

Figure 423

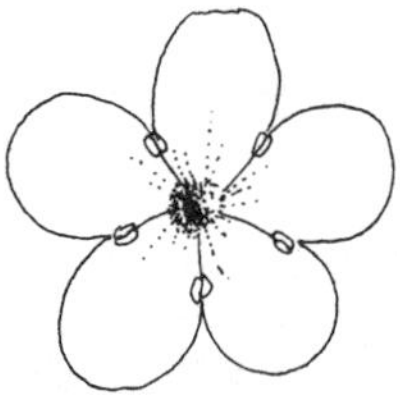

Figure 424

Hastate. Arrowhead-shaped, but with the basal lobes turned outward rather than downward; halberd-shaped. Figure 425. (compare **sagittate**)

Hastiform. See **hastate**.

Haustorium (pl. **haustoria**). A specialized

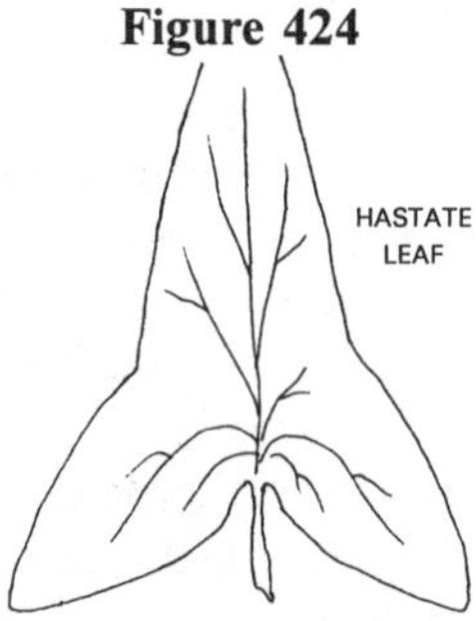

Figure 425

root-like organ used by parasitic plants to draw nourishment from host plants.

Head. A dense cluster of sessile or subsessile flowers; the involucrate inflorescence of the Compositae (Asteraceae). Figures 426 and 427.

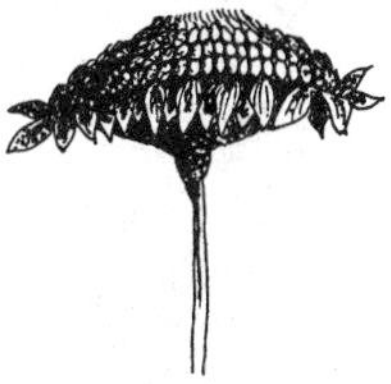

Figure 426

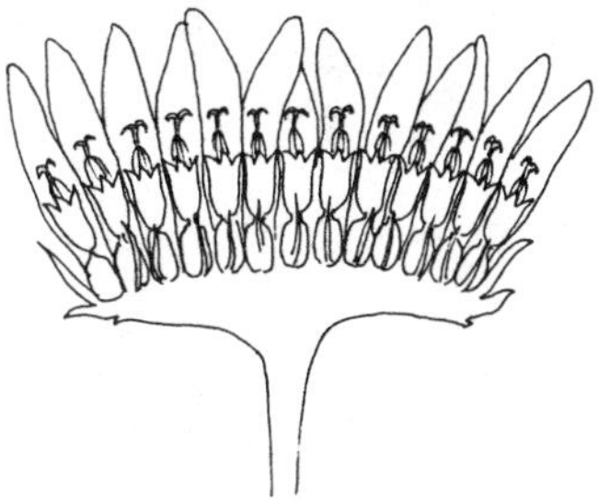

Figure 427

Heartwood. The innermost, usually somewhat darker wood of a woody stem. Figure 428.

Helicoid. Coiled like a spiral or helix, as in some one-sided cymose inflorescences in the Boraginaceae. Figure 429.

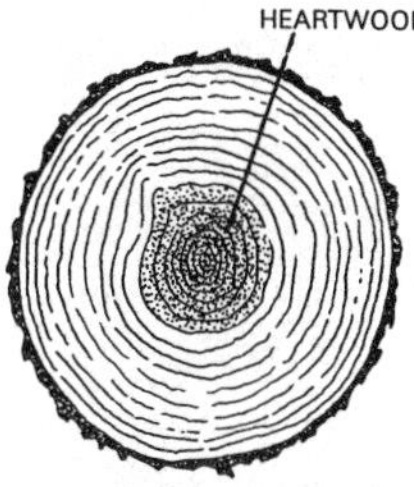

Figure 428

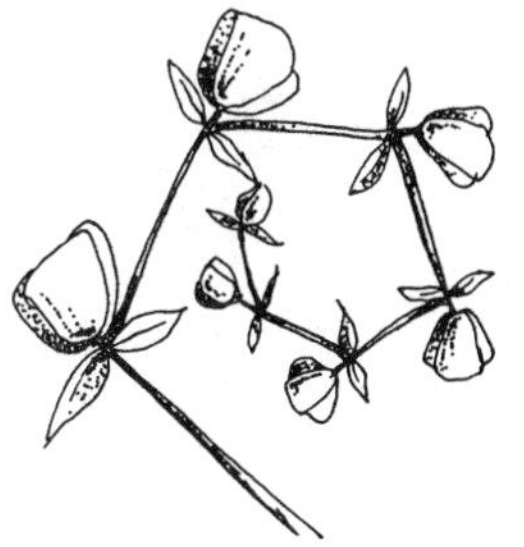

Figure 429

Helmet. See **hood**.

Hemi- (prefix). Meaning half.

Hemianatropous ovule. See **hemitropous ovule**.

Hemispheric. Shaped like half of a sphere.

Hemispherical. See **hemispheric**.

Hemispheroidal. Shaped like half of a spheroid.

Hemitropous ovule. An ovule which is half inverted so that the funiculus is attached near the middle with the micropyle at a right angle. Figure 430.

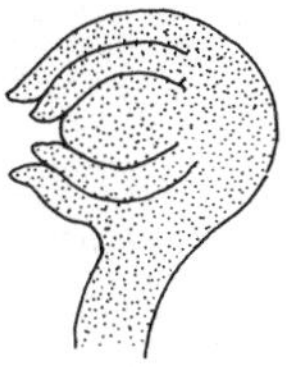

Figure 430

Herb. A plant without a persistent aboveground woody stem, the stems dying back to the ground at the end of the growing season.

Herbaceous. With the characteristics of an herb; not woody.

Herbage. The non-reproductive parts of the plant; the non-woody stems, leaves, and roots of a plant.

Herbarium. A collection of dried plant specimens.

Hermaphrodite. A hermaphroditic plant.

Hermaphroditic. With pistils and stamens in the same flower; bisexual; monoclinous; perfect. Figure 431.

Hesperidium. A fleshy berrylike fruit with a tough rind, as a lemon or orange. Figure 432.

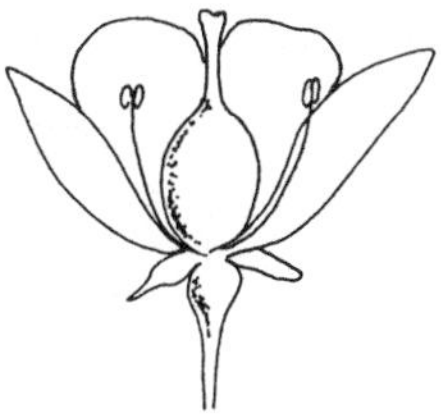

Figure 431

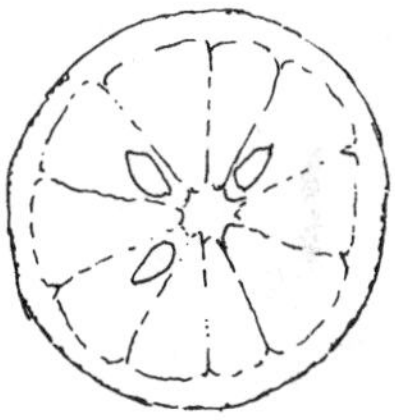

Figure 432

Hetero- (prefix). Meaning different or other.

Heterogamous. With flowers of differing sex.

Heterogamy. Having heterogamous flowers.

Heterogonous. With two or more different kinds of perfect flowers on different individuals of the same species, the kinds of flowers differing in the relative length of the pistils and stamens. (compare **homogonous**)

Heterogony. State of being heterogonous.

Heteromerous. With a variable number of parts, as in a flower with a different number of members in each floral whorl. Figure 433.

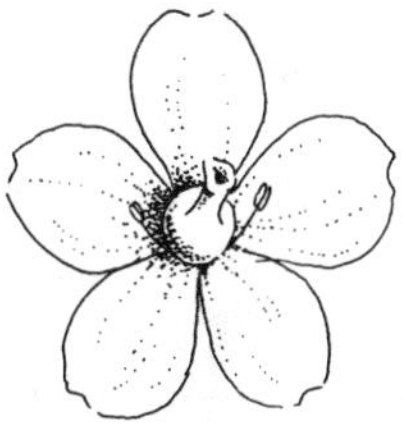

Figure 433

Heteromorphic. Of more than one kind or form.

Heteromorphous. See **heteromorphic**.

Heterophyllous. With different kinds of leaves on the same plant. Figure 434.

Heterosporous. Having spores of two different kinds, microspores and megaspores. Figure 435.

Figure 434

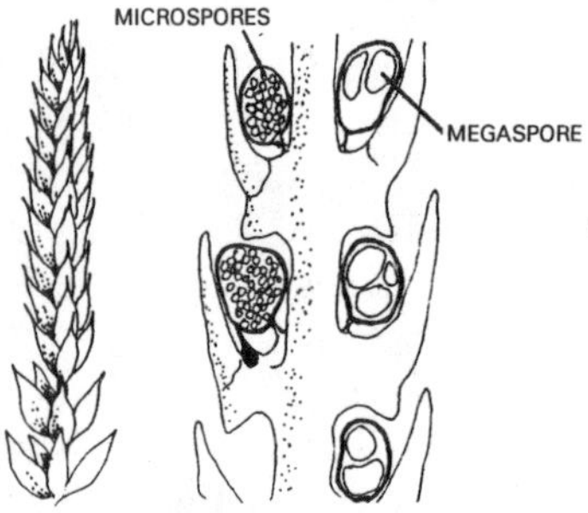

Figure 435

Heterostylic. With styles of different lengths in flowers of the same species. Figures 436 and 437.

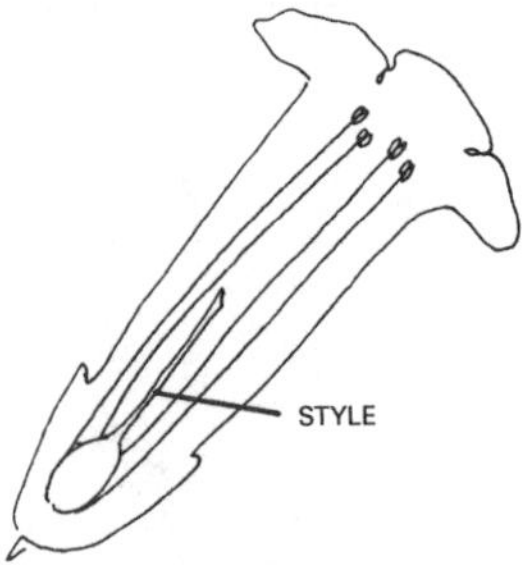

Figure 436

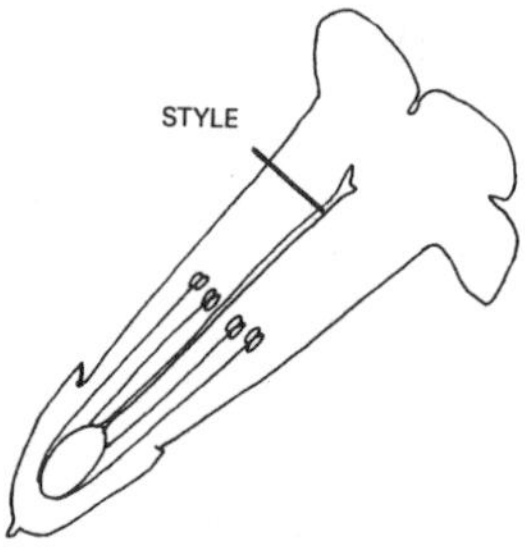

Figure 437

Heterostylous. See **heterostylic.**

Hexa- (prefix). Meaning six.

Hexamerous. With parts arranged in sets or multiples of six. Figure 438.

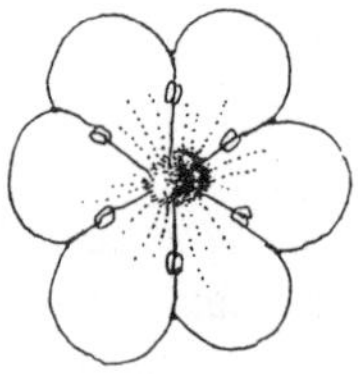

Figure 438

Hexaploid. With six full sets of chromosomes in each cell.

Hibernal. Flowering or appearing in the winter.

Hilum. A scar on a seed indicating its point of attachment. Figures 439 and 440; a scar indicating the point of attachment of the ovary in grasses.

Hip. A berry-like structure composed of an enlarged hypanthium surrounding numerous achenes, as in roses. Figures 441 and 442.

Hippocrepiform. Horseshoe-shaped. Figure 443.

Hirsute. Pubescent with coarse, stiff hairs. Figure 444.

Hirsutulous. Pubescent with very small, coarse, stiff hairs. Figure 445.

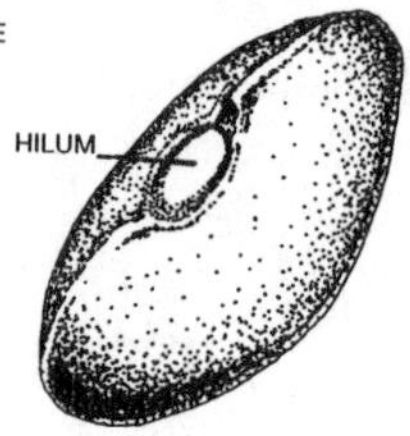

Figure 439

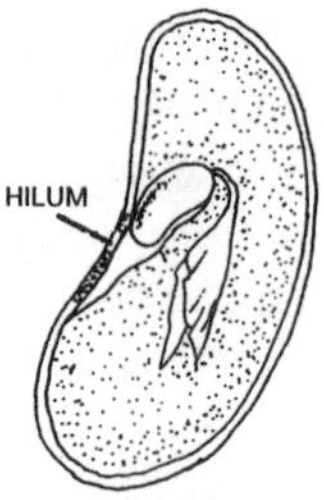

Figure 440

Figure 441

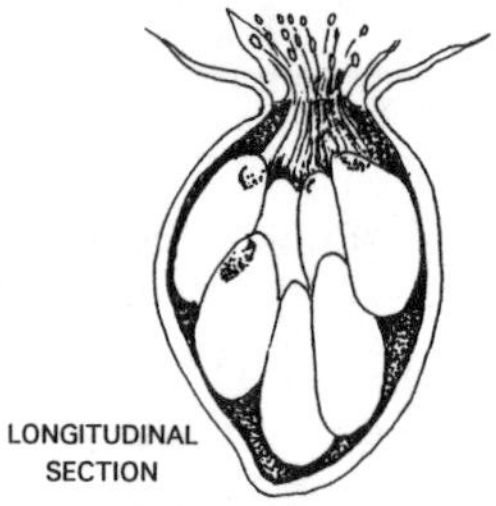

Figure 442

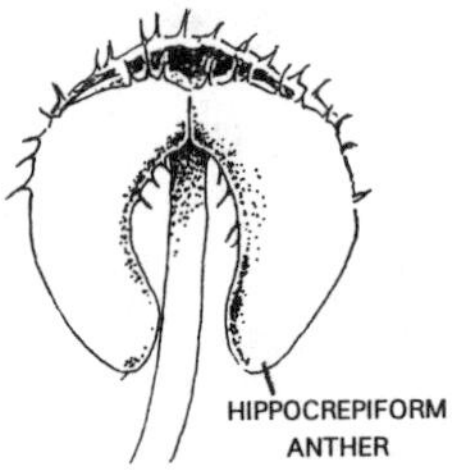

Figure 443

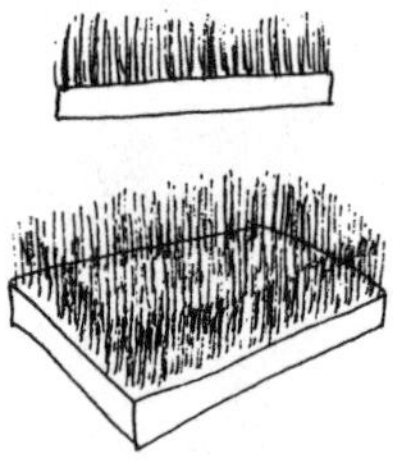

Figure 444

Hirtellate. See **hirsutulous.**

Hirtellous. See **hirsutulous.**

Hispid. Rough with firm, stiff hairs. Figure 446.

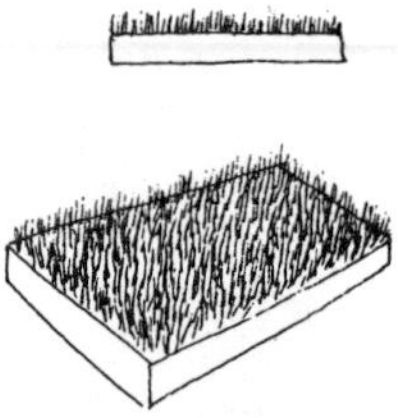

Figure 445

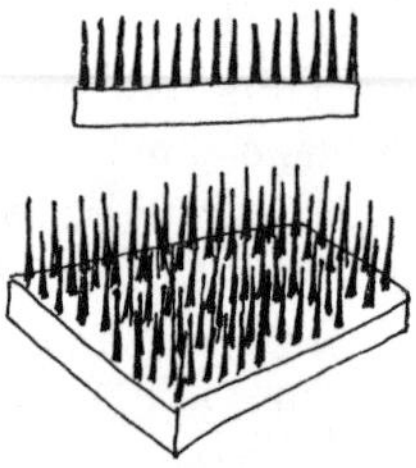

Figure 446

Hispidulous. Minutely hispid. Figure 447.

Hoary. With gray or white short, fine hairs. Figure 448.

Holosericeous. Covered with fine, silky hairs.

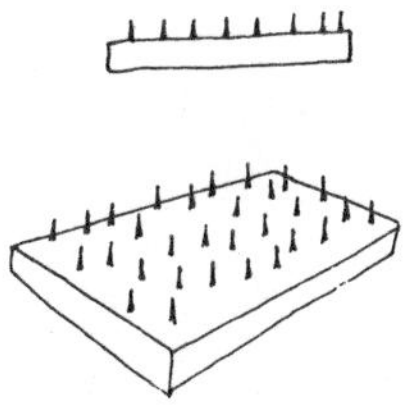

Figure 447

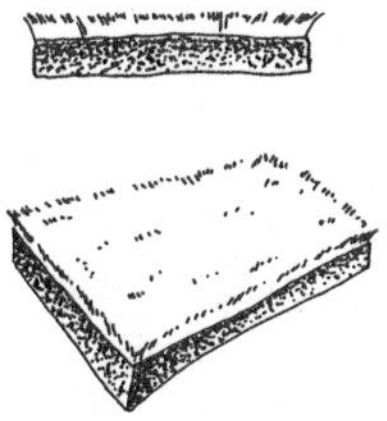

Figure 448

Figure 449.

Homo- (prefix). Meaning the same.

Homochromous. Being all of one color, as the flower heads of some Compositae (Asteraceae).

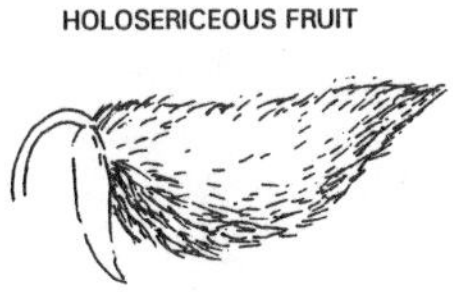

Figure 449

Homogamous. With flowers of the same sex; with the pistils and stamens maturing at the same time. (compare **dichogamous**)

Homogamy. Having homogamous flowers.

Homogeneous. With parts all of the same kind; uniform.

Homogonous. With perfect flowers which do not differ in the relative length of the pistils and stamens. (compare **heterogonous**)

Homogony. State of being homogonous.

Homomorphic. All of the same kind or form.

Homomorphous. See **homomorphic**.

Homosporous. Having spores of one kind.

Homostylic. With styles of more or less constant length in flowers of the same species.

Homostylous. See **homostylic**.

Hood. A hollow, arched covering, as the upper petal in *Aconitum*. Figures 450 and 451.

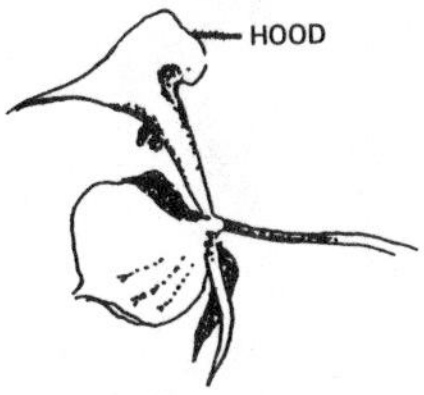

Figure 450

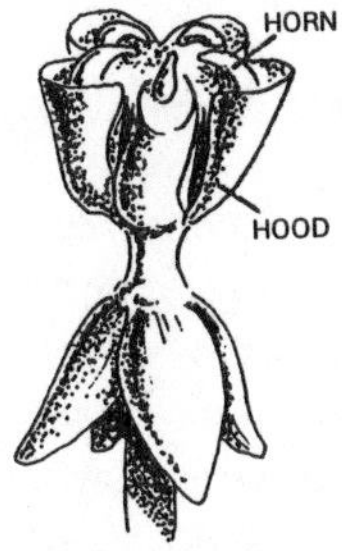

Figure 451

Hooked. Hook-shaped; bent like a hook. Figure 452.

Horn. A tapering projection resembling the horn of a cow. Figure 451.

Host. A plant providing nourishment to a parasite.

Humifuse. Spreading along the ground. Figure 453.

Figure 452

Figure 453

Hyaline. Thin, membranous and translucent or transparent.

Hybrid. The offspring from a cross between parent plants of different varieties, subspecies, species, or genera.

Hybrid swarm. Hybrid plants which are back-crossing to the parents and crossing with themselves, so that there is a continuous intergradation of forms in the population.

Hydathode. An opening which exudes water, usually from a leaf.

Hydrophyte. A plant growing in water. (compare **mesophyte** and **xerophyte**)

Hygroscopic. Absorbing moisture from the air and sometimes swelling, shrinking, or changing position due to changes in moisture content.

Hypanthium. A cup-shaped extension of the floral axis usually formed from the union of the basal parts of the calyx, corolla, and androecium, commonly surrounding or enclosing the pistils. Figure 454.

Figure 454

Hypanthodium. An inflorescence with flowers borne on the walls of a capitulum, as in *Ficus*. Figure 455.

Hypo- (prefix). Meaning beneath or under.

Hypocotyl. That portion of the embryonic stem below the cotyledons. Figures 456 and 457.

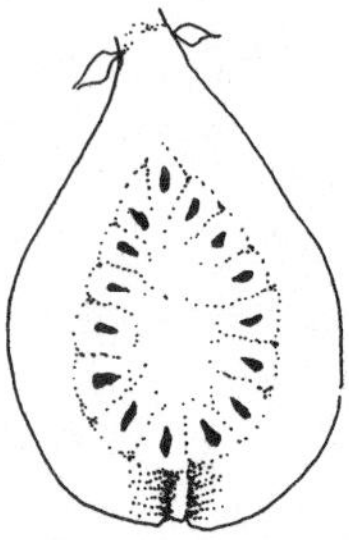

Figure 455

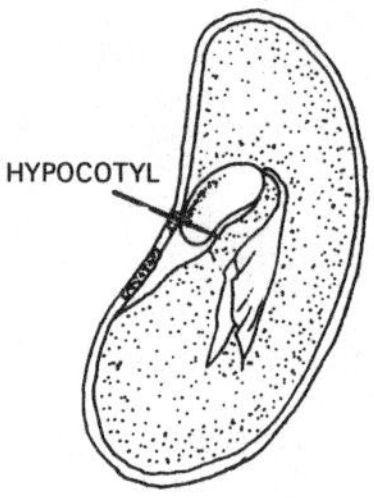

Figure 456

Hypocrateriform. Platter-shaped.

Hypogeal. See **hypogeous.**

Hypogaeous. See **hypogeous.**

Hypogeous. Beneath the ground; said of seedling germination in which the cotyledons remain beneath the ground.

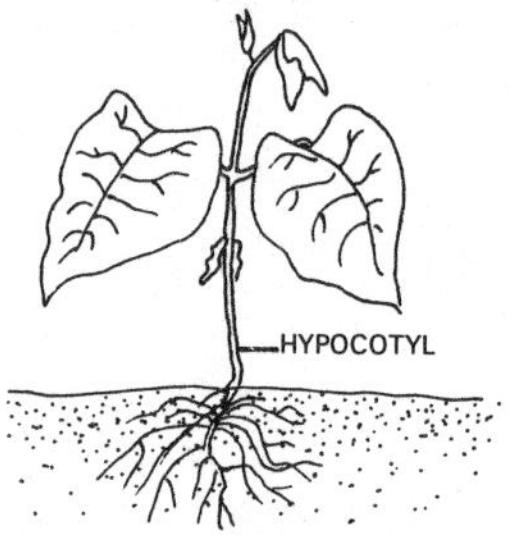

Figure 457

Hypogynous. With stamens, petals, and sepals attached below the ovary, the ovary superior to the other floral parts. Figure 458.

Imbricate. Overlapping like tiles or shingles on a roof. Figure 459.

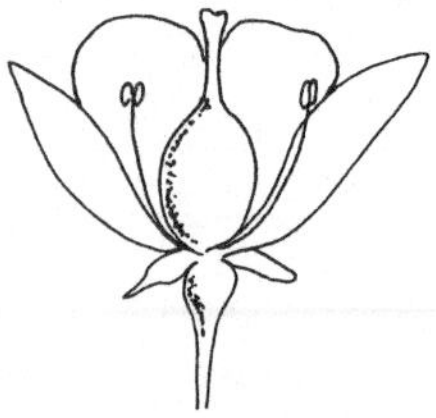

Figure 458

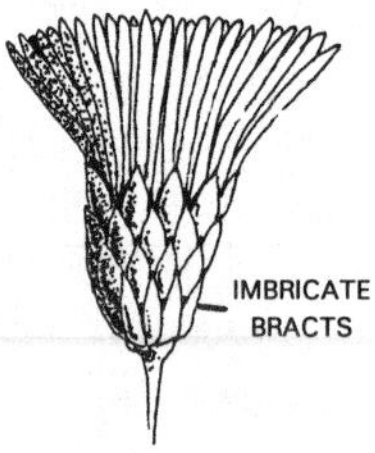

Figure 459

Immersed. Growing under water.

Imparipinnate. Odd-pinnate; unequally pinnate. Figure 460.

Imperfect. With either stamens or pistils, but not both; unisexual. Figure 461.

Implicate. Twisted together; intertwined. Figure 462.

Impressed. Situated below the surface, as in some leaf veins. Figure 463.

Figure 460

Figure 461

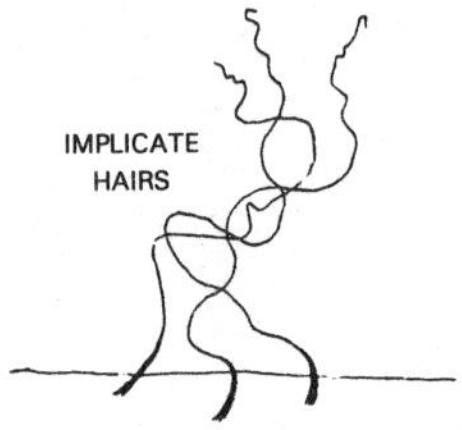

Figure 462

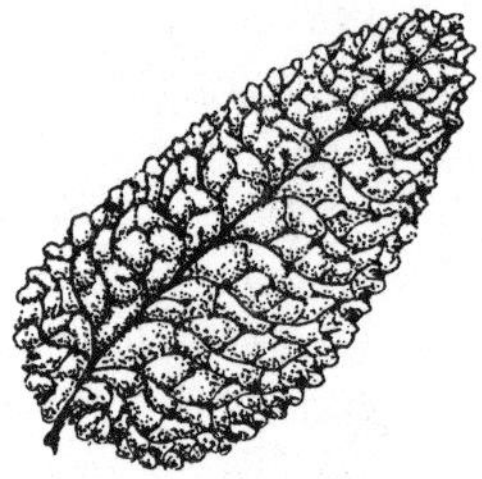

Figure 463

Inaequilateral. See **inequilateral.**

Incanous. With a whitish pubescence. Figure 464.

Incised. Cut sharply, deeply, and usually irregularly. Figure 465.

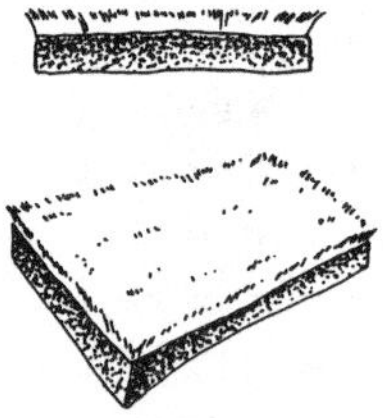

Figure 464

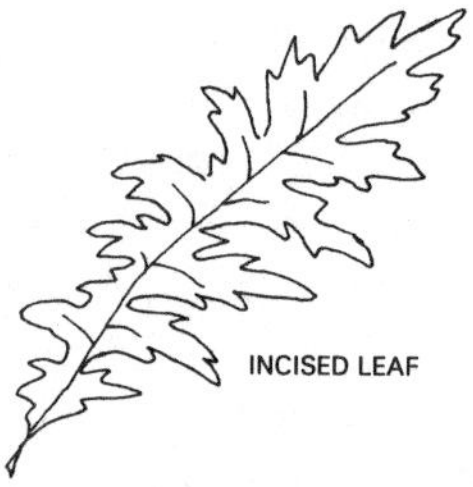

Figure 465

Inclined. Rising upward at a moderate angle. Figure 466.

Included. Not projecting beyond the surrounding parts, as stamens contained within a corolla; not excluded. Figure 467.

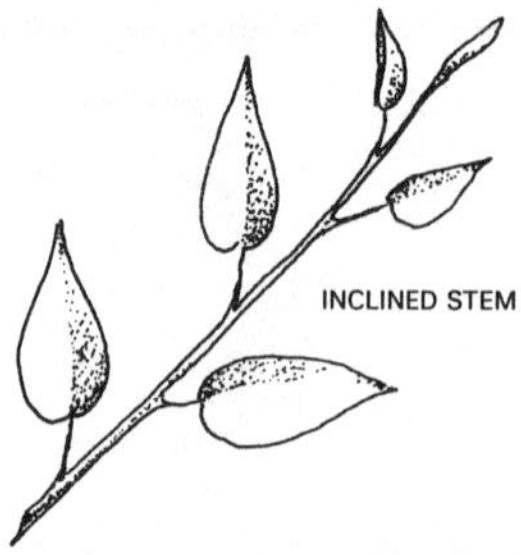

Figure 466

Incomplete. Lacking an expected part or series of parts, as in a flower lacking one of the floral whorls (i.e. sepals, petals, stamens, or pistils).

Incrassate. Thickened or swollen. Figure 468.

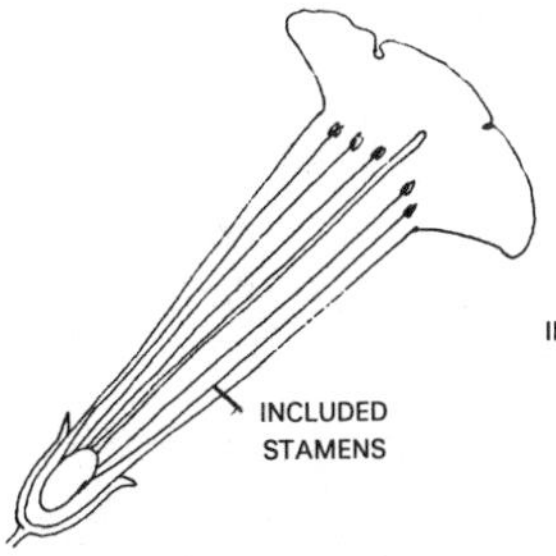

Figure 467

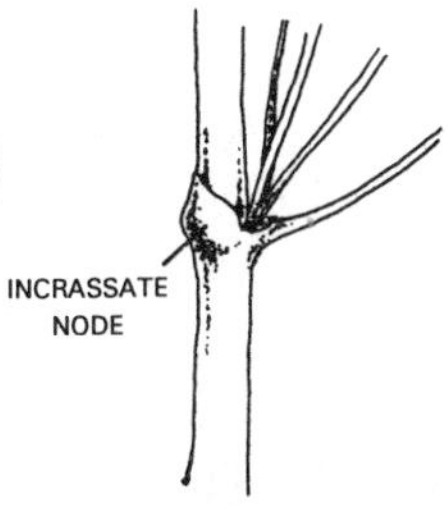

Figure 468

Incumbent cotyledons. Cotyledons lying against the radicle along the back of one of the cotyledons. Figure 469. (compare **accumbent cotyledons**)

Incurved. Curved inward; curved toward the base or apex. Figure 470.

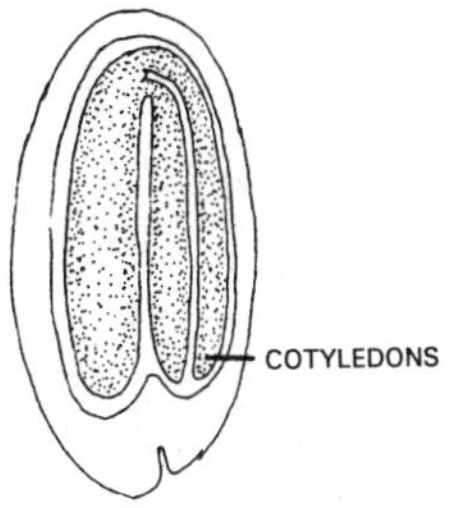

Figure 469

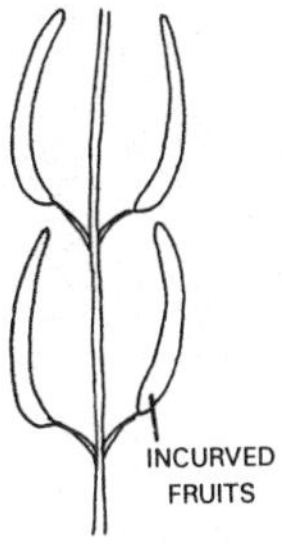

Figure 470

Indehiscent. Not opening at maturity along definite lines or by pores.

Indeterminate. Describes an inflorescence in which the lower or outer flowers bloom first, allowing indefinite elongation of the main axis. Figure 471.

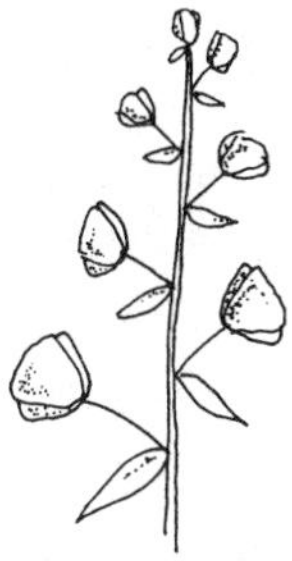

Figure 471

Indigenous. Native to a particular area; not introduced.

Indument. The epidermal coverings of a plant, collectively.

Induplicate. With the petals or sepals edge to edge along their entire length, the margins rolled inward. Figure 472.

Indurate. Hardened.

Indusium (pl. **indusia**). A thin epidermal outgrowth from a fern leaf that covers the sorus. Figure 473.

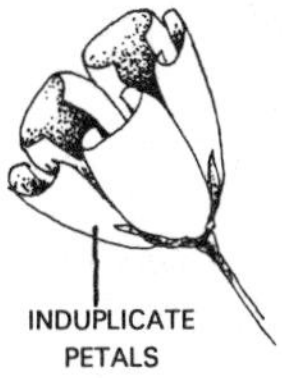

Figure 472

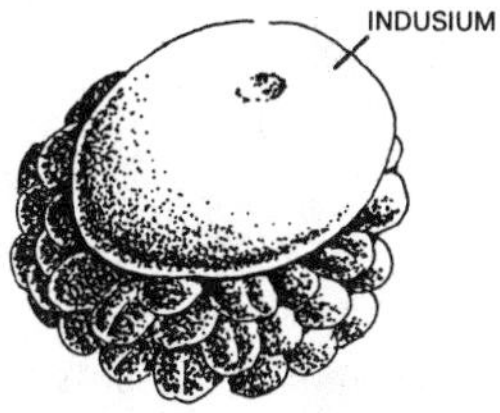

Figure 473

Inequilateral. With sides of unequal shape and length. Figure 474.

Inermous. See **unarmed**.

Inferior. Attached beneath, as an ovary that is attached beneath the point of attachment of the other floral whorls which appear, therefore, to arise from the top of the ovary. Figure 475.

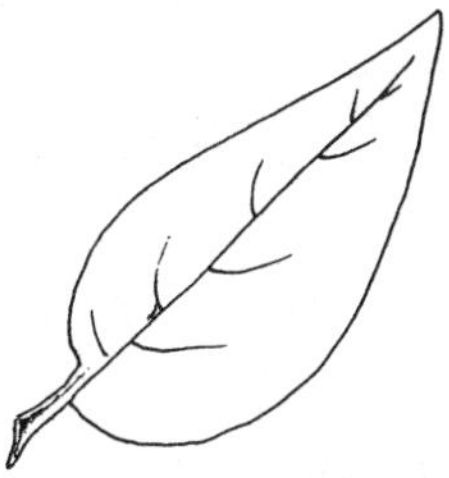

Figure 474

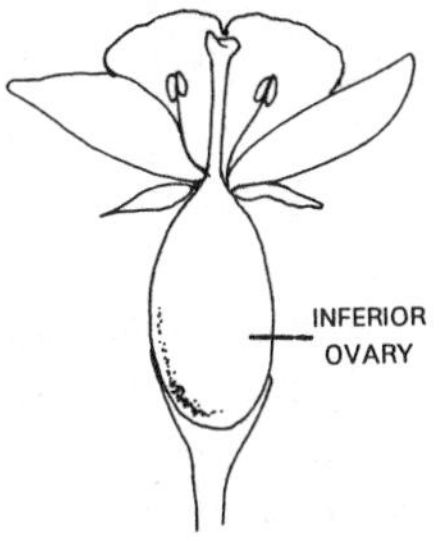

Figure 475

Infertile. Sterile or inviable.

Inflated. Swollen or expanded; bladdery. Figure 476.

Inflexed. Bent or turned downward or inward, toward the axis. Figure 477.

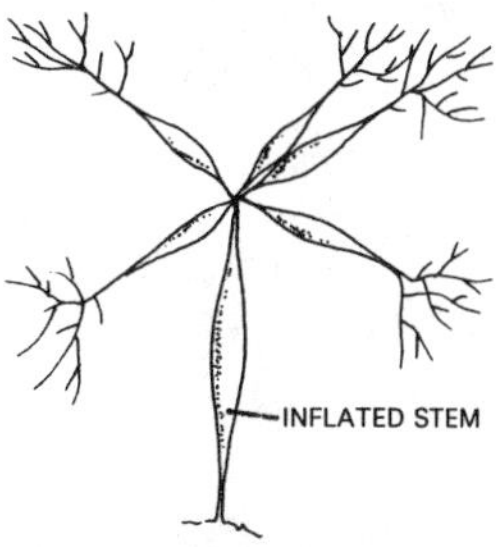

Figure 476

Figure 477

Inflorescence. The flowering part of a plant; a flower cluster; the arrangement of the flowers on the flowering axis. Figure 478.

Infra- (prefix). Meaning below or beneath.

Infra-axillary. Below the axil.

Inframedial. Below the middle.

Infrastaminal. Below the stamens.

Figure 478

Infrastipular. Below the stipules.

Infundibuliform. Funnel-shaped. Figure 479.

Innate. Borne at the apex, as an anther at the apex of the filament. Figure 480.

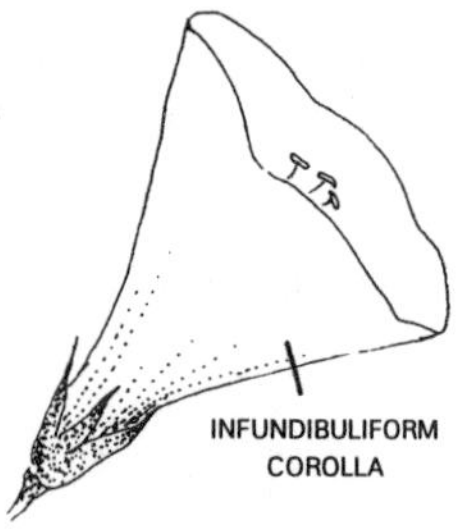

Figure 479 **Figure 480**

Innocuous. Harmless; lacking thorns or spines.

Innovation. A short, basal offset from the base of a stem. Figure 481.

Inodorous. Without an odor.

Inrolled. Curled or rolled inward; involute. Figure 482.

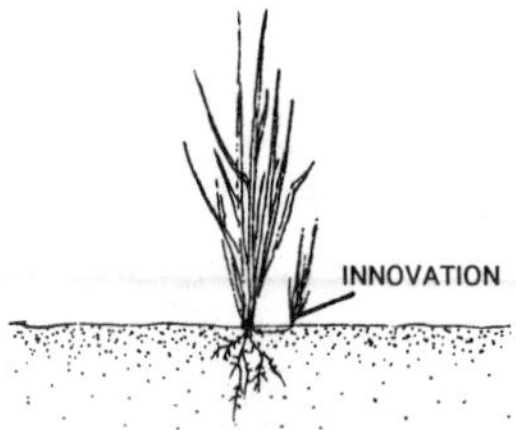

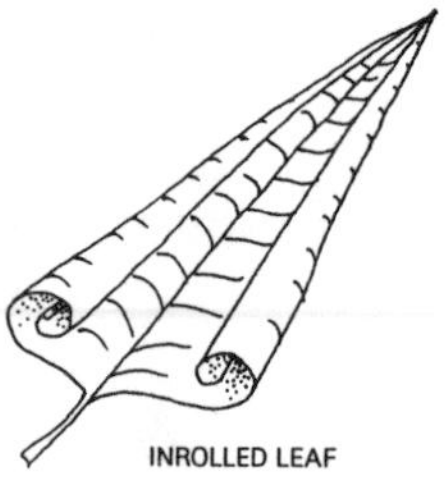

Figure 481 **Figure 482**

Insectivorous. Capturing and digesting insects.

Inserted. Attached to or growing out of. Figure 483.

Insipid. Lacking taste or flavor.

Integument. The covering of the ovule which will become the seed coat. Figure 484.

Inter- (prefix). Meaning between or among.

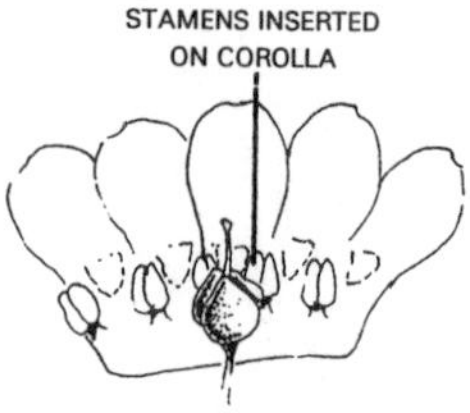

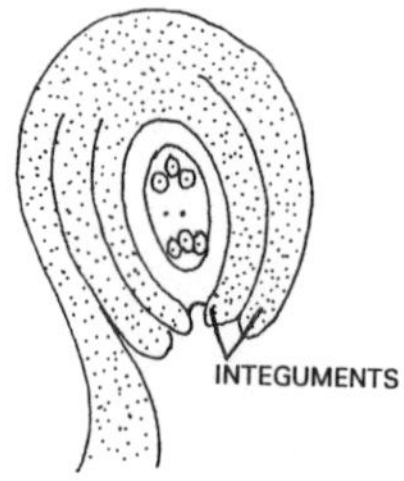

Figure 483 **Figure 484**

Intercostal. Situated between the ribs or nerves. Figure 485.

Internerve. The space between two nerves. Figure 485.

Internode. The portion of a stem between two nodes. Figure 486.

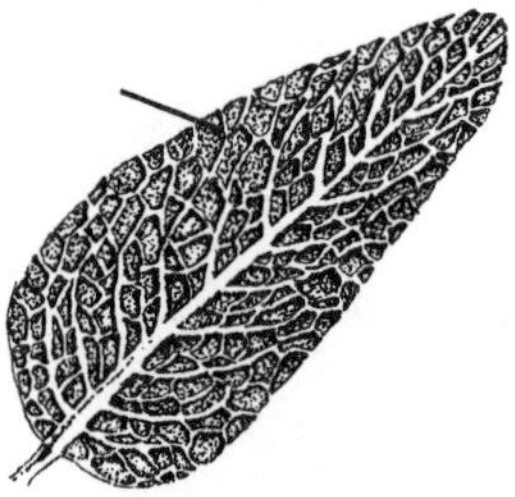

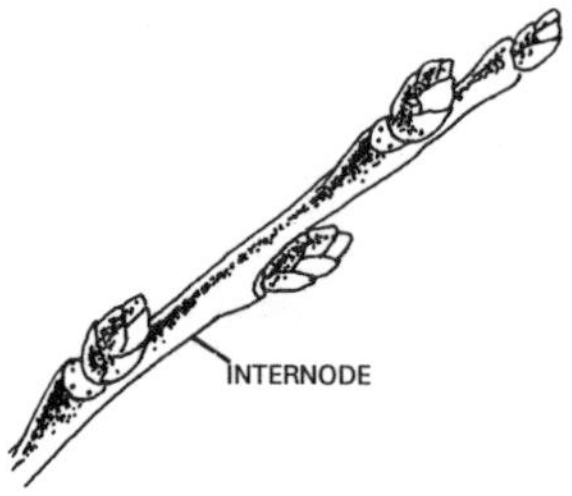

Figure 485 **Figure 486**

Interpetiolar. Between the petioles.

Interrupted. Not continuous.

Interruptedly pinnate. Pinnate with leaflets of various sizes intermixed. Figure 487.

Intervenous. Pertaining to the spaces between veins. Figure 485.

Intine. The inner layer of the two-layered wall of a pollen grain. Figure 488.

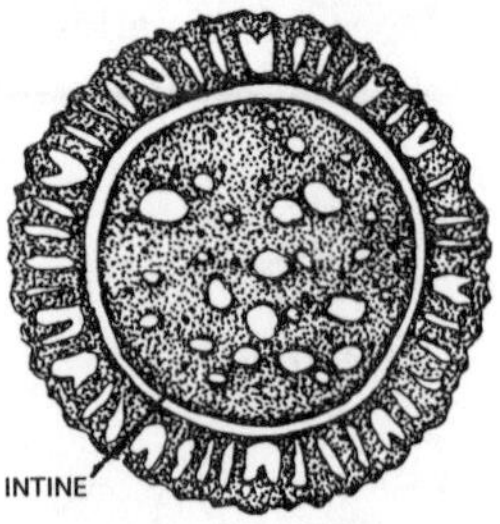

Figure 487 **Figure 488**

Intolerant. Not surviving well under a dense forest canopy.

Intra- (prefix). Meaning within.

Intrastaminal. Within the androecium.

Intricate. Tangled together. Figure 489.

Introduced. Brought in intentionally from another area; not native.

Introgression. Flow of genetic material between taxa.

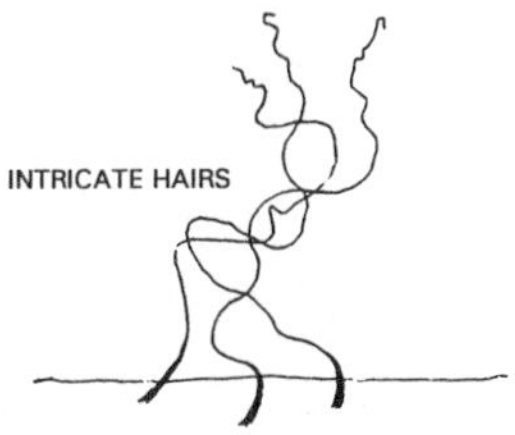

Figure 489

Introrse. Turned inward, toward the axis. Figure 490. (compare **extrorse**)

Intrusion. Protrusion into, as placentae into the cell of an ovary. Figure 491.

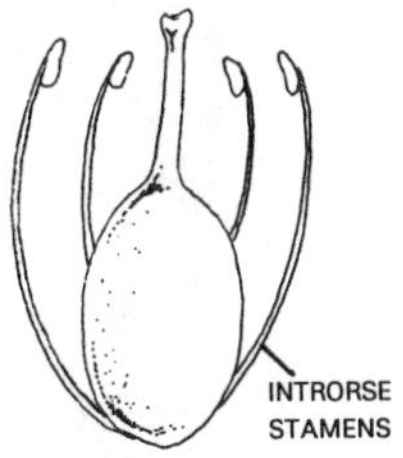

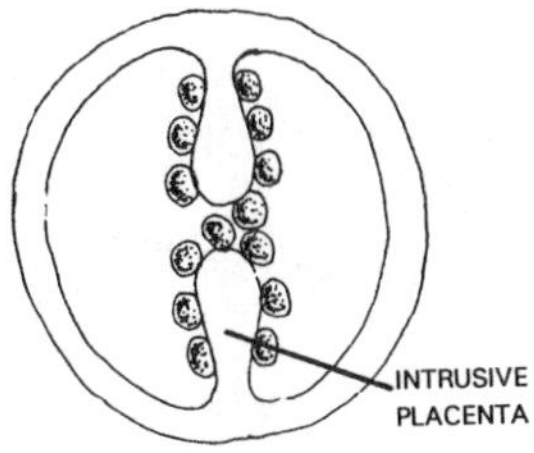

Figure 490 **Figure 491**

Inverted. Positioned opposite the typical direction; reversed.

Investing. Covering or surrounding.

Involucel. A small involucre; a secondary involucre, as in the bracts of the secondary umbels in the Umbelliferae (Apiaceae). Figure 492.

Involucral. Of or pertaining to an involucre.

Involucrate. With an involucre. Figure 493.

Involucre. A whorl of bracts subtending a flower or flower cluster. Figure 493.

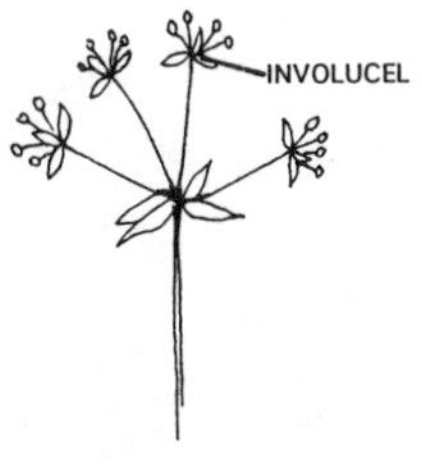

Figure 492 **Figure 493**

Involucrum (pl. **involucra**). See **involucre**.

Involute. With the margins rolled inward toward the upper side. Figure 494. (compare **revolute**)

Iridescent. Displaying many colors, as in a rainbow.

Irregular. Bilaterally symmetrical; said of a flower in which all parts are not similar in size and arrangement on the receptacle. Figure 495. (compare **regular**, and see **zygomorphic**)

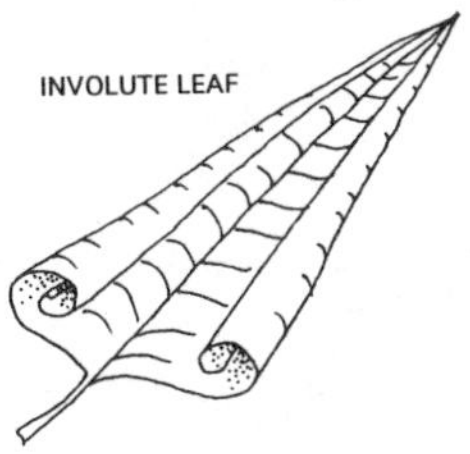

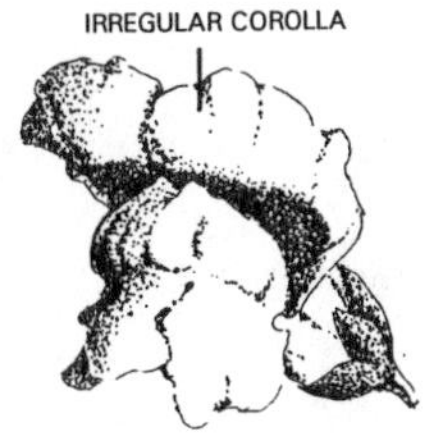

Figure 494 **Figure 495**

Isomerous. With an equal number of parts, as in a flower with an equal number of members in each floral whorl.

Joint. The section of a stem from which a leaf or branch arises; a node, especially on a grass stem. Figure 496.

Jointed. Having nodes or points of articulation, as in the stems of *Opuntia*. Figure 497.

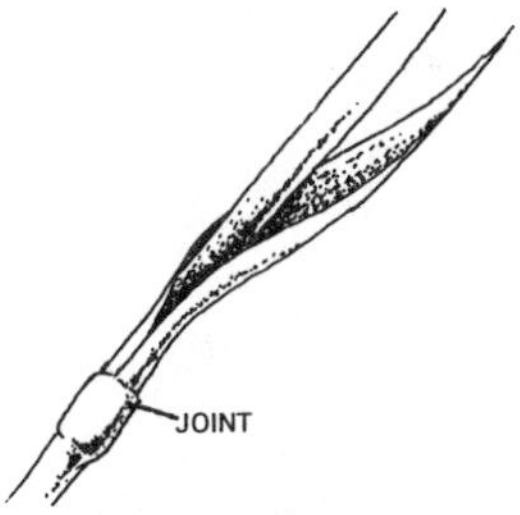

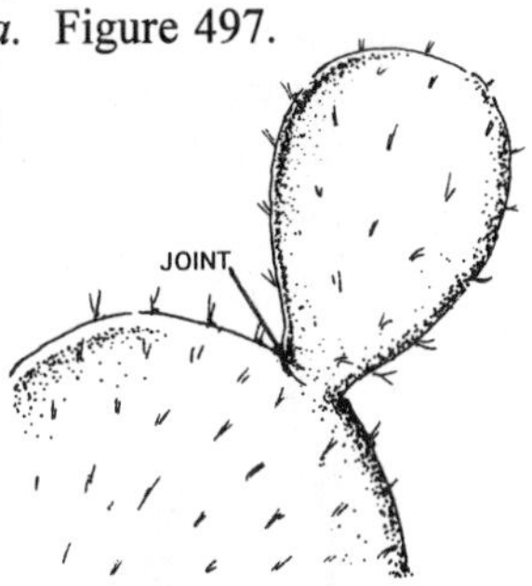

Figure 496 **Figure 497**

Jugate. With parts in pairs, as the leaflets of a pinnate leaf. Figure 498.

Karyotype. All of the chromosomes within the nucleus, especially the size, shape, and number of these chromosomes.

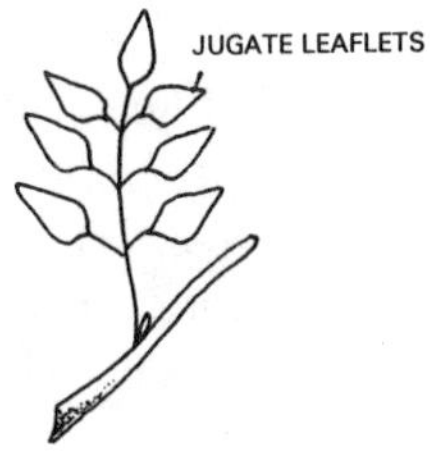

Figure 498

Keel. A prominent longitudinal ridge, like the keel of a boat; the two lower united petals of a papillonaceous flower. Figures 499 and 500.

Keeled. Ridged, like the keel of a boat. Figure 499.

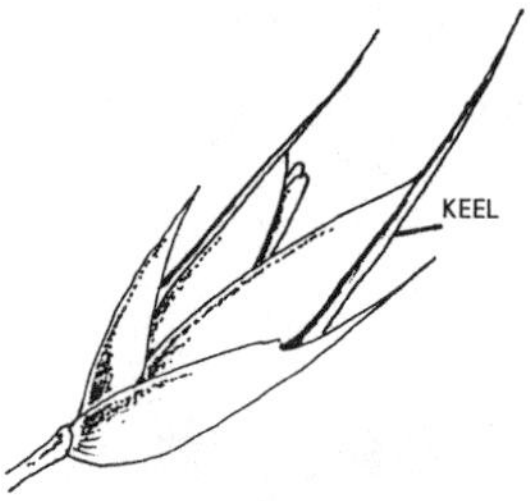

Figure 499

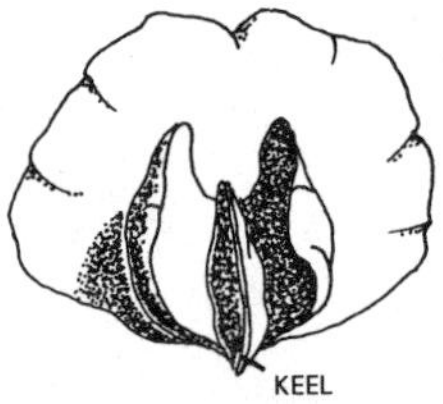

Figure 500

Krummholz. Literally crooked forest; the low wind-contorted forest at timberline.

Labellum. Lip; the exceptional petal of an orchid blossom. Figure 501.

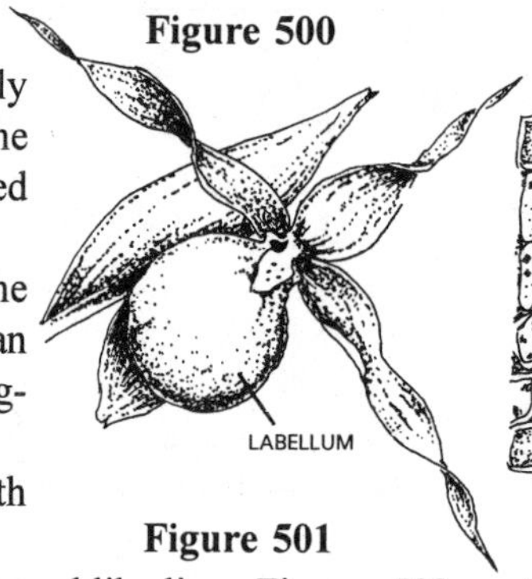

Figure 501

Labiate. Lipped; with parts which are arranged like lips or shaped like lips. Figures 502 and 503; of or pertaining to a member of the Labiatae (Lamiaceae).

Labium (pl. **labia**). The lower lip of a bilabiate corolla. Figures 502 and 503.

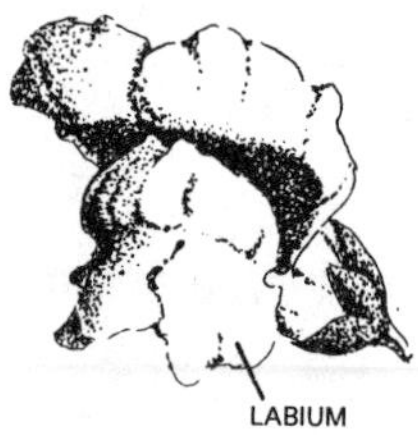

Figure 502

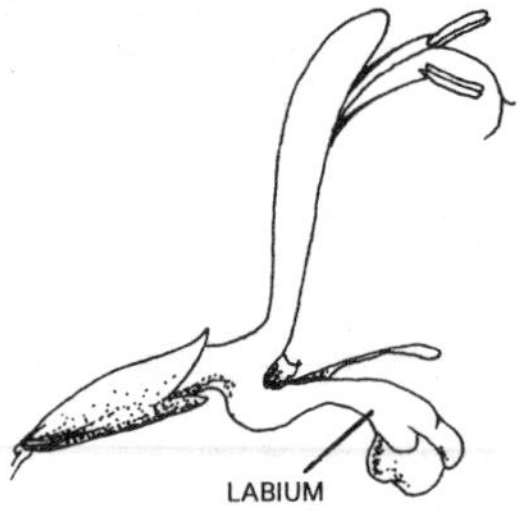

Figure 503

Lacerate. Cut or cleft irregularly, as if torn. Figure 504.

Laciniate. Cut into narrow, irregular lobes or segments. Figure 505.

Lactiferous. Bearing or containing a milky latex.

Lacuna (pl. **lacunae**). An empty air space or gap within a tissue. Figure 506.

Lacustrine. Growing around lakes.

Laevigate. Lustrous; shining.

Lamella (pl. **lamellae**). An erect scale inserted on the petal in some corollas and forming part of the corona; a flat plate or ridge. Figure 507.

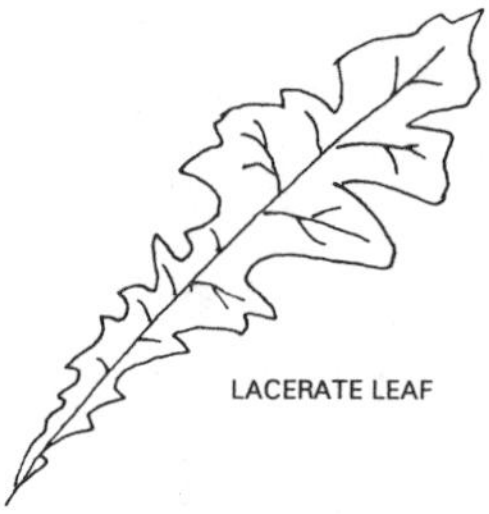

Figure 504

Figure 505

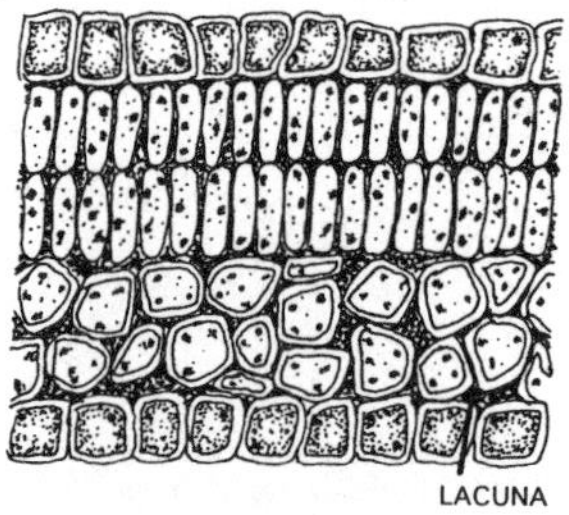

Figure 506

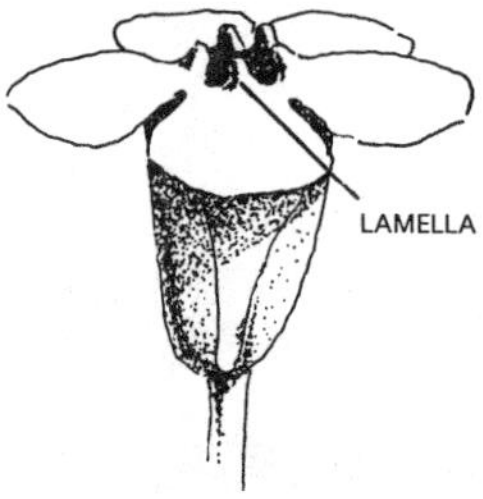

Figure 507

Lamellar. Of or pertaining to lamellae. Figure 507; with lamellae; plate-like. Figure 508.

Lamellate. See **lamellar**.

Lamina. The expanded portion, or blade, of a leaf or petal. Figures 509 and 510.

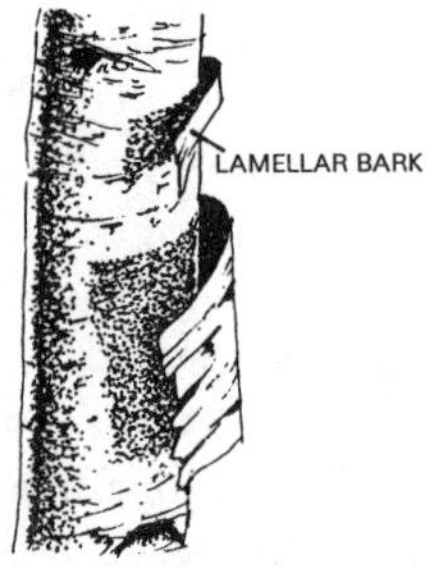

Figure 508

Laminar. Thin, flat, and expanded, as the blade of a leaf. Figure 509.

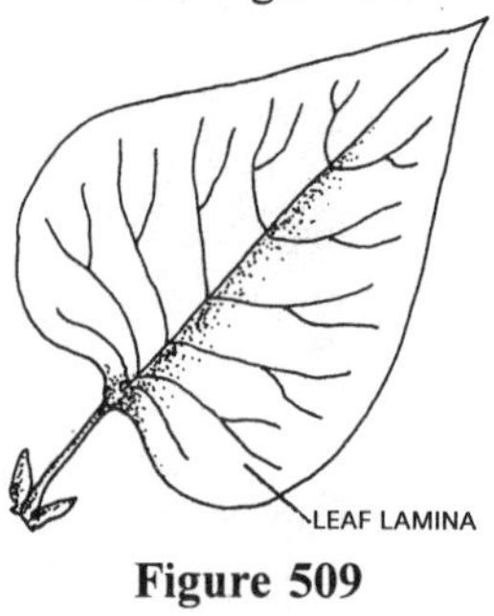

Figure 509

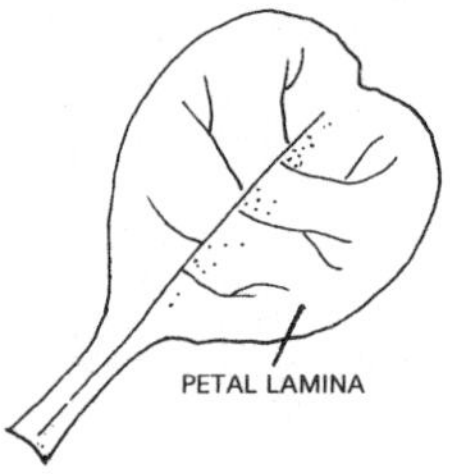

Figure 510

Laminate. With plates or layers; separating into

plates or layers. Figure 508.

Lanate. Woolly; densely covered with long tangled hairs. Figure 511.

Lanceolate. Lance-shaped; much longer than wide, with the widest point below the middle. Figure 512.

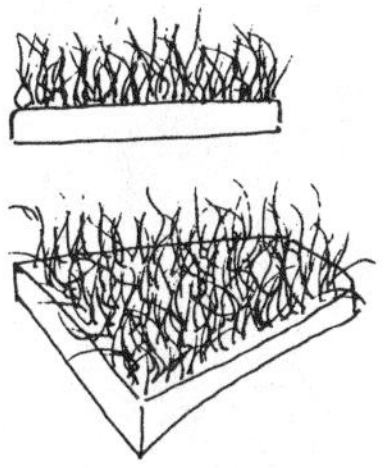

Figure 511

Figure 512

Lanuginose. See **lanuginous**.

Lanuginous. Downy or woolly; with soft downy hairs. Figure 513.

Lanugo. A covering of soft downy hairs. Figure 513.

Lanulose. Diminutive of lanate; minutely woolly. Figure 514.

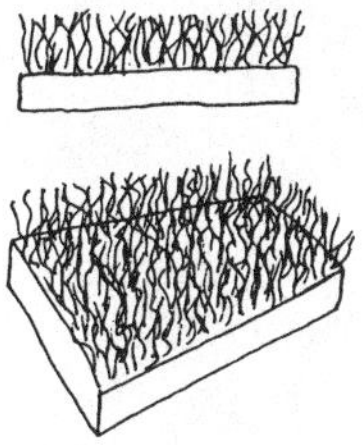

Figure 513

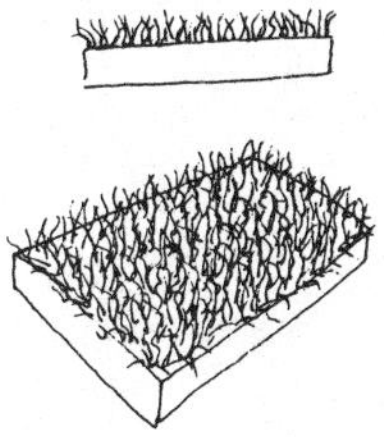

Figure 514

Lateral. Borne on or at the side. Figure 515.

Latex. A milky sap.

Laticifer. A tube or channel containing latex.

Laticiferous. Bearing or containing latex.

Latrorse. Dehiscing longitudinally and laterally. Figure 516.

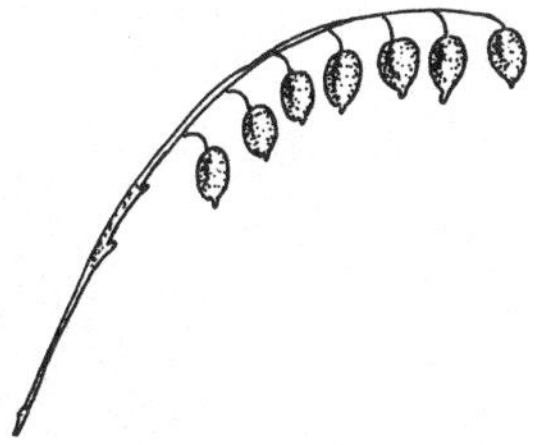

Figure 515

Lax. Loose; with parts open and spreading, not compact. Figure 517.

Leaf. An expanded, usually photosynthetic organ of a plant. Figure 518.

Leaflet. A division of a compound leaf. Figure 519.

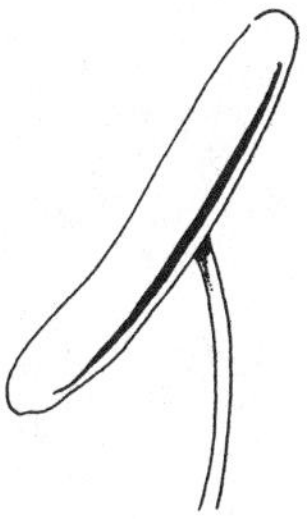

Figure 516

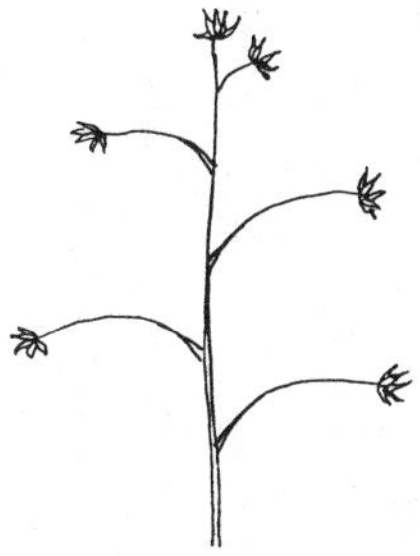

Figure 517

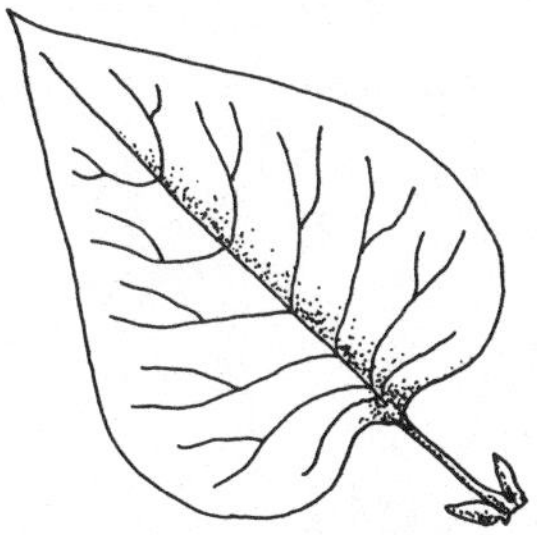

Figure 518

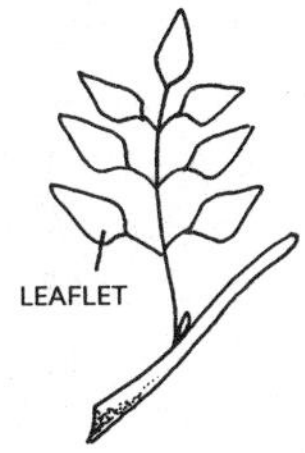

Figure 519

Leaf scar. The scar remaining on a twig after a leaf falls. Figures 520 and 521.

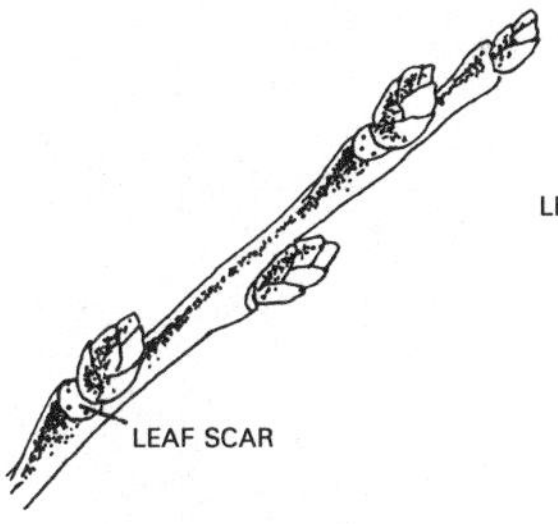

Figure 520

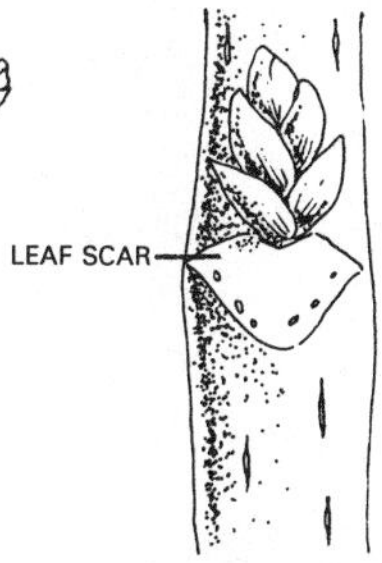

Figure 521

Legume. A dry, dehiscent fruit derived from a single carpel and usually opening along two lines of dehiscence, as a pea pod Figure 522; a plant belonging to the Leguminosae (Fabaceae) family.

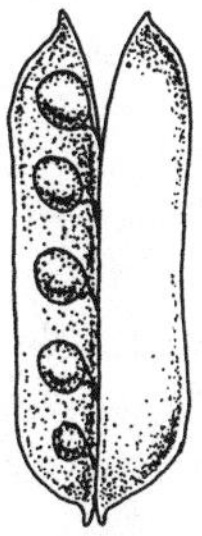

Figure 522

Lemma. The lower of the two bracts (lemma and palea) which subtend a grass floret, often partially surrounding the palea. Figure 523.

Lenticel. A slightly raised, somewhat corky, often lens-shaped area on the surface of a young stem. Figure 524.

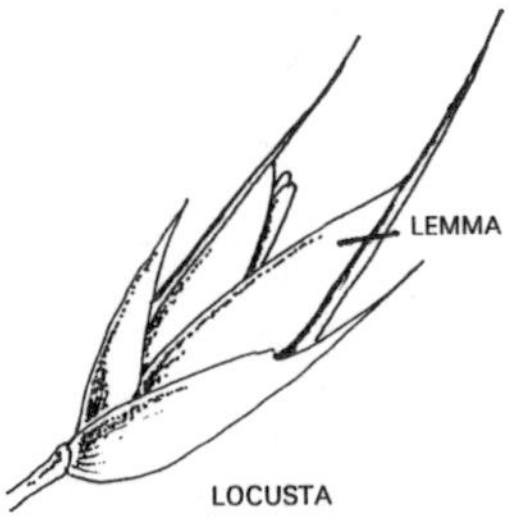

Figure 523

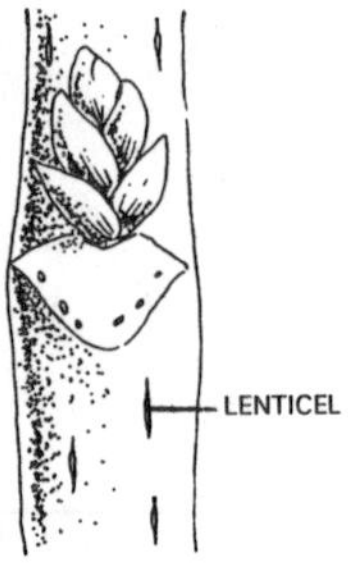

Figure 524

Lenticular. Lentil-shaped (lens-shaped); biconvex. Figure 525.

Lentiginous. See **scurfy**.

Lepidote. Covered with small, scurfy scales. Figure 526.

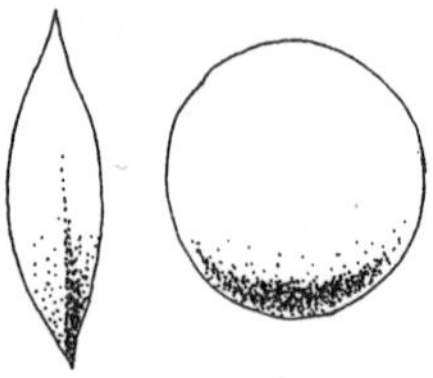

Figure 525

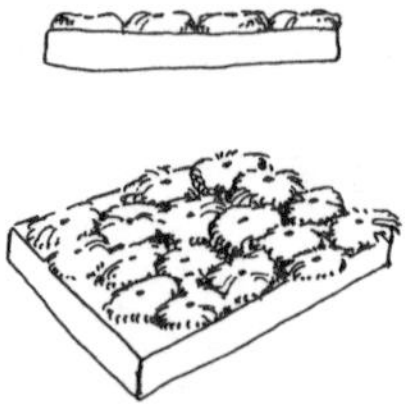

Figure 526

Liana. A woody, climbing vine.

Ligneous. Woody.

Lignified. See **ligneous**.

Lignose. See **ligneous**.

Ligula. See **ligule**.

Ligulate. With a ligule; strap-shaped. Figure 527.

Ligule. A tongue-shaped or strap-shaped organ. Figure 527; the flattened part of the ray corolla in the Compositae (Asteraceae). Figure 528; the membranous appendage arising from the inner surface of the leaf at the junction with the leaf sheath in many grasses and some sedges. Figure 529; a tongue-like projection borne at the base of the leaves above the sporangia in *Isoetes*. Figure 530.

Limb. The expanded part of a petal or leaf. Figure 531; the expanded part of a sympetalous corolla. Figure 532.

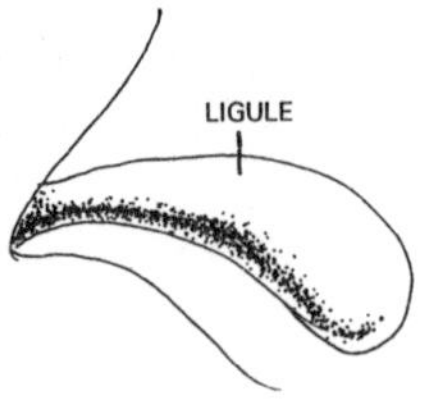

Figure 527

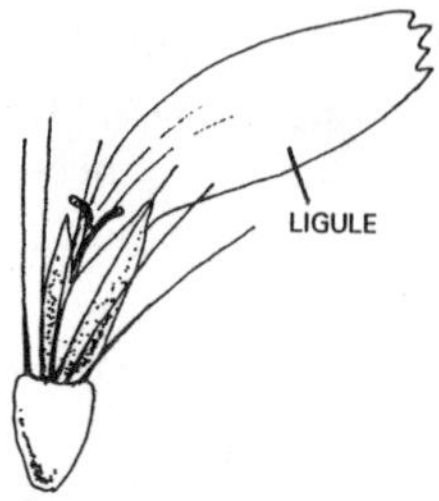

Figure 528

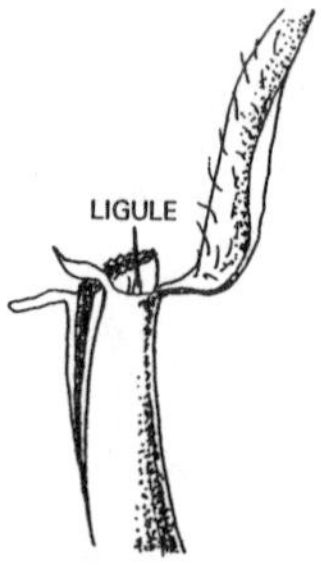

Figure 529

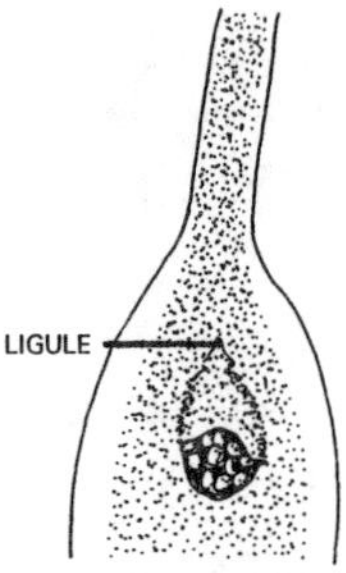

Figure 530

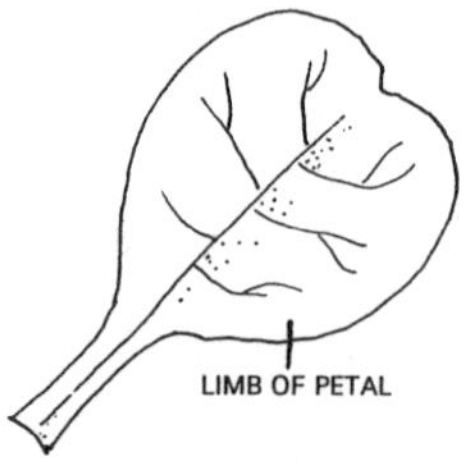

Figure 531

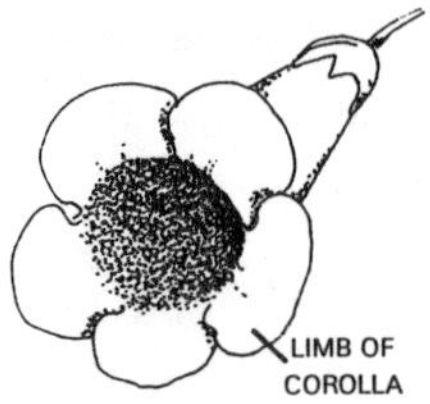

Figure 532

Limbate. Bordered, as in a leaf or flower in which one color forms an edging or margin around another. Figure 533.

Linear. Resembling a line; long and narrow with more or less parallel sides. Figure 534.

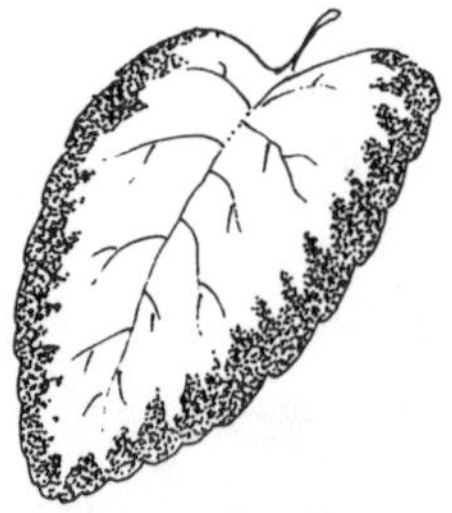

Figure 533

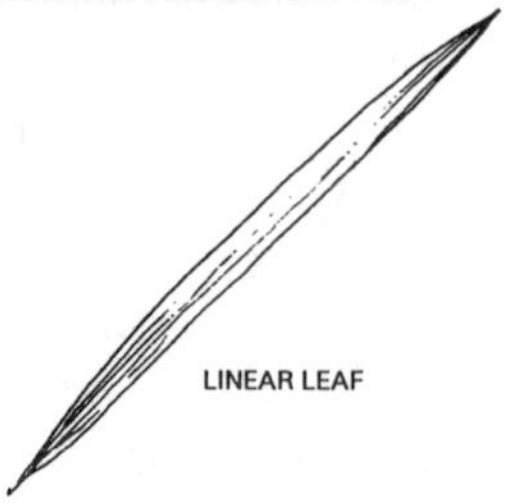

Figure 534

Lineate. Marked with lines.

Linguiform. See **lingulate**.

Lingulate. Tongue-shaped. Figure 527.

Lip. One of the two projections or segments of an irregular, two-lipped corolla or calyx. Figure 535; a labium; the exceptional petal of an orchid blossom. Figure 536.

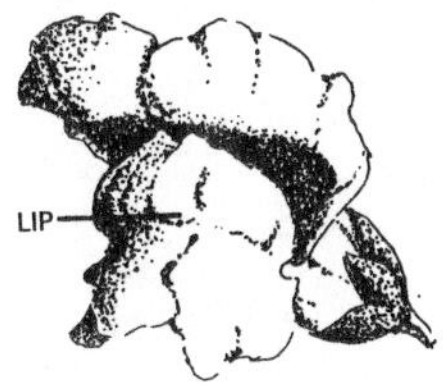

Figure 535

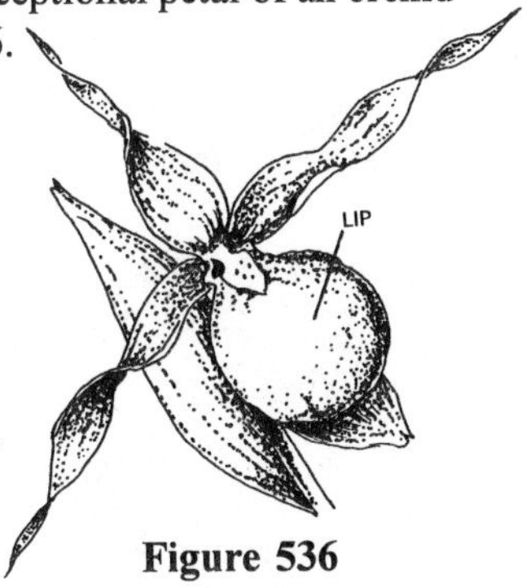

Figure 536

Litoral. See **littoral**.

Littoral. Growing along the shore.

Livid. Pale grayish-blue.

Lobate. In the form of a lobe; lobed. Figure 537.

Lobe. A rounded division or segment of an organ, as of a leaf. Figure 537.

Lobed. Bearing lobes which are cut less than half way to the base or midvein. Figure 537.

Lobulate. With lobules. Figure 538.

Lobule. A small lobe; a lobelike subdivision of a lobe. Figure 538.

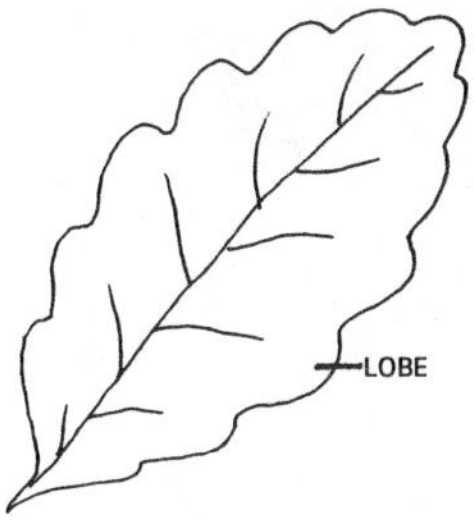

Figure 537

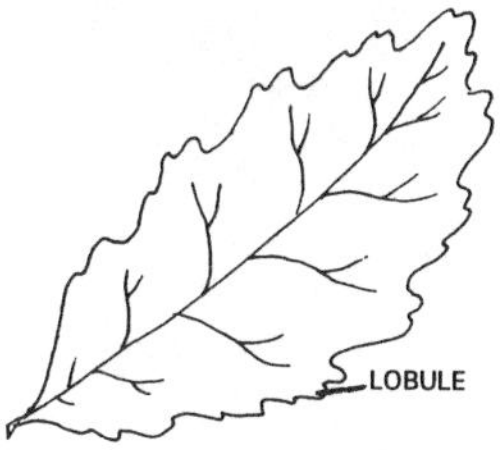

Figure 538

Locular. With locules. Figure 539.

Locule. The chamber or cavity ("cell") of an organ, as in the cell of an ovary containing the seed or the pollen bearing compartment of an anther. Figure 539.

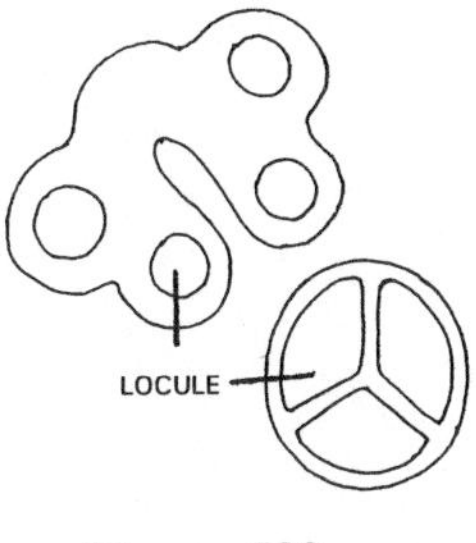

Figure 539

Loculicidal. Dehiscing through the locules of a fruit rather than through the septa. Figure 540. (compare **septicidal** and **poricidal**)

Loculus (pl. **loculi**). See **locule**.

Locusta. The spikelet of grasses. Figure 523.

Lodicule. Paired, rudimentary scales at the base of the ovary in grass flowers. Figure 541.

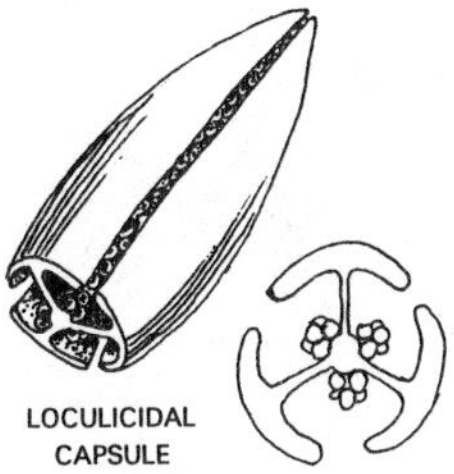

Figure 540

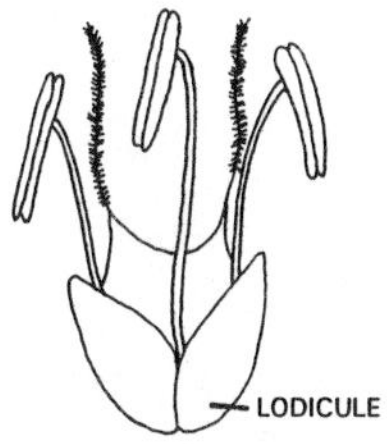

Figure 541

Loment. A legume which is constricted between the seeds. Figure 542.

Lomentaceous. Lomentlike; with loments.

Lomentum (pl. **lomenta**). See **loment**.

Longitudinal. Along the long axis of an organ. Figures 543 and 544.

Lunate. Crescent-shaped. Figure 545.

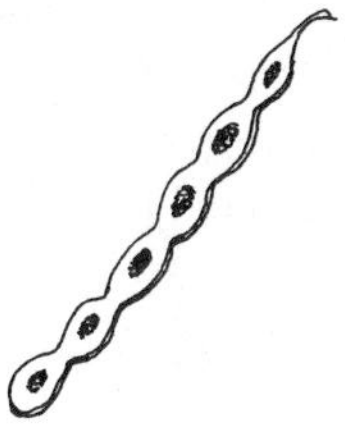

Figure 542

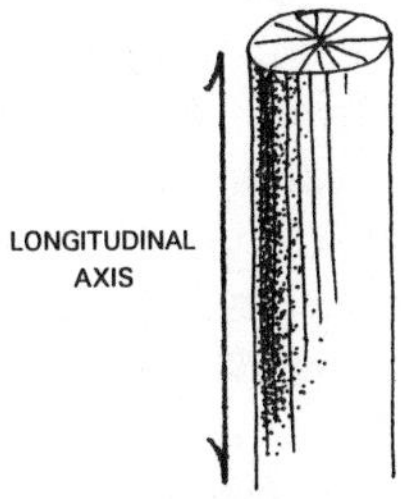

Figure 543

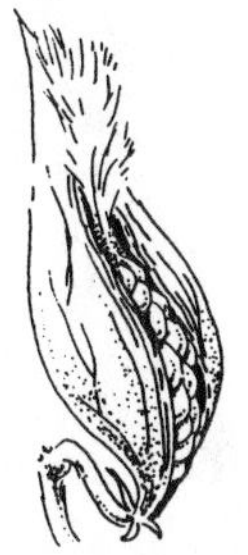

Figure 544

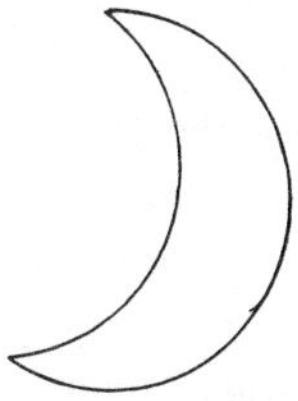

Figure 545

Lunulate. Diminutive of **lunate**.

Lustrous. Shiny or glossy.

Luteous. See **lutescent**.

Lutescent. Yellowish.

Lyrate. Lyre-shaped; pinnatifid, with the terminal lobe large and rounded and the lower lobes much smaller. Figure 546.

Lysigenous. Formed by the dissolution of tissue.

Machaerantheroid. Involucral bracts with recurved tips. Figure 547.

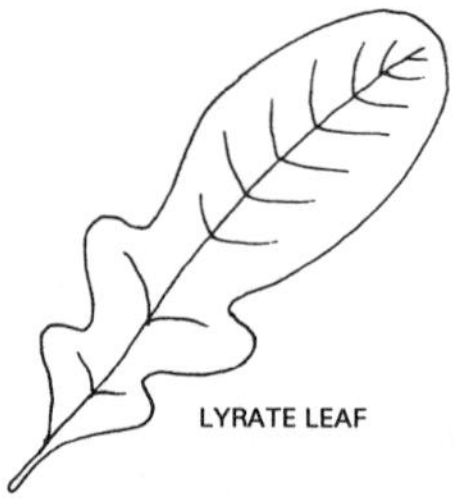

Figure 546

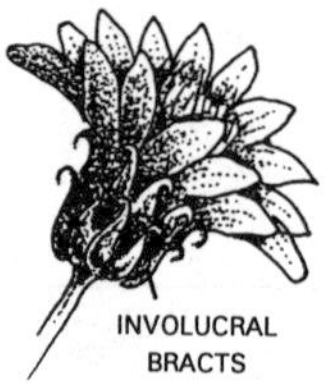

Figure 547

Macro- (prefix). Meaning large.

Macrocladous. With long branches.

Macrophyll. The relatively large, expanded leaf of higher vascular plants. Figure 548. (compare **microphyll**)

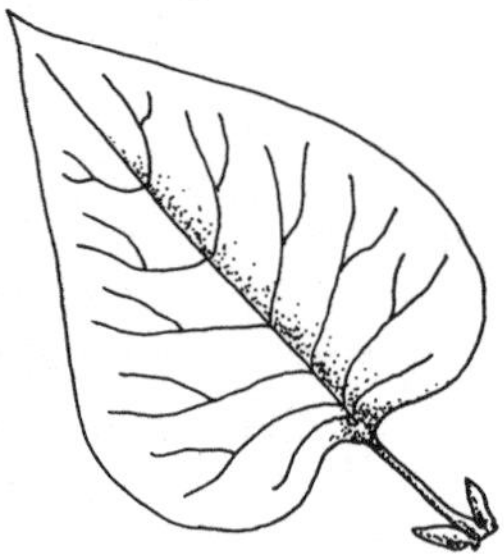

Figure 548

Macrophyllous. With large leaves or leaflets; with macrophylls.

Macrosporangium. See **megasporangium**.

Macrospore. See **megaspore**.

Macrosporophyll. See **megasporophyll**.

Macrostylous. With a long style.

Macula (pl. **maculae**). A spot or blotch. Figure 549.

Maculate. Spotted or blotched. Figure 549.

Malacophyllous. With soft leaves.

Malodorous. Having a disagreeable odor.

Figure 549

Malpighiaceous hair. See **malpighian hair**.

Malpighian hair. Straight hairs tapering to both free ends and attached near the middle; pick-shaped. Figure 550. (same as **dolabriform**)

Malvaceous. Mallowlike.

Mammiform. Breast-shaped.

Mammilla (pl. **mammillae**). A nipple-like protuberance. Figure 551.

Mammillate. With nipple-like protuberances. Figure 551.

Mammose. See **mammillate**.

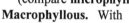

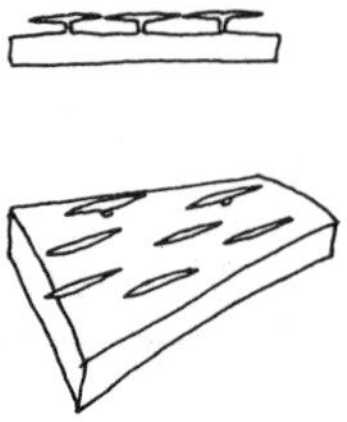

Figure 550

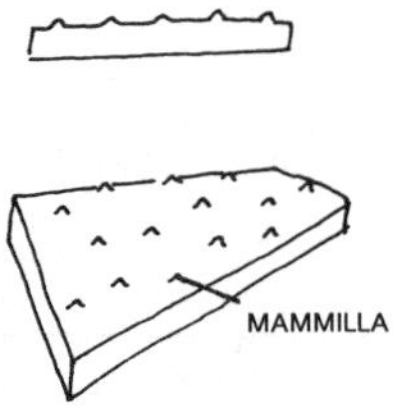

Figure 551

Manicate. With a thick, interwoven pubescence. Figure 552.

Many. In botanical descriptions, this term usually means more than ten.

Marcescent. Withering but persistent, as the sepals and petals in some flowers or the leaves at the base of some plants. Figure 553.

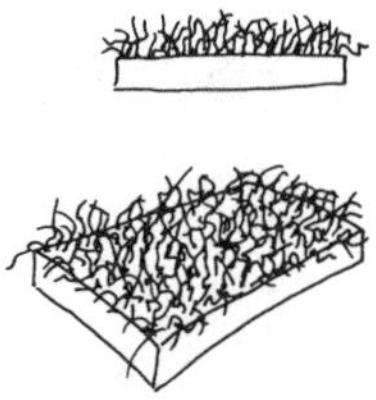

Figure 552

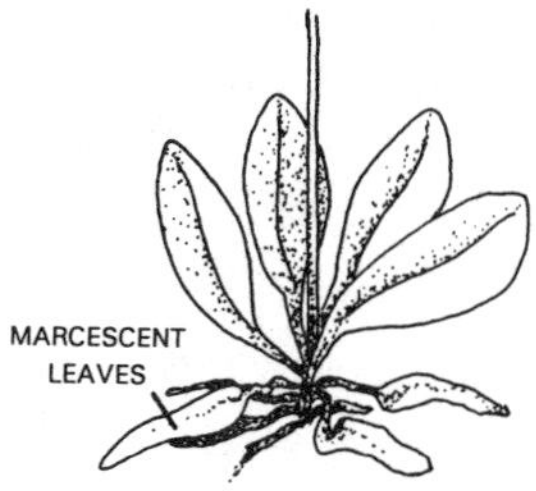

Figure 553

Margin. The edge, as in the edge of a leaf blade. Figure 554.

Marginal placentation. Ovules attached to the juxtaposed margins of a simple pistil. Figure 555.

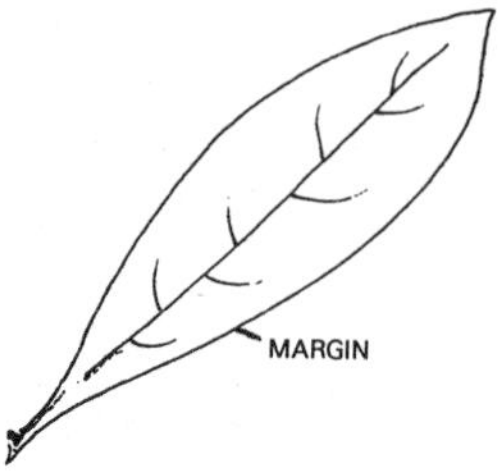

Figure 554

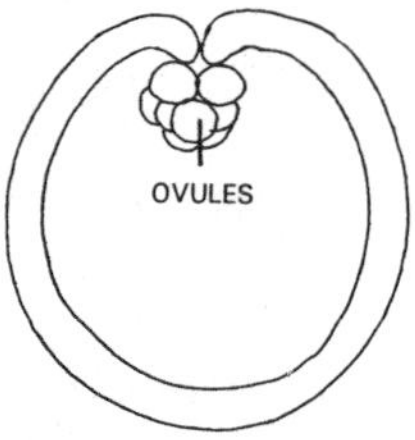

Figure 555

Marginate. With a distinct margin.

Maritime. Growing near the sea and often being saltwater tolerant.

Masked. See **personate**.

Massula. See **pollinium**.

Mast. Nuts used for food, particularly acorns and beechnuts.

Matutinal. Functioning in the morning, as in flowers which open in the morning.

Mauve. Bluish or pinkish purple.

Mealy. With the consistency of meal; powdery, dry, and crumbly. Figure 556.

Medial. Of the middle; situated in the middle.

Median. See **medial**.

Mega- (prefix). Meaning large.

Megaphyllous. With large leaves.

Figure 556

Megasporangium (pl. **megasporangia**). A spore-producing structure (sporangium) which bears megaspores. Figure 557.

Megaspore. A female spore which will give rise to the female gametophyte. Figure 557.

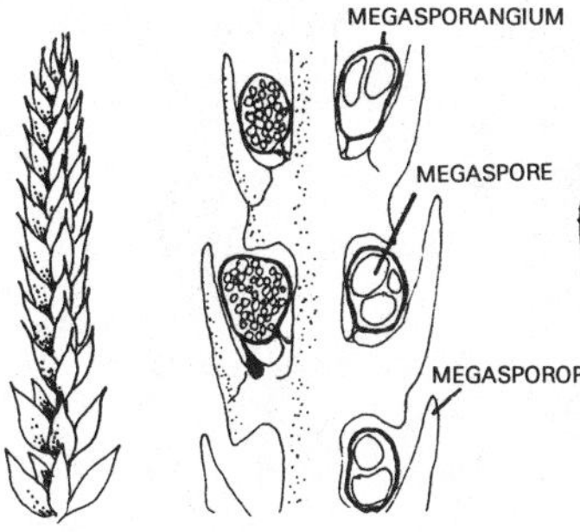

Figure 557

Megasporophyll. A modified leaf which bears one or more megasporangia. Figure 557.

Melanophyllous. With dark leaves.

Melanoxylon. Dark wood.

Membranaceous. See **membranous**.

Membranous. Thin, soft, flexible, and more or less translucent, like a membrane.

Meniscoid. Concavo-convex; one side concave and the other convex. Figure 558.

Mephitic. Having a strong, disagreeable odor.

Mericarp. A section of a schizocarp; one of the two halves of the fruit in the Umbelliferae (Apiaceae). Figure 559.

Meristem. Undifferentiated, actively dividing tissues at the growing tips of shoots and roots.

Figure 558

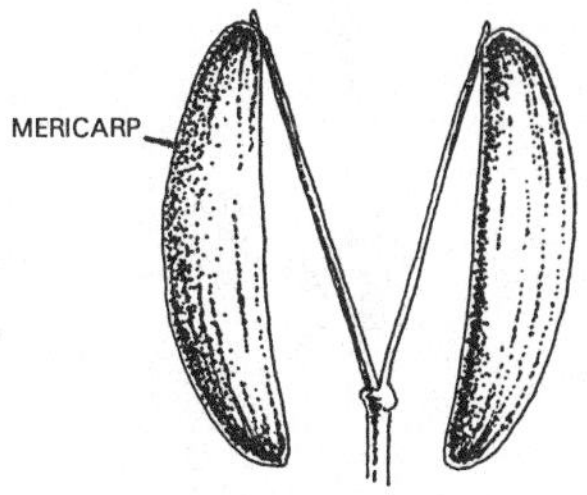

Figure 559

Meristematic. Of or pertaining to the meristem.

-merous (suffix). Meaning parts of a set. A 5-merous corolla would have five petals.

Mesic. Moist.

Meso- (prefix). Meaning middle.

Mesocarp. The middle layer of the pericarp of a fruit. Figure 560. (compare **endocarp** and **exocarp**)

Mesocotyl. That portion of the embryonic stem between the coleoptile and the scutellum in a grass embryo. Figure 561.

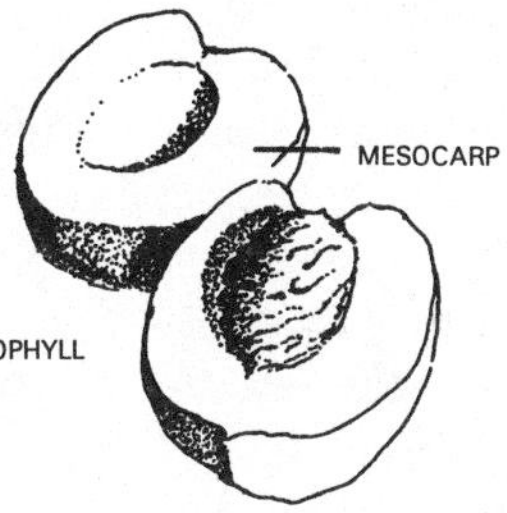

Figure 560

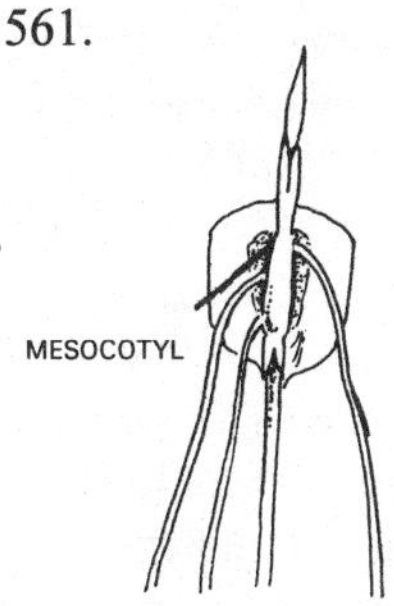

Figure 561

Mesophyll. The central tissues of a leaf between the upper and lower epidermis. Figure 562.

Mesophyte. A plant growing in average moisture conditions. (compare **hydrophyte** and **xerophyte**)

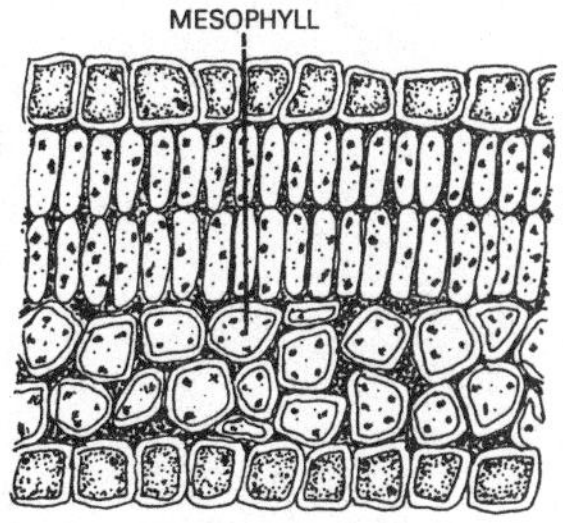

Figure 562

Metandry. Female flowers maturing before the male flowers; protogyny.

Micro- (prefix). Meaning small.

Microphyll. The relatively small, narrow, single-veined leaf of some lower vascular plants. Figures 563 and 564. (compare **macrophyll**)

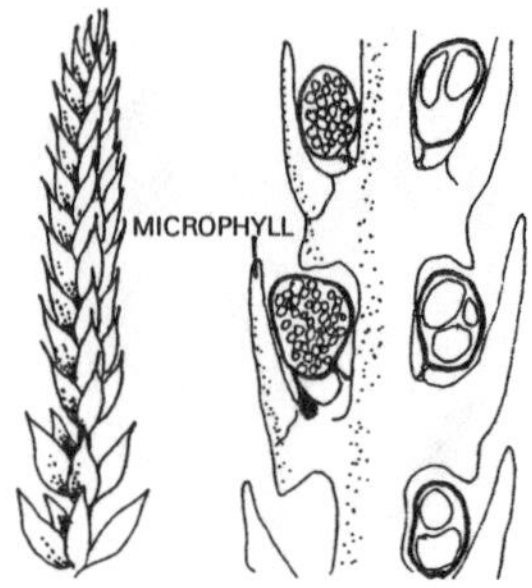

Figure 563

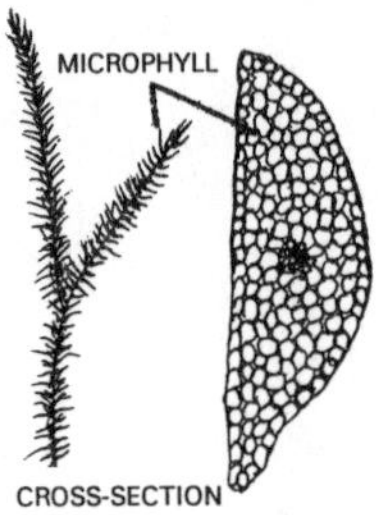

Figure 564

Micropyle. The opening in the integuments of the ovule. Figure 565.

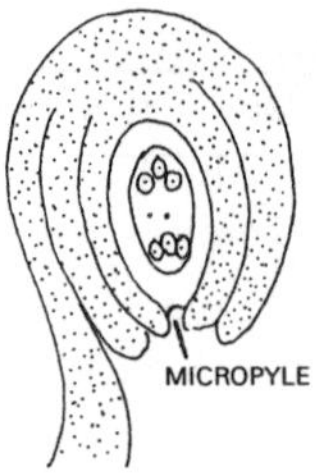

Figure 565

Microsporangium (pl. **microsporangia**). A spore-producing structure (sporangium) which bears microspores. Figure 563.

Microspore. A male spore which will give rise to the male gametophyte. Figure 563.

Microsporophyll. A modified leaf which bears one or more microsporangia; a stamen. Figure 563.

Midlobe. The central lobe. Figure 566.

Midnerve. The central nerve. Figure 567.

Midrib. The central rib or vein of a leaf or other organ. Figure 567.

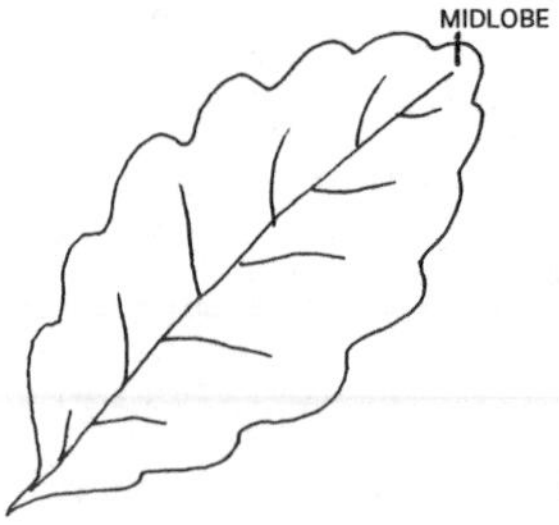

Figure 566

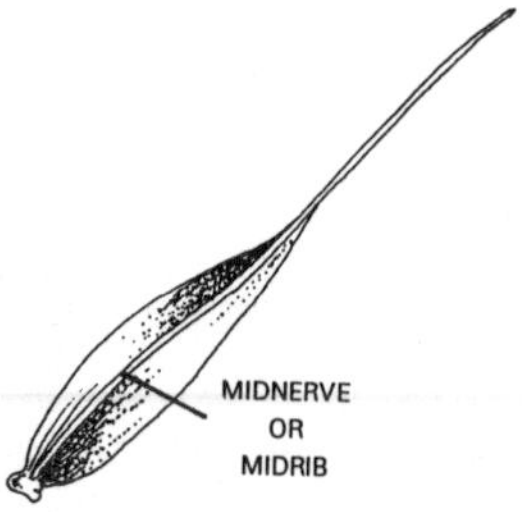

Figure 567

Midvein. The central vein. Figure 568.

Mitriform. Shaped like a mitra. Figure 569.

Mixed bud. A bud which produces both leaves and flowers.

Mixed inflorescence. An inflorescence with both racemose and cymose portions.

Molendinaceous. With large, winglike developments. Figure 570.

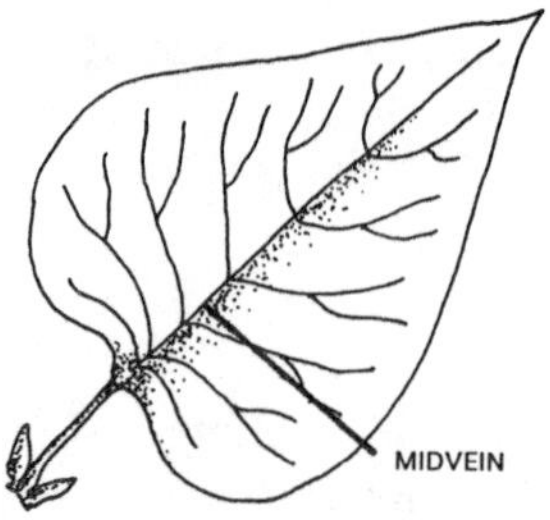

Figure 568

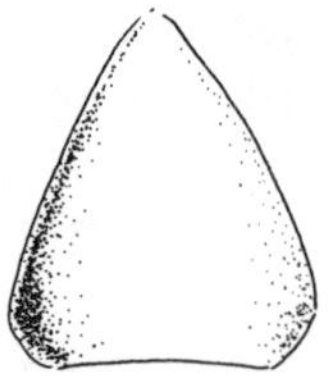

Figure 569

Monad. A single individual that is free from other such individuals rather than being united into a group.

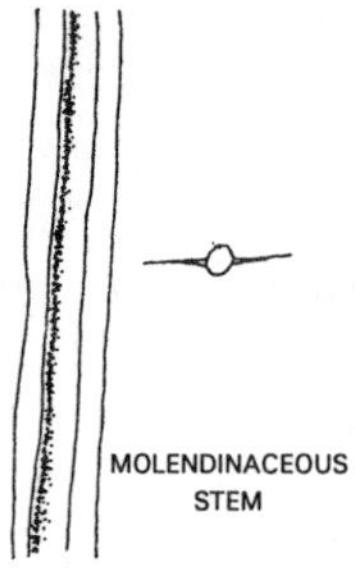

Figure 570

Monadelphous. Stamens united by the filaments and forming a tube around the gynoecium. Figures 571 and 572.

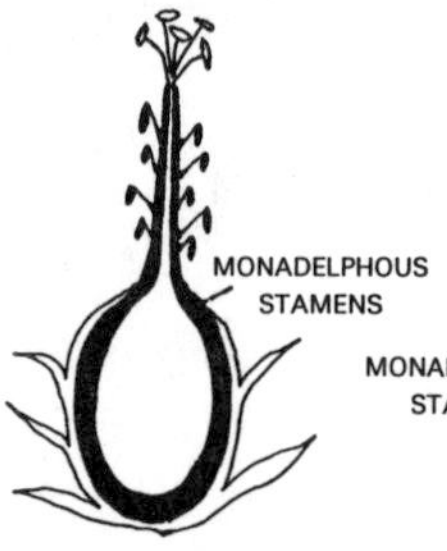

Figure 571

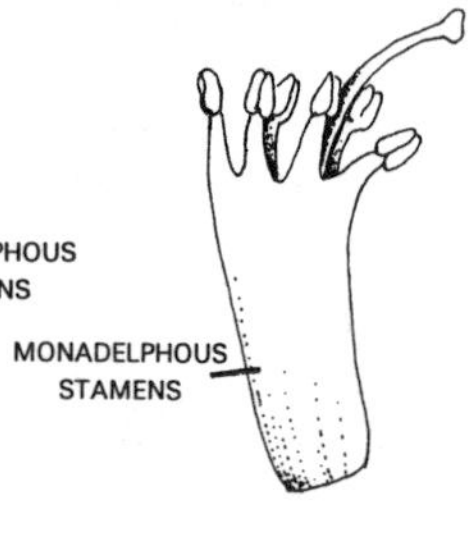

Figure 572

Monandrous. With a single stamen.

Monanthous. One-flowered.

Monecious. See **monoecious**.

Moniliform. Cylindrical and constricted at regular intervals, giving a beaded necklace-like appearance. Figure 573.

Mono- (prefix). Meaning one.

Figure 573

Monocarpic. Flowering and bearing fruit only once and then dying. The term may be applied to perennials, biennials, or annuals.

Monocarpous. With one carpel.

Monocephalous. With one head.
Monochasial. With the form of a monochasium.
Monochasium. A type of cymose inflorescence with only a single main axis. Figure 574.
Monochlamydeous. With only one type of perianth member. Figure 575.

Figure 574

Figure 575

Monochromatic. Of a single color.
Monoclinous. With pistils and stamens in the same flower; perfect. Figure 576.
Monocotyledon. Plants with a single seed leaf, or cotyledon. Figure 577.
Monocotyledonous. With a single cotyledon. Figure 577.

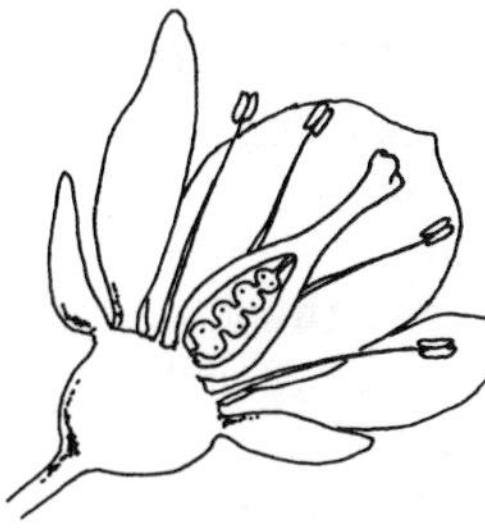

Figure 576

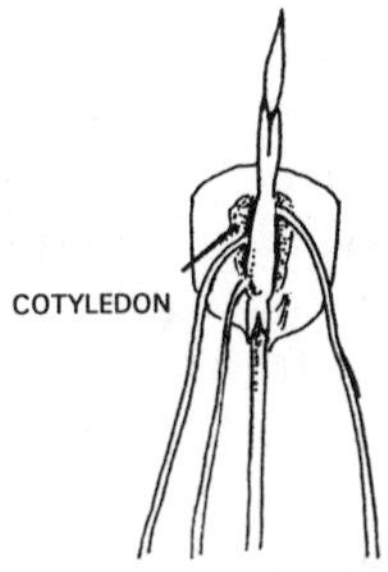

Figure 577

Monocyclic. With a single whorl. Figure 578.
Monodynamous. With one stamen distinctly larger than the others. Figure 579.

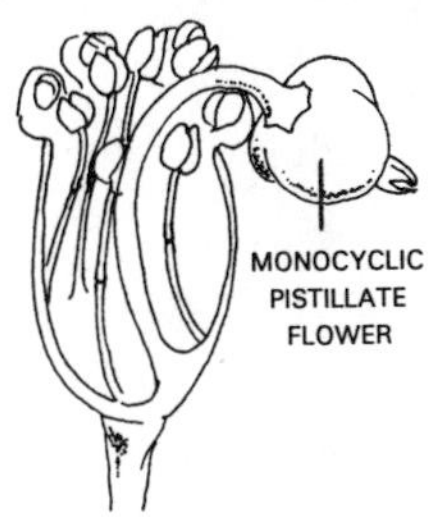

Figure 578

Figure 579

Monoecious. Flowers imperfect, the staminate and pistillate flowers borne on the same plant. (compare **dioecious**)
Monogynous. With one carpel.
Monolocular. With a single cell or chamber; unilocular. Figure 580.
Monomerous. With a single member, as in a floral whorl with only one part. Figure 578.
Monomorphic. With a single form; all alike in appearance.

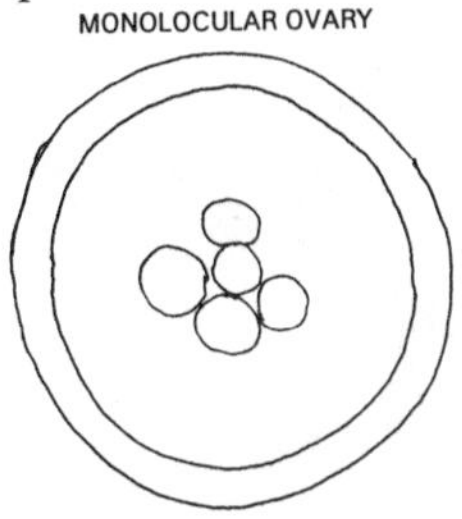

Figure 580

Monopetalous. See **sympetalous** or **gamopetalous**.
Monophyllous. Of a single leaf; with simple leaves; said of plants that have simple leaves though their relatives have compound leaves.
Monopodial. Of or pertaining to a monopodium; with the branches arising from a single main axis.
Monopodium (pl. **monopodia**). A single main axis giving rise to lateral branches. Figure 581. (compare **sympodium**)
Monopterous. With a single wing. Figure 582.

Figure 581

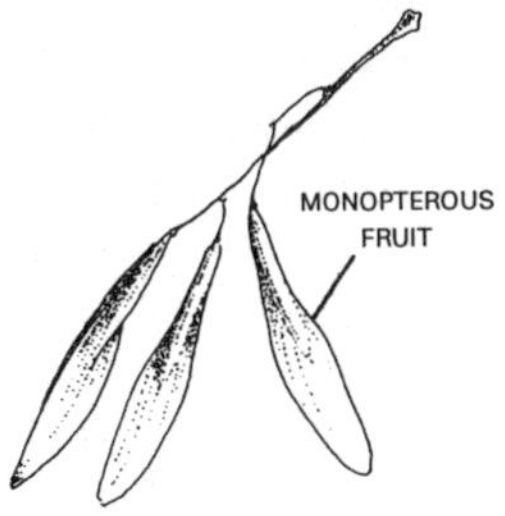

Figure 582

Monosepalous. See **gamosepalous**.
Monospermous. One-seeded.
Monostachous. With flowers arranged in a single spike.
Monostichous. In a single vertical rank or row. Figure 583.

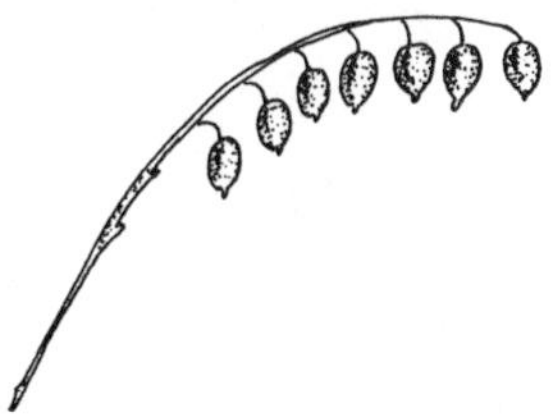

Figure 583

Monostylous. With a single style.

Monosymmetrical. Bilaterally symmetrical; zygomorphic. Figure 584.

Monotrichous. With a single bristle. Figure 585.

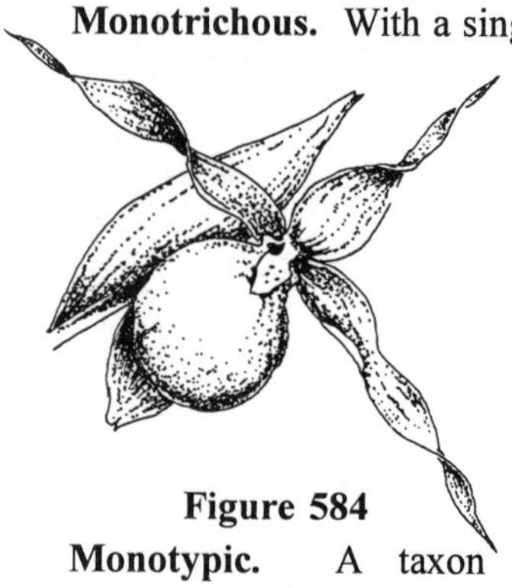

Figure 584

Figure 585

Monotypic. A taxon with only a single representative, as a genus with a single species or a family with a single genus.

Montane. A plant growing in the mountains.

Moschate. With a musky scent.

Mottled. With colored spots or blotches. Figure 586.

Mucilaginous. Slimy and moist; mucilage-like.

Figure 586

Mucro. A short, sharp, abrupt point, usually at the tip of a leaf or other organ. Figure 587.

Mucronate. Tipped with a short, sharp, abrupt point (mucro). Figure 587.

Mucronulate. Tipped with a very small mucro. Figure 588.

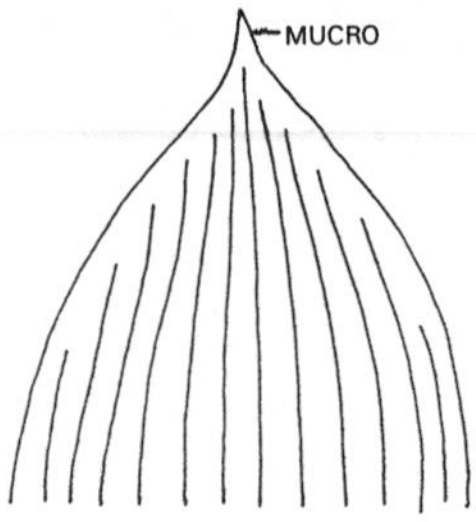

Figure 587

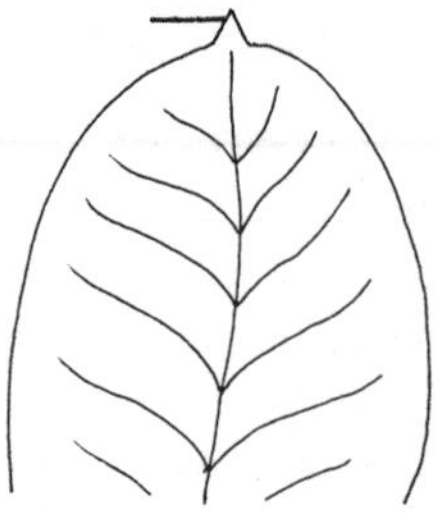

Figure 588

Multi- (prefix). Meaning many.

Multiciliate. With many cilia. Figure 589.

Multicipital. Many-headed, as the crown of a root divided into a number of caudices. Figure 590.

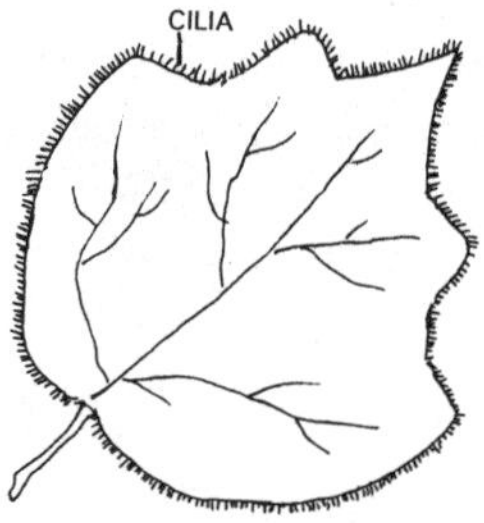

Figure 589

Figure 590

Multicostate. With many ribs. Figure 591.

Multifid. Cleft into many narrow segments or lobes. Figure 592.

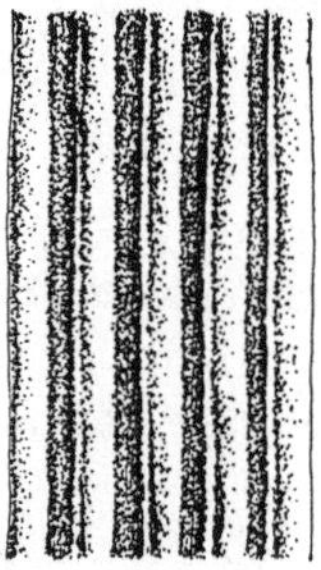

Figure 591

Figure 592

Multiflorous. Bearing many flowers.

Multifoliate. Bearing many leaves or leaflets.

Multiparous. A cyme with many lateral axes.

Multipartite. See **multifid**.

Multiperennial. See **pliestesial**.

Multiple fruit. A fruit formed from several separate flowers crowded on a single axis, as a mulberry or pineapple. Figure 593.

Multiplicate. Repeatedly folded. Figure 594.

Figure 593

Figure 594

Multiradiate. With many rays. Figure 595.

Multiseptate. With many septae or partitions. Figure 596.

Muricate. Rough with small, sharp projections or points. Figure 597.

Murication. A small, sharp projection or point. Figure 597.

Muriculate. Very finely muricate. Figure 598.

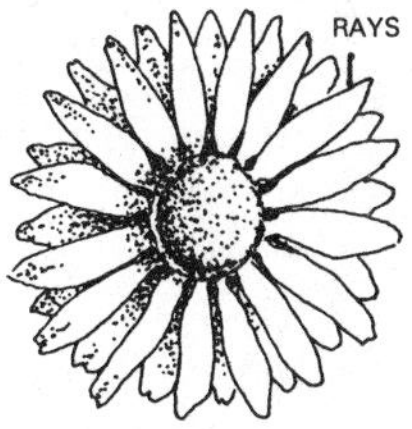

Figure 595

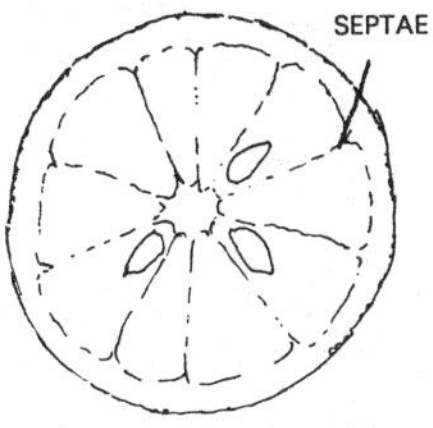

Figure 596

Figure 597

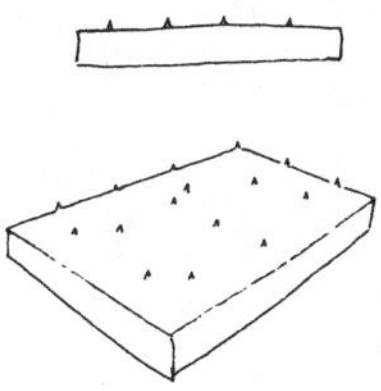

Figure 598

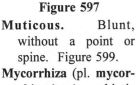

Muticous. Blunt, without a point or spine. Figure 599.

Mycorrhiza (pl. **mycorrhizae**). A symbiotic relationship between a fungus and the root of a plant.

Mycotrophic. Modified by a mycorrhizal relationship.

Figure 599

Myochrous. Mouse-colored.

Nacreous. With a pearly luster; pearlescent.

Naked. Lacking hairs, structures, or appendages typically present, as in a flower lacking a perianth; nude.

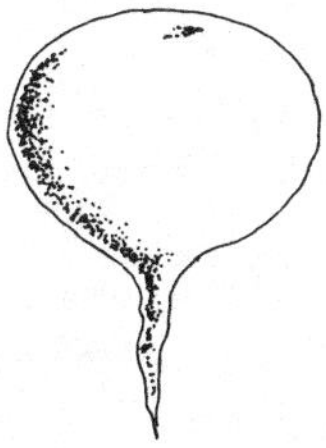

Figure 600

Naked bud. A bud lacking scales.

Napaceous. See **napiform**.

Napiform. Turnip-shaped. Figure 600.

Nascent. Beginning to develop, but not yet fully formed.

Natant. Floating in water, immersed.

Naturalized. Plants introduced from elsewhere, but now established.

Naucum. The soft, fleshy part of a drupe. Figure 601.

Nautiloid. Spiral-shaped, like a *Nautilus* shell. Figure 602.

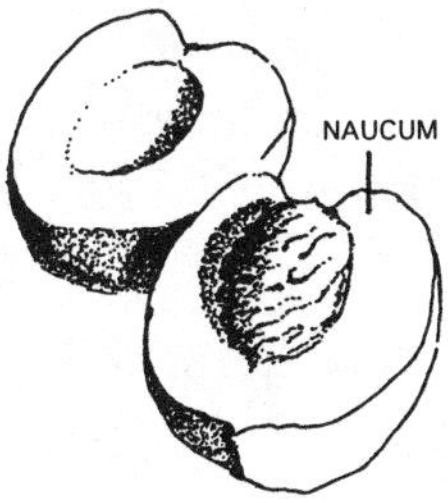

Figure 601

Figure 602

Navicular. Boat-shaped. Figure 603.

Nebulose. Indistinct, as in a fine, diffuse inflorescence. Figure 604.

Figure 603

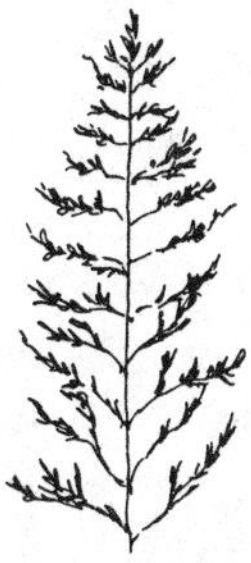

Figure 604

Nectar. A sugary, sticky fluid secreted by many plants.

Nectar gland. See **nectary**.

Nectar guides. Lines or spots, often invisible except in ultraviolet light, directing pollinators toward the nectaries. Figure 605.

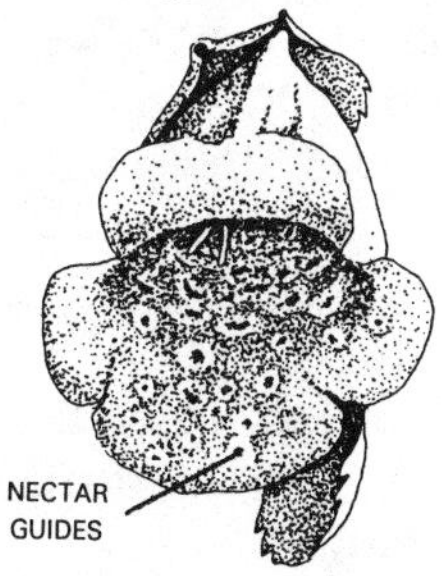

Figure 605

Nectariferous. With nectar.

Nectary. A tissue or organ which produces nectar. Figure 606.

Needle. A slender, needle-shaped leaf, as in the pinaceae. Figure 607.

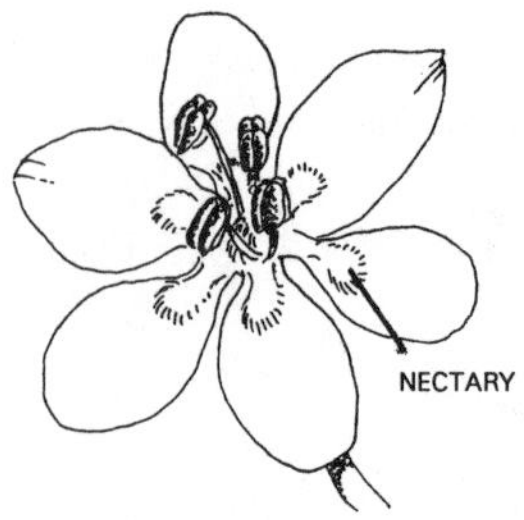

Figure 606

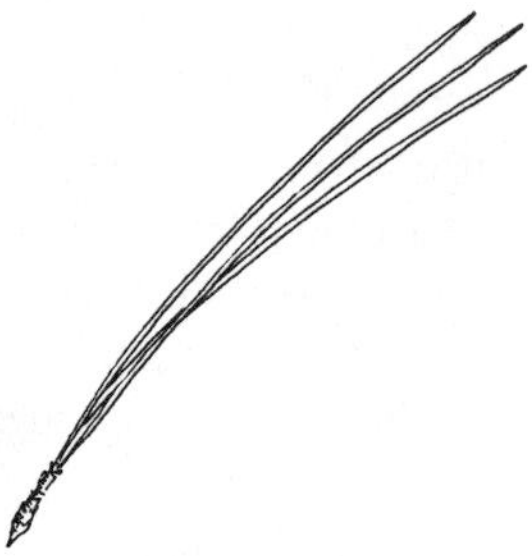

Figure 607

Nema (pl. **nemata**). A filament or thread.

Nephroid. Kidney-shaped; reniform. Figure 608.

Nervation. The arrangement of nerves or veins in an organ. Figure 609.

Nerve. A prominent, simple vein or rib of a leaf or other organ. Figure 609.

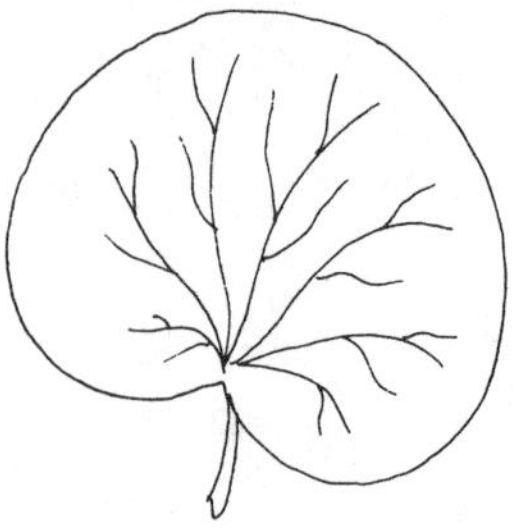

Figure 608

Figure 609

Nerviform. Resembling a nerve.

Nervose. With prominent nerves. Figure 609.

Netted. See **net-veined**.

Net-veined. In the form of a network; reticulate. Figure 609.

Neuter. Lacking functional stamens or pistils.

Neutral. See **neuter**.

Nidulent. Lying within a cavity; embedded within a pulp. Figure 610.

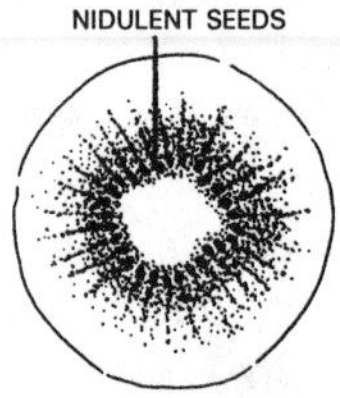

Figure 610

Nigrescent. Blackish.

Nitid. Lustrous; shining.

Niveous. White.

Nocturnal. Functioning at night, as in flowers which open at night.

Nodal. Of, on, or pertaining to a node. Figure 611.

Nodding. Bent to one side and downward. Figure 612.

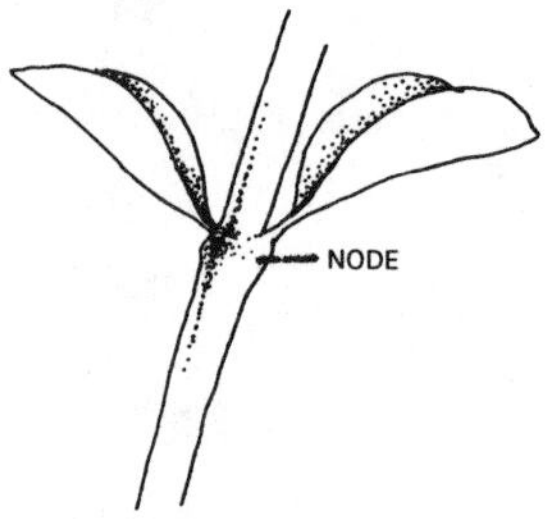

Figure 611

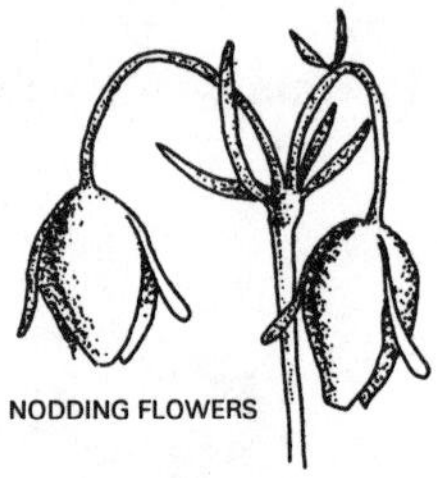

Figure 612

Node. The position on the stem where leaves or branches originate. Figure 611.

Nodiferous. With nodes. Figure 611.

Nodiform. See **nodulose**.

Nodose. With knobs or nodules. Figures 613 and 614; with nodes. Figure 611.

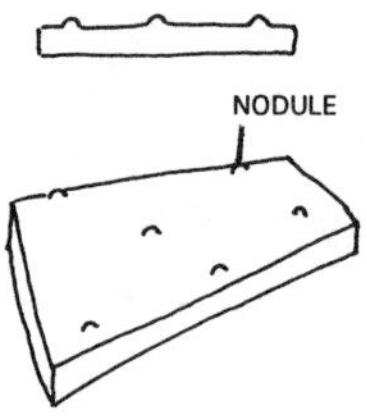

Figure 613

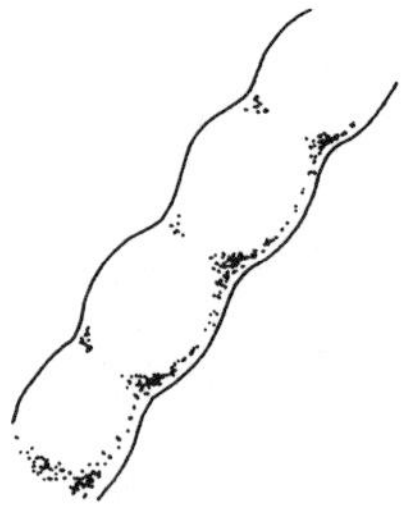

Figure 614

Nodule. A swelling or knob. Figure 613.

Nodulose. With minute knobs or nodules. Figure 615.

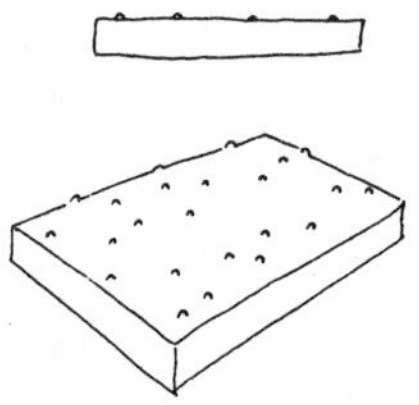

Figure 615

Nomad. A plant growing in pastures.

Nomophilous. Growing in pastures.

Notate. Marked with lines or spots. Figure 616.

Figure 616

Nucamentaceous. Catkinlike; indehiscent.

Nucamentum. A catkin or ament. Figure 617.

Nucellus. The part of the ovule just beneath the integuments and surrounding the female gametophyte. Figure 618.

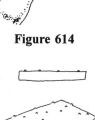

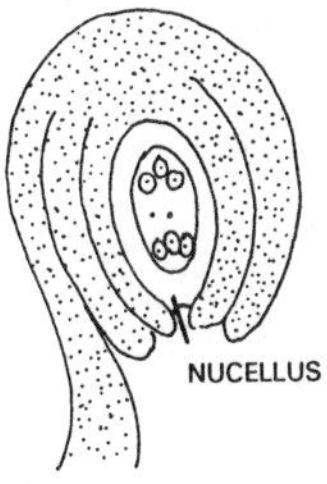

Figure 617 **Figure 618**

Nuciferous. Bearing nuts.

Nude. See **naked**.

Nudicaul. With leafless stems.

Numerous. In botanical descriptions, this term usually means more than ten.

Nut. A hard, dry, indehiscent fruit, usually with a single seed. Figure 619.

Nutant. Drooping; nodding. Figure 612.

Nutlet. A small nut; one of the lobes or sections of the mature ovary of some members of the Boraginaceae, Verbenaceae, and Labiatae (Lamiaceae). Figure 620.

Nux. A nut. Figure 619.

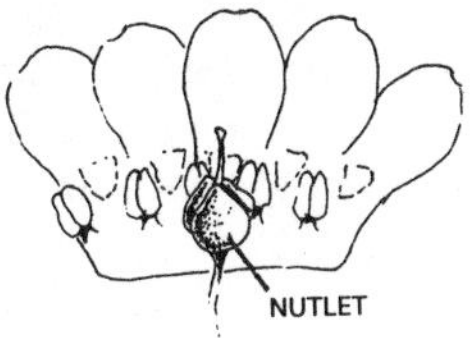

Figure 619 **Figure 620**

Nyctanthous. Night-flowering.

Nyctigamous. Opening at night.

Nyctitropic. Movement or positioning of plant organs at night that is unlike those occurring during the day.

Ob- (prefix). Meaning inversion; in a reverse direction.

Obclavate. Club-shaped, with the attachment at the broad end. Figure 621.

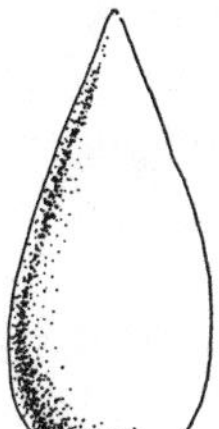

Figure 621

Obcompressed. Compressed opposite the usual way, as in a structure which is flattened dorso-ventrally when similar structures are flattened laterally. Figure 622.

Obconic or **obconical.** Conical or cone-shaped, with the attachment at the narrow end. Figure 623.

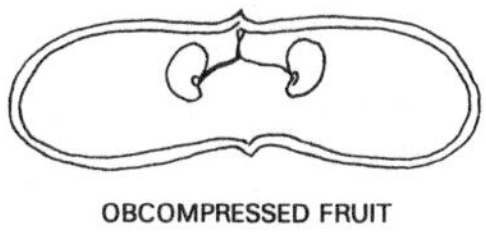

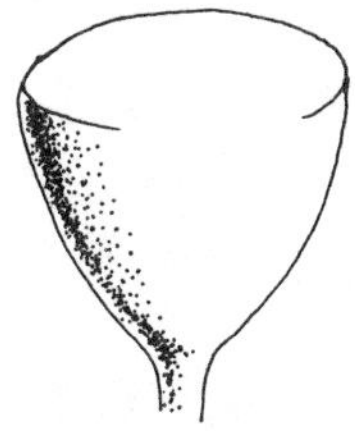

Figure 622 **Figure 623**

Obcordate. Inversely cordate, with the attachment at the narrower end; sometimes refers to any leaf with a deeply notched apex. Figures 624 and 625.

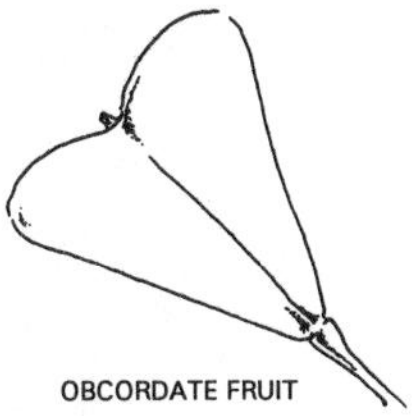

Figure 624 **Figure 625**

Obcordiform. See **obcordate**.

Obdeltoid. Deltoid, with the attachment at the pointed end. Figure 626.

Obdiplostemonous. Having two whorls of stamens, the outer whorl opposite the petals and the inner whorl opposite the sepals. Figure 627.

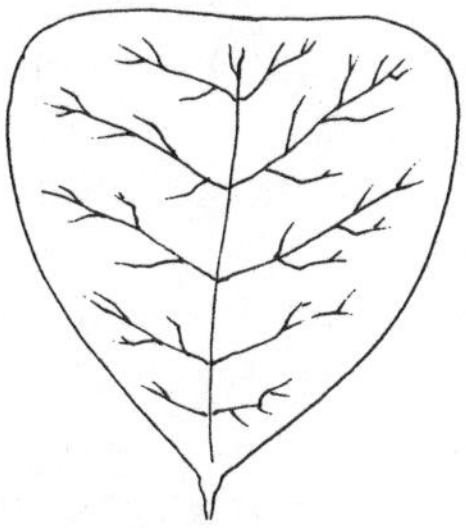

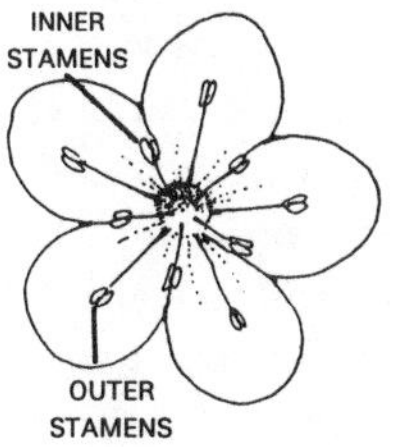

Figure 626 **Figure 627**

Obelliptic or **obelliptical.** Almost elliptic, but

with the distal end somewhat larger than the proximal end. Figure 628.

Oblanceolate. Inversely lanceolate, with the attachment at the narrower end. Figure 629.

Figure 628

Figure 629

Oblate. Spheroidal and flattened at the poles. Figure 630.

Obligate. Restricted to particular conditions or circumstances, as a parasite incapable of independent survival.

Oblique. With unequal sides, especially a leaf base. Figure 631; slanting.

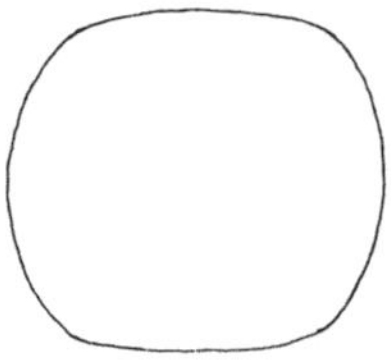

Figure 630

Oblong. Two to four times longer than broad with nearly parallel sides. Figure 632.

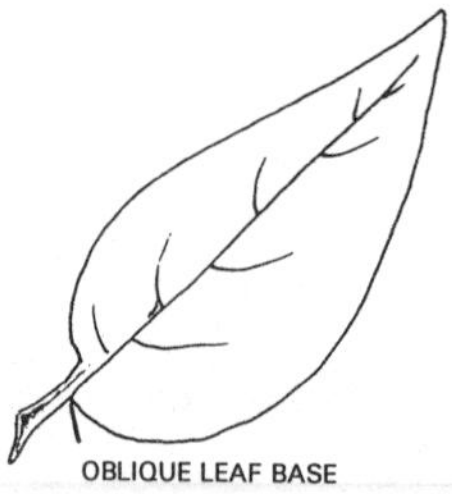

Figure 631

Figure 632

Obovate. Inversely ovate, with the attachment at the narrower end. Figure 633.

Obovoid. Inversely ovoid, with the attachment at the narrower end. Figure 634.

Obpyramidal. Inversely pyramidal, with the attachment at the narrower end. Figure 635.

Obsolescent. See **obsolete**.

Obsolete. An organ or structure which is much reduced and likely nonfunctional, though believed at one time to have been more perfectly formed; vestigial. Figure 636.

Figure 633

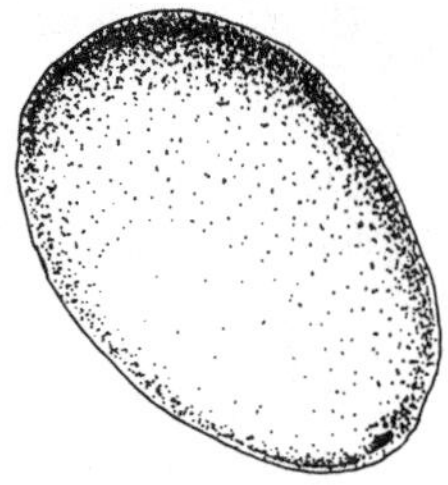

Figure 634

Figure 635

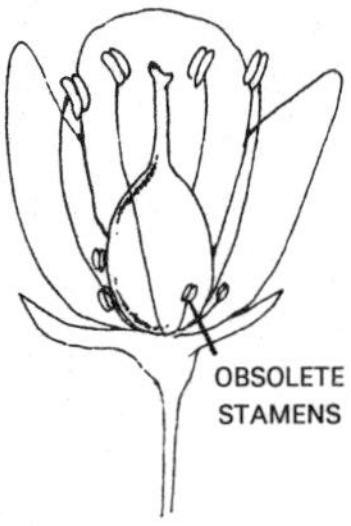

Figure 636

Obturator. A small glandular structure attached to the pollinia of members of the Asclepiadaceae and Orchidaceae which closes the opening to the anther. Figure 637.

Obtuse. Blunt or rounded at the apex; with the sides coming together at the apex at an angle greater than 90 degrees. Figure 638.

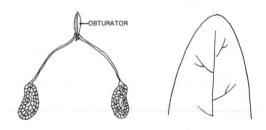

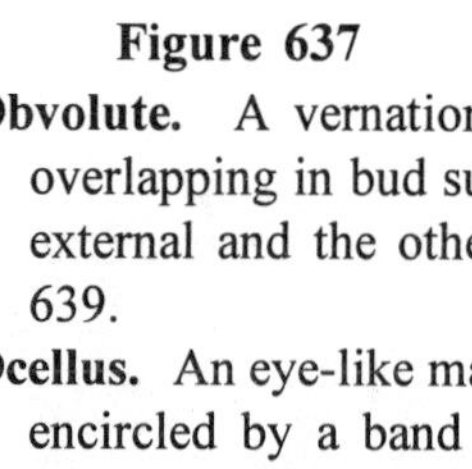

Figure 637

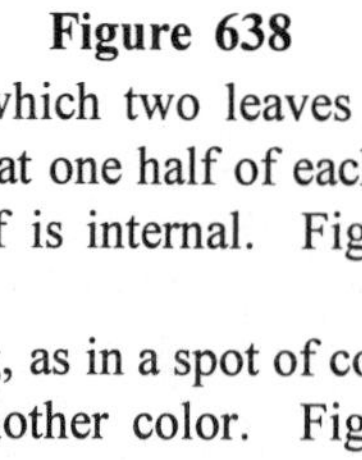

Figure 638

Obvolute. A vernation in which two leaves are overlapping in bud such that one half of each is external and the other half is internal. Figure 639.

Ocellus. An eye-like marking, as in a spot of color encircled by a band of another color. Figure 640.

Ochraceous. Ochre-colored.

Ochrea (pl. **ochreae**). See **ocrea**.

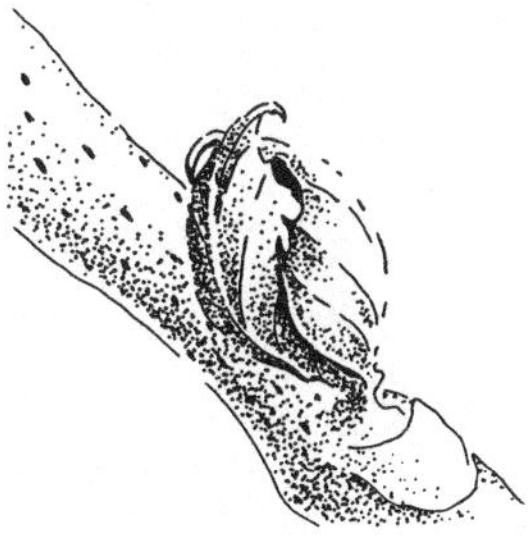

Figure 639

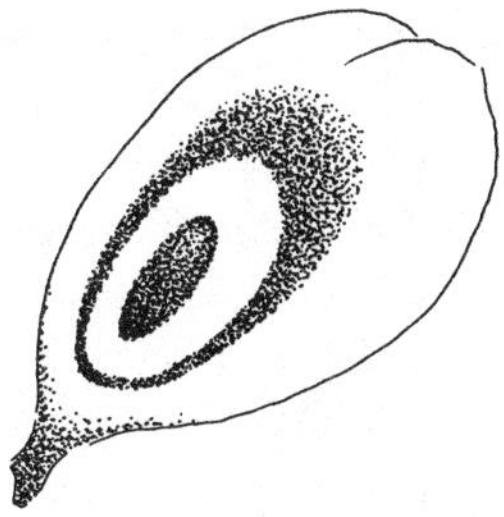

Figure 640

Ochreate. See **ocreate**.
Ochroleucous. Yellowish white; cream-colored.
Ocrea (pl. **ocreae**). A sheath around the stem formed from the stipules, as in many members of the Polygonaceae. Figure 641.
Ocreate. With sheathing stipules. Figure 641.
Ocreola (pl. **ocreolae**). A minute stipular sheath around the secondary divisions of the inflorescence in some members of the Polygonaceae. Figure 642.
Ocreolate. With minute sheathing stipules; often applied to bract bases. Figure 642.

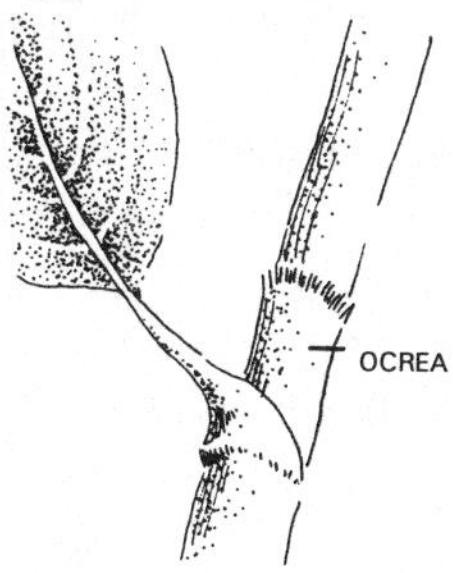

Figure 641

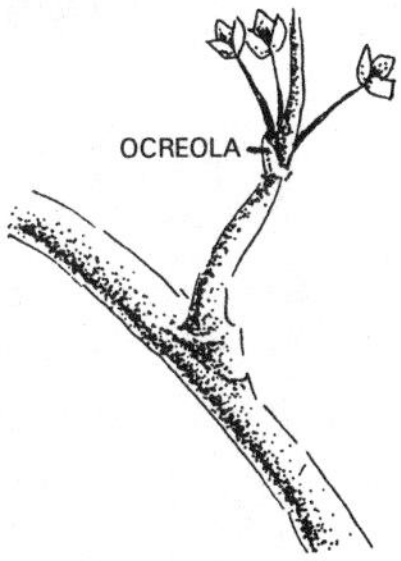

Figure 642

Octandrous. With eight stamens. Figure 643.
Octogynous. With eight pistils or styles. Figure 643.
Octolocular. With eight locules. Figure 644.

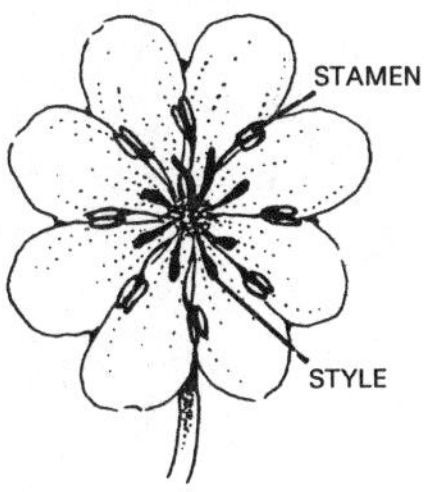

Figure 643

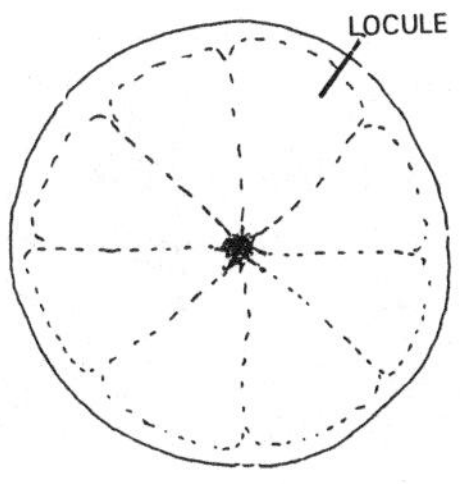

Figure 644

Octopetalous. With eight petals. Figure 643.
Octoradiate. With eight ray flowers. Figure 645.
Octosepalous. With eight sepals. Figure 646.

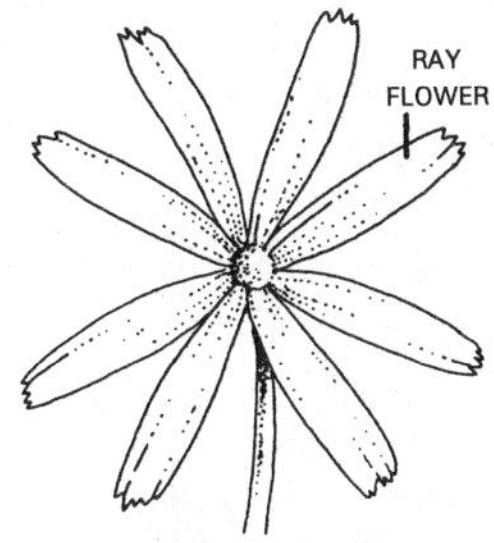

Figure 645

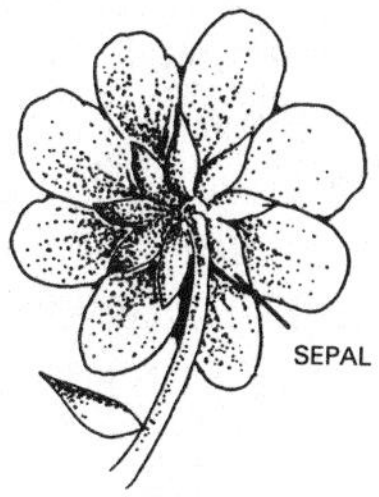

Figure 646

Octostemonous. With eight stamens. Figure 643.
Octostichous. In eight ranks or rows.
Odd-pinnate. Pinnately compound with a terminal leaflet rather than a pair of leaflets or a tendril, so that there is an odd number of leaflets. Figure 647.

Figure 647

Odoriferous. With a distinct odor.
Offset. A short, often prostrate, shoot originating near the ground at the base of another shoot. Figure 648.
Offshoot. A shoot or branch arising from a main stem. Figure 649.

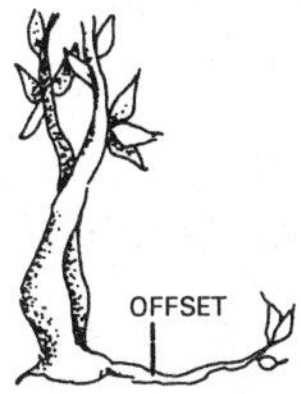

Figure 648

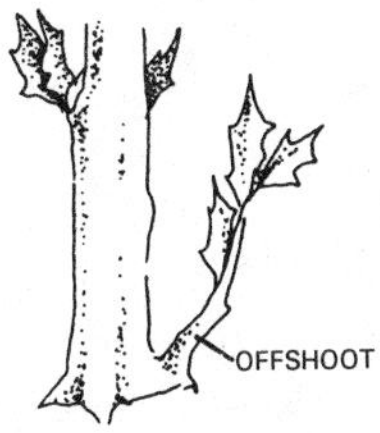

Figure 649

Oil tube. Narrow ducts in the walls of the fruit of many members of the Umbelliferae (Apiaceae) containing volatile oils. Figure 650.
Oleaginous. Oily; oil-producing.
Oleiferous. Oil-bearing.
Oligandrous. With few stamens.
Oligocarpic. See **oligocarpous**.
Oligocarpous. Bearing less than the typical

amount of fruit.

Oligomerous. With less than the typical number of parts.

Oligophyllous. With few leaves.

Oligospermous. With few seeds.

Olivaceous. Olive green; olive-like.

Operculate. With an operculum. Figure 651.

Operculum. A small lid, such as the deciduous cap of a circumscissile capsule. Figure 651.

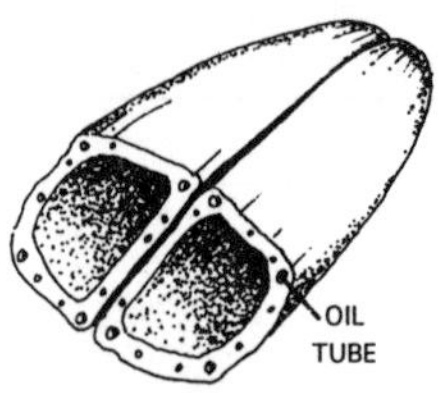

Figure 650

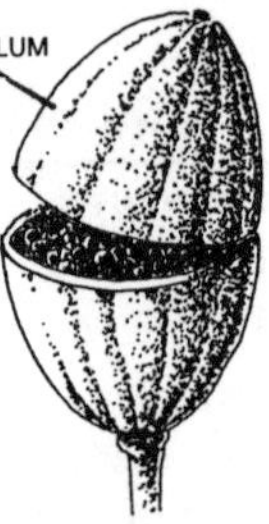

Figure 651

Opposite. Borne across from one another at the same node, as in a stem with two leaves per node. Figure 652; borne over, or on the same radius, as other organs rather than between other organs, as a stamen in front of a petal. Figure 653. (compare **alternate**)

Figure 652

Figure 653

Oppositiflorous. With opposite pedicels or peduncles. Figure 652.

Oppositifolious. With opposite leaves. Figure 652.

Orbicular. Approximately circular in outline. Figure 654. (compare **spherical**)

Figure 654

Orbiculate. See **orbicular**.

Orchioid. Orchid-like.

Organ. A plant part with a specific function, as a leaf.

Orifice. An opening or mouth, as the mouth-like opening of a tubular corolla. Figure 655.

Figure 655

Ornithophilous. Pollinated by birds.

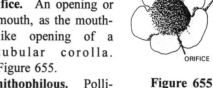

Orophilous. Growing in mountainous areas.

Orthocladous. With straight branches. Figure 656.

Orthopterous. Straight-winged. Figure 657.

Figure 656

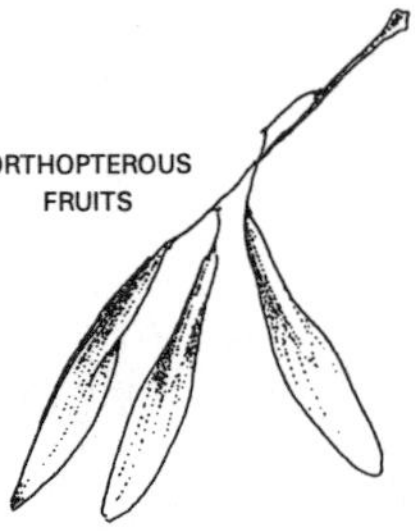

Figure 657

Orthostichous. With parts arranged in straight ranks or rows. Figure 658.

Orthotropic. Of, pertaining to, or exhibiting an essentially vertical growth habit. Figure 659.

Figure 658

Figure 659

Orthotropous ovule. An ovule which is straight and erect. Figure 660.

Osseous. Bony.

Ossiculus. The stone or pit of a drupe; a pyrene. Figure 661.

Ossified. Becoming bony.

Outcross. To transfer pollen from the anthers of the flowers of one plant to the stigma of the

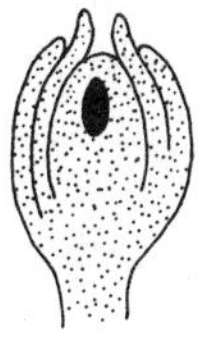

Figure 660

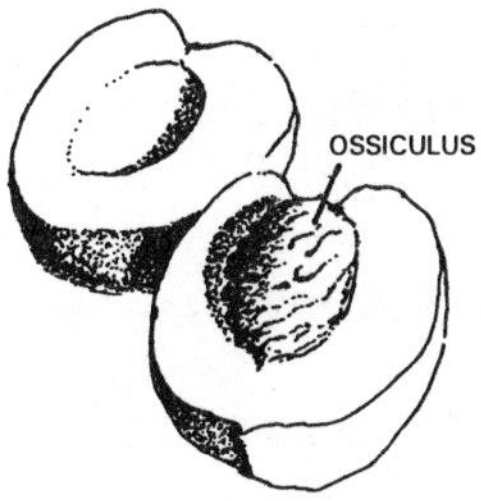

Figure 661

flower of another plant.

Oval. Broadly elliptic, the width over one-half the length. Figure 662.

Ovary. The expanded basal portion of the pistil that contains the ovules. Figure 663.

Figure 662

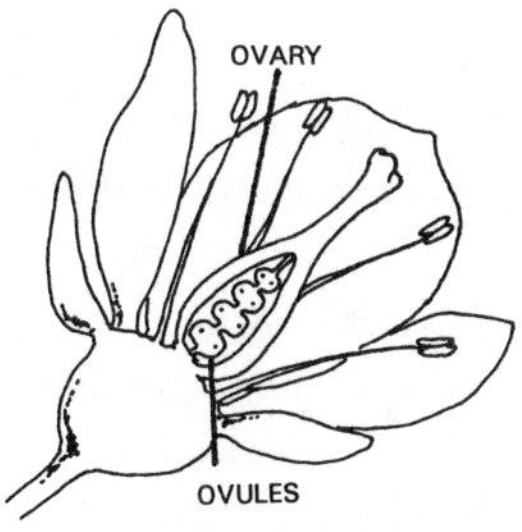

Figure 663

Ovate. Egg-shaped in outline and attached at the broad end (applied to plane surfaces). Figure 664. (compare **ovoid**)

Ovoid. Egg-shaped (applied to three-dimensional structures). Figure 665.

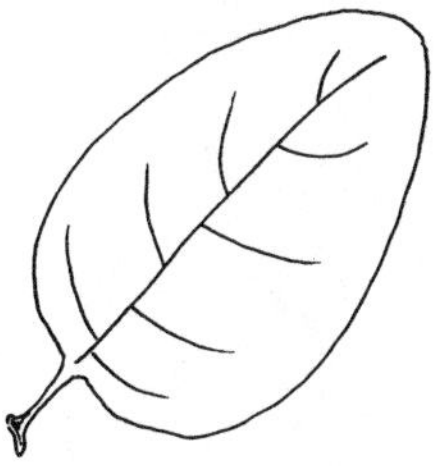

Figure 664

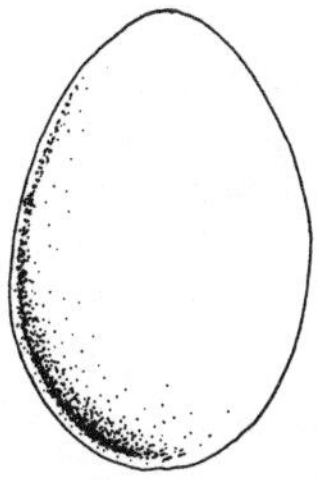

Figure 665

Ovulate. Producing ovules.

Ovule. An immature seed; the megasporangium and surrounding integuments of a seed plant. Figures 666 and 663.

Ovuliferous. Bearing ovules.

Oxylophyte. A plant growing on acidic soils.

Pachycladous. With thick branches.

Pagina. The blade of a leaf. Figure 667.

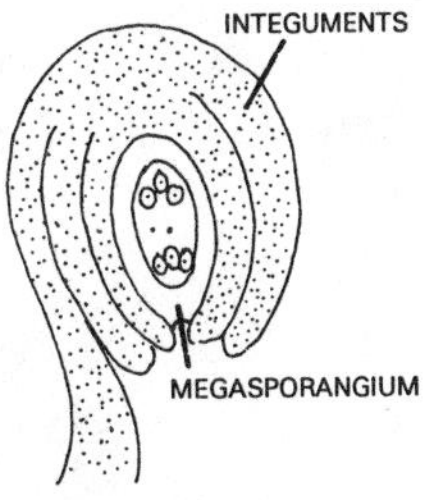

Figure 666

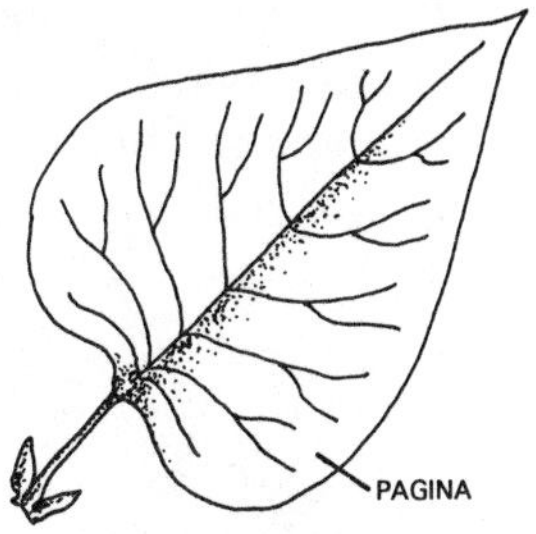

Figure 667

Palate. A raised appendage on the lower lip of a corolla which partially or completely closes the throat. Figure 668.

Palea (pl. **paleae**). A chaffy scale or bract; the uppermost of the two bracts (lemma and palea) which subtend a grass floret, often partially surrounded by the lemma. Figure 669.

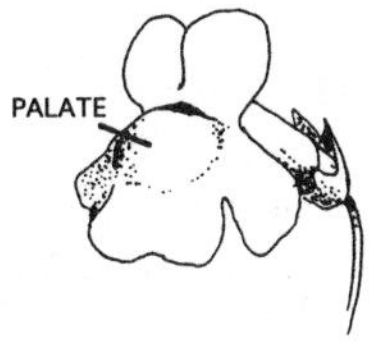

Figure 668

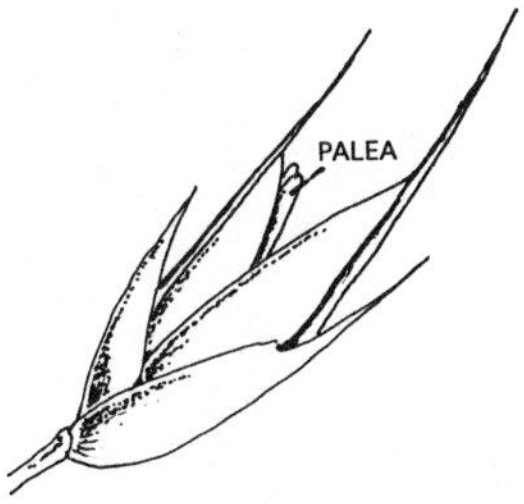

Figure 669

Paleaceous. Chaffy; with chaffy scales. Figure 670.

Paleola (pl. **paleolae**). A tiny palea; a lodicule. Figure 671.

Paleolate. With a lodicule. Figure 671.

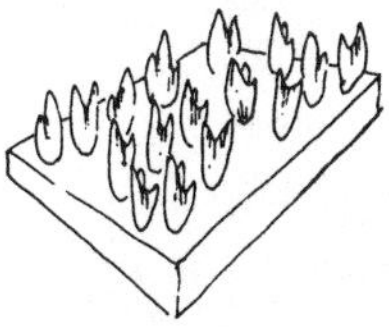

Figure 670

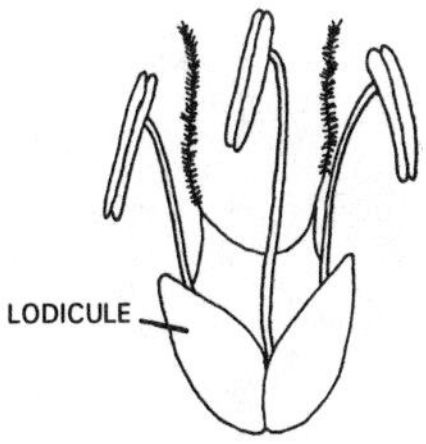

Figure 671

Paleomorphic. Lacking symmetry.

Palet. See **palea**.

Pallid. Pale.

Palmate. Lobed, veined, or divided from a common point, like the fingers of a hand. Figure 672. (compare **pinnate**)

Palmate-pinnate. With the primary leaflets palmately arranged and the secondary leaflets pinnately arranged. Figure 673.

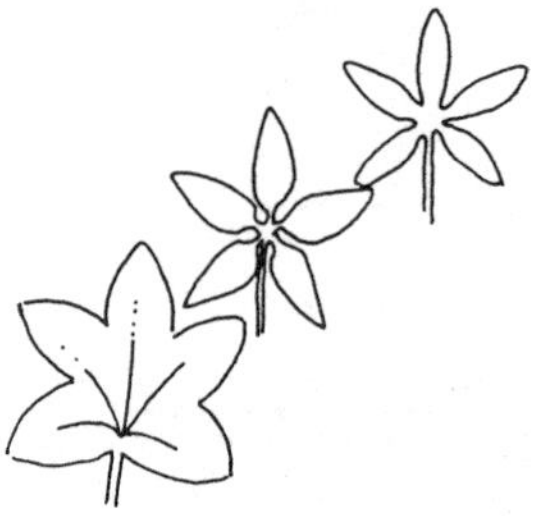

Figure 672

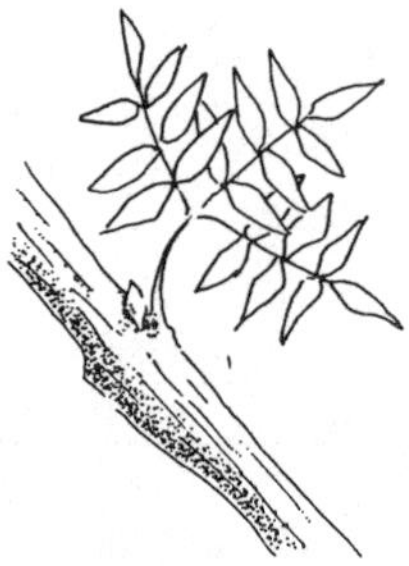

Figure 673

Palmatifid. Palmately cleft or lobed. Figure 674.

Palmatisect. Palmately divided. Figure 675.

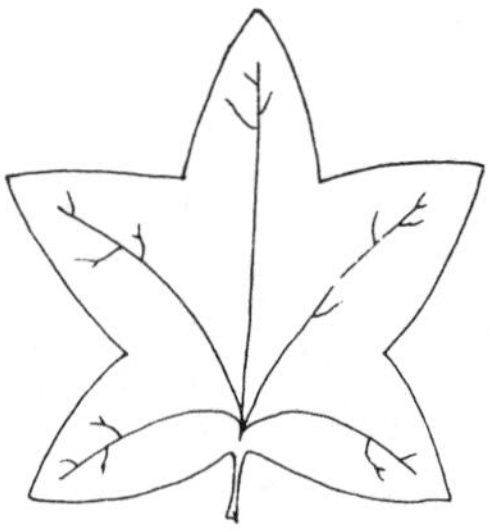

Figure 674

Figure 675

Paludose. Growing in wet meadows or marshes. (same as **palustrine**)

Palustrine. See **paludose.**

Pampiniform. Tendril-like.

Pampinus. A tendril. Figure 676.

Pandurate. Fiddle-shaped. Figure 677.

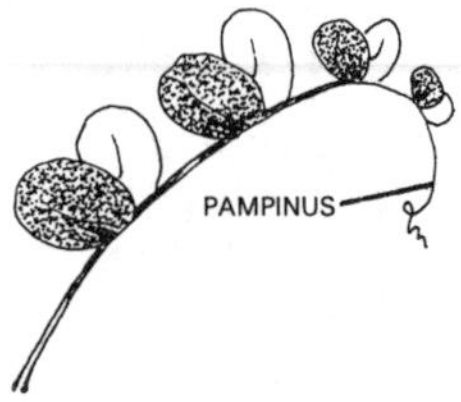

Figure 676

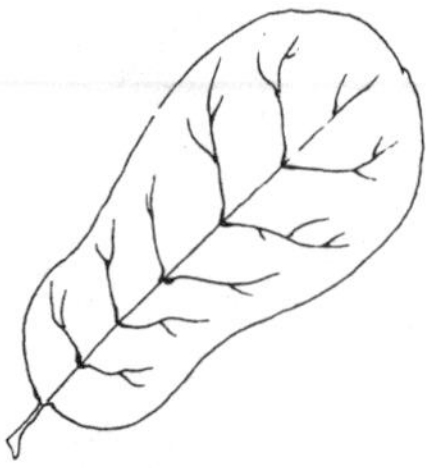

Figure 677

Panduriform. See **pandurate.**

Panicle. A branched, racemose inflorescence with flowers maturing from the bottom upwards. Figure 678.

Paniculate. Having flowers in panicles. Figure 678.

Paniculiform. An inflorescence with the general appearance, but not necessarily the structure, of a true panicle.

Panniform. See **pannose.**

Pannose. Covered with a short, dense, felt-like tomentum. Figure 679.

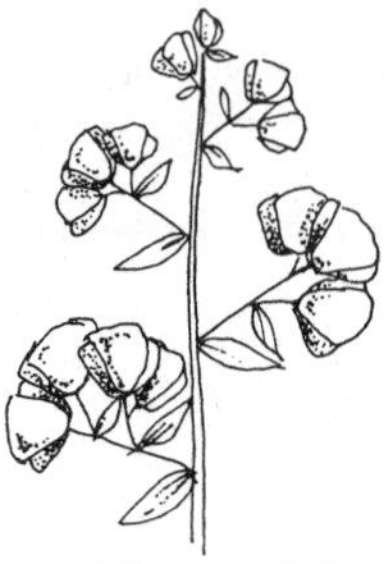

Figure 678

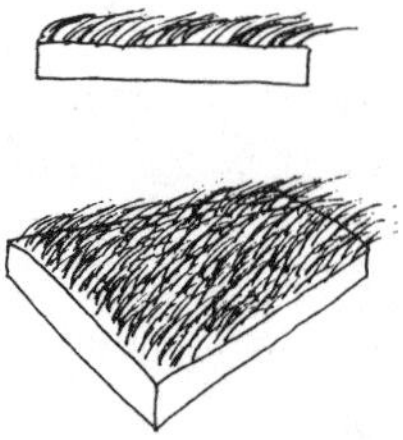

Figure 679

Papaveraceous. Resembling a poppy; a member of the Papaveraceae.

Papilionaceous. Butterflylike, as the irregular corolla of a pea, with a banner petal, two wing petals, and a keel petal. Figure 680.

Papilla (pl. **Papillae**). A short, rounded nipplelike bump or projection. Figure 681.

Papillary. Resembling papillae.

Papillate. Having papillae. Figure 681.

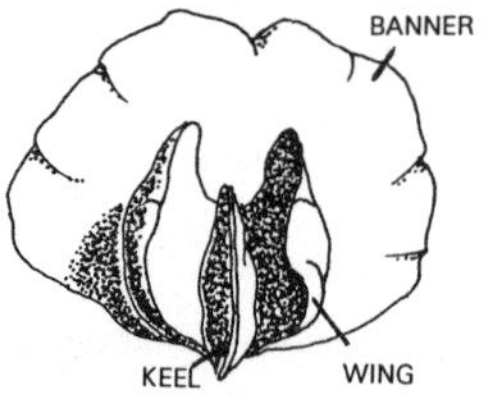

Figure 680

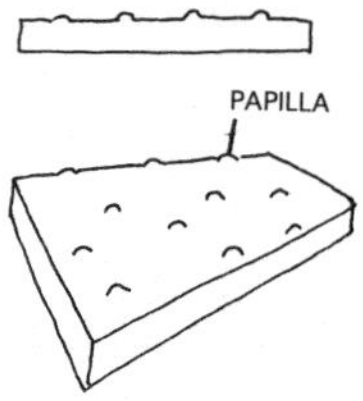

Figure 681

Papillose. Having minute papillae. Figure 682.

Papillose-hispid. With stiff hairs borne on swollen, nipplelike bases. Figure 683.

Pappiferous. Pappus bearing. Figure 684.

Pappose. Pappus-bear-

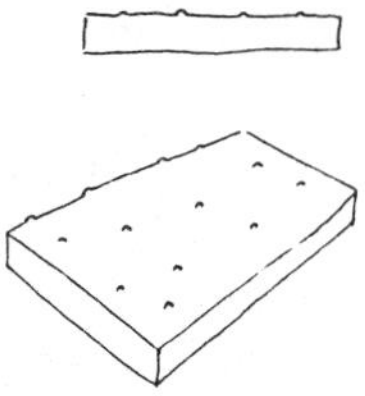

Figure 682

ing. Figure 684.

Pappus. The modified calyx of the Compositae (Asteraceae), consisting of awns, scales, or bristles at the apex of the achene. Figure 684.

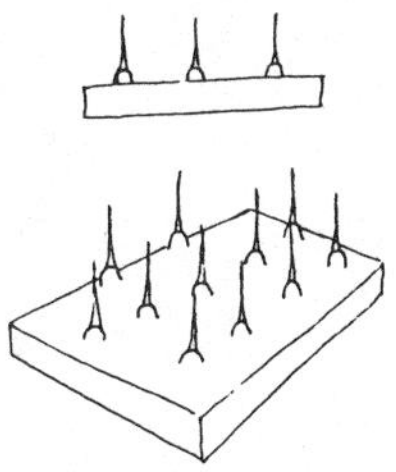

Figure 683

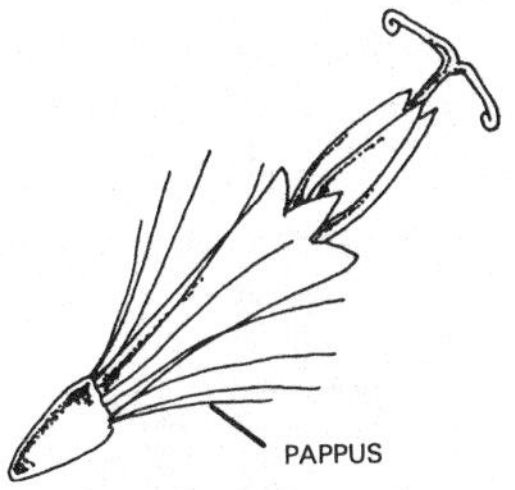

Figure 684

Papyraceous. Papery in texture and usually color.

Parallelodromous. See **parallel-veined**.

Parallel-veined. With the main veins parallel to the leaf axis or to each other. Figure 685. (compare **net-veined**)

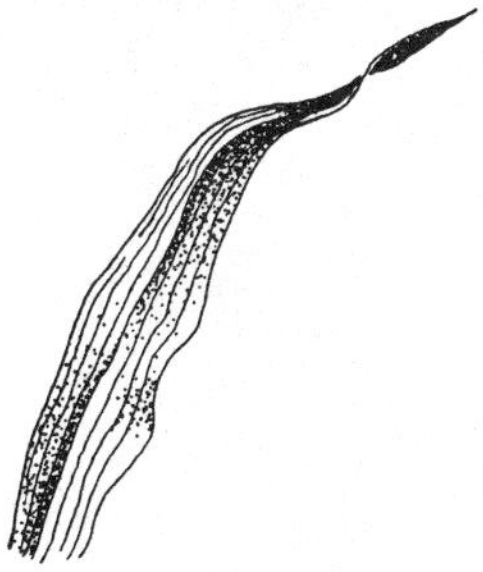

Figure 685

Paraphysis (pl. **paraphyses**). A sterile filament occurring among the sporangia of some ferns. Figure 686.

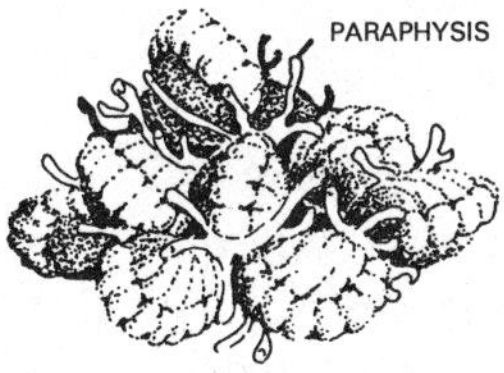

Figure 686

Parasite. An organism that obtains its food or water, at least partly, from a host organism. (compare **epiphyte**)

Parietal. Positioned along the edges or wall, rather than on the axis.

Parietal placentation. Ovules attached to the walls of the ovary. Figure 687.

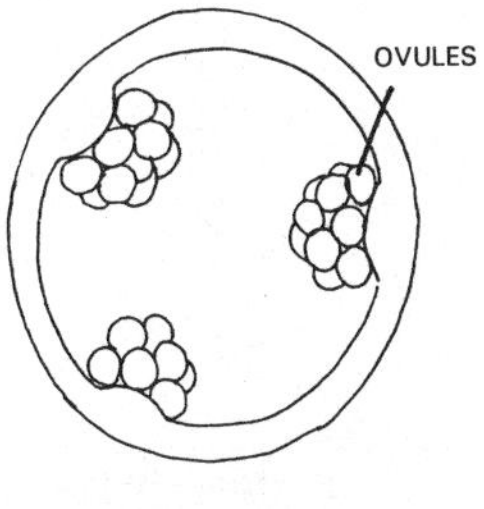

Figure 687

Paripinnate. Even-pinnate; lacking a terminal leaflet. Figure 688.

Parted. Deeply cleft, usually more than half the distance to the base or midvein. Figure 689.

Figure 688

Figure 689

Parthenocarpy. Development of a fruit without fertilization or seed production.

Parthenogenesis. Development from the egg without fertilization.

Parti-colored. Of different colors; variegated.

Patelliform. Shaped like a kneecap. Figure 690.

Patent. Spreading or expanded. Figure 691.

Patulous. Open or spreading. Figure 691.

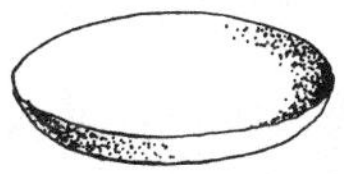

Figure 690

Figure 691

Pectinate. Comb-like; with close, regularly spaced divisions, appendages, or hairs, often in a single row, like the teeth of a comb. Figure 692.

Pedate. Palmately divided, with the lateral lobes 2-cleft. Figure 693.

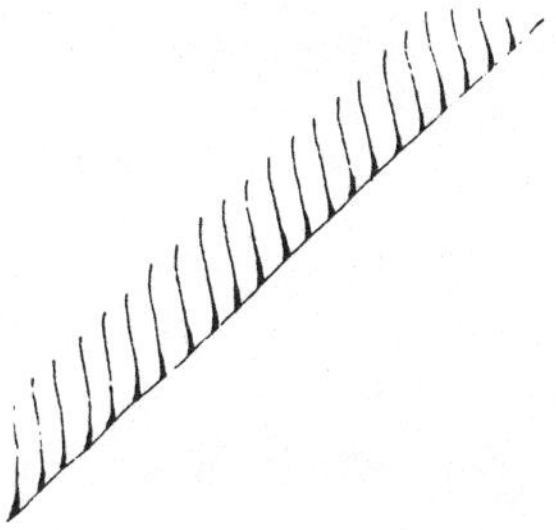

Figure 692

Figure 693

Pedatifid. Pedately cleft. Figure 694.

Pedicel. The stalk of a single flower in an inflorescence, or of a grass spikelet. Figure 695.

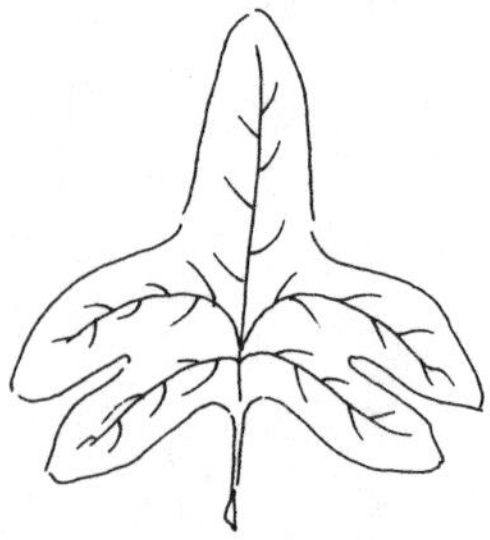

Figure 694

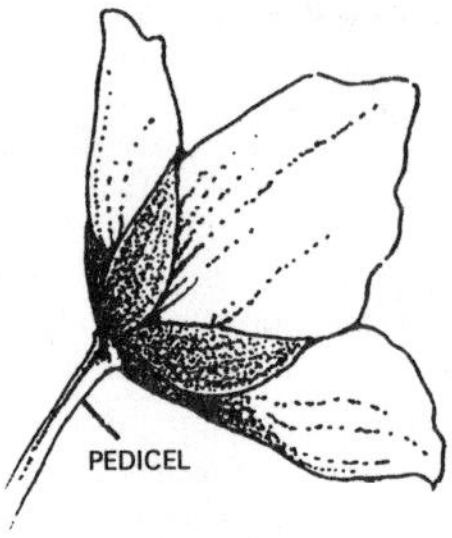

Figure 695

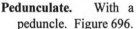

Pedicellate. With a pedicel. Figure 695.
Pedicle. See **pedicel.**
Pediculus. See **pedicel.**
Peduncle. The stalk of a solitary flower or of an inflorescence. Figure 696.
Pedunculate. With a peduncle. Figure 696.

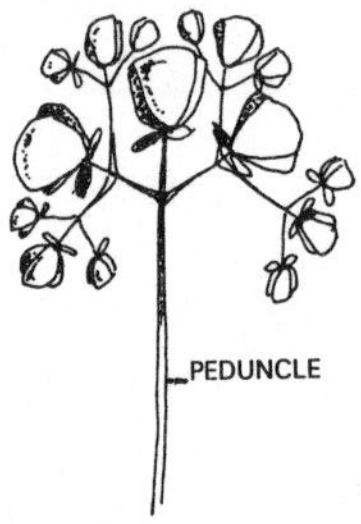

Figure 696

Pellicle. A thin, membranous or skinlike covering.
Pellucid. Transparent or translucent.
Peloria. Radial symmetry in flowers normally bilaterally symmetrical.
Peltafid. Peltate and divided into segments. Figure 697.
Peltate. Shield-shaped; a flat structure borne on a stalk attached to the lower surface rather than to the base or margin. Figure 698.

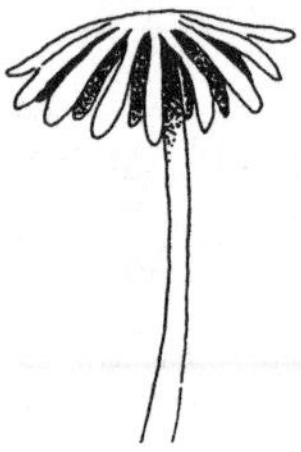

Figure 697

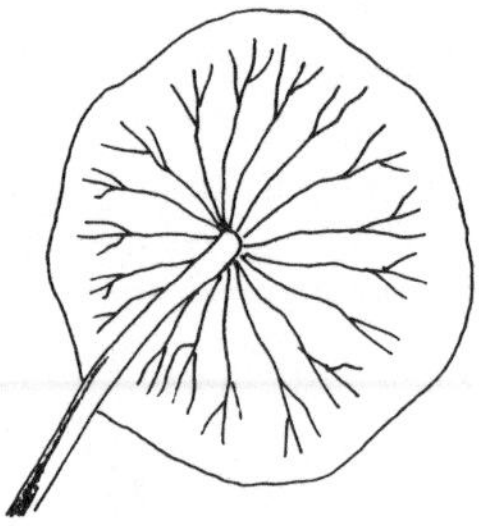

Figure 698

Peltiform. See **peltate.**
Pendent. See **pendulous.**
Pendulous. Hanging or drooping downward. Figure 699.
Penicil. A brushlike tuft of short hairs. Figure 700.
Penicillate. With a tuft of short hairs at the end, like a brush; with a penicil or penicils. Figure 700.
Pennate. See **pinnate.**
Pennatifid. See **pinnatifid.**
Penni-parallel. See **penniveined.**
Penniveined. With parts arising parallel to one another from a main axis, like the veins of a feather. Figure 701.
Penta- (prefix). Meaning five.
Pentacamerous. With five locules. Figure 702.

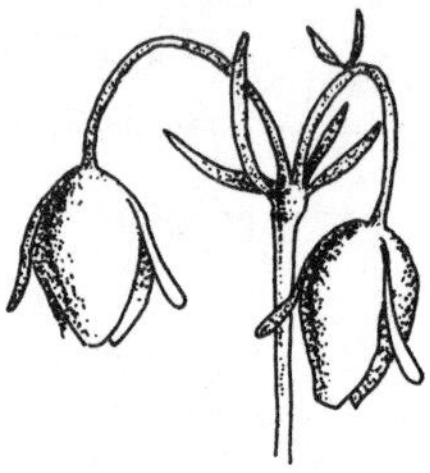

Figure 699

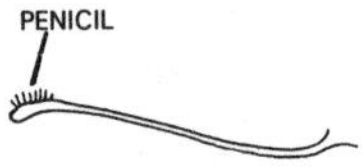

Figure 700

Figure 701

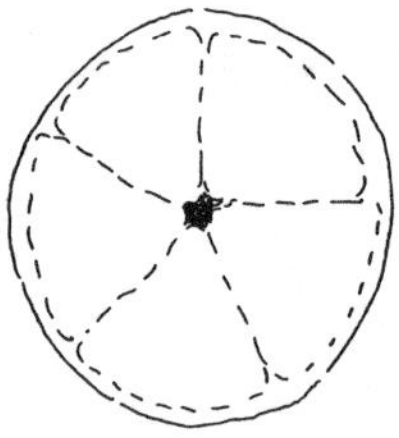

Figure 702

Pentacarpellary. With five carpels. Figure 703.
Pentacyclic. With five whorls.
Pentadactylous. Divided into five fingerlike segments. Figure 704.

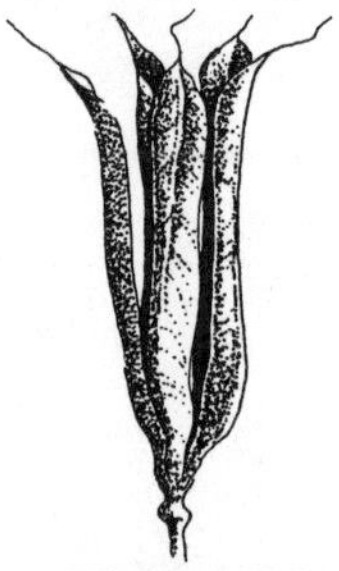

Figure 703

Figure 704

Pentadelphous. With the stamens arranged into five groups or clusters.
Pentagonal. Five-angled. Figure 705.
Pentagynous. With five pistils or styles. Figure

703.

Pentamerous. With parts arranged in sets or multiples of five. Figure 706.

Pentandrous. With five stamens. Figure 706.

Pentapetalous. With five petals. Figure 706.

Pentapterous. With five wings.

Pentasepalous. With five sepals. Figure 706.

Pentastichous. In five vertical rows or ranks.

Pepo. A fleshy, indehiscent, many-seeded fruit with a tough rind, as a melon or a cucumber. Figure 707.

Perennate. To renew, as when lateral shoots arise from a caudex.

Perennial. A plant that lives three or more years.

Perfect. With both male and female reproductive organs (stamens and pistils); bisexual. Figure 708.

Perfoliate. A leaf with the margins entirely surrounding the stem, so that the stem appears to pass through the leaf. Figure 709.

Figure 705

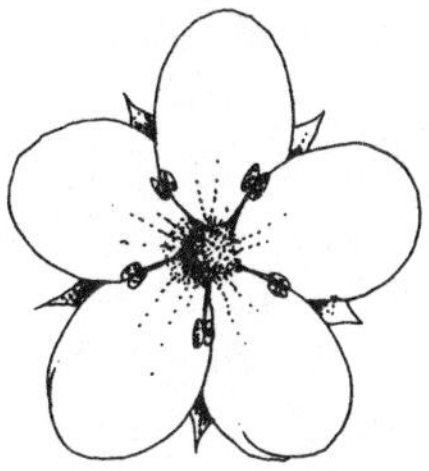

Figure 706

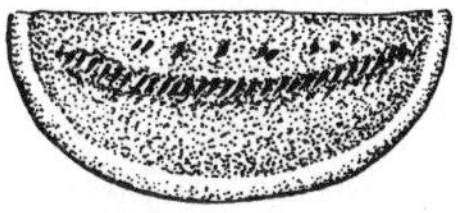

Figure 707

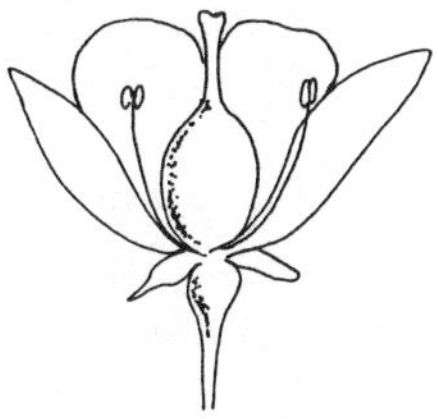

Figure 708

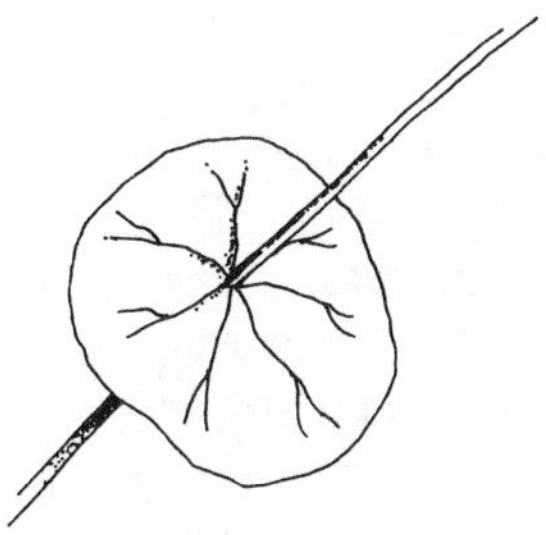

Figure 709

Perforate. With holes or perforations. Figure 710.

Perianth. The calyx and corolla of a flower, collectively, especially when they are similar in appearance. Figure 711.

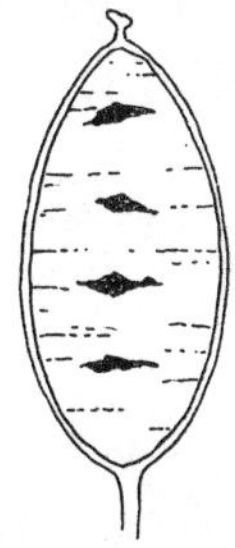

Figure 710

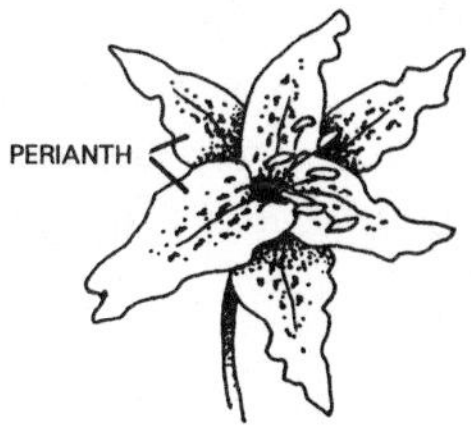

Figure 711

Pericarp. The wall of the fruit. Figure 712.

Periclinium. An involucre. Figure 713.

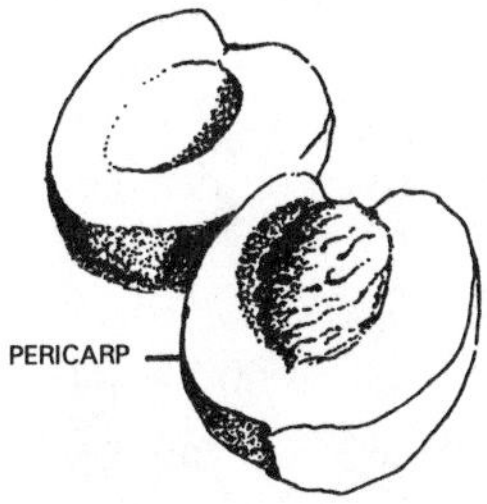

Figure 712

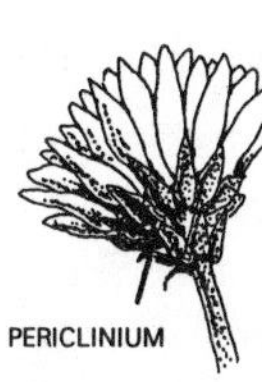

Figure 713

Peridroma. The rachis of a fern frond. Figure 714.

Perigynium (pl. **perigynia**). A scalelike bract enclosing the pistil in *Carex*. Figure 715.

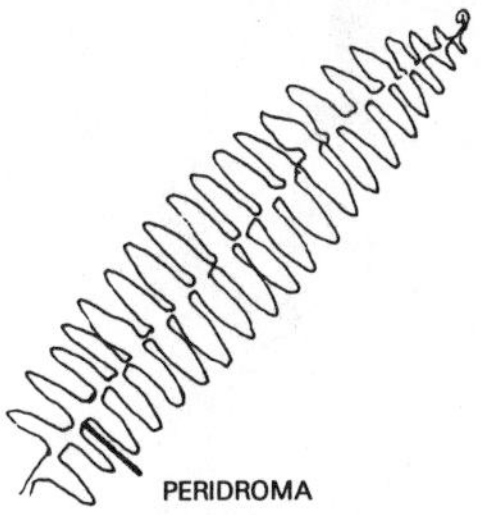

Figure 714

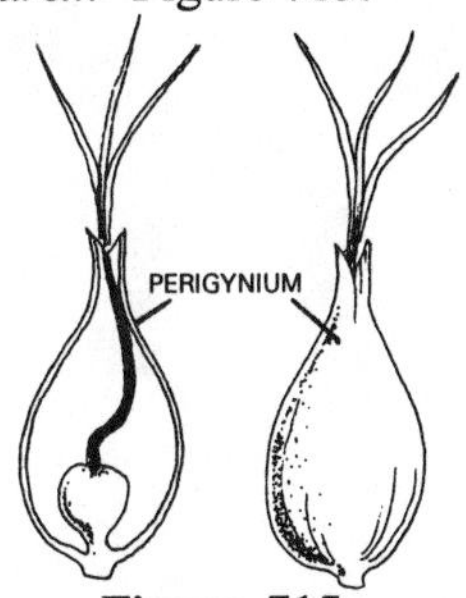

Figure 715

Perigynous. With stamens, petals, and sepals borne on a calyx tube (hypanthium) surrounding, but not actually attached to, the superior ovary. Figure 716.

Peripheral. Outside of

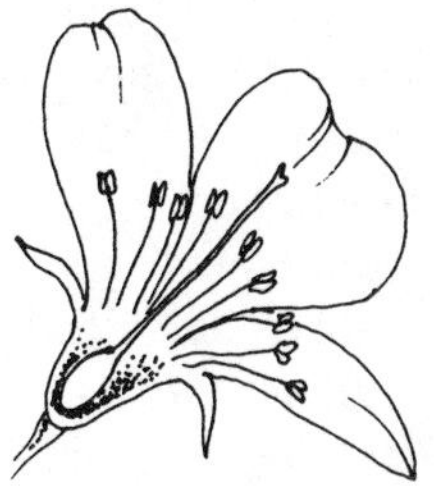

Figure 716

or external to.

Peripterous. With a surrounding border or wing. Figure 717.

Perisperm. Food storage tissue in some seeds, arising from the nucellus.

Perispore. A membrane surrounding a spore, as in *Equisetum* spores. Figure 718.

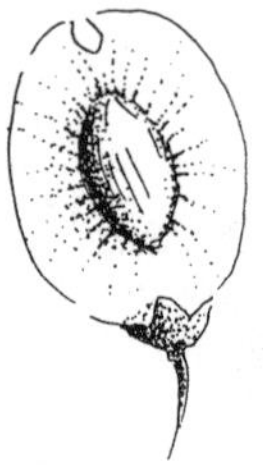

Figure 717

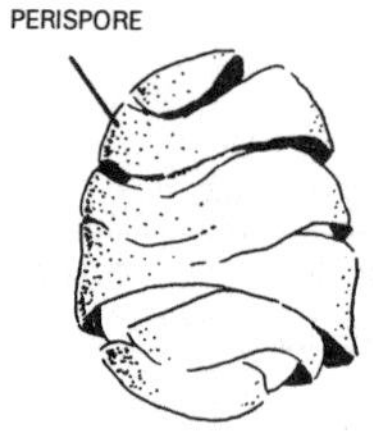

Figure 718

Pernicious. Harmful, destructive, or deadly in nature.

Persicicolor. Peach-colored.

Persistent. Remaining attached after similar parts are normally dropped, after the function has been completed.

Personate. Two-lipped, with the throat closed by a prominent projection (palate); masked. Figure 719.

Perspicous. Transparent.

Perula. A leaf-bud scale. Figure 720.

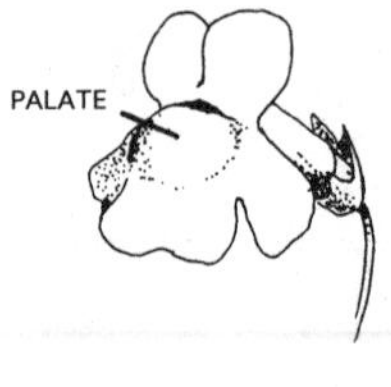

Figure 719

Figure 720

Perulate. Bearing scales. Figure 720.

Petal. An individual segment or member of the corolla, usually colored or white. Figure 721.

Petalantherous. Of a stamen with a petaloid filament. Figure 722.

Petaliferous. Bearing petals. Figure 721.

Petaline. Of or pertaining to a petal; petaloid.

Petalode. An organ (usually a stamen) which resembles a petal. Figure 722.

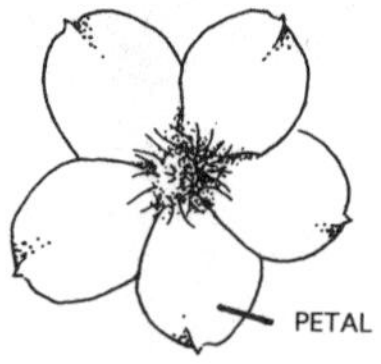

Figure 721

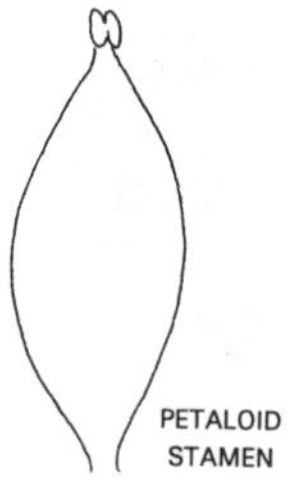

Figure 722

Petalody. A condition in which various organs in a flower, such as stamens, become petals or become petaloid, as in some double flowers.

Petaloid. Petal-like in appearance.

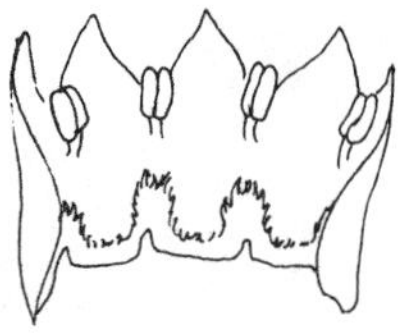

Figure 723

Petalostemonous. With the staminal filaments fused to the corolla and the anthers free. Figure 723.

Petalous. With petals. Figure 721.

Petiolar. Pertaining to the petiole; growing from the petiole. Figure 724.

Figure 724

Petiolate. With a petiole. Figure 724.

Petiole. A leaf stalk. Figure 724.

Petioled. See **petiolate**.

Petioliform. Resembling a petiole.

Petiolulate. With a petiolule. Figure 725.

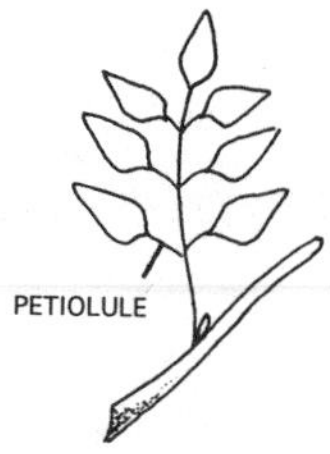

Figure 725

Petiolule. The stalk of a leaflet of a compound leaf. Figure 725.

Phaenantherous. With stamens exserted from the corolla. Figure 726.

Phaenocarpous. With the carpel (fruit) free from the surrounding floral parts. Figure 727.

Phalange. Two or more stamens joined by their filaments. Figure 728.

Phalanx. See **Phalange.**

Phanerogam. A plant which produces seeds. (compare **cryptogam**)

Phloem. The food conducting tissue of vascular plants; bark. Figure 729.

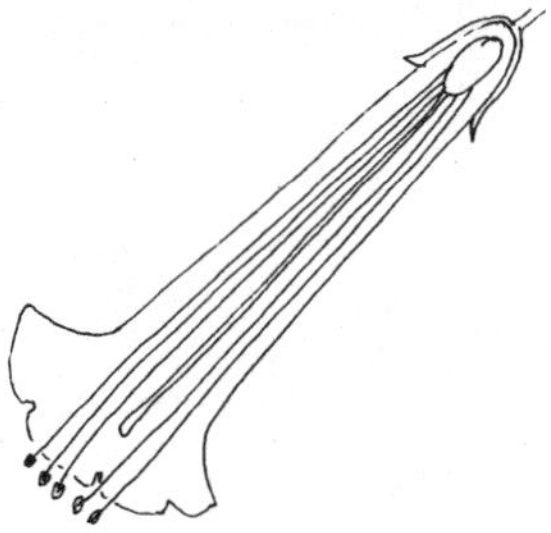

Figure 726

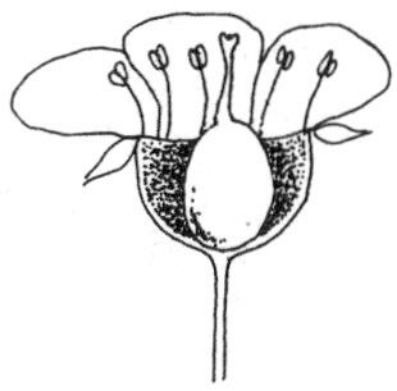

Figure 727

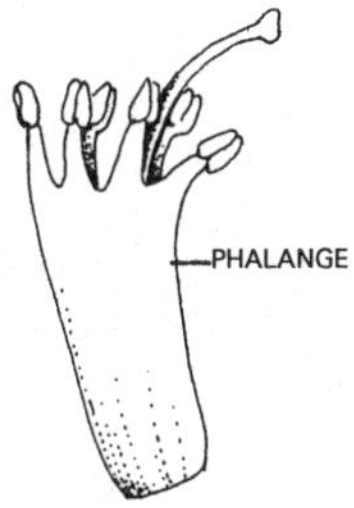

Figure 728

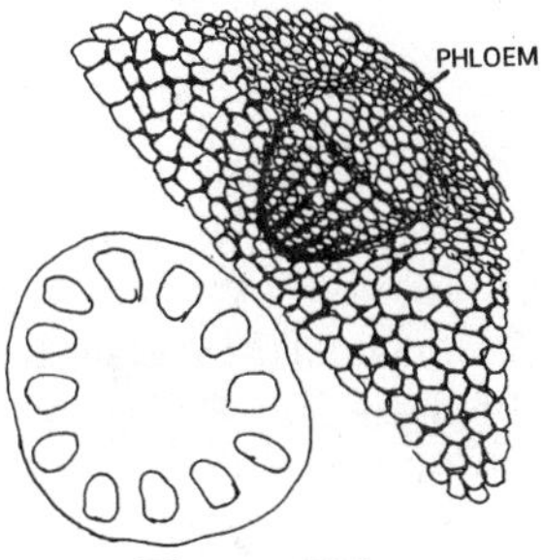

Figure 729

Phoeniceous. Purple-red.

Phoranthium. The receptacle of the flower head of the Compositae (Asteraceae). Figure 730.

Phreatophyte. A plant with its root system typically in soil saturated with water.

Phyllary. An involucral bract of the Compositae (Asteraceae). Figure 731.

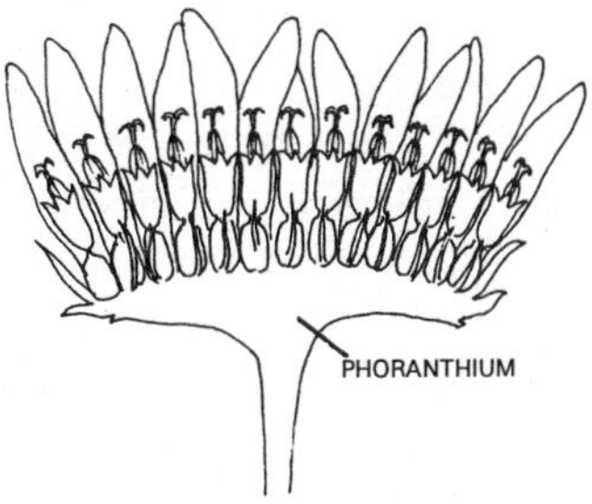

Figure 730

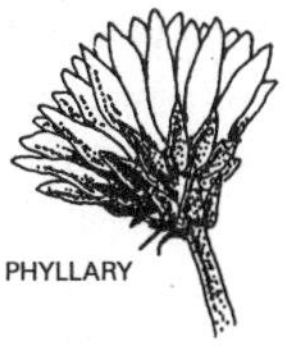

Figure 731

Phylloclad. See **phylloclade.**

Phylloclade. Part of a stem with the form and function of a leaf. Figure 732. (same as **cladophyll**)

Phyllode. An expanded, leaflike petiole lacking a true leaf blade. Figure 733.

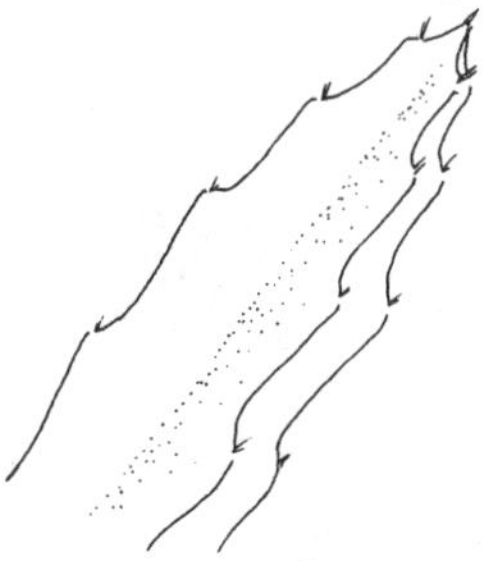

Figure 732

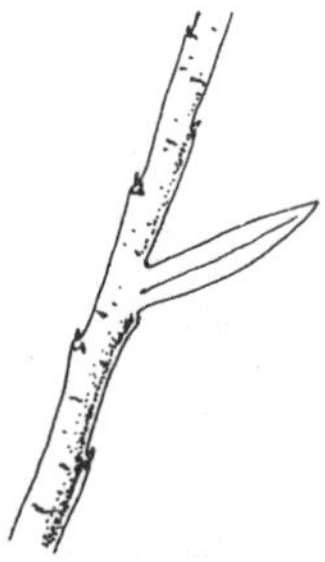

Figure 733

Phyllodium (pl. **phyllodia**). See **phyllode.**

Phylloid. Leaflike.

Phyllome. A leaf. Figure 734.

Phyllopode. The dilated leaf base of an *Isoetes* leaf. Figure 735.

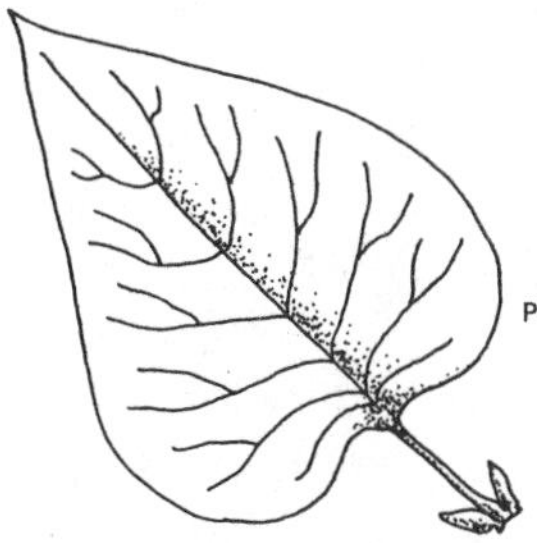

Figure 734

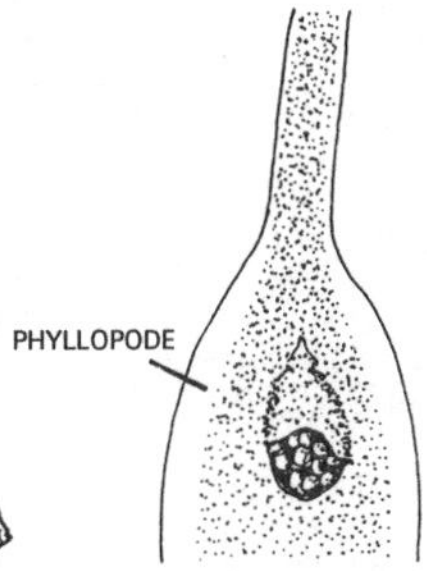

Figure 735

Phyllopodic. With the lowest leaves well developed, not reduced to scales. Figure 736. (compare **aphyllopodic**)

Phyllotaxis. See **phyllotaxy.**

Phyllotaxy. The arrangement of leaves on a stem. When expressed as a fraction, the numerator indicates the number of turns around the stem, and the denominator indicates the number of internodes between two leaves in direct vertical alignment on the stem.

Figure 736

Phytomere. A section of a grass shoot including an internode, the leaf and a portion of the node at the top of the internode, and a portion of the node at the bottom of the internode. Figure 737.

Pileate. With a cap. Figure 738.

Piliferous. Tipped with a fine hairlike structure.

Figure 737

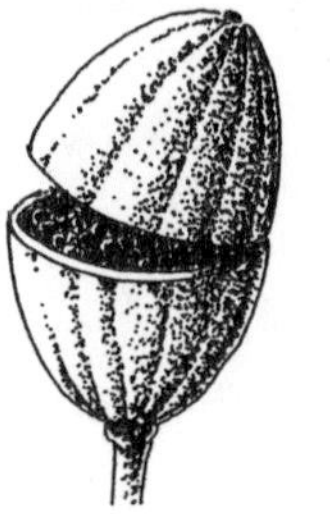

Figure 738

Figure 739.

Piliform. With the form of a hair.

Piloglandulose. With glandular hairs. Figure 740.

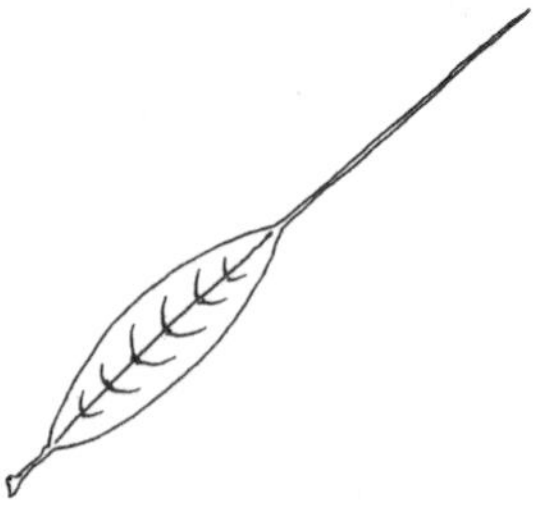

Figure 739

Figure 740

Pilose. Bearing long, soft, straight hairs. Figure 741.

Pilosulose. Bearing minute, long, soft, straight hairs. Figure 742.

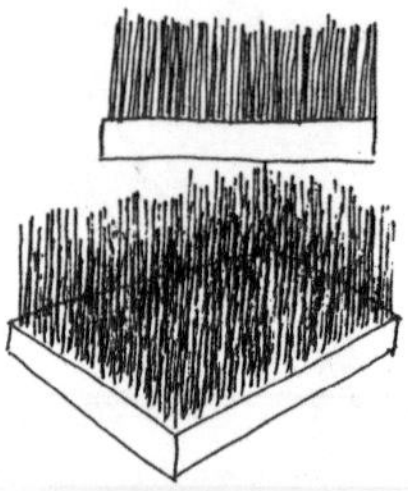

Figure 741

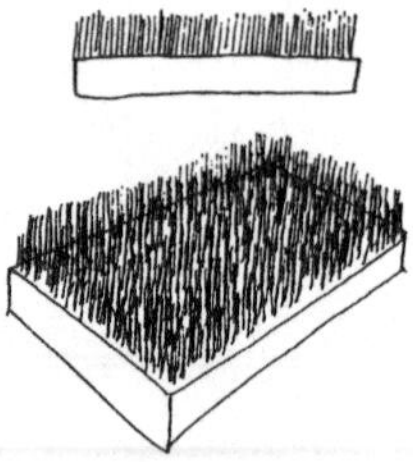

Figure 742

Pilosulous. See **pilosulose.**

Pin. A heterostylic flower with a fairly long style and short stamens. Figure 743. (compare **thrum**)

Pinna (pl. **pinnae**). One of the primary divisions or leaflets of a pinnate leaf. Figure 744.

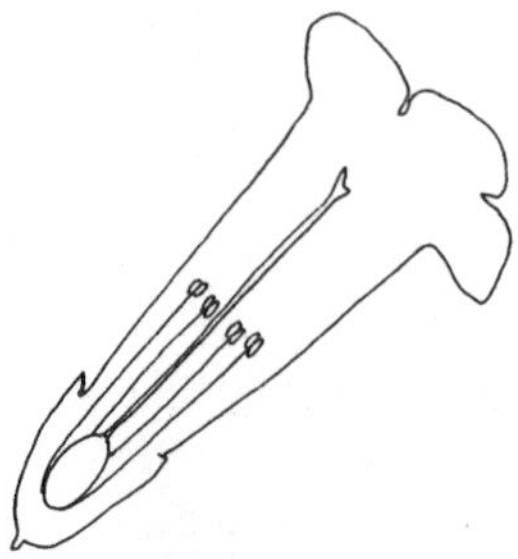

Figure 743

Pinnate. A compound leaf with leaflets arranged on opposite sides of an elongated axis. Figure 744.

Pinnatifid. Pinnately cleft or lobed half the distance or more to the midrib, but not reaching the midrib. Figure 745.

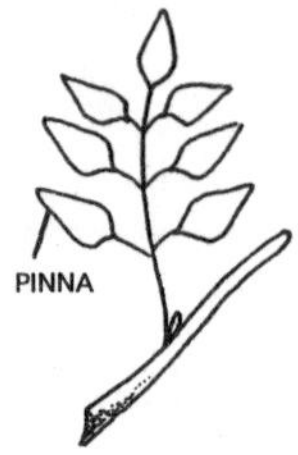

Figure 744

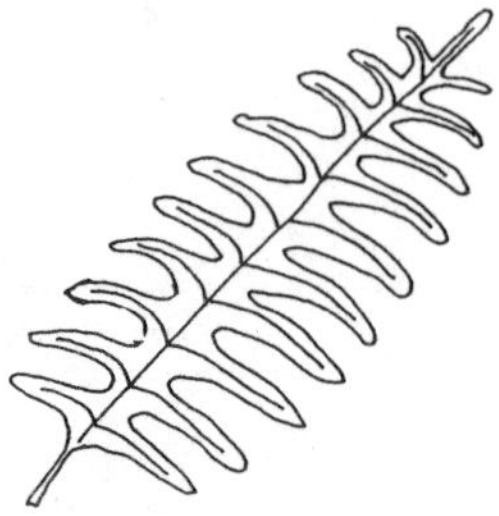

Figure 745

Pinnatilobate. With pinnately arranged lobes. Figure 746.

Pinnation. Pinnate condition or development.

Pinnatipartite. Pinnately parted. Figure 747.

Pinnatisect. Pinnately cleft to the midrib. Figure 747.

Figure 746

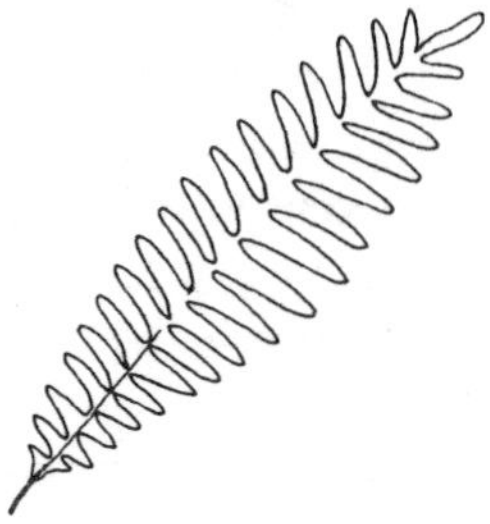

Figure 747

Pinninerved. Pinnately veined. Figure 748.

Pinnipalmate. Intermediate between pinnate and palmate, as in a leaf with the first pair of veins larger and more distinctive than the others. Figure 749.

Figure 748

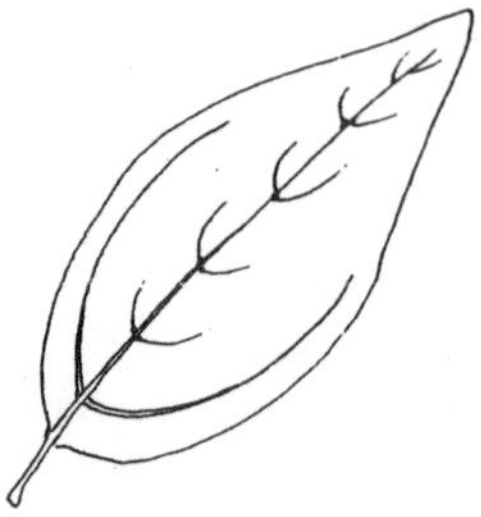

Figure 749

Pinnule. The pinnate division of a pinna in a bipinnately compound leaf, or the ultimate divisions of a leaf which is more than twice pinnately compound. Figure 750.

Pip. A small seed of a fleshy fruit. Figure 751.

Pippin. A seed. Figure 752.

Piriform. Pear-shaped. Figure 753.

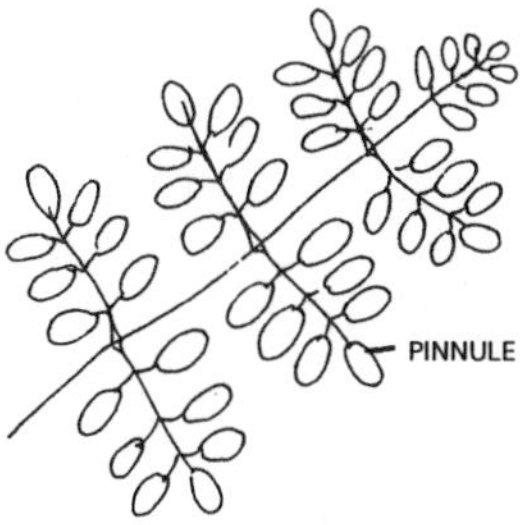

Figure 750

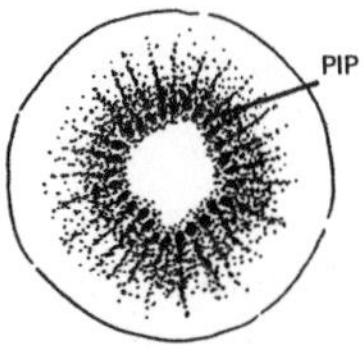

Figure 751

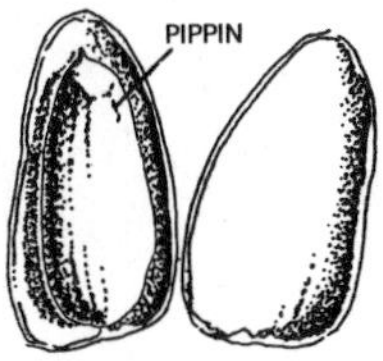

Figure 752

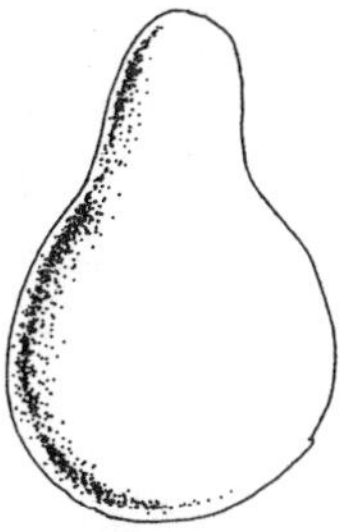

Figure 753

Pisaceous. Pea-green.

Pisiferous. Bearing peas.

Pisiform. Pea-shaped. Figure 754.

Pistil. The female reproductive organ of a flower, typically consisting of a stigma, style, and ovary. Figure 755. (compare **gynoecium**)

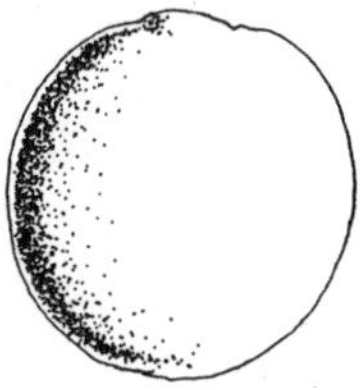

Figure 754

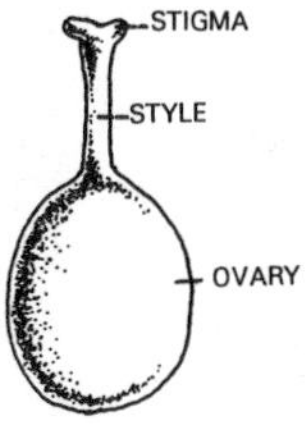

Figure 755

Pistillate. Bearing a pistil or pistils, but lacking stamens. Figure 756. (same as **carpellate**; compare **pistillate**)

Pit. A small depression. Figure 757; the stony endocarp of a drupe, as in a peach or cherry. Figure 758; the small openings in the walls of tracheids and vessel elements which allow water to move from cell to cell. Figure 759.

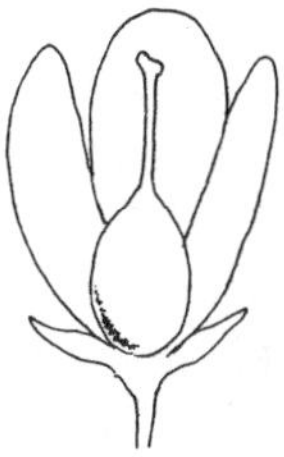

Figure 756

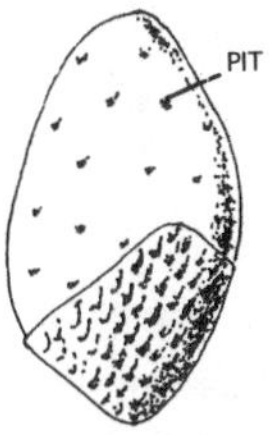

Figure 757

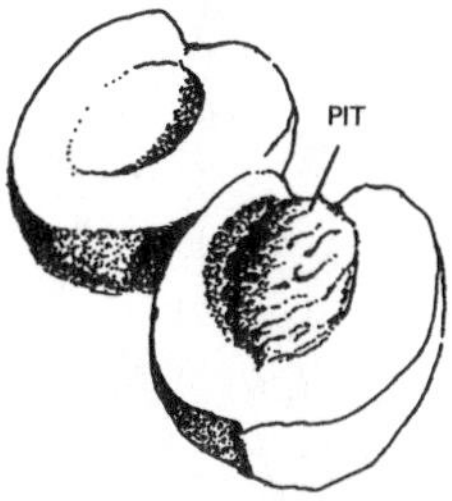

Figure 758

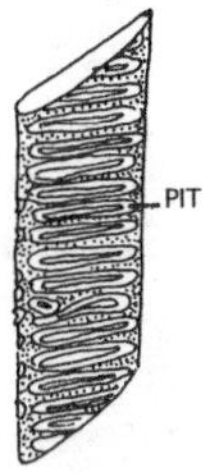

Figure 759

Pith. The spongy, parenchymatous central tissue in some stems and roots. Figure 760.

Pitted. With small pits or depressions. Figure 757.

Placenta (pl. **placentae**). The portion of the ovary bearing ovules. Figure 761.

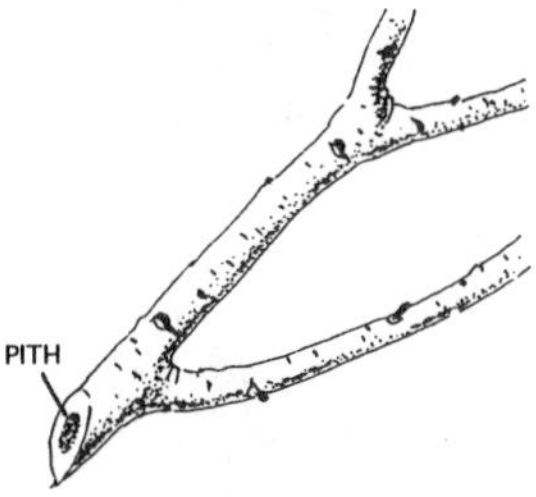

Figure 760

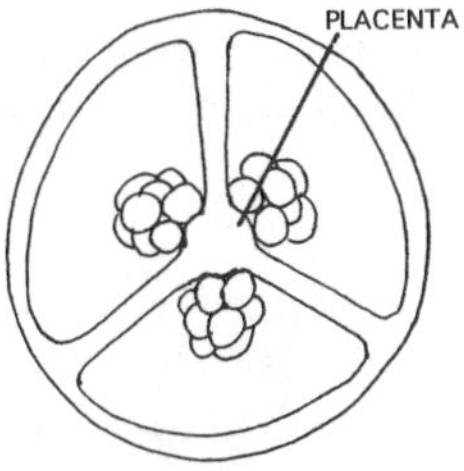

Figure 761

Placentation. The arrangement or configuration of the placentae. (see **axile**, **basal**, **free central**, and **parietal** placentation)

Plait. A fold or pleat, as in some corollas. Figure 762.

Plane. With a flat surface.

Plano-convex. Flat on one side and convex on the

other. Figure 763.

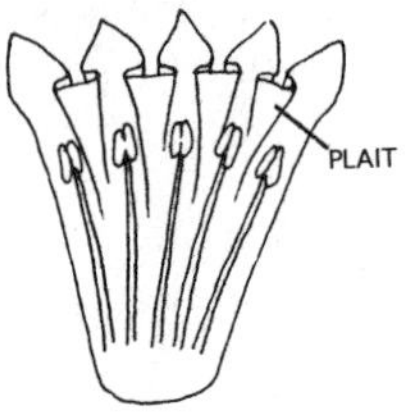

Figure 762 **Figure 763**

Pleiochasium. A cymose inflorescence with more than two branches from the main axis. Figure 764.

Figure 764

Pleiomery. The condition of having more than the usual number of floral whorls.

Pleiopetalous. With many petals.

Pleiosepalous. With many sepals.

Pleiospermous. With many seeds.

Plicate. Plaited or folded, as a folding fan. Figure 762.

Pliestesial. Living several years before flowering and fruiting, and then dying, as in *Agave*.

Plococarpium. A fruit consisting of follicles around an axis. Figure 765.

Plumbeous. Lead-colored.

Plumose. Feathery; with hairs or fine bristles on both sides of a main axis, as a plume. Figure 766.

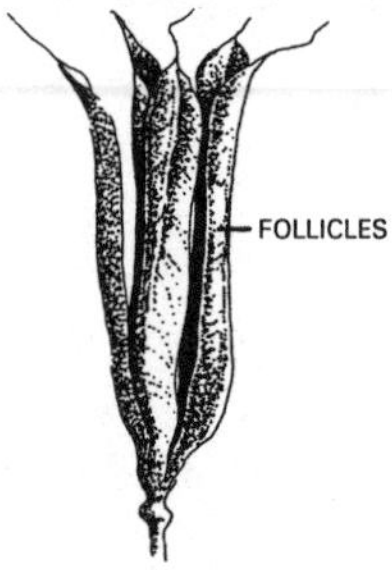

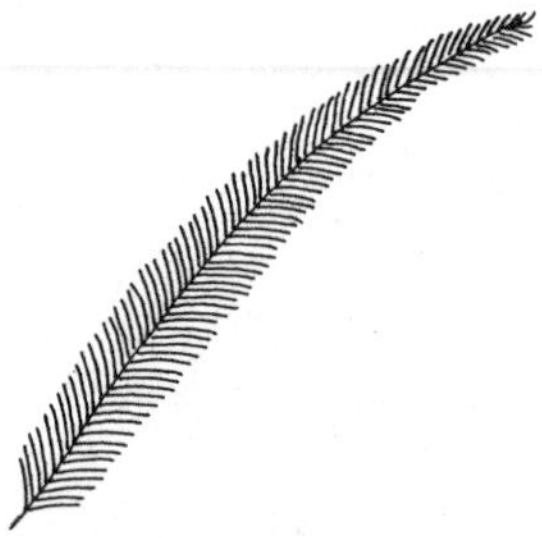

Figure 765 **Figure 766**

Plumule. The portion of the embryo above the point of attachment of the cotyledon(s) which gives rise to the shoot. Figure 767. (same as **epicotyl**)

Pluri- (prefix). Meaning many or several.

Pluricellular. Of many cells. Figure 768.

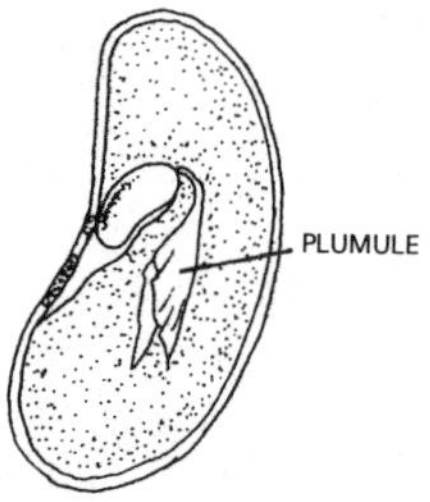

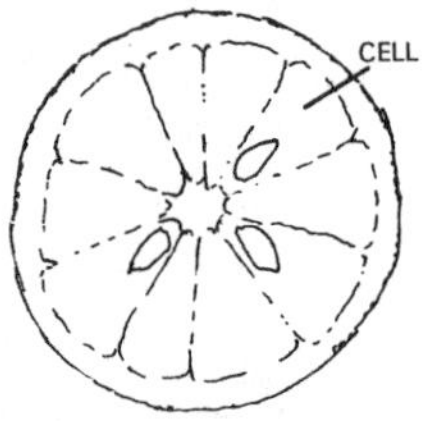

Figure 767 **Figure 768**

Pluricipital. With many heads, as in a highly branched caudex. Figure 769.

Plurilocular. See **pluricellular**.

Pluriseriate. In many series or rows. Figure 770.

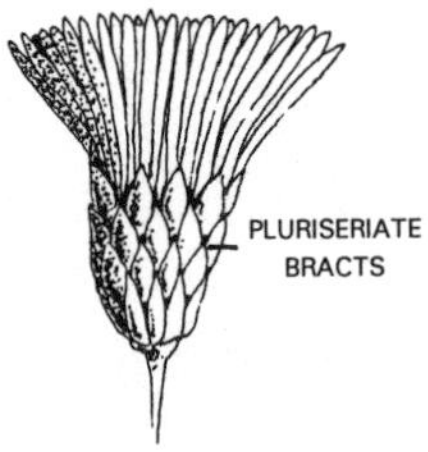

Figure 769 **Figure 770**

Pod. Any dry, dehiscent fruit, especially a legume or follicle. Figure 771.

Podocarp. A fruit borne on a stipe. Figure 772.

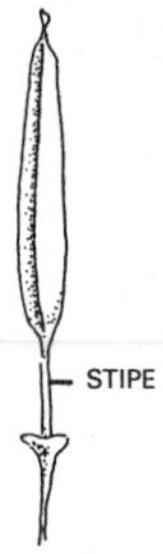

Figure 771 **Figure 772**

Podogyne. See **carpopodium**.

Pollen. The mature microspores or developing male gametophytes of a seed plant, produced in the microsporangium of a gymnosperm or in the anther of an angiosperm. Figure 773.

Pollination. The transfer of pollen from the anther to the stigma.

Polliniferous. Bearing pollen.

Pollinium (pl. **pollinia**). A mass of waxy pollen grains transported as a unit in many members of the Orchidaceae and Asclepiadaceae. Figure 774.

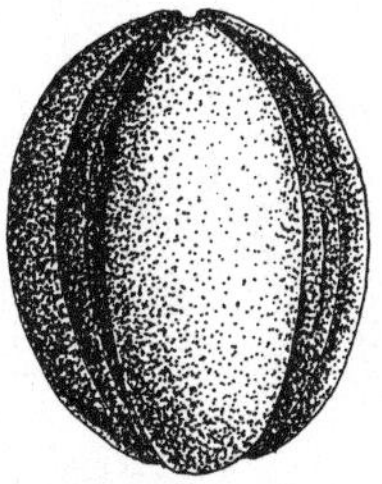

Figure 773

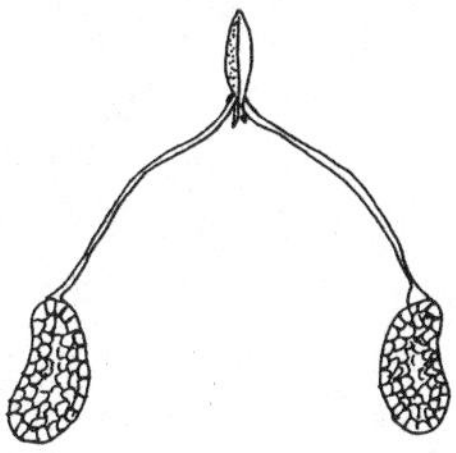

Figure 774

Poly- (prefix). Meaning many.

Polyadelphous. Borne in several distinct groups, as the stamens of some flowers. Figure 775. (compare **monadelphous** and **diadelphous**)

Figure 775

Polyandrous. With many stamens (usually more than ten). Figure 776.

Polyanthous. With many flowers, especially when clustered together in an involucre. Figure 777.

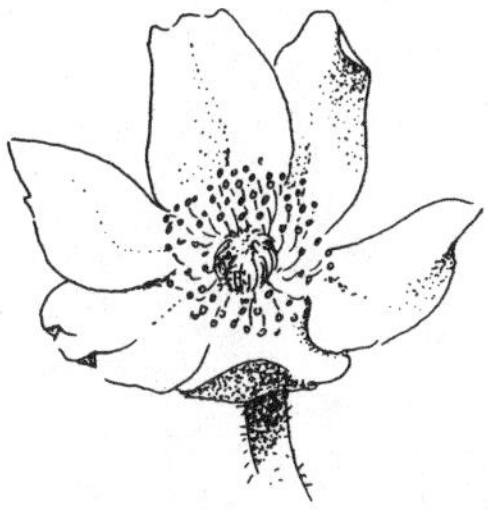

Figure 776

Figure 777

Polycarpic. See **perennial**.

Polycarpous. With many carpels. Figure 778.

Polycephalous. With many flower heads. Figure 779.

Polychasium. A cymose inflorescence in which each axis produces more than two lateral axes. Figure 780.

Polychrome. Many colored.

Figure 778

Figure 779

Polycyclic. With many whorls.

Polygamo-dioecious. Mostly dioecious, but with some perfect flowers.

Polygamo-monoecious. Mostly monoecious, but with some perfect flowers.

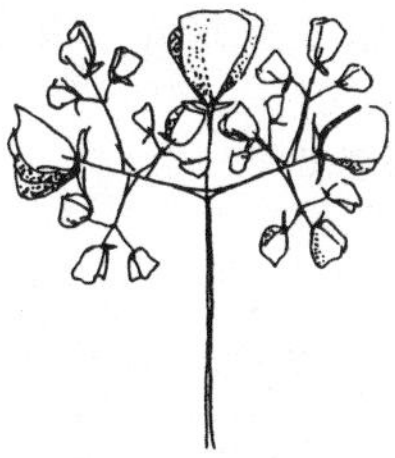

Figure 780

Polygamous. With unisexual and bisexual flowers on the same plant.

Polygonal. Many-angled.

Polygynous. With many pistils or styles. Figure 778.

Polymerous. With many parts, as in a floral whorl with many members.

Polymorphic. Variable; with many forms.

Polymorphous. See **polymorphic**.

Polypetalous. A corolla of completely separate petals. Figure 781. (same as **apopetalous**; compare **gamopetalous** and **sympetalous**)

Polyploid. With three or more complete sets of chromosomes in each cell.

Polysepalous. A calyx of separate sepals. Figure 782. (compare **synsepalous** and **gamosepalous**)

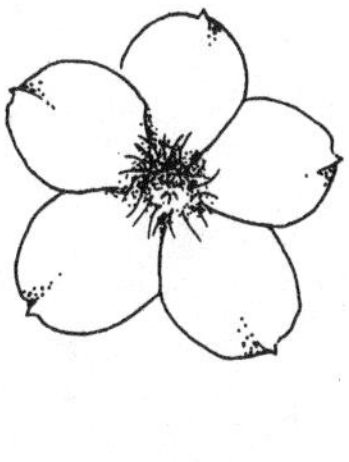

Figure 781

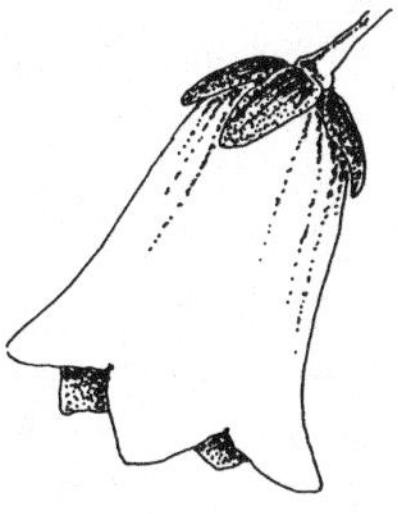

Figure 782

Polystachyous. With many spikes.

Polystemonous. With many stamens. Figure 783.

Polystichous. Arranged into several rows.

Pomaceous. Of or relating to a pome.

Pome. A fleshy, indehiscent fruit derived from an inferior, compound ovary, consisting of a modified floral tube surrounding a core, as in an apple. Figure 784.

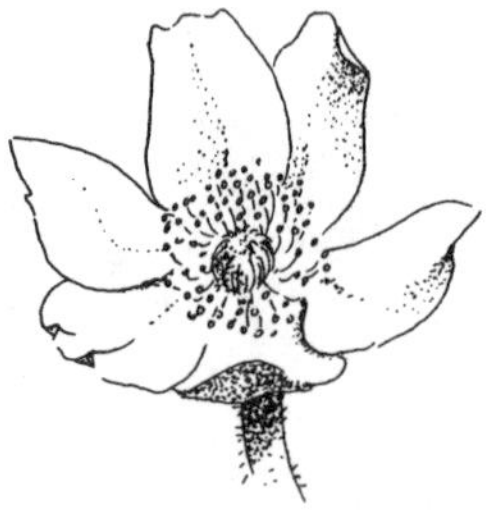

Figure 783

Porandrous. With anthers opening by pores. Figure 785.

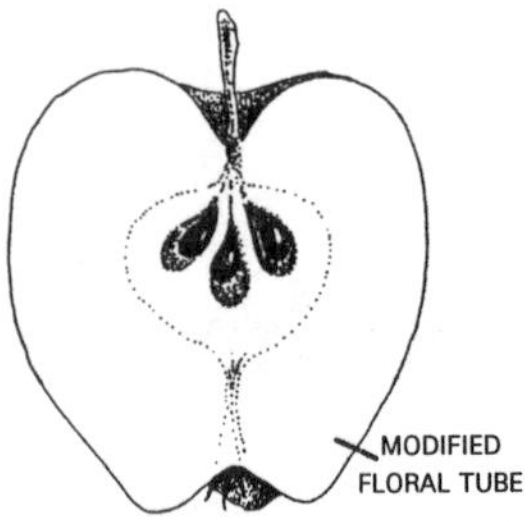

Figure 784

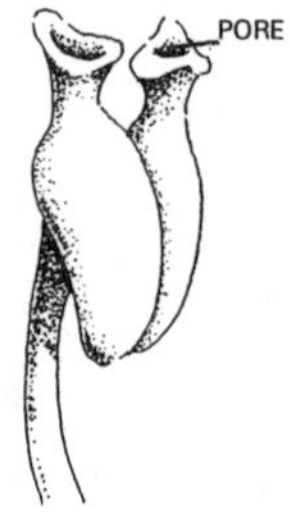

Figure 785

Pore. A small opening. Figure 786.

Poricidal. Opening by pores, as in a poppy capsule. Figure 786.

Porose. With pores. Figure 786.

Porrect. Extended forward; resembling a parrot beak. Figure 787.

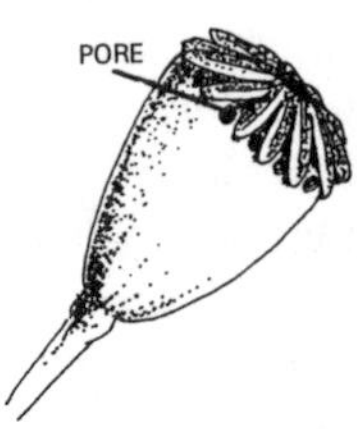

Figure 786

Posterior. At the back; on the side toward the axis, as the upper lip of a bilabiate corolla. Figure 788. (compare **anterior**)

Praemorse. Terminating abruptly, as if bitten off. Figure 789.

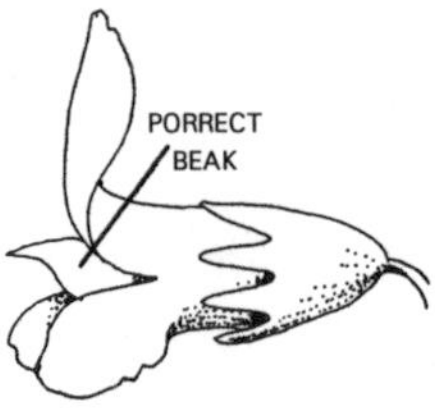

Figure 787

Precocious. Developing or appearing very early; with the flowers developing before the leaves.

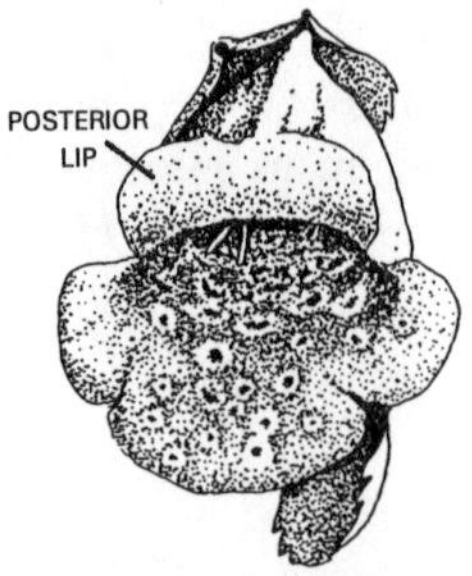

Figure 788

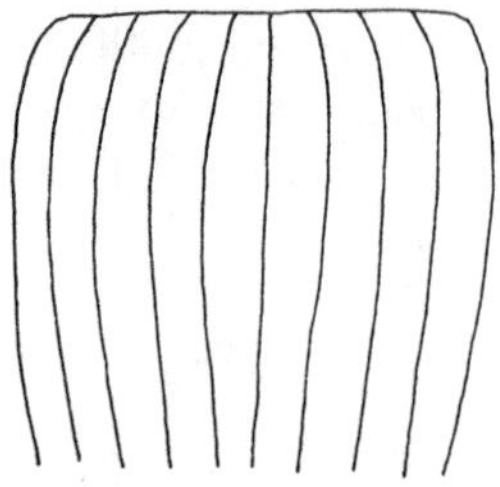

Figure 789

Prehensile. Adapted for grasping, as in a tendril. Figure 790.

Prevernal. Pre-spring.

Prickle. A small, sharp outgrowth of the epidermis or bark. Figure 791. (compare **spine** and **thorn**)

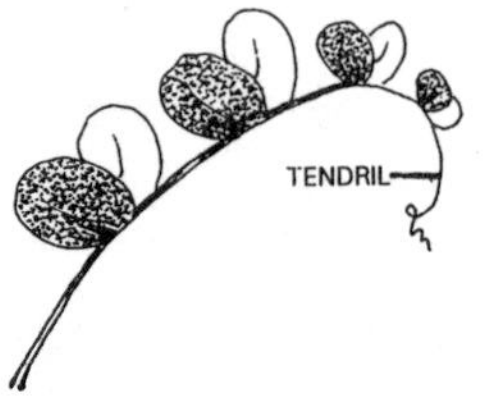

Figure 790

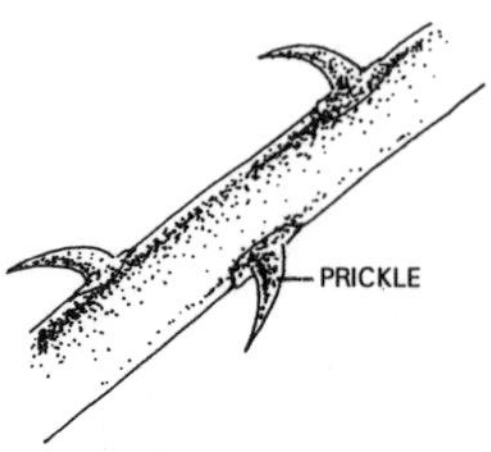

Figure 791

Primary. First, as the first leaflet of a compound leaf. Figure 792.

Primine. The outer integument layer of an ovule. Figure 793.

Figure 792

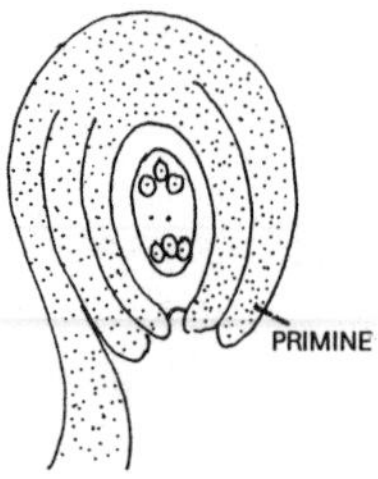

Figure 793

Primocane. The first-year, usually flowerless, cane (shoot) of *Rubus*. (compare **floricane**)

Prismatic. With sharp, definite angles and flat sides, like a prism; brilliant. Figure 794.

Process. An outgrowth or appendage. Figures 795 and 796.

Procumbent. Lying or trailing on the ground, but

not rooting at the nodes. Figure 797.

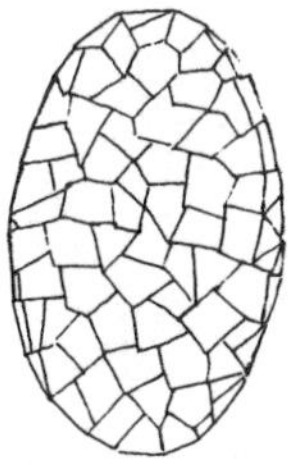

Figure 794

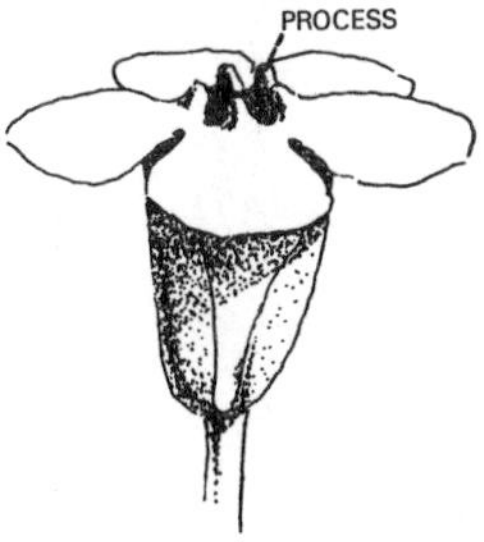

Figure 795

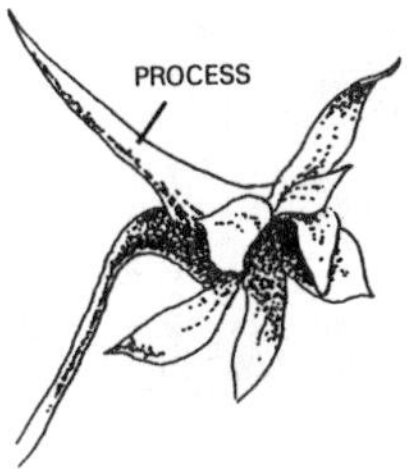

Figure 796

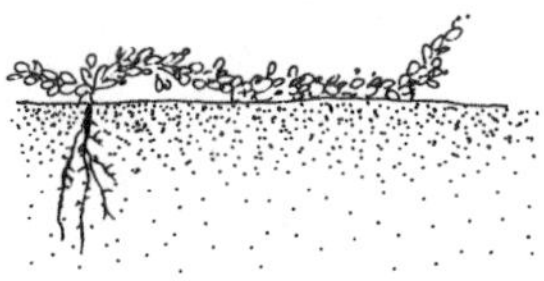

Figure 797

Projected. Extending outward. Figure 798.

Proliferous. Bearing plantlets or bulblets, usually from the leaves. Figure 799.

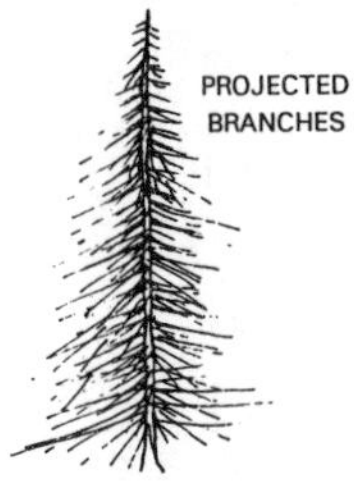

Figure 798

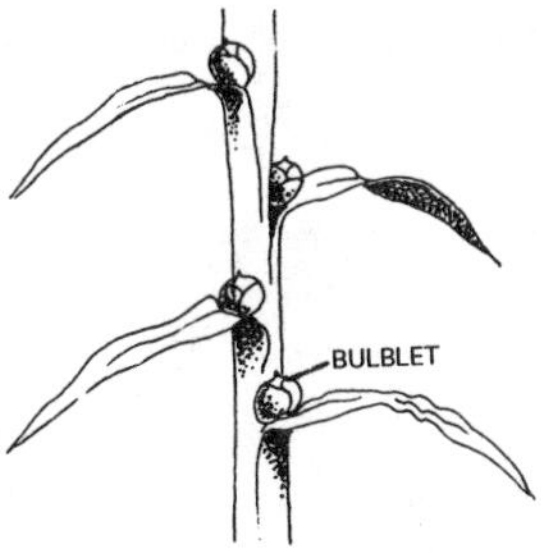

Figure 799

Prominent. Standing out from the surrounding surface, as raised veins on the surface of a leaf. Figure 800.

Propagule. A structure, such as a seed or spore, which gives rise to a new plant. Figure 801.

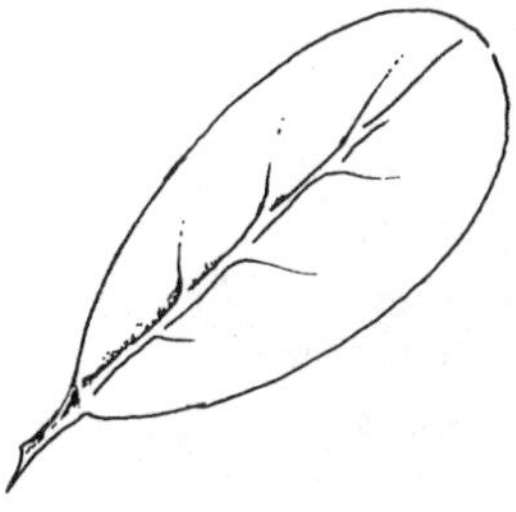

Figure 800

Prophyll. One of the paired bracteoles subtending the flowers in some *Juncus* species. Figure 802.

Prophyllum. See **prophyll**.

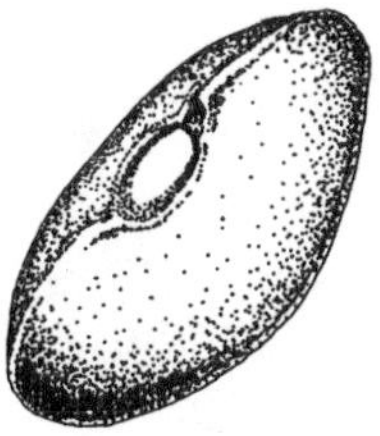

Figure 801

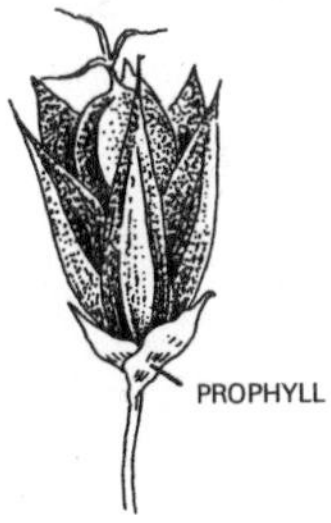

Figure 802

Prop root. Adventitious roots arising from lower nodes and providing support to a stem. Figure 803.

Prostrate. Lying flat on the ground. Figure 804.

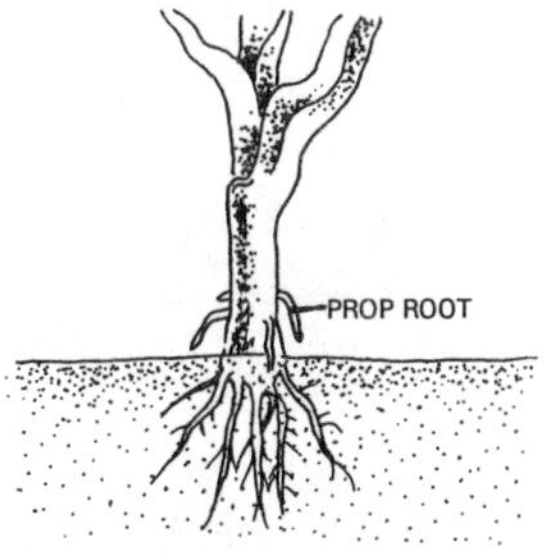

Figure 803

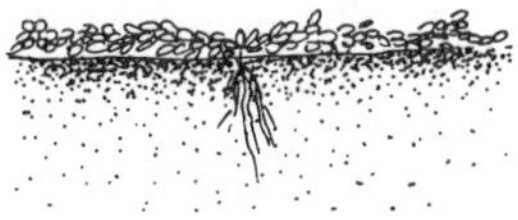

Figure 804

Protandry (adj. **protandrous**). The anthers releasing pollen before the stigma is receptive.

Proterandry (adj. **proterandrous**). See **protandry**.

Proteranthy (adj. **proteranthous**). With the flowers developing before the leaves.

Proterogyny (adj. **proterogynous**). See **protogyny**.

Prothallium. See **prothallus**.

Prothallus (pl. **prothallia**). The small, usually flat, thallus-like growth germinating from a spore; the gametophyte generation in the alternation of generations. Figure 805.

Figure 805

Protogyny (adj. **protogynous**). The stigma receptive before the anthers release pollen.

Protostele. A stele with a solid core of vascular tissue, lacking a pith. Figure 806.

Protuberance. A rounded bulge, swelling, or projection. Figure 807.

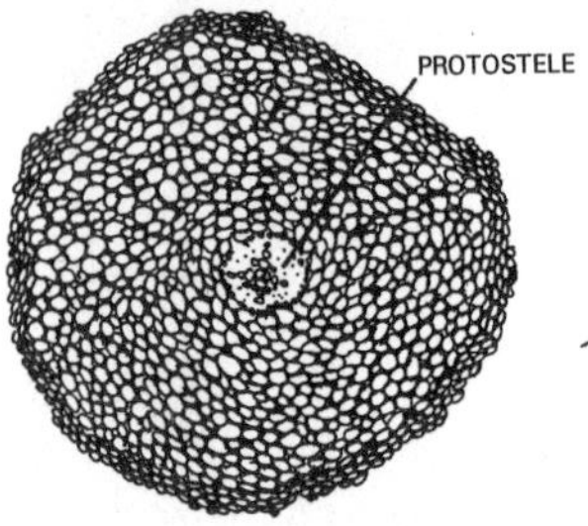

Figure 806

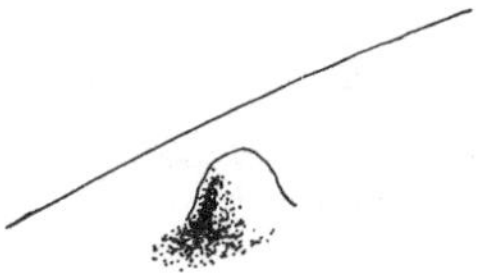

Figure 807

Proximal. Toward the base, or the end of the organ by which it is attached. Figure 808. (compare **distal**)

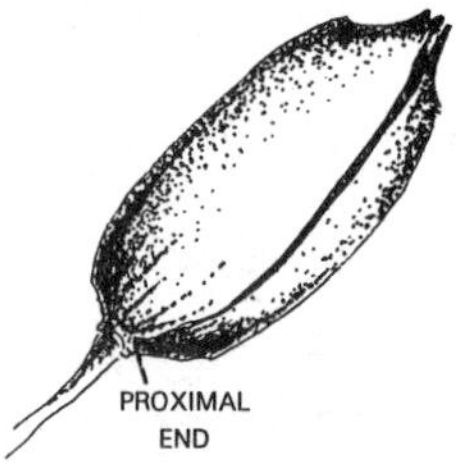

Figure 808

Pruinate. See **pruinose.**

Pruinose. With a waxy, powdery, usually whitish coating (bloom) on the surface; conspicuously glaucous, like a prune.

Pruniform. Plum-shaped. Figure 809.

Psammophyte. A plant growing in sand.

Pseudanthium. A compact inflorescence of many small flowers which simulates a single flower. Figure 810.

Figure 809

Figure 810

Pseudo- (prefix). Meaning false.

Pseudocarp. A fruit which develops from the receptacle rather than from the ovary, as in a pome. Figure 811.

Pseudofasciculate. Closely clustered, but not actually joined into a bundle. Figure 812.

Pseudomonomerous. A structure which appears to be simple, though actually derived from the fusion of separate structures, as a pistil which appears to be composed of a single carpel, though actually composed of two or more carpels.

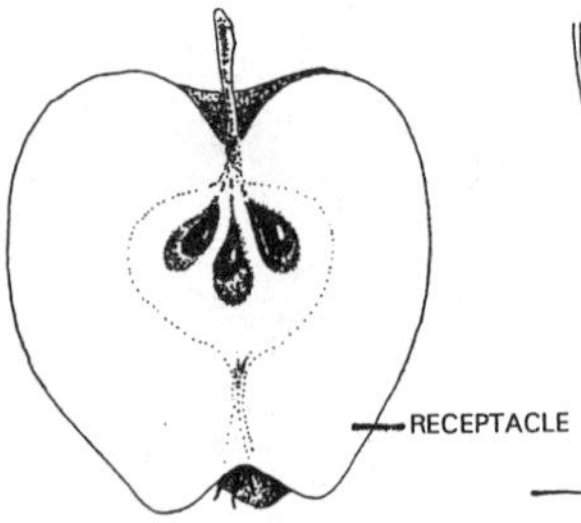

Figure 811

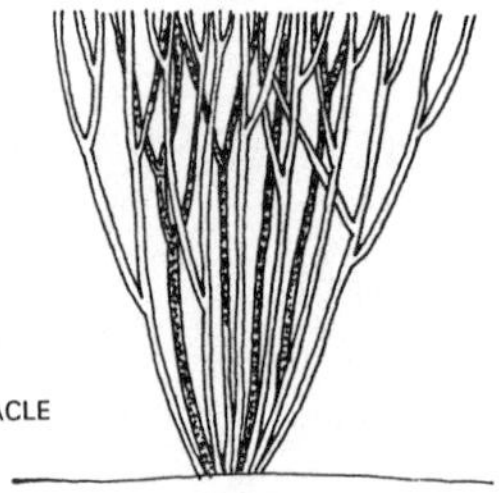

Figure 812

Pseudoscape. A false scape, where not all of the leaves are truly basal in origin though, superficially, they appear to be so. Figure 813.

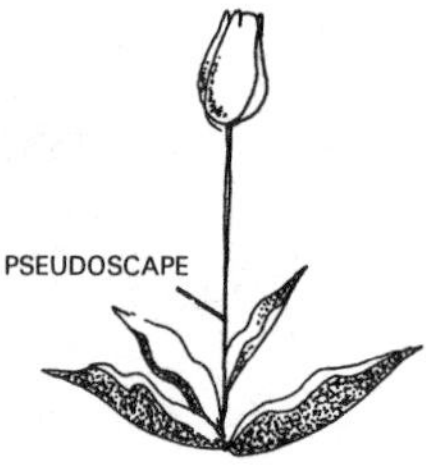

Figure 813

Pseudoverticillate. Not actually whorled, but appearing so.

Pterocarpous. With winged fruits. Figure 814.

Pterocaulous. With winged stems. Figure 815.

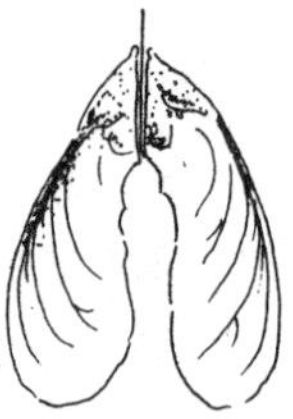

Figure 814

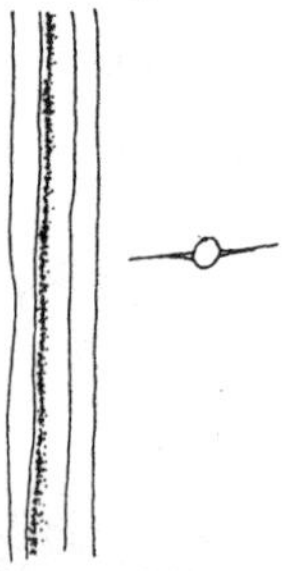

Figure 815

Pterospermous. With winged seeds. Figure 816.

Pterygopous. With winged peduncles.

Puberulence. Fine, short hairs. Figure 817.

Puberulent. Minutely pubescent; with fine, short hairs. Figure 817.

Puberulous. See **puberulent.**

Figure 816

Pubescence. Hairiness; short, soft hairs. Figure 818.

Pubescent. Covered with short, soft hairs. Figure 818; bearing any kind of hairs.

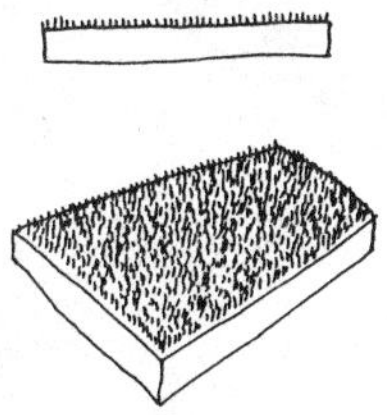

Figure 817

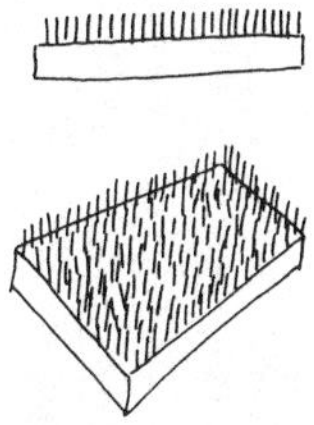

Figure 818

Pulveraceous. See **pulverulent**.

Pulverulent. Appearing dusty or powdery.

Pulvinate. Cushion-like or mat-like. Figure 819.

Pulviniform. See **pulvinate**.

Pulvinus (pl. **pulvini**). A swelling or enlargement at the base of a petiole or petiolule. Figure 820.

Figure 819

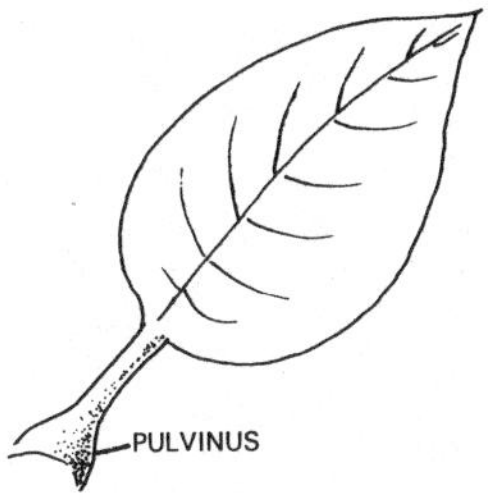

Figure 820

Punctate. Dotted with pits or with translucent, sunken glands or with colored dots. Figure 821.

Puncticulate. Minutely punctate. Figure 822.

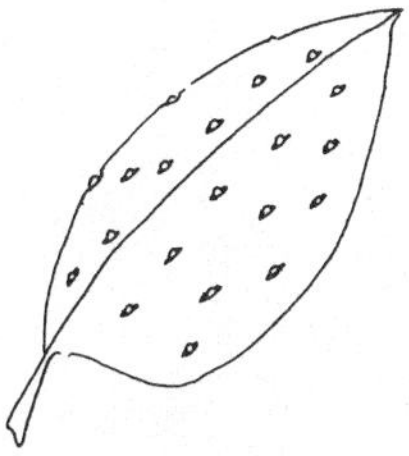

Figure 821

Figure 822

Punctiform. Reduced to a point.

Pungent. Tipped with a sharp, rigid point. Figure 823; with a sharp, acrid odor or taste.

Puniceous. Crimson colored.

Purpurescent. Becoming purplish.

Pustular. See **pustulose**.

Pustulate. See **pustulose**.

Pustule. Small blisterlike elevations. Figure 824.

Pustuliferous. See **pustulose**.

Pustulose. With small blisters or pustules, often at the base of a hair. Figure 824.

Figure 823

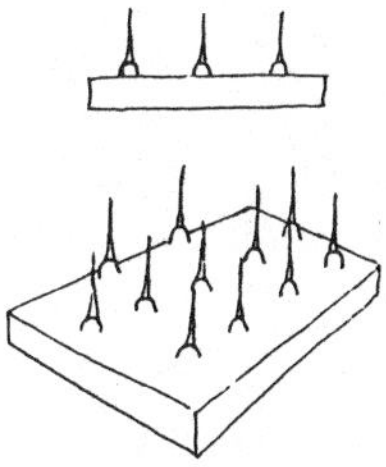

Figure 824

Putamen. The hard stony endocarp of some fruits. Figure 825; a nut shell.

Pyramidal. Tetrahedral; pyramid-shaped. Figure 826.

Pyrene. The stone or pit of a drupe or drupelet. Figure 825.

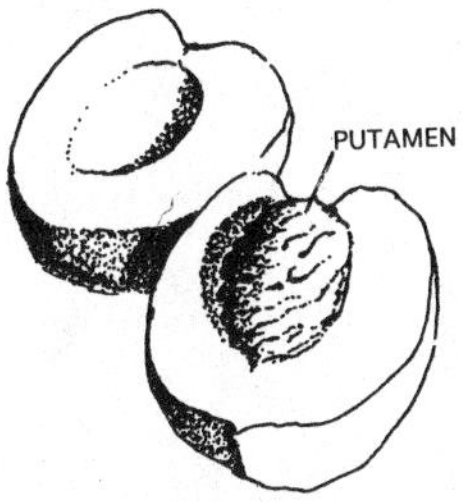

Figure 825

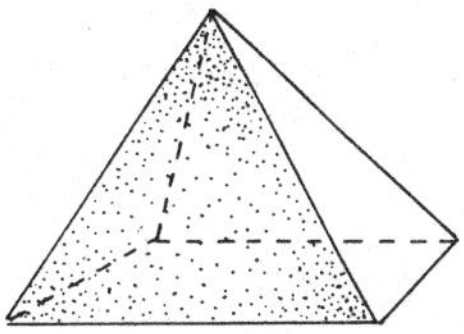

Figure 826

Pyriform. Pear-shaped. Figure 827.

Pyxidate. With a pyxis. Figure 828.

Pyxidium. See **pyxis**.

Pyxis. A circumscissile capsule, the top coming off as a lid. Figure 828.

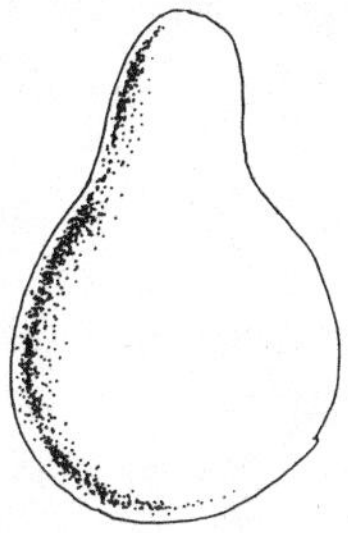

Figure 827

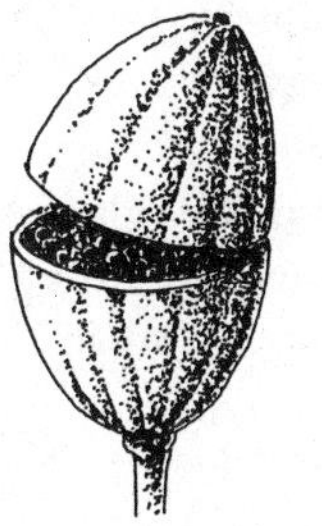

Figure 828

Quadrangular. Four-angled. Figure 829.

Quadrangulate. See **quadrangular**.

Quadrate. Square; rectangular.

Quadri- (prefix). Meaning four.

Quadrifoliate. With four leaves or four leaflets. Figure 830.

Quadrilateral. With four sides. Figure 831.

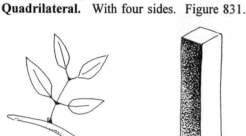

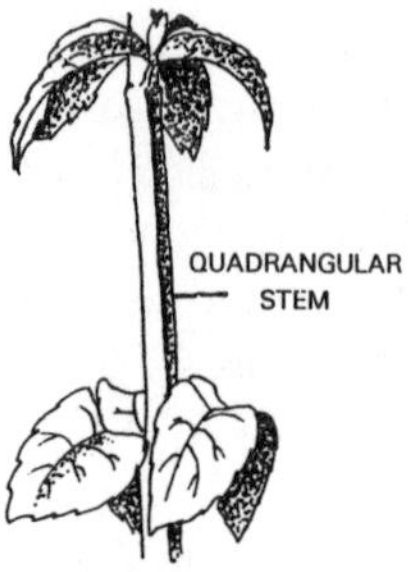

Figure 829

Figure 830

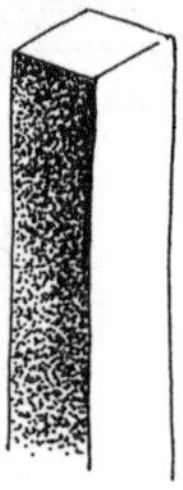

Figure 831

Quadripinnatifid. Four times pinnately cleft. Figure 832.

Quinary. In fives.

Quinate. Five-parted. Figure 833.

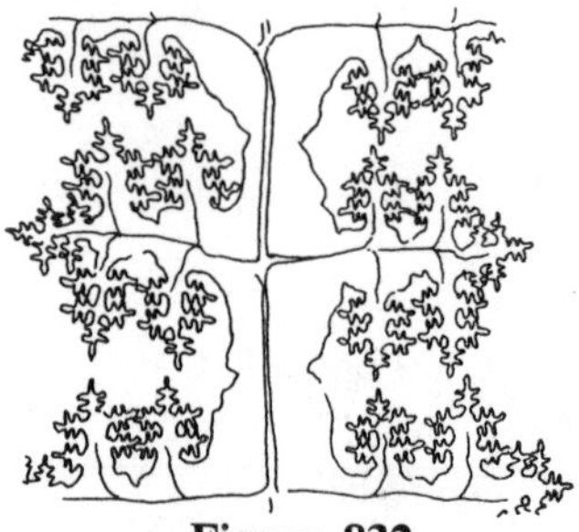

Figure 832

Figure 833

Quincuncial. With a five-ranked leaf arrangement. Figure 834.

Quinquecostate. With five ribs. Figure 835.

Quinquefarious. Arranged in five ranks. Figure 836.

Quinquefoliate. With five leaves or five leaflets. Figure 837.

Quinquejugate. Arranged in five pairs. Figure 838.

Quinquelocular. With five cells or locules. Figure 839.

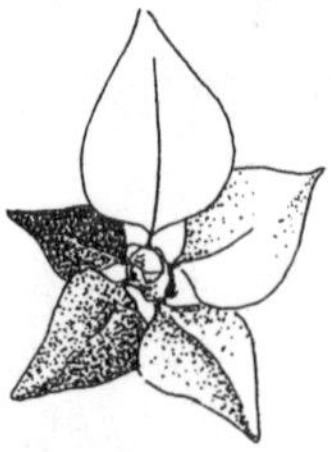

Figure 834

Figure 835

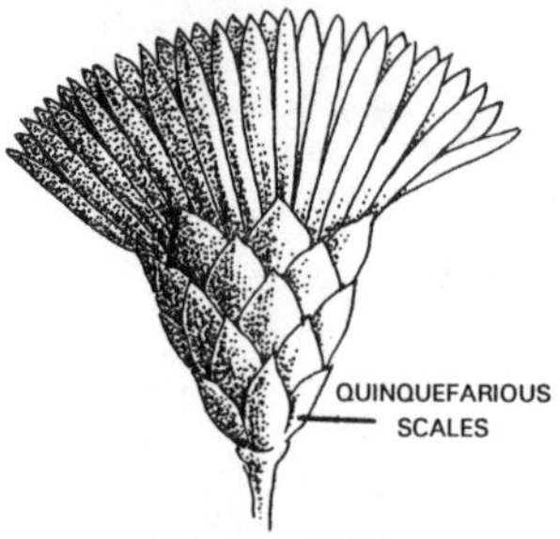

Figure 836

Figure 837

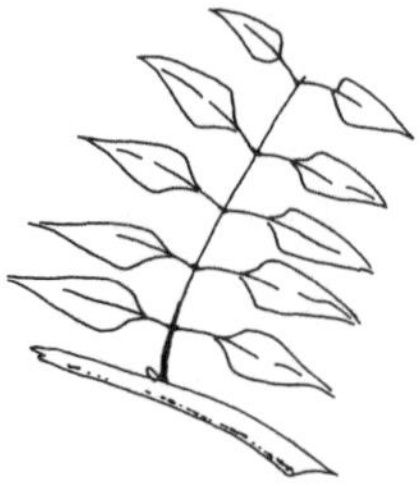

Figure 838

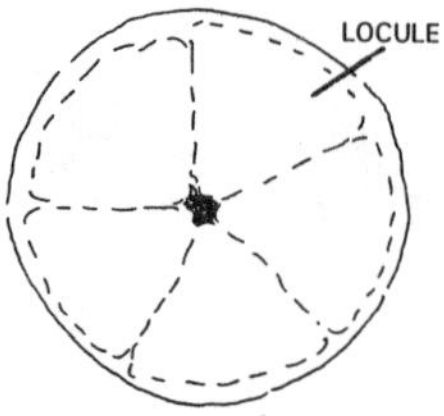

Figure 839

Quinquenerved. With five main nerves. Figure 840.

Quinquepartite. Divided into five parts. Figure 841.

Figure 840

Figure 841

Raceme. An unbranched, elongated inflorescence with pedicellate flowers maturing from the bottom upwards. Figure 842.

Racemiferous. See **racemose.**

Racemiform. An inflorescence with the general appearance, but not necessarily the structure, of a true raceme.

Racemose. Having flowers in racemes. The term is sometimes used in the same sense as **racemiform**.

Figure 842

Rachilla. The axis of a grass or sedge spikelet; a small rachis. Figure 843.

Rachis. The main axis of a structure, such as a compound leaf or an inflorescence. Figure 844.

Figure 843

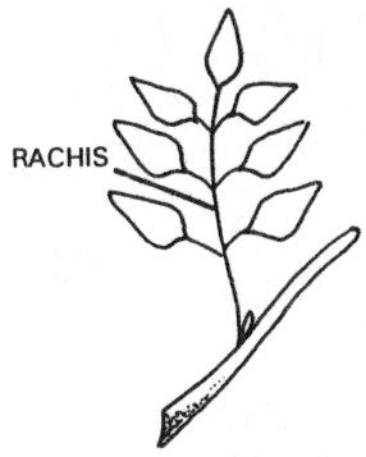

Figure 844

Radial. With structures radiating from a central point, as spokes on a wheel. Figure 845; the lateral spines of a cactus.

Radiant. See **radiate**.

Radiate. With parts spreading from a central point. Figure 846; in the Compositae (Asteraceae), with some of the flowers of the involucrate head ligulate (the petals united into a straplike corolla). Figure 847.

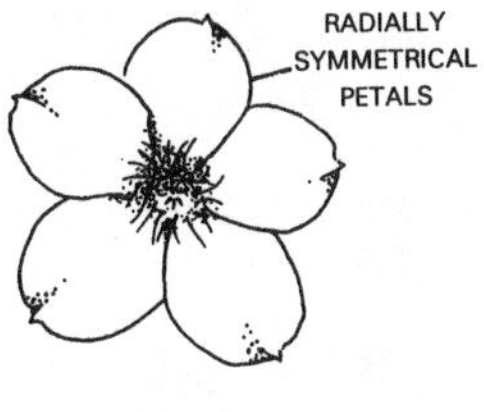

Figure 845

Radical. Pertaining to the root; arising from, or near, the roots.

Radicant. A root arising from the node of a prostrate stem. Figure 848.

Radicicolous. With the flower positioned directly upon the root crown. Figure 849.

Radicle. The part of the plant embryo which will develop into the primary root. Figure 850.

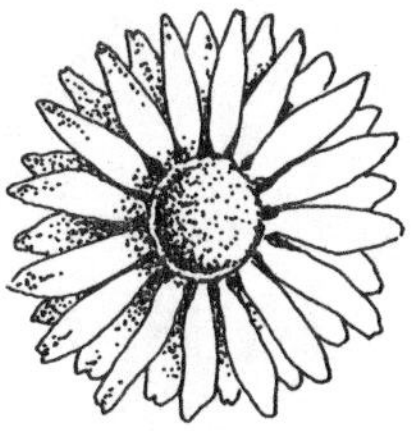

Figure 846

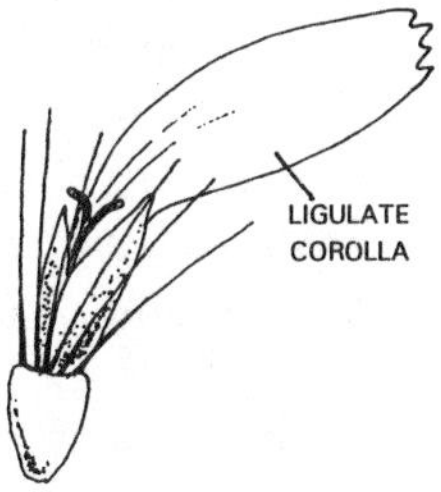

Figure 847

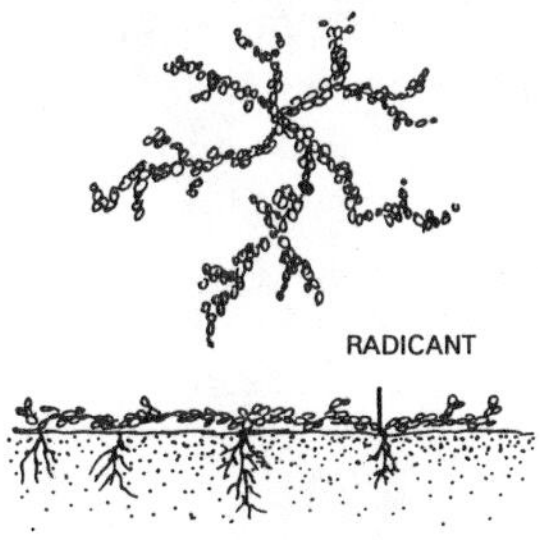

Figure 848

Figure 849

Ramal. See **rameal**.

Rameal. Pertaining to the branches.

Ramentaceous. Having ramentum. Figure 851.

Ramentum. The flattened, scaly outgrowths on the epidermis of the stem and leaves of some ferns. Figure 851.

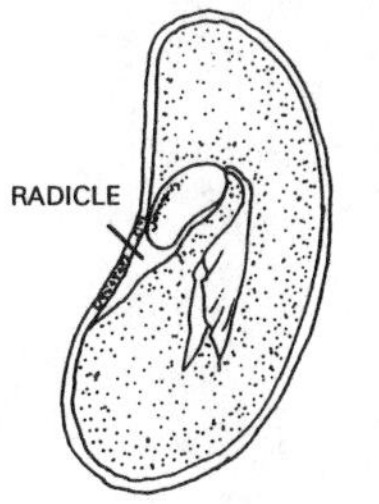

Figure 850

Figure 851

Ramification. The arrangement of branching parts.

Ramiform. Branchlike in form; branched.

Ramose. With many branches; branching. Figure 852.

Ramous. See **ramose**.

Ramulose. See **ramose**.

Figure 852

Range. The area of distribution of a plant.

Rank. A vertical row, as in a plant with 2-ranked leaves arranged into two rows. Figure 853.

Ranked. Arranged into vertical rows. Figure 853.

Figure 853

Raphal. Of or pertaining to the raphe.

Raphe (pl. **raphae**). A ridge on the seed formed by the portion of the funiculus fused to the seed coat. Figure 854.

Rapiformis. Turnip-shaped. Figure 855.

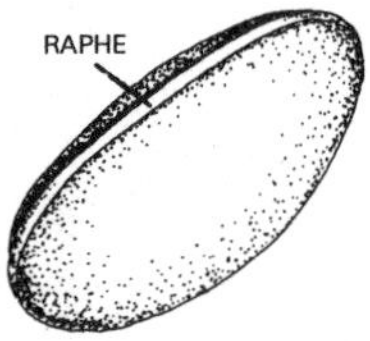

Figure 854

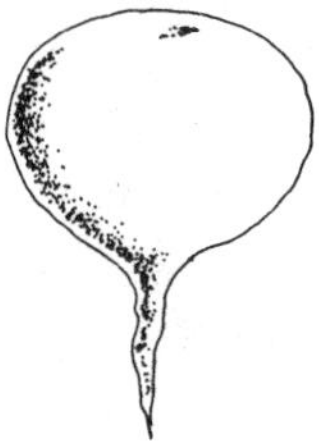

Figure 855

Ratoon. A shoot arising from the root of a plant that has been cut down. Figure 856.

Ray. The straplike portion of a ligulate flower (or the ligulate flower itself) in the Compositae (Asteraceae). Figure 857; a branch of an umbel. Figure 858.

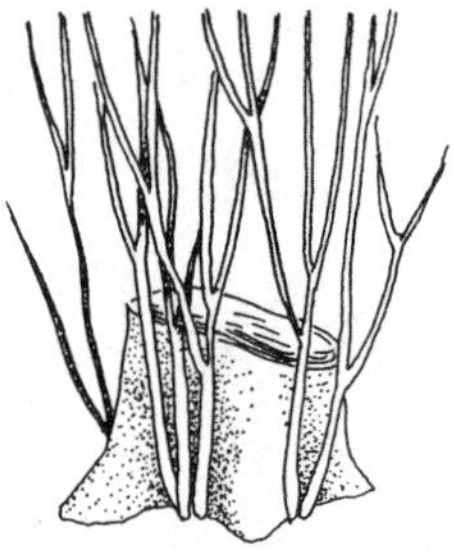

Figure 856

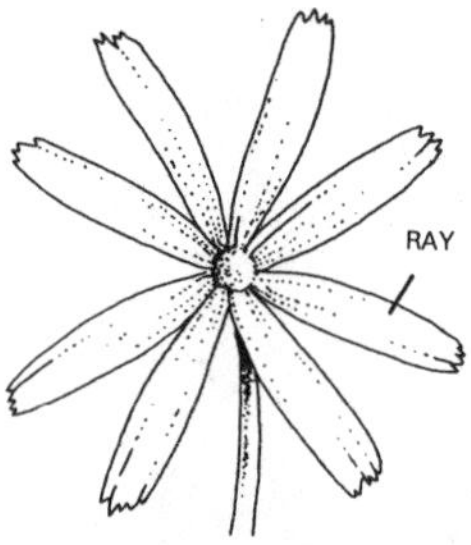

Figure 857

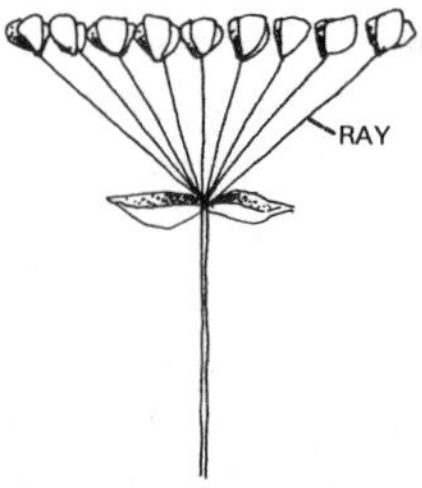

Figure 858

Ray flower. A ligulate flower of the Compositae (Asteraceae). Figure 859.

Receptacle. The portion of the pedicel upon which the flower parts are borne. Figure 860; in the Compositae (Asteraceae), the part of the peduncle where the flowers of the head are borne. Figure 861.

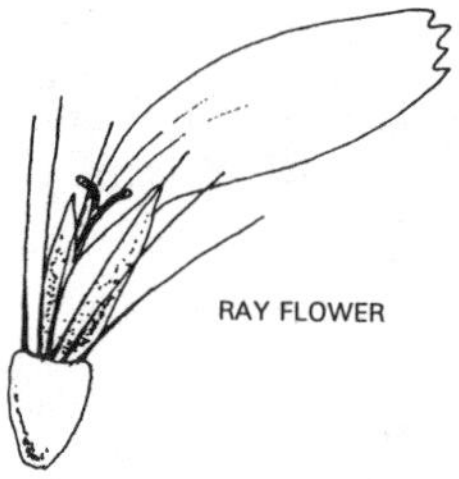

Figure 859

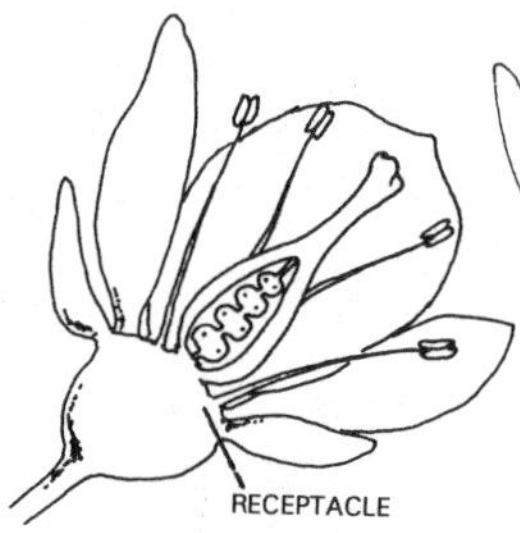

Figure 860

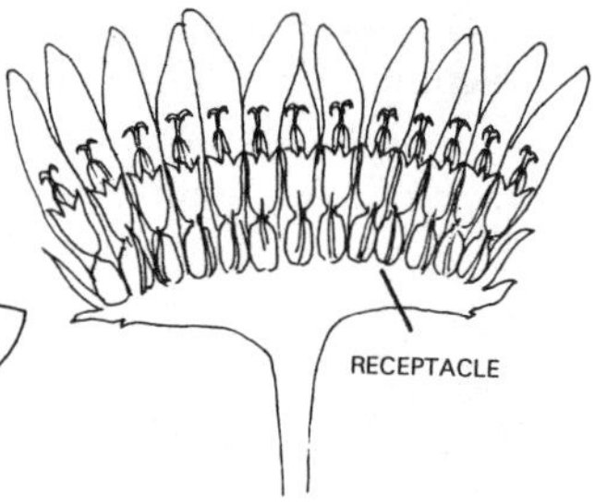

Figure 861

Reclinate. Bent abruptly downward. Figure 862.

Reclining. Bending or curving downward; lying upon something and being supported by it. Figure 863.

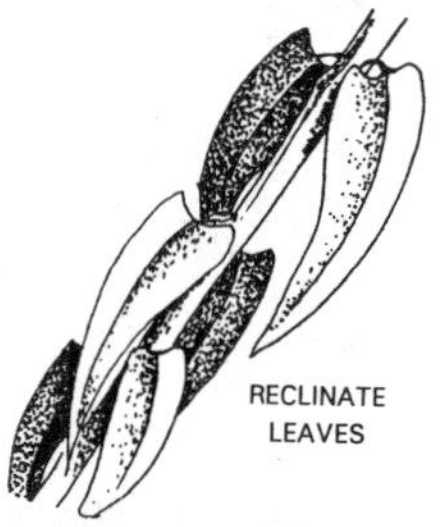

Figure 862

Figure 863

Recumbent. Leaning or resting on the ground; prostrate. Figure 864.

Recurved. Curved backward. Figure 865.

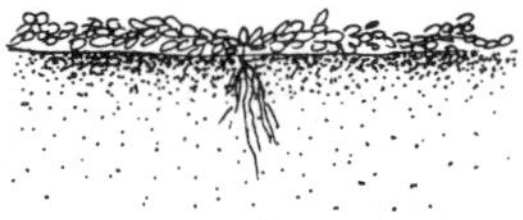

Figure 864

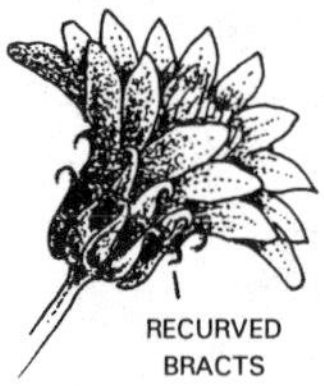

Figure 865

Reduced. Diminished in size.
Reflexed. Bent backward or downward. Figure 866.
Refoliate. To produce leaves again, as after rain, wind, or disease.
Refracted. Bent backward from the base. Figure 862.

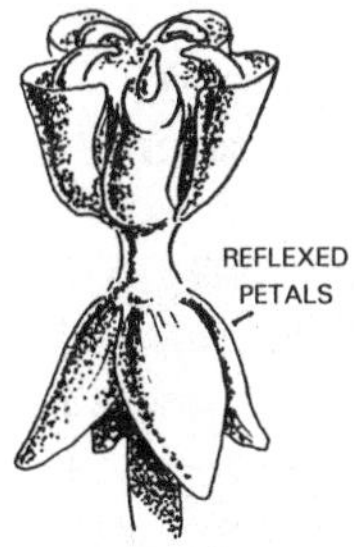

Figure 866

Regma (pl. **regmata**). A dry fruit of three or more carpels which separate at maturity; a type of schizocarp. Figure 867.
Regular. Radially symmetrical; said of a flower in which all parts are similar in size and arrangement on the receptacle. Figure 868. (compare **irregular**, and see **actinomorphic**)

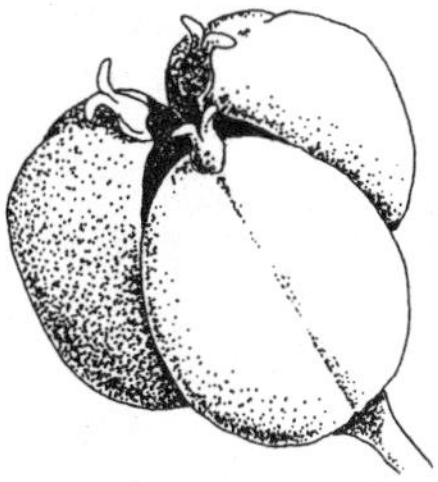
Figure 867

Figure 868

Regularly. Evenly or uniformly.
Relict. A plant which has survived from a past geologic epoch.
Remote. Distantly spaced.
Reniform. Kidney-shaped. Figure 869.
Repand. With a slightly wavy or weakly sinuate margin; undulate. Figure 870.

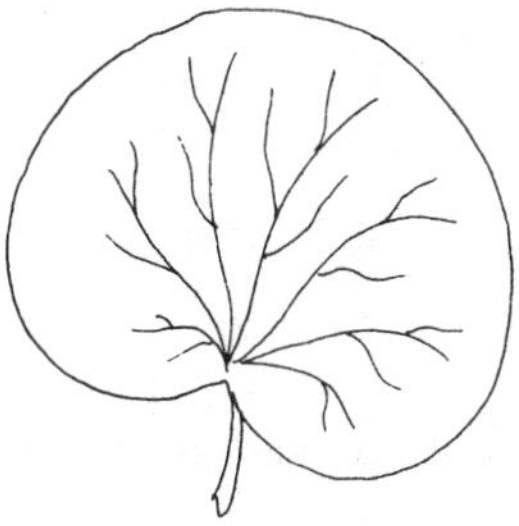
Figure 869

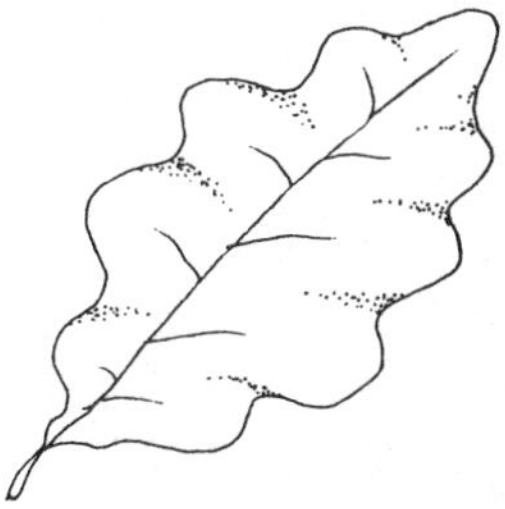
Figure 870

Repent. Prostrate; creeping. Figure 871.
Replicate. Folded backward. Figure 862.
Replum. Partition or septum between the two valves or compartments of silicles or siliques in the Cruciferae (Brassicaceae). Figure 872.

Figure 871

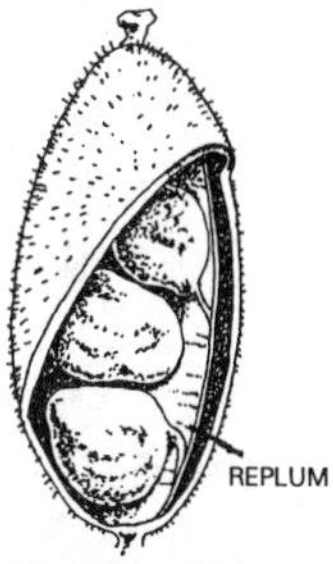

Figure 872

Reptant. See **repent.**
Resiniferous. See **resinous.**
Resinous. Bearing resin and often, therefore, sticky.
Resupinate. Upside down due to twisting of the pedicel, as the flowers of some orchids. Figure 873.

Figure 873

Reticulate. In the form of a network; net-veined. Figure 874.
Reticulum (pl. **reticula**). A network of veins or fibers. Figure 874.
Retinaculum. The gland attached to the pollinia of the Orchidaceae. Figure 875.

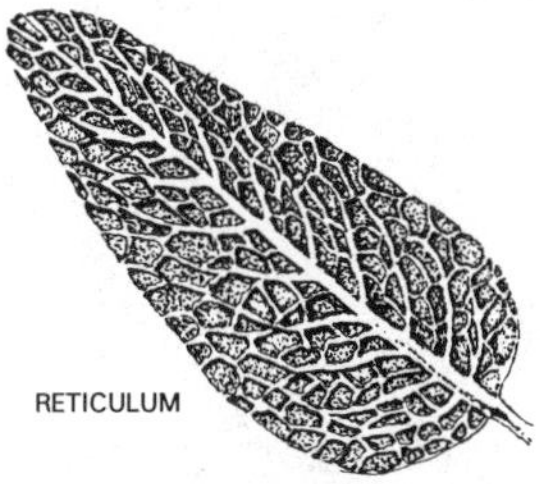

Figure 874

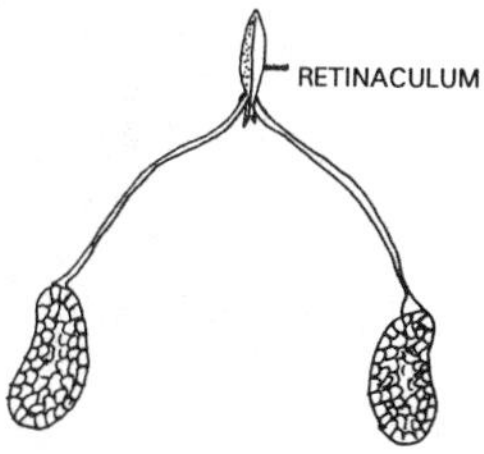

Figure 875

Retrocurved. See **recurved.**
Retroflexed. See **reflexed.**
Retrorse (Adv. **retrorsely**). Directed downward or backward. Figure 876. (compare **antrorse**)
Retuse. With a shallow notch in a round or blunt apex. Figure 877.
Revolute. With the margins rolled backward

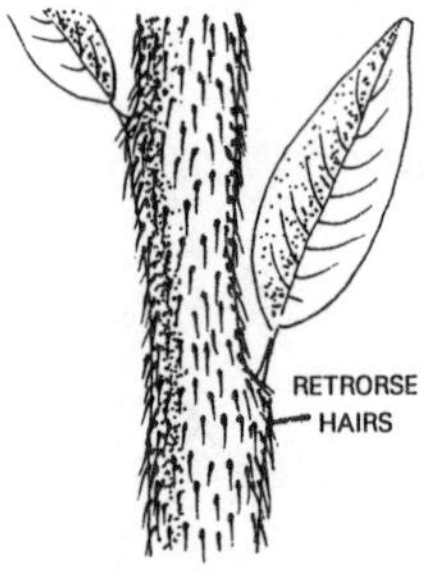

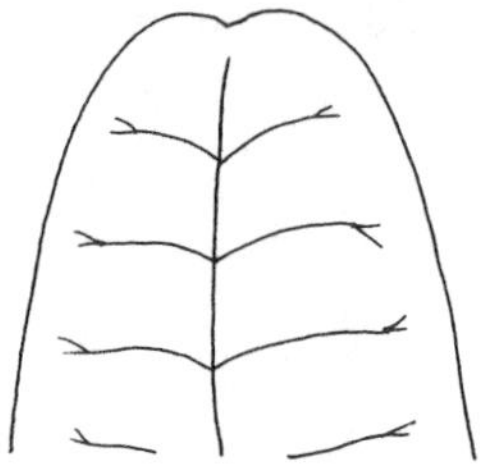

Figure 876 Figure 877

toward the underside. Figure 878. (compare **involute**)

Rhabdocarpous. With long rod-shaped fruits. Figure 879.

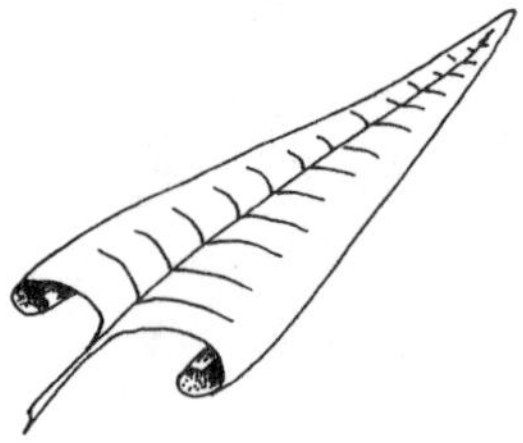

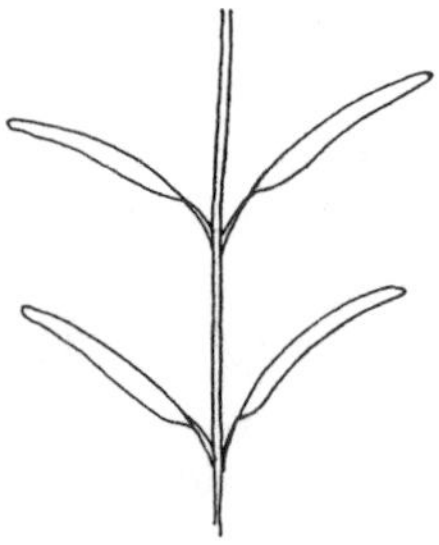

Figure 878 Figure 879

Rhachilla. See **rachilla**.

Rhachis. See **rachis**.

Rhaphe. See **raphe**.

Rhipidium. A flattened, fan-shaped cyme. Figure 880.

Rhizanthous. With the flowers arising so close to the ground that they appear to be arising from the root. Figure 881.

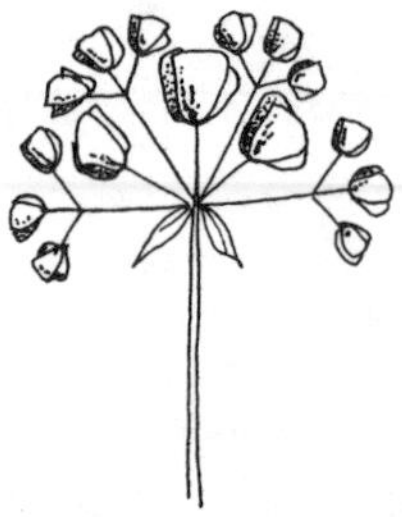

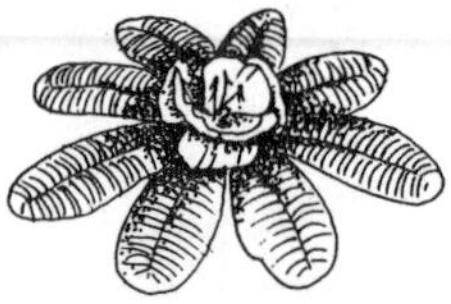

Figure 880 Figure 881

Rhizocarpic. With the roots living for several to many years and the stems dying each year.

Rhizocarpous. See **rhizocarpic**.

Rhizogenic. Root producing.

Rhizoid. A rootlike structure lacking conductive tissues (xylem and phloem).

Rhizomatous. With rhizomes. Figure 882.

Rhizome. A horizontal underground stem; rootstock. Figure 882.

Rhizomorphous. Root-like in appearance.

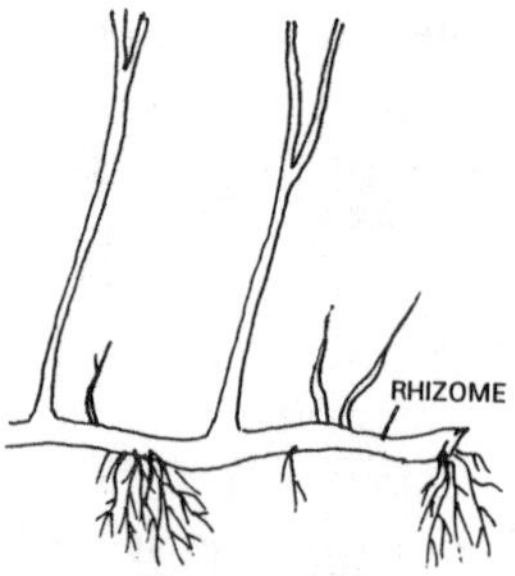

Figure 882

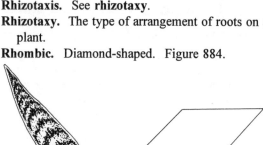

Rhizophyllous. With roots arising from the leaves. Figure 883.

Rhizotaxis. See **rhizotaxy**.

Rhizotaxy. The type of arrangement of roots on a plant.

Rhombic. Diamond-shaped. Figure 884.

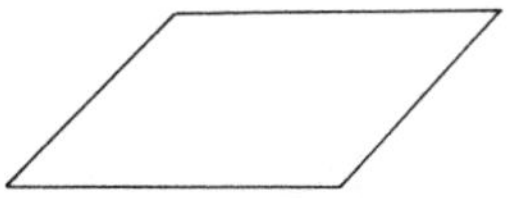

Figure 883 Figure 884

Rhomboid. See **rhomboidal**.

Rhomboidal. Quadrangular, nearly rhombic, with obtuse lateral angles. Figure 885.

Rib. A main longitudinal vein in a structure, as in a leaf. Figure 886.

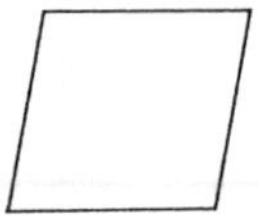

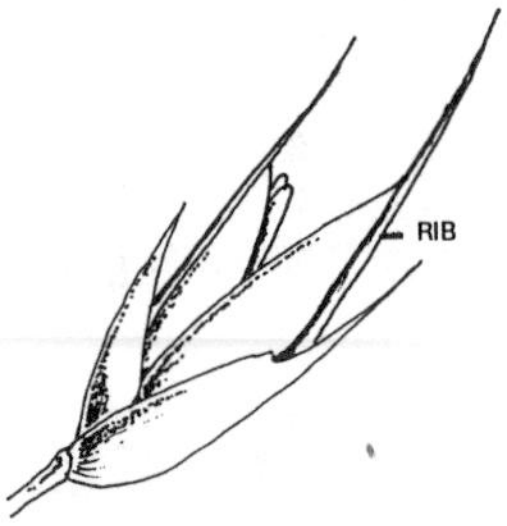

Figure 885 Figure 886

Ribbed. With prominent ribs or veins. Figures 886 and 887.

Rictus. The mouth of a bilabiate corolla. Figure 888.

Rigescent. Becoming rigid.

Rigid. Stiff and inflexible.

Rimose. With fissures or cracks, as in the bark of

some trees. Figure 889.

Rimous. See **rimose**.

Rind. A thick outer covering, as in the tough outer layer of a pepo. Figure 890.

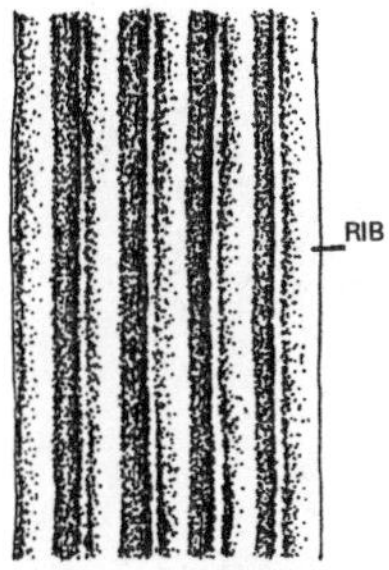

Figure 887

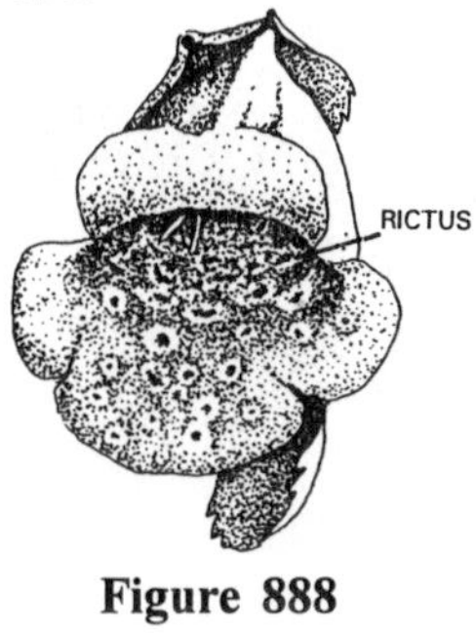

Figure 888

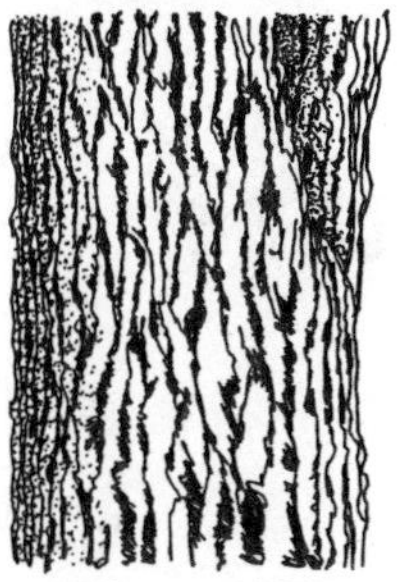

Figure 889

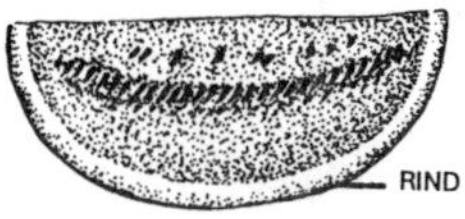

Figure 890

Ringent. Gaping; with widely spreading lips, as in some corollas. Figure 891.

Riparian. Growing along the banks of streams, springs, or seeps.

Riparious. See **riparian**.

Ripe. Fully developed and mature.

Rivulose. With meandering channels or marked with sinuous lines resembling a rivulet. Figure 892.

Root. That portion of the plant axis lacking nodes and leaves and usually found below ground. Figure 893.

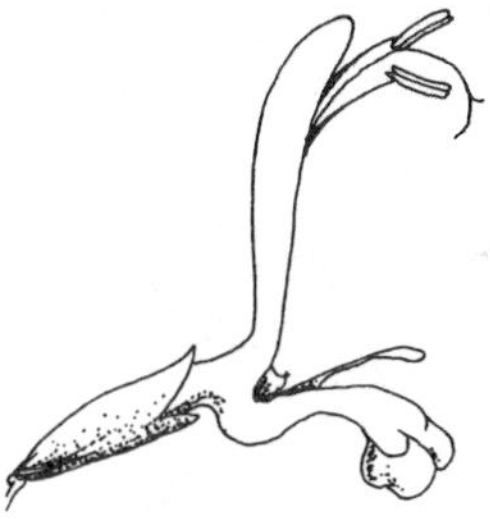

Figure 891

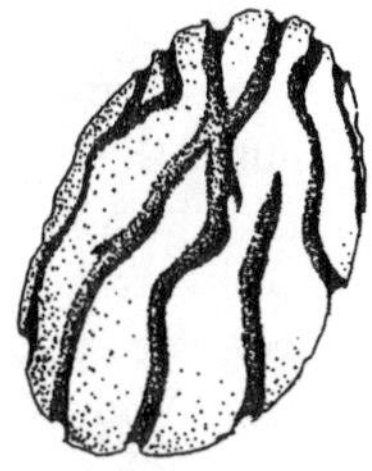

Figure 892

Rootlet. A small root. Figure 893.

Rootstock. See **rhizome**.

Roridulate. With a covering of waxy platelets, appearing moist.

Roseate. Tinged with red; rosy.

Rosette. A dense radiating cluster of leaves (or other organs), usually at or near ground level. Figure 894.

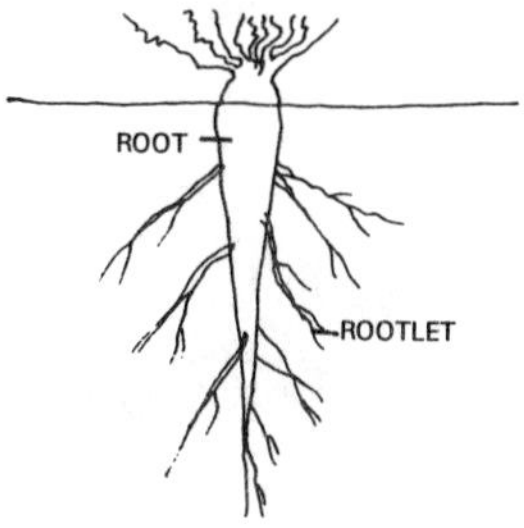

Figure 893

Rostellate. With a tiny, short, stout, terminal beak. Figure 895.

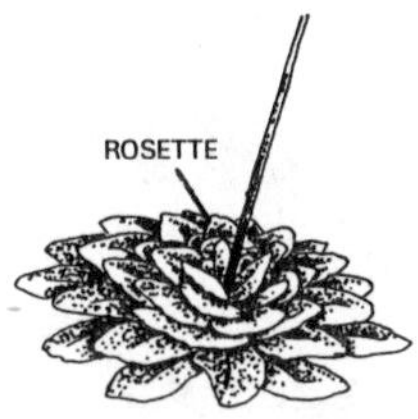

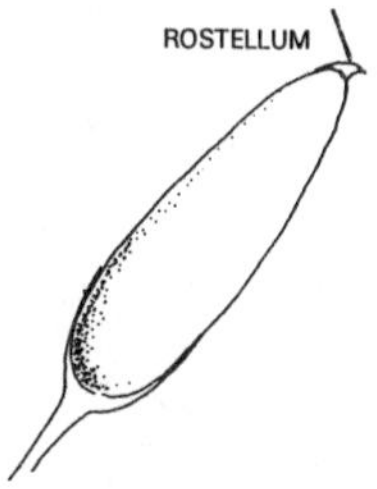

Figure 894 Figure 895

Rostellum. A small beak. Figure 895; an extension from the upper edge of the stigma in orchids. Figure 896.

Rostrate. With a short, stout, terminal beak. Figure 897.

Rostrum. A beaklike structure. Figure 897.

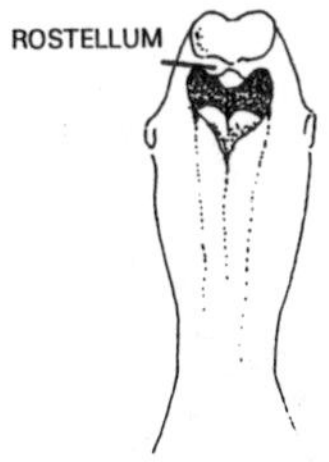

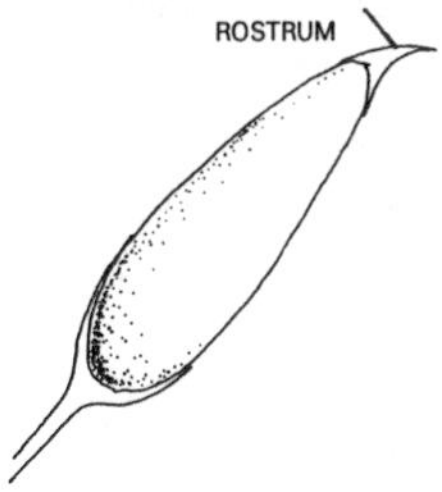

Figure 896 Figure 897

Rosulate. With the leaves arranged in basal rosettes, the stem very short or lacking. Figure 898.

Rotate. Disc-shaped; flat and circular, as a sympetalous corolla with widely spreading lobes and little or no tube. Figure 899.

Rotund. Round or rounded in outline. Figure 900.

Rotundifolious. With round leaves. Figure 901.

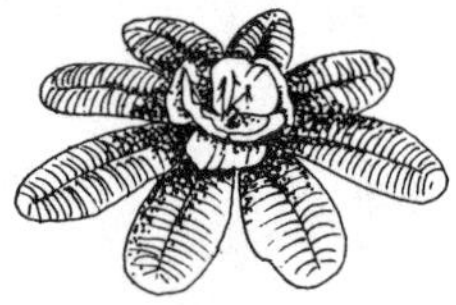

Figure 898

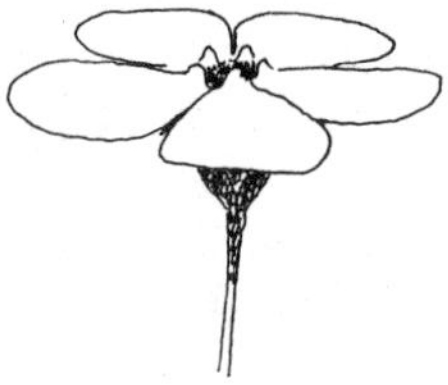

Figure 899

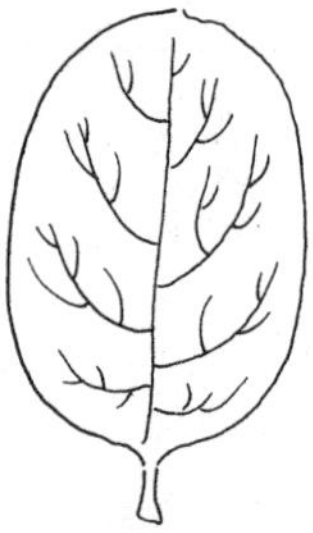

Figure 900

Figure 901

Rubescent. Becoming red or reddish.
Rubiginose. See **rubiginous**.
Rubiginous. Rust-colored.
Ruderal. Growing in disturbed habitats; weedy.
Rudimentary. Imperfectly developed; vestigial. Figure 902.
Rufescent. See **rubescent**.
Rufous. Reddish-brown.
Rufus. See **rufous**.
Ruga (pl. **rugae**). A fold or wrinkle. Figure 903.
Rugate. See **rugose**.
Rugose. Wrinkled. Figure 903.

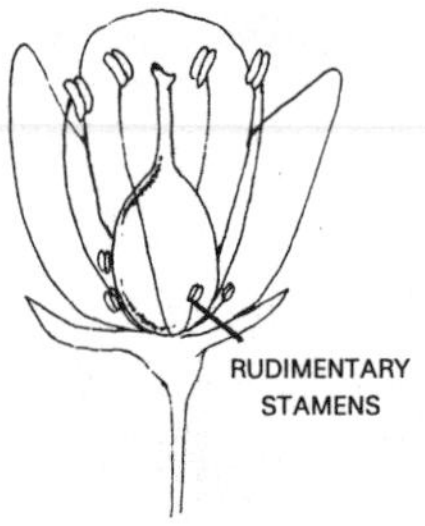

Figure 902

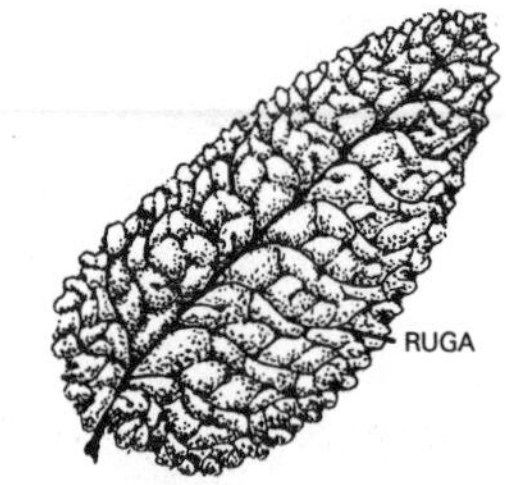

Figure 903

Rugulose. Slightly wrinkled. Figure 904.
Ruminate. Roughly wrinkled, as if chewed. Figure 905.
Runcinate. Sharply pinnatifid or cleft, the segments directed downward. Figure 906.
Runner. A slender stolon or prostrate stem rooting at the nodes or at the tip. Figure 907.

Figure 904

Figure 905

Figure 906

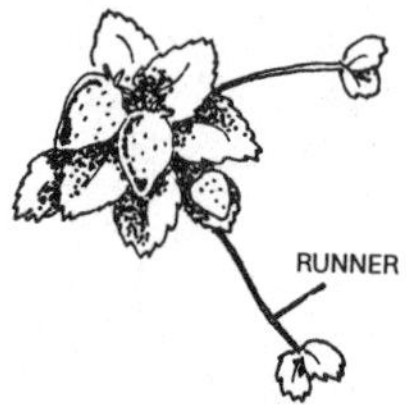

Figure 907

Ruptile. Dehiscing irregularly. Figure 908.
Rushlike. Grasslike in appearance, with inconspicuous flowers. Figure 909.

Figure 908

Figure 909

Sac. A bag-shaped compartment, as the cavity of an anther. Figure 910.
Saccate. With a sac, or in the shape of a sac; bag-shaped. Figure 911.
Sacciform. See **saccate**.
Sacculate. With a saccule, or in the shape of a saccule. Figure 912.
Saccule. A very small sac or cavity. Figure 912.
Sacculus (pl. **sacculi**). See **saccule**.
Sagittate. Arrowhead-shaped, with the basal lobes directed downward. Figure 913. (compare

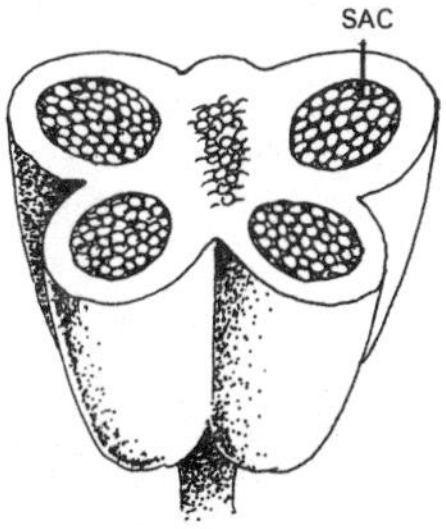

Figure 910

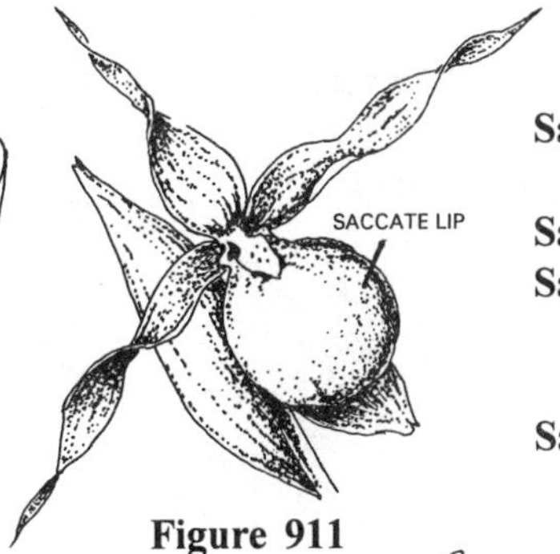

Figure 911

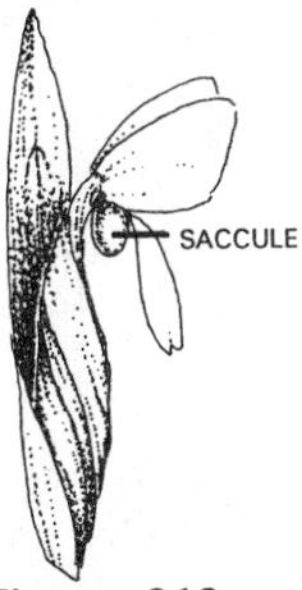

Figure 912

Figure 913

hastate)

Sagittiform. See **sagittate**.

Salient. Projecting outward. Figure 914.

Salverform. With a slender tube and an abruptly spreading, flattened limb. Figure 915.

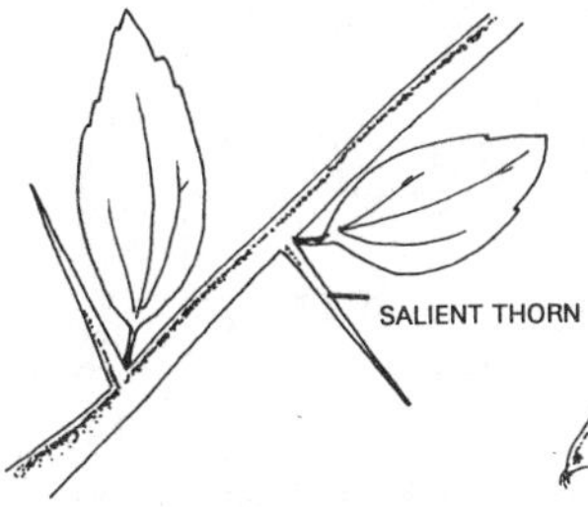

Figure 914

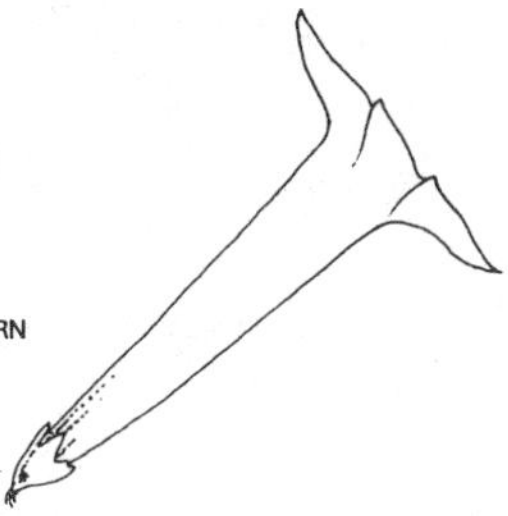

Figure 915

Samara. A dry, indehiscent, winged fruit. Figure 916.

Samaroid. Samara-like.

Sanguine. Blood red.

Sanguineous. Blood red.

Sap. The juice of a plant; the fluids circulated throughout a plant.

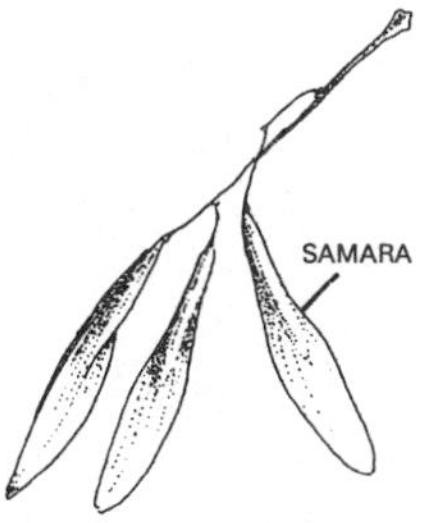

Figure 916

Sapid. With an agreeable taste.

Saponaceous. Soapy, as in a substance or object slippery to the touch.

Sapor. The flavor or taste of a plant or plant substance.

Saprobe (Adj. **saprobic**). See **saprophyte**.

Saprophyte (Adj. **saprophytic**). A plant living on dead organic matter, lacking chlorophyll. (compare **parasite**)

Sapwood. The outer, newer, usually somewhat lighter, wood of a woody stem; the wood that is actively transporting water; alburnum. Figure 917.

Sarcocarp. The fleshy portion (mesocarp) of a fleshy fruit. Figure 918.

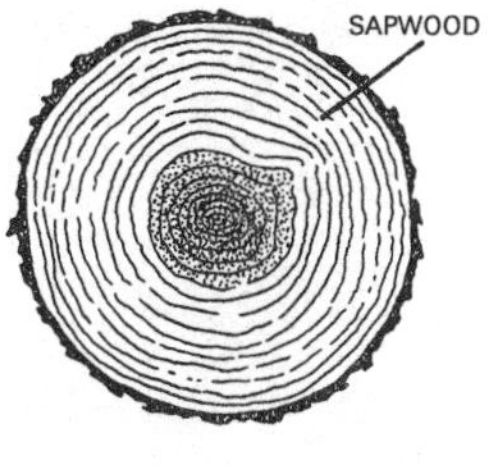

Figure 917

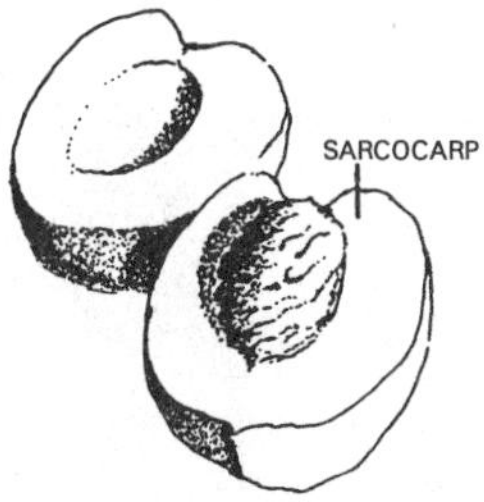

Figure 918

Sarcocaulous. With fleshy stems. Figure 919.

Sarcous. Fleshy. Figure 920.

Figure 919

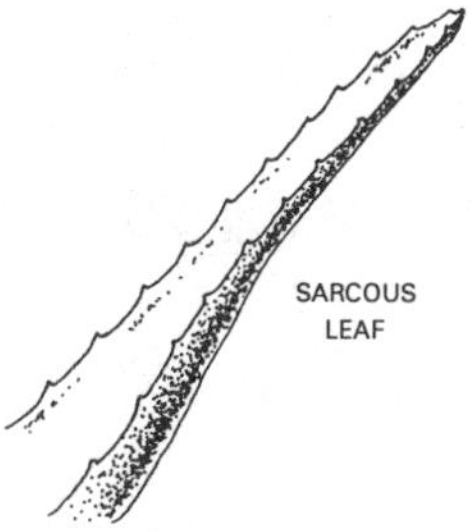

Figure 920

Sarment. A long, slender runner. Figure 921.

Sarmentose. With long, slender runners. Figure 921.

Scaberulent. See **scaberulose**.

Scaberulose. Slightly rough to the touch, due to the structure of the epidermal cells, or to the presence of short stiff hairs. Figure 922.

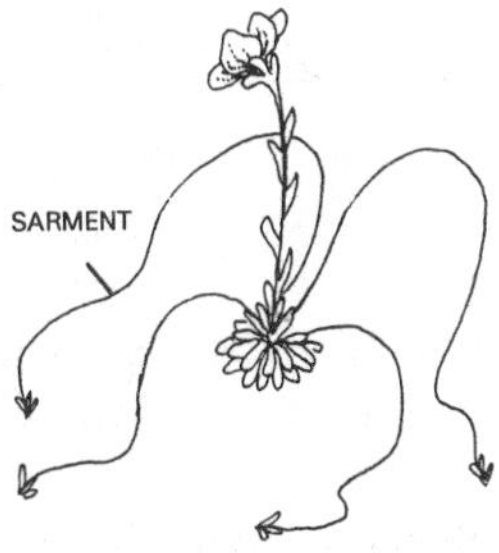

Figure 921

Scaberulous. See **scaberulose.**

Scabrellate. See **scaberulose.**

Scabrid. Roughened.

Scabridulous. Minutely roughened.

Scabrous. Rough to the touch, due to the structure of the epidermal cells, or to the presence of short stiff hairs. Figure 923.

Figure 922

Scalariform. Ladder-like. Figure 924.

Figure 923

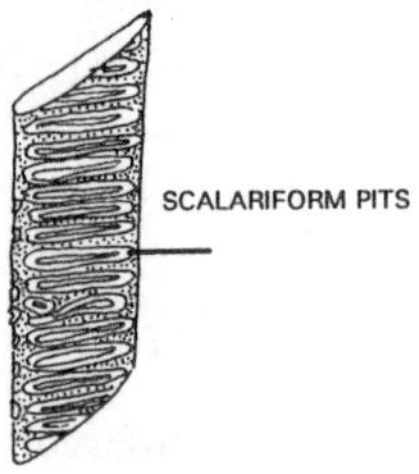

Figure 924

Scale. Any thin, flat, scarious structure. Figures 925 and 926.

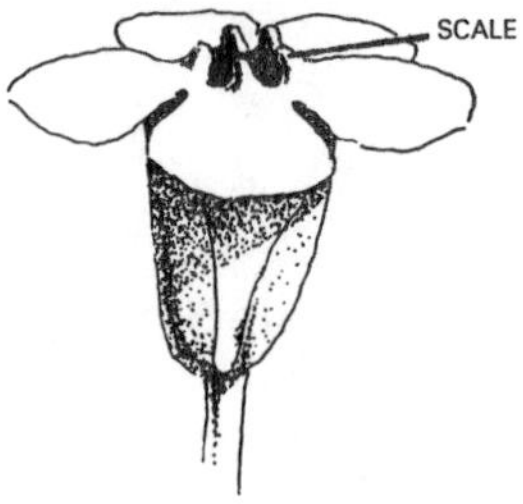

Figure 925

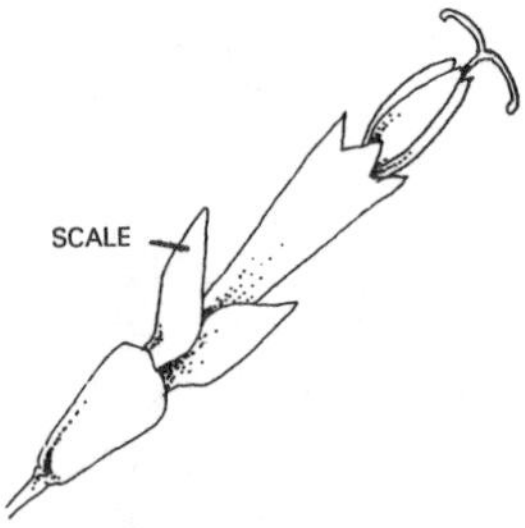

Figure 926

Scandent. Climbing. Figure 927.

Scape. A leafless peduncle arising from ground level (usually from a basal rosette) in acaulescent plants. Figure 928.

Scaphoid. Boat-shaped. Figure 929.

Scapiflorous. See **scapose.**

Scapiform. Scapelike but not entirely leafless. Figure 930.

Scapose. With flowers borne on a scape; scapelike. Figure 928.

Scar. The mark left on a seed after detachment from the placenta. Figure 931; the mark left on a stem after leaf abscission. Figure 932.

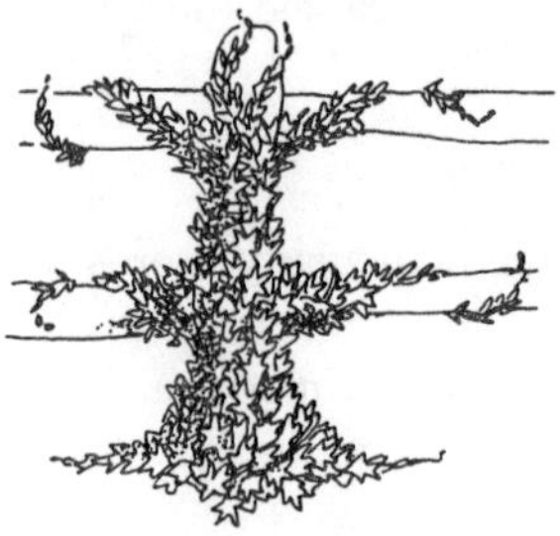

Figure 927

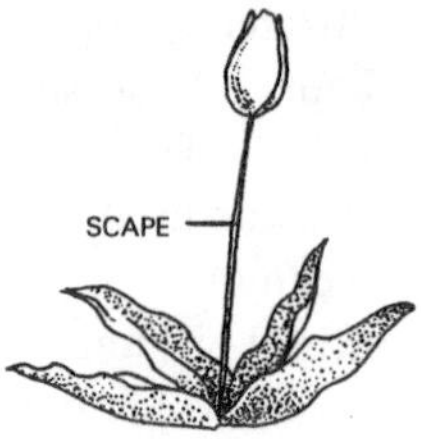

Figure 928

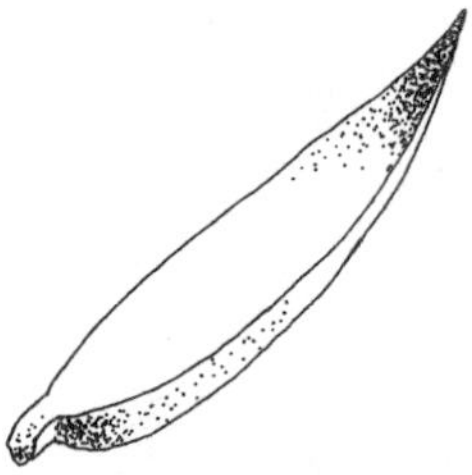

Figure 929

Figure 930

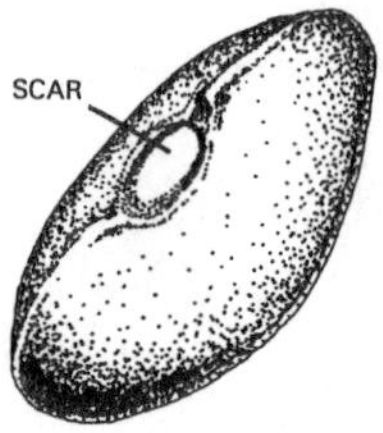

Figure 931

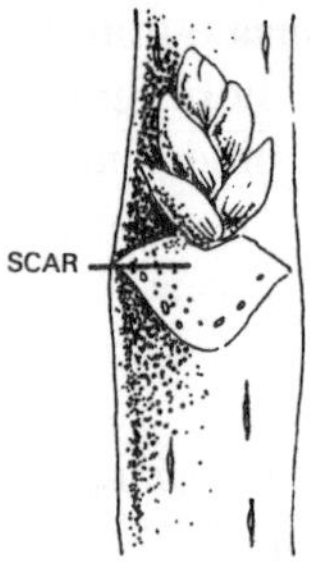

Figure 932

Scarious. Thin, dry, and membranous in texture, not green. Figure 933.

Scattered. Irregularly, and usually sparsely, arranged. Figure 934.

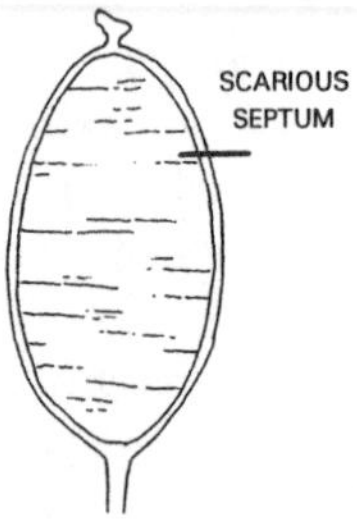

Figure 933

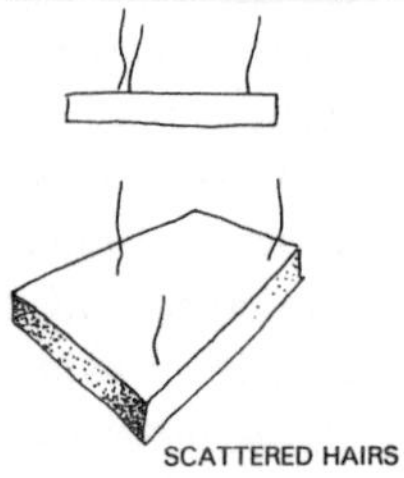

Figure 934

Schizocarp. A dry, indehiscent fruit which splits into separate one-seeded segments (carpels) at maturity. Figure 935.

Schizogenous. Formed by the splitting or separation of tissue.

Schizopetalous. With cut petals. Figure 936.

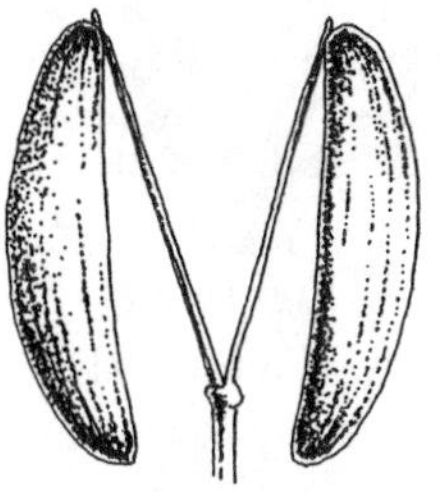

Figure 935

Figure 936

Scissile. Splitting easily.

Sciuroid. Shaped like the tail of a squirrel, as in some grass inflorescences. Figure 937.

Scleranthium. An achene enclosed within a hardened calyx tube. Figure 938.

Figure 937

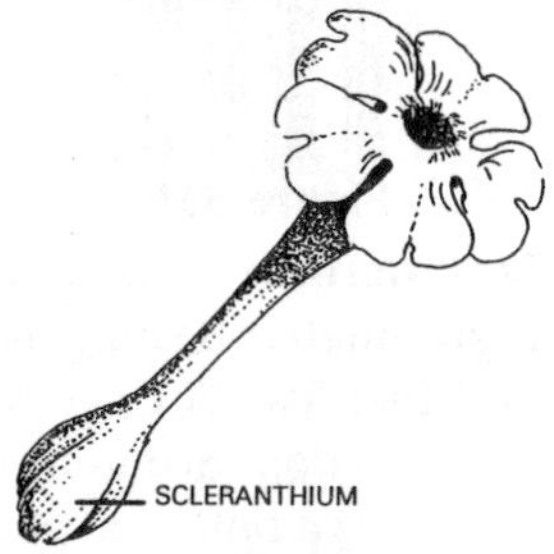

Figure 938

Scleroid. See **sclerotic**.

Sclerophyll. A stiff, firm leaf which retains its stiffness even when wilted.

Sclerophyllous. With stiff, firm leaves; with sclerophylls.

Sclerosis. A hardening or thickening of tissue due to lignification.

Sclerotic. Hardened or thickened.

Sclerous. See **sclerotic**.

Scobiform. Sawdust-like in appearance.

Scobina. The zigzag rachilla of some grass spikelets. Figure 939.

Scobinate. With a roughened surface, as though rasped. Figure 940.

Figure 939

Scorpioid. Shaped like a scorpion's tail, as in some coiled cymes. Figure 941; a determinate inflorescence with a zigzag rachis. Figure 942.

Scrobiculate. Pitted or furrowed. Figure 943.

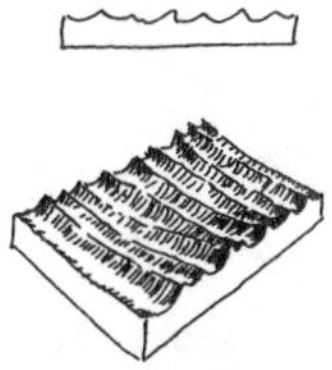

Figure 940

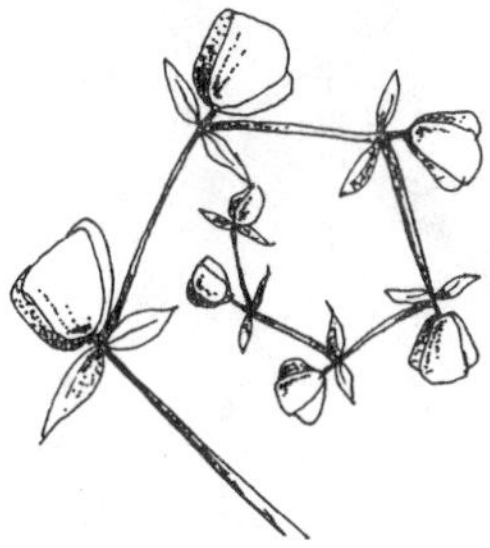

Figure 941

Figure 942

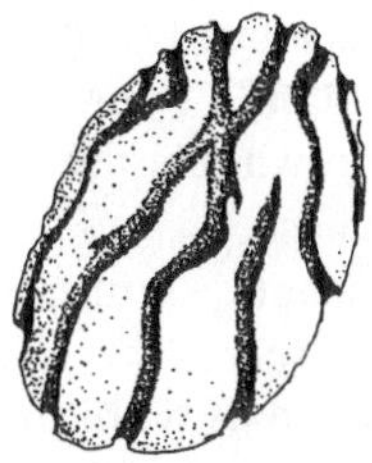

Figure 943

Scrotiform. Scrotum-like in appearance. Figure 944.

Scurf. Small branlike scales. Figure 945.

Scurfy. Covered with small, branlike scales. Figure 945.

Figure 944

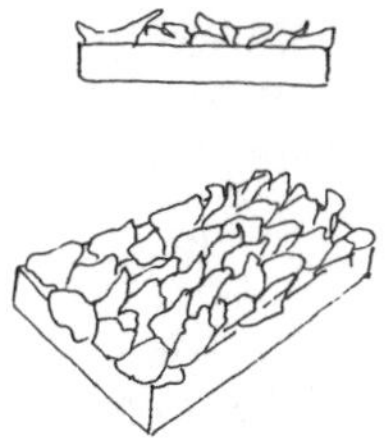

Figure 945

Scutate. Shaped like a small shield. Figure 946.

Scutellate. With scutella; saucer-shaped or shield-shaped. Figure 946.

Scutelliform. Saucer-shaped or shield-shaped. Figure 946.

Scutellum (pl. **scutella**). A small platelike or shieldlike structure, as in some monocot seeds. Figures 947 and 948.

Scutum. An expanded style tip, as in *Asclepias*. Figure 949.

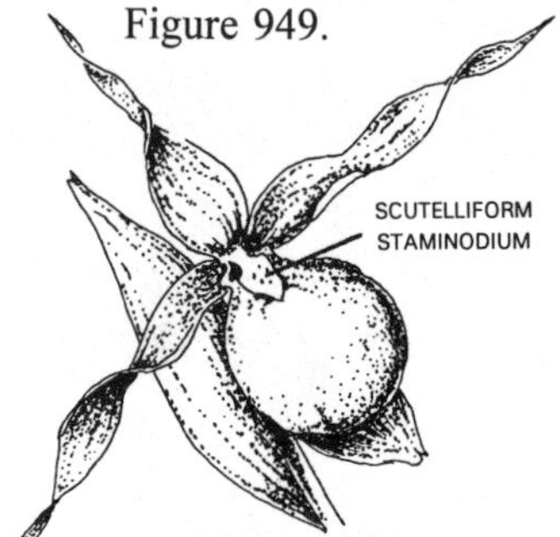

Figure 946

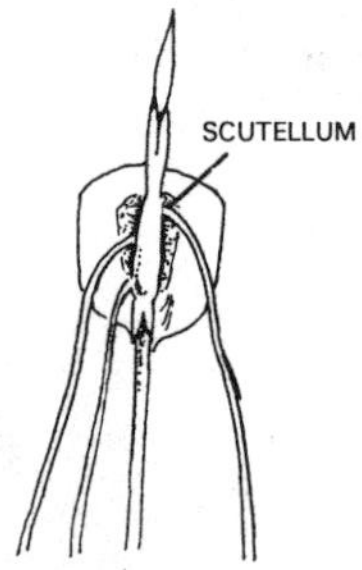

Figure 947

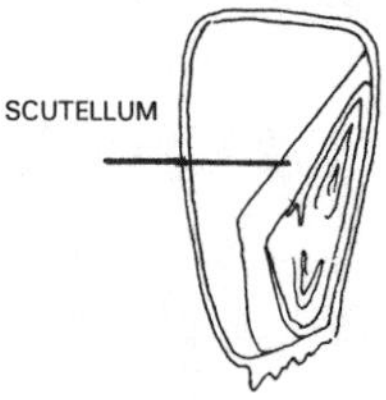

Figure 948

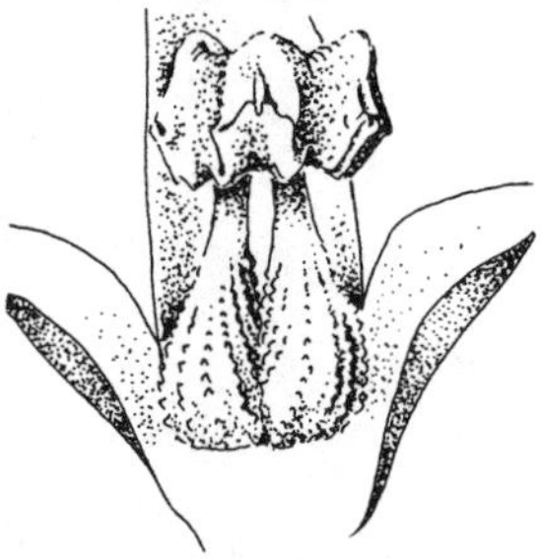

Figure 949

Sebaceous. Tallowy or fatty.

Secondary leaflet. The leaflet below the terminal leaflet. Figure 950.

Secondary peduncle. An inflorescence branch. Figure 951.

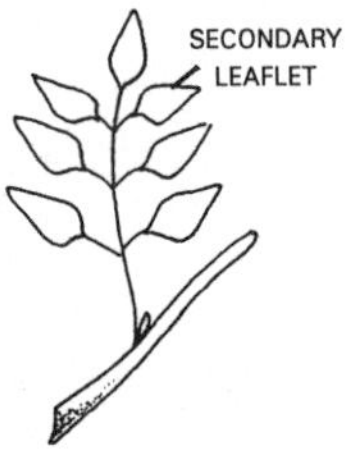

Figure 950

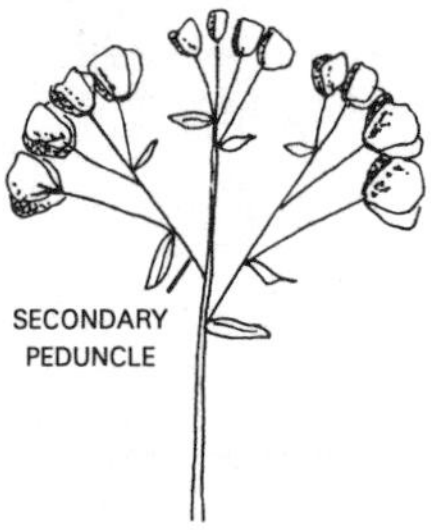

Figure 951

Seculate. Sickle-shaped. Figure 952.

Secund. Arranged on one side of the axis only. Figure 953.

Secundine. The inner integument of the ovule. Figure 954.

Seed. A ripened ovule. Figure 955.

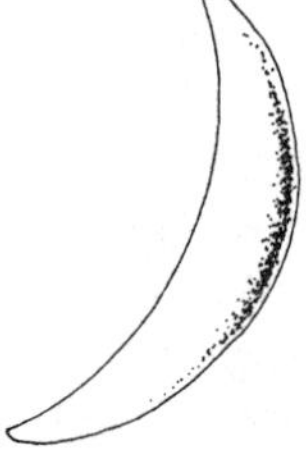

Figure 952

Seed coat. The outer covering of the seed, from the integuments of the ovule. Figure 955.

Seed leaf. A cotyledon. Figure 956.

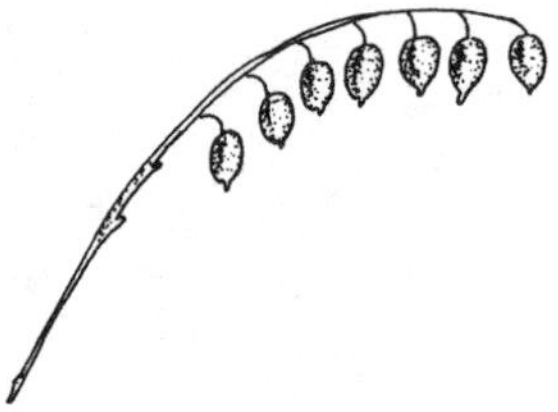

Figure 953

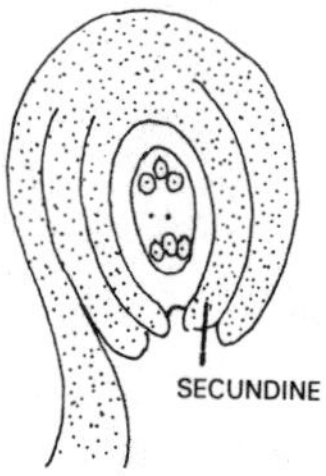

Figure 954

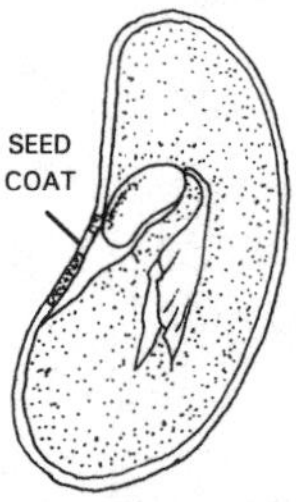

Figure 955

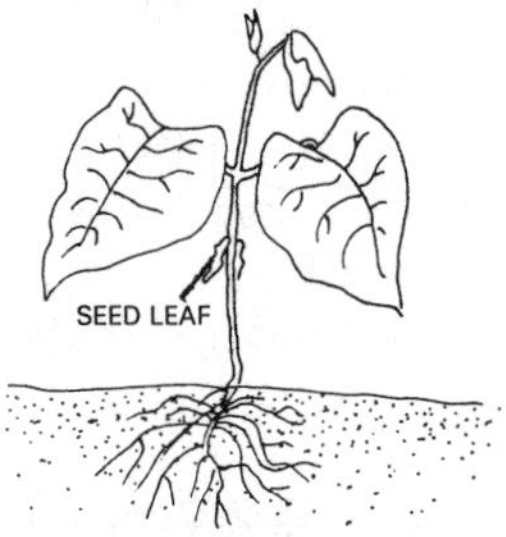

Figure 956

Seed stalk. The funiculus. Figure 957.

Segment. A section or division of an organ. Figure 958.

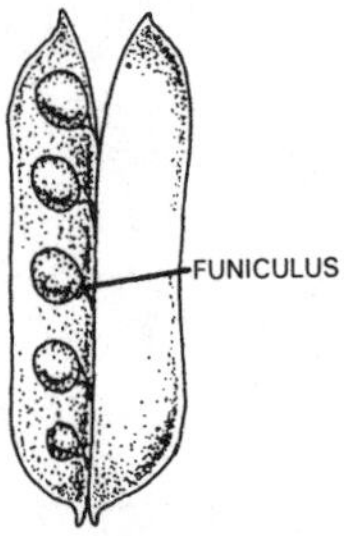

Figure 957

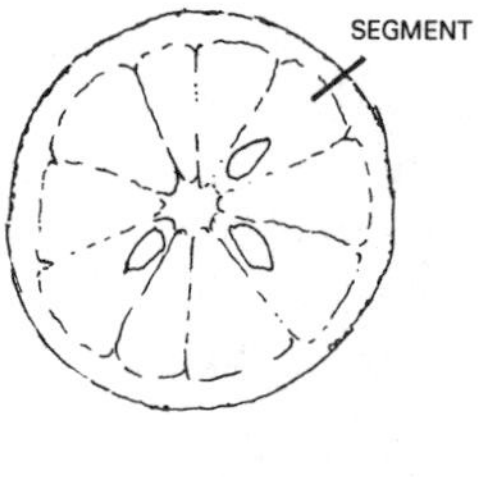

Figure 958

Sejugous. With six pairs of leaflets. Figure 959.

Seleniferous. Bearing selenium.

Selenophyte. A plant that grows on seleniferous soils and takes up selenium from these soils.

Figure 959

Self-pollination. Transfer of pollen from the

anthers to the stigma of the same flower or to the stigma of another flower on the same plant.

Semen. A seed. Figure 960.

Semi- (prefix). Half; partly or almost.

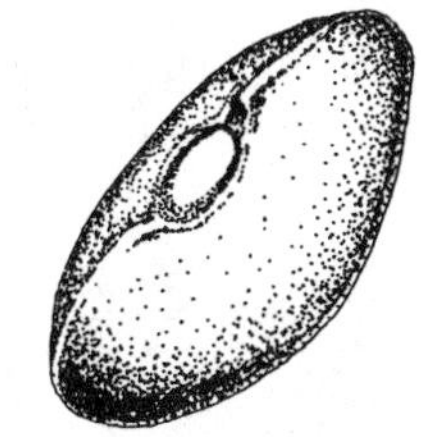

Figure 960

Semicarpous. With ovaries of carpels partly fused, the styles and stigmas separate. Figure 961.

Seminiferous. Seed-bearing.

Semitropical. See **subtropical**.

Semperflorous. Flowering throughout the year.

Sensitive. Responsive to touch.

Sepal. A segment of the calyx. Figure 962.

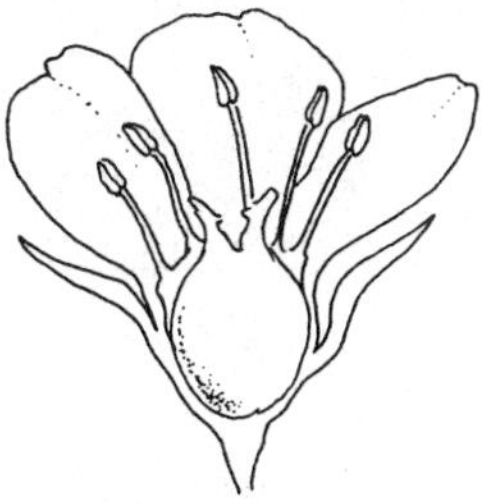

Figure 961

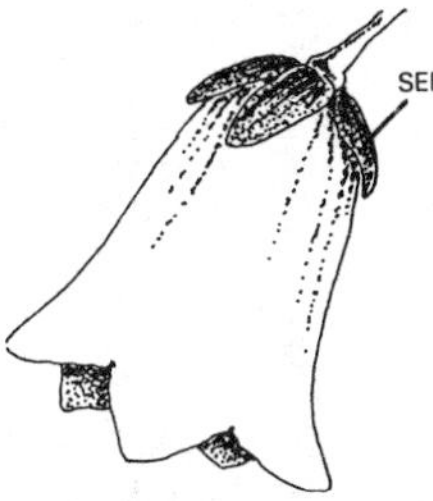

Figure 962

Sepaloid. Sepallike in color and texture.

Septate. Divided by one or more partitions. Figure 963.

Septentate. With parts in sevens. Figure 964.

Figure 963

Figure 964

Septicidal. Dehiscing through the septa and between the locules. Figure 965. (compare **loculicidal** and **poricidal)**

Septifolious. With seven leaves or seven leaflets. Figure 964.

Septifragal. Separation of the valves from the septae at dehiscence. Figure 966.

Septum (pl. **septa**). A partition, as the partitions separating the locules of an ovary. Figure 967.

Seriate. Arranged in rows or series. Figure 968.

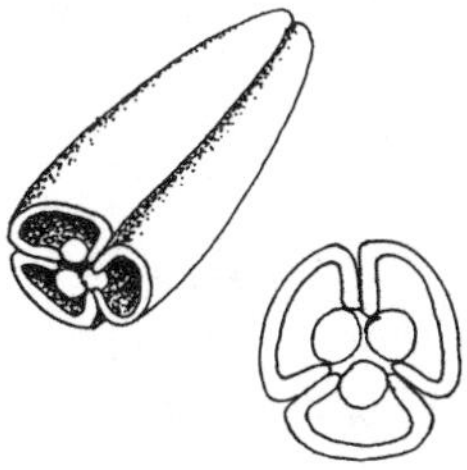

Figure 965

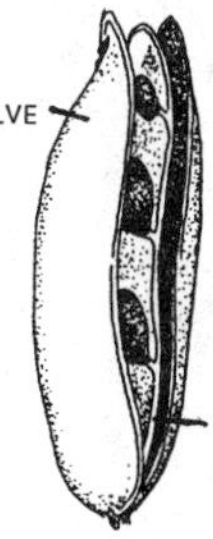

Figure 966

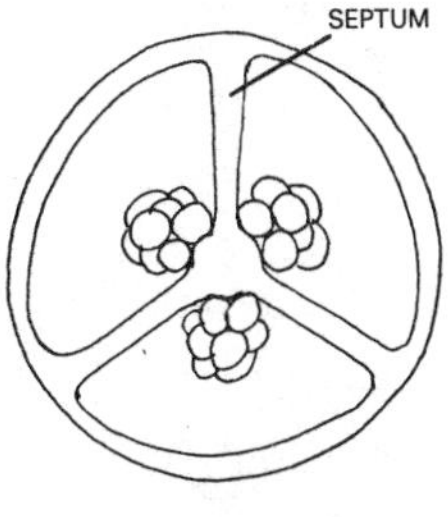

Figure 967

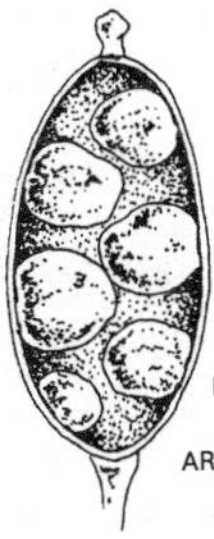

Figure 968

Sericeous. Silky, with long, soft, slender, somewhat appressed hairs. Figure 969.

Serotinal. See **serotinous**.

Serotinous. Late; late in flowering or leafing; with flowers developing after the leaves are fully developed.

Serra. A tooth of a serrate leaf. Figure 970.

Serrate. Saw-like; toothed along the margin, the sharp teeth pointing forward. Figure 970.

Serration. A serrated margin. Figure 970; one of the teeth along a serrated margin; a serrated condition.

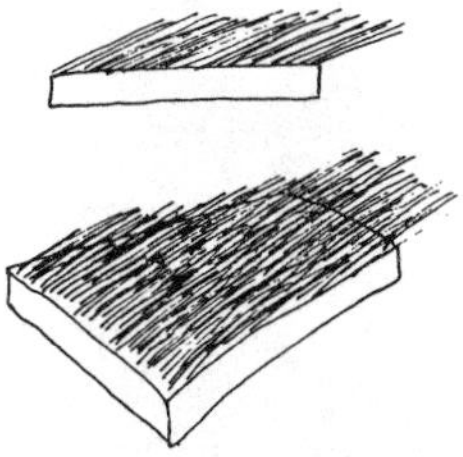

Figure 969

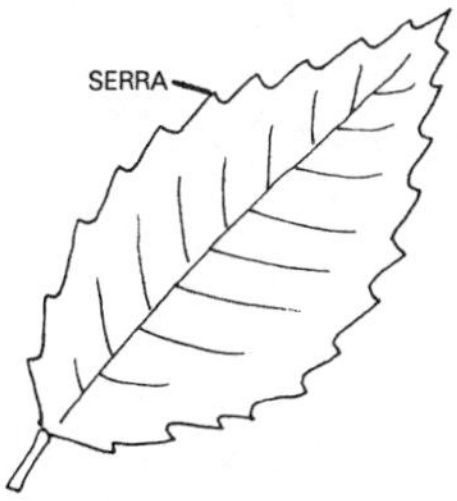

Figure 970

Serriform. See **serrate**.

Serrulate. Toothed along the margin with minute, sharp, forward-pointing teeth. Figure 971.

Serrulation. A serrulate margin. Figure 971; one of the teeth along a serrulate margin; a serrulate condition.

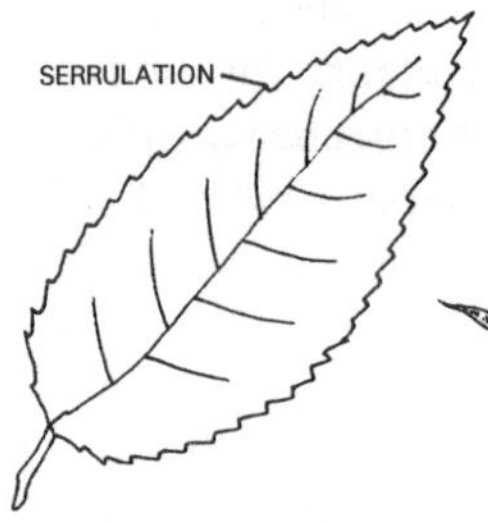

Figure 971

Sessile. Attached directly, without a supporting stalk, as a leaf without a petiole. Figure 972.

Seta (pl. **setae**). A bristle. Figure 973.

Setaceous. Bristle-like; with bristles. Figure 973.

Setiferous. Bristle-bearing. Figure 973.

Setiform. See **setaceous**.

Setose. Covered with bristles. Figure 973.

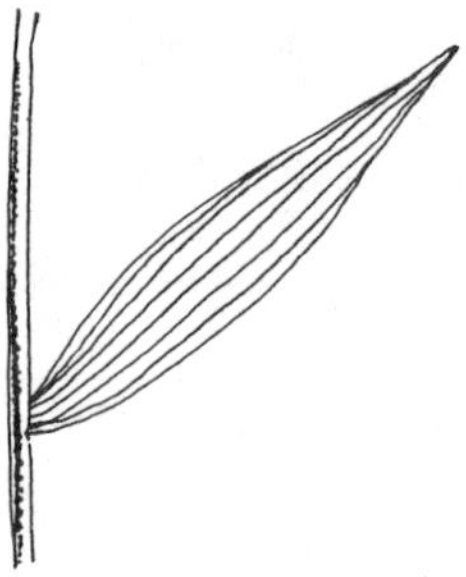

Figure 972

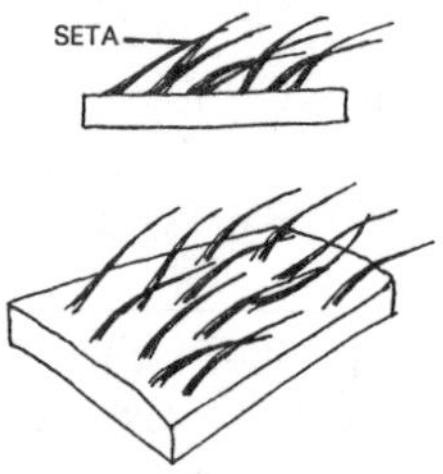

Figure 973

Setulose. Covered with minute bristles. Figure 974.

Sheath. The portion of an organ which surrounds, at least partly, another organ, as the leaf base of a grass surrounds the stem. Figure 975.

Sheathing. Forming a sheath, as the leaf base of a grass forms a sheath as it surrounds the stem. Figure 975.

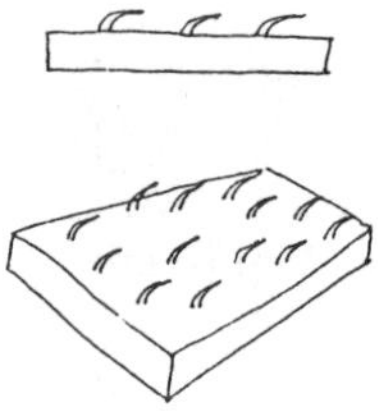

Figure 974

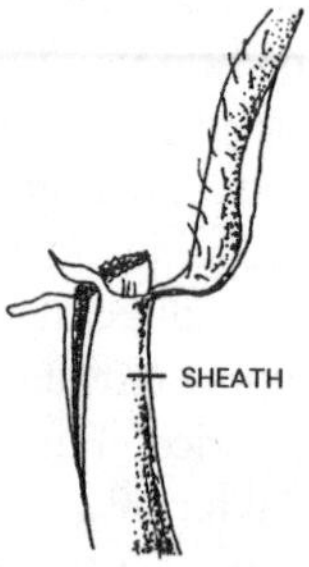

Figure 975

Shield. The staminode of *Cypripedium*. Figure 976; the outermost portion of a cone scale of a conifer cone. Figure 977.

Shoot. A young stem or branch. Figure 978.

Shrub. A woody plant, with several stems, that is shorter than a typical tree. Figure 979.

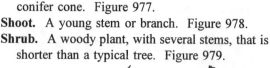

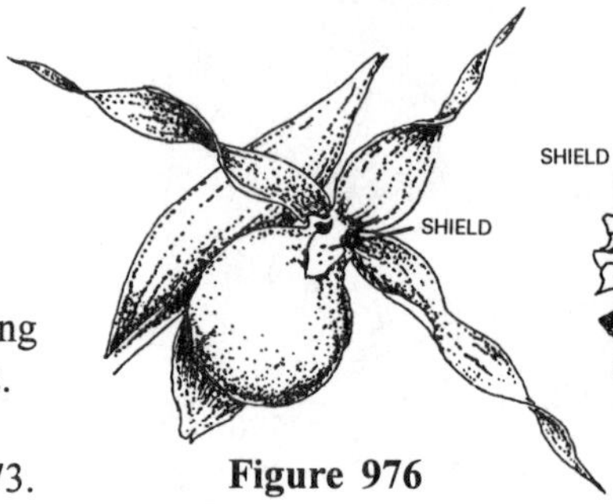

Figure 976

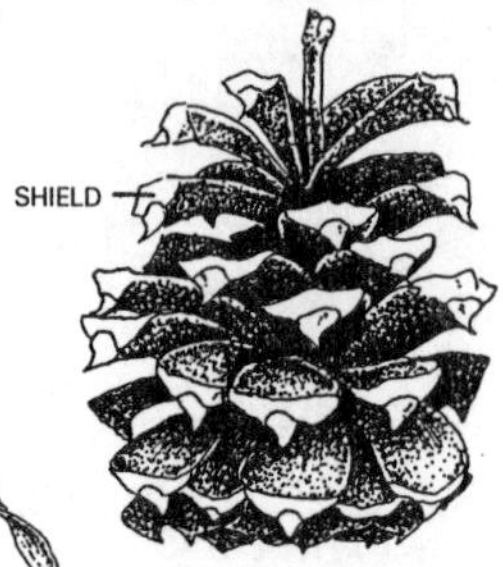

Figure 977

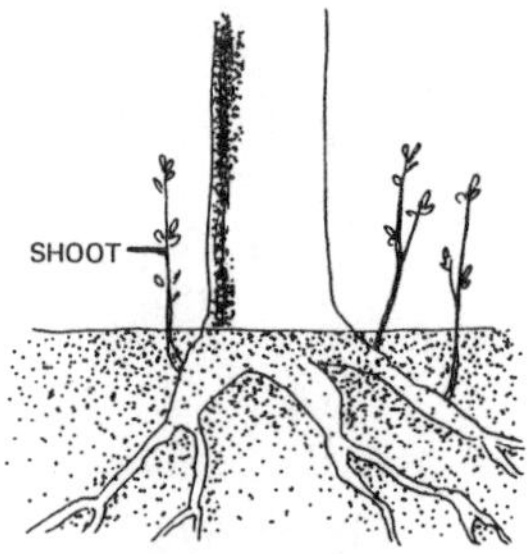

Figure 978

Figure 979

Sigmoid. S-shaped; doubly curved, like the letter S. Figure 980.

Siliceous. Relating to, or containing silica.

Silicious. See **siliceous**.

Silicle. A dry, dehiscent fruit of the Cruciferae (Brassicaceae), typically less than twice as long as wide, with two valves separating from the persistent placentae and septum (replum). Figures 981 and 982.

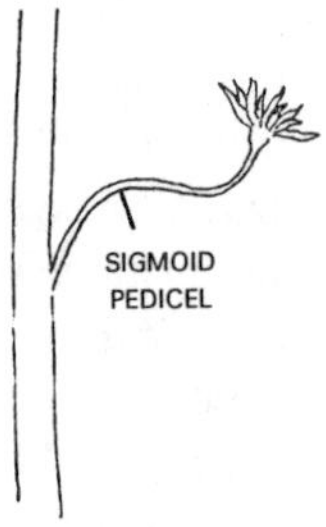

Figure 980

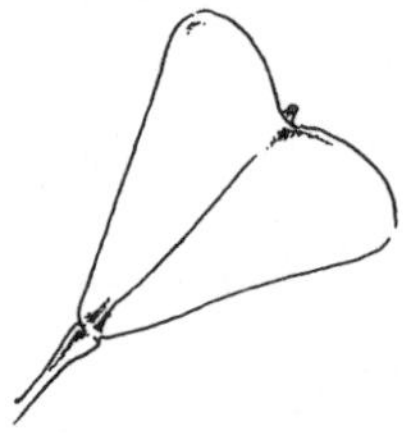

Figure 981

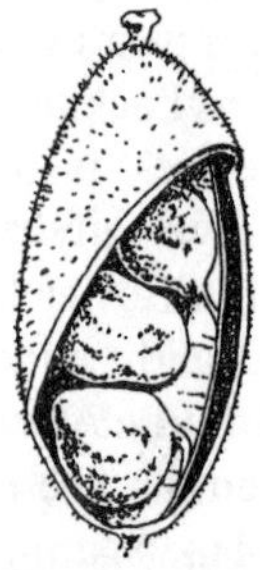

Figure 982

Silique. A dry, dehiscent fruit of the Cruciferae (Brassicaceae), typically more than twice as long as wide, with two valves separating from the persistent placentae and septum (replum). Figure 983.

Silk. The hairlike styles in maize. Figure 984.

Figure 983

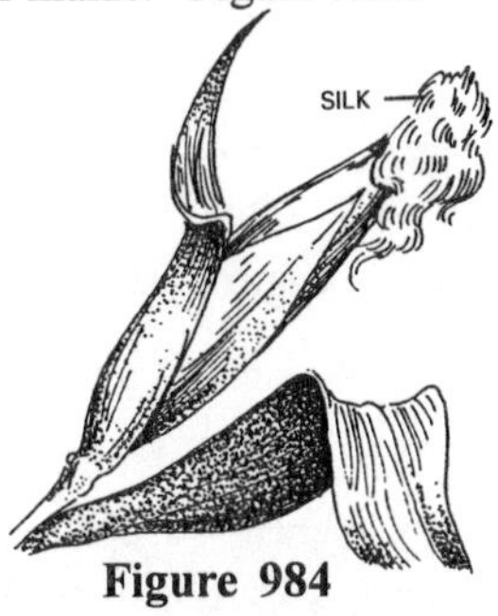

Figure 984

Silky. Silklike in appearance or texture; sericeous.

Simple. Undivided, as a leaf blade which is not separated into leaflets (though the blade may be deeply lobed or cleft). Figures 985 and 986; single, as a pistil composed of only one carpel. Figure 987; unbranched, as a stem or hair. Figure 988.

Figure 985

Figure 986

Figure 987

Figure 988

Sinistrorse. Turned to the left or spirally arranged to the left, as in the leaves on some stems. Figure 989. (compare **dextrorse**)

Sinuate. With a strongly wavy margin. Figure 990. (compare **undulate** or **repand**)

Sinuous. Of a wavy or serpentine form. Figure 991.

Sinus. The cleft, depression, or recess between two lobes of an expanded organ such as a leaf or petal. Figure 992.

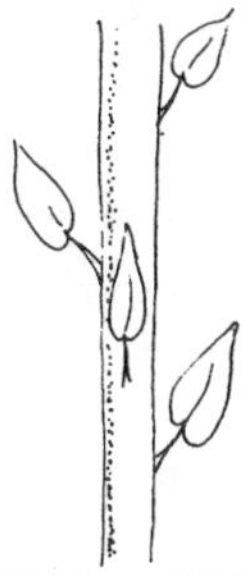

Figure 989

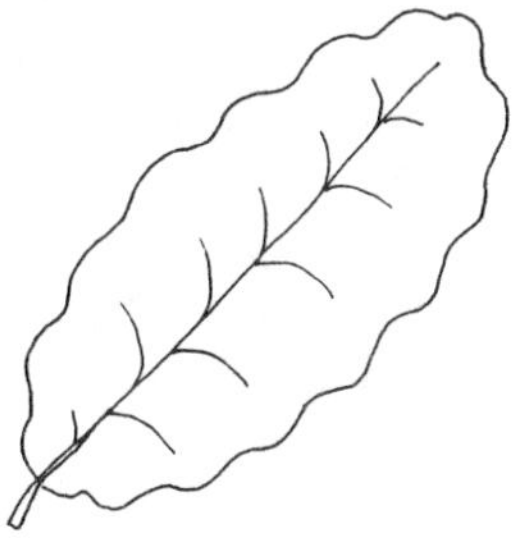

Figure 990

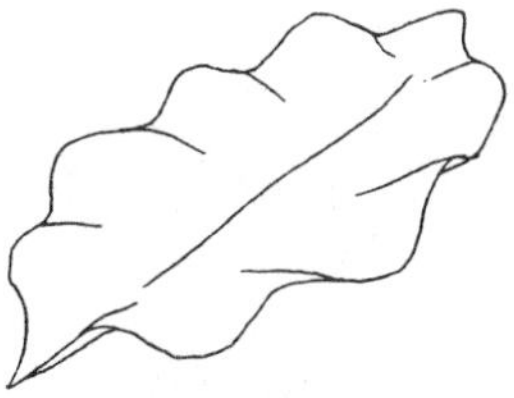

Figure 991

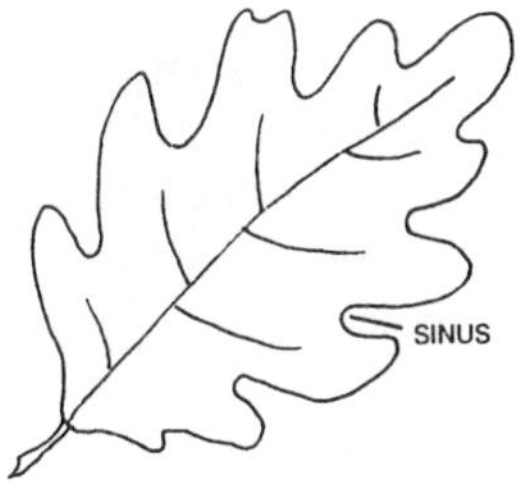

Figure 992

Smooth. With an even surface; not rough to the touch.

Sobol. Elongated caudex branches; a shoot arising from the base of a stem or from the rhizome. Figure 993.

Sobole. See **sobol**.

Soboliferous. Of or pertaining to sobols; bearing sobols. Figure 993.

Sole. That end of the carpel most distant from the apex. Figure 994.

Solitary. Occurring singly and not borne in a cluster or group. Figure 995.

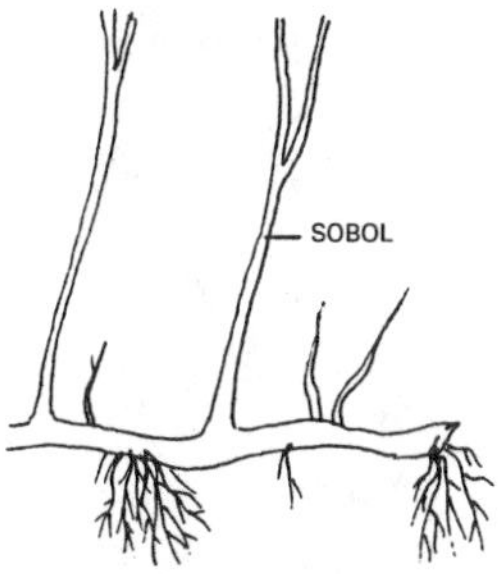

Figure 993

Figure 994

Somatic. Pertaining to or of the body, as all of the cells of a plant except the egg and sperm.

Sordid. Of a dull, dingy, or muddy color.

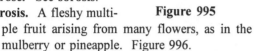

Sorose. See **sorosis**.

Sorosis. A fleshy multiple fruit arising from many flowers, as in the mulberry or pineapple. Figure 996.

Sorus (pl. **sori**). A cluster of sporangia on the surface of a fern leaf. Figure 997.

Figure 995

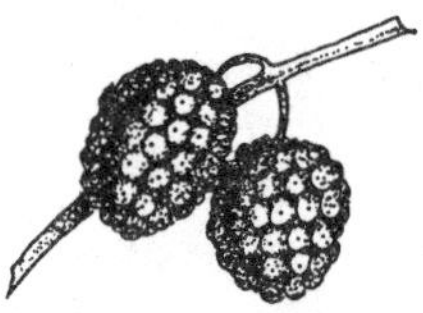

Figure 996

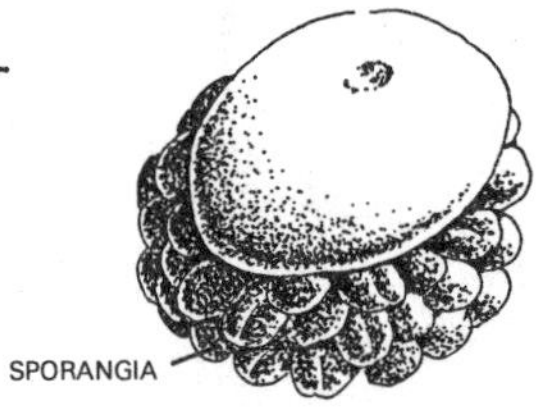

Figure 997

Spadiceous. Spadix bearing. Figure 998; spadix-like.

Spadix. A spike with small flowers crowded on a thickened axis. Figure 998.

Spathaceous. Spathe bearing. Figure 998; spathe-like.

Spathe. A large bract or pair of bracts subtending and often enclosing an inflorescence. Figure 998.

Spathella. A glume in a grass flower. Figure 999.

Spathellula. A palea in a grass flower. Figure 999.

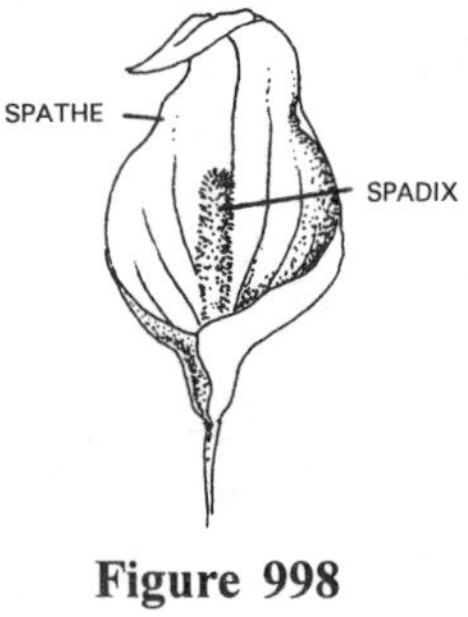

Figure 998

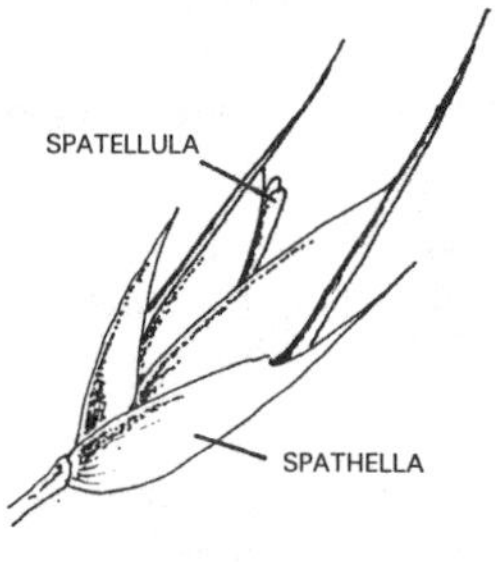

Figure 999

Spathulate. See **spatulate**.

Spatulate. Like a spatula in shape, with a rounded blade above gradually tapering to the base. Figure 1000.

Speiranthy. The condition of having twisted flowers. Figure 1001.

Figure 1000

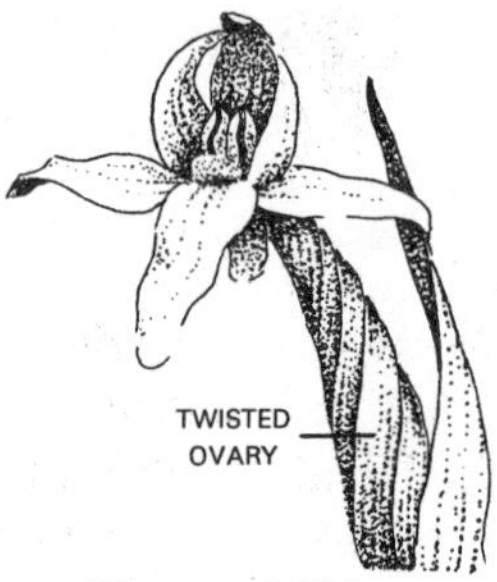

Figure 1001

Spermatophyte. Plants reproducing by seeds.

Spermophyte. See **spermatophyte**.

Sphenoid. Wedge-shaped; cuneate. Figure 1002.

Spherical. A three-dimensional, isodiametrical structure, round in outline. Figure 1003. (same as **globose**)

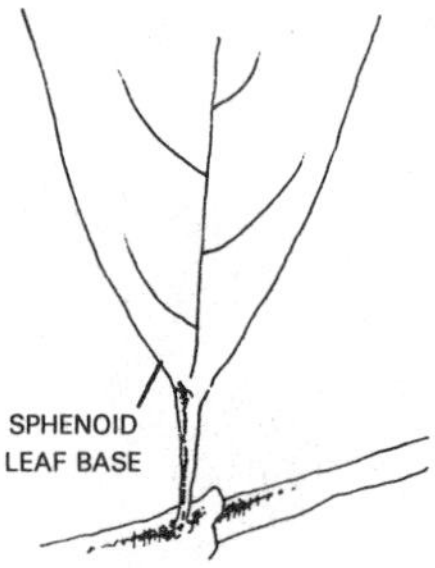

Figure 1002

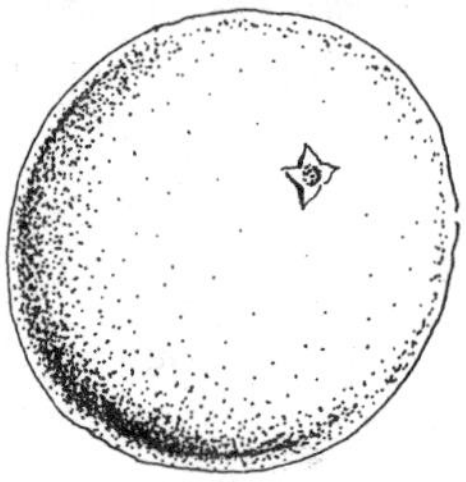

Figure 1003

Spheroidal. Almost spherical, but elliptical in cross section. Figure 1004.

Spicate. Arranged in a spike. Figure 1005.

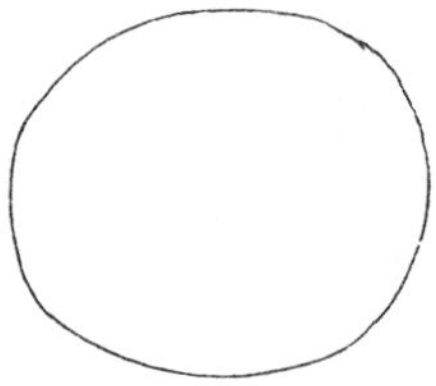

Figure 1004

Figure 1005

Spiciform. An inflorescence with the general appearance, but not necessarily the structure, of a true spike.

Spicula (pl. **spiculae**). See **spicule**.
Spicular. See **spiculate**.
Spiculate. Spiculelike; bearing spicules. Figure 1006.
Spicule. A short, pointed, epidermal projection. Figure 1006.
Spiculose. See **spiculate**.

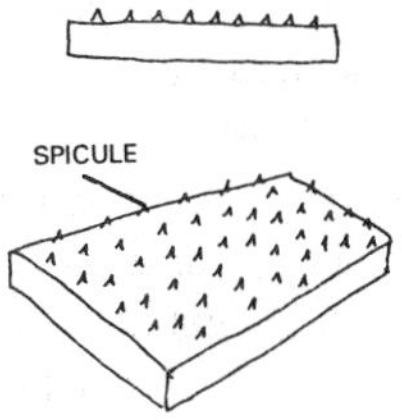

Figure 1006

Spiculum (pl. **spiculae**). See **spicule**.
Spike. An unbranched, elongated inflorescence with sessile or subsessile flowers or spikelets maturing from the bottom upwards. Figure 1007. (compare **raceme**)
Spikelet. A small spike or secondary spike; the ultimate flower cluster of grasses and sedges, consisting of 1-many flowers subtended by two bracts (glumes). Figure 1008.

Figure 1007

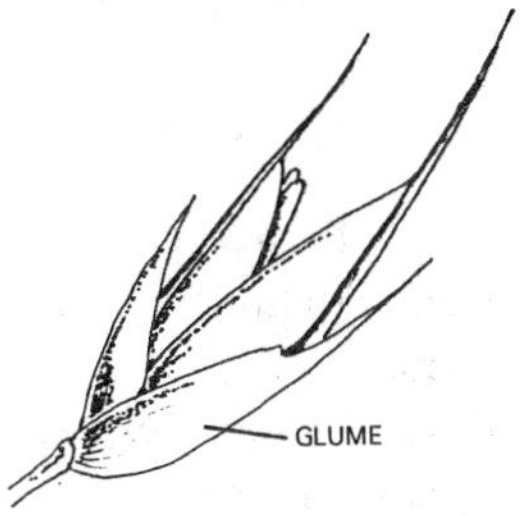

Figure 1008

Spindle-shaped. Broadest near the middle and tapering toward both ends. Figure 1009. (see **fusiform**)
Spine. A stiff, slender, sharp-pointed structure arising from below the epidermis, representing a modified leaf or stipule; any structure with the appearance of a true spine. Figure 1010.

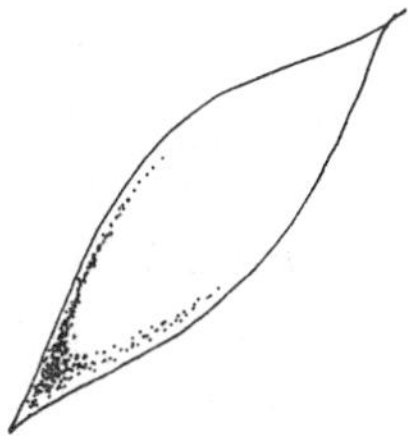
Figure 1009

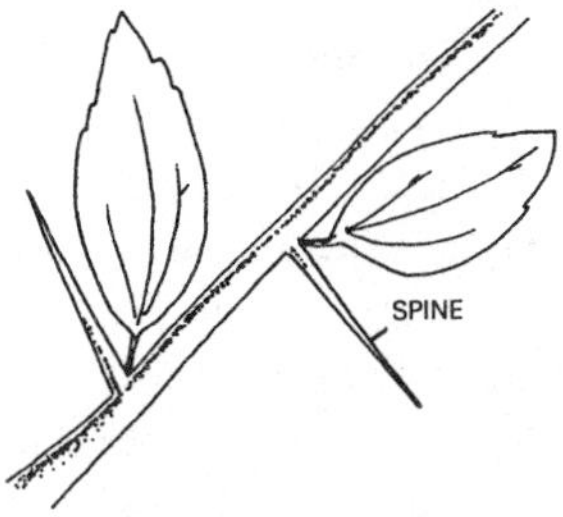

Figure 1010

Spinescent. Bearing a spine or a spinelike point at the tip; bearing spines. Figure 1010.
Spiniferous. See **spinose**.
Spinose. Bearing spines. Figure 1011.
Spinous. See **spinose**.
Spinule. A small spine. Figure 1012.
Spinulose. Bearing spinules. Figure 1012.
Spiny. With spines. Figure 1011.

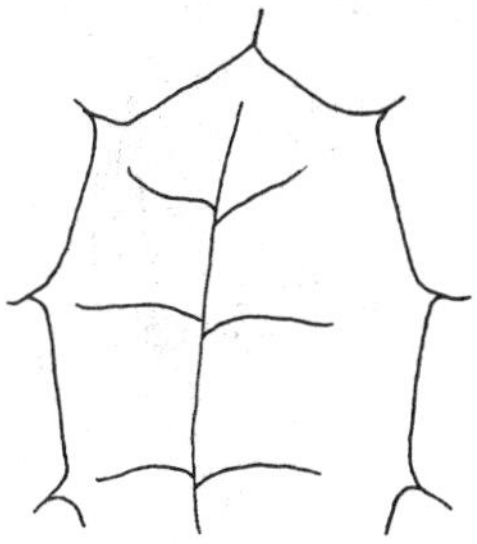
Figure 1011

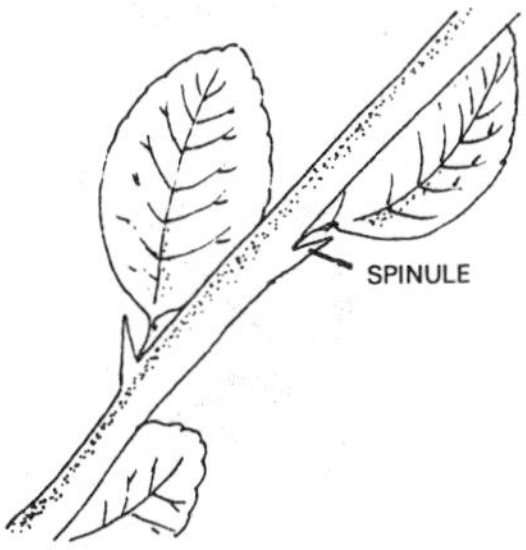

Figure 1012

Spongiose. Soft and spongy.
Sporadic. Occurring in a scattered distribution rather than in a continuous range.
Sporangiophore. A stalk bearing sporangia. Figure 1013.

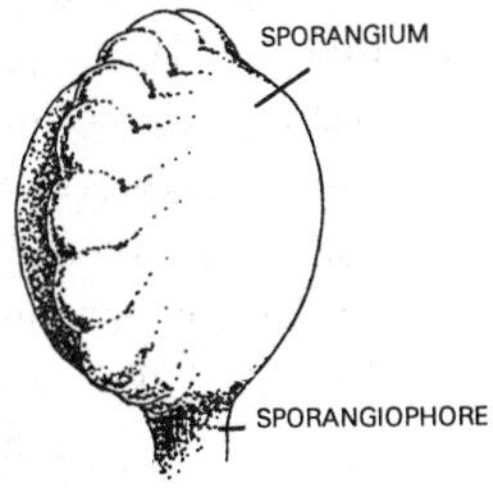

Figure 1013

Sporangium (pl. **sporangia**). A spore-bearing case or sac. Figures 1013, 1014 and 1015.

Figure 1014

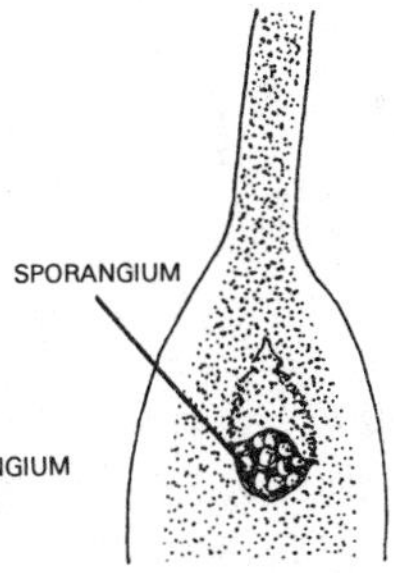

Figure 1015

Spore. A reproductive cell resulting from meiotic cell division in a sporangium, representing the first cell of the gametophyte generation. Figure 1016.
Sporiferous. Bearing spores.
Sporocarp. A specialized structure containing sporangia. Figure 1017.

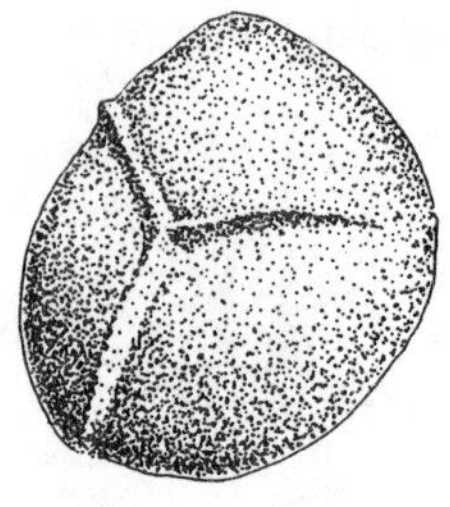

Figure 1016

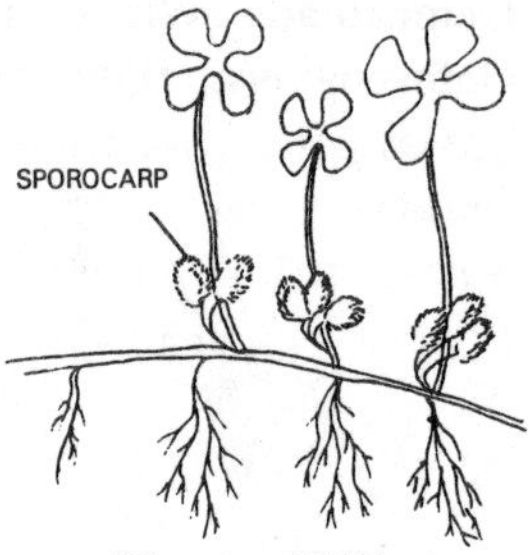

Figure 1017

Sporophyll. A sporangium-bearing leaf, often modified in structure. Figure 1018.

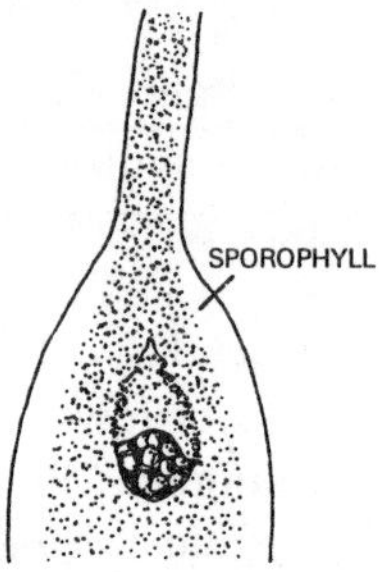

Figure 1018

Sporophyte. The diploid (2n), spore-producing generation of the plant reproductive cycle, the dominant and conspicuous plant in the vascular plants. (compare **gametophyte**)

Sprawling. Bending or curving downward; lying upon something and being supported by it. Figure 1019.

Spray. A slender shoot or branch with its leaves, flowers, or fruits. Figure 1020.

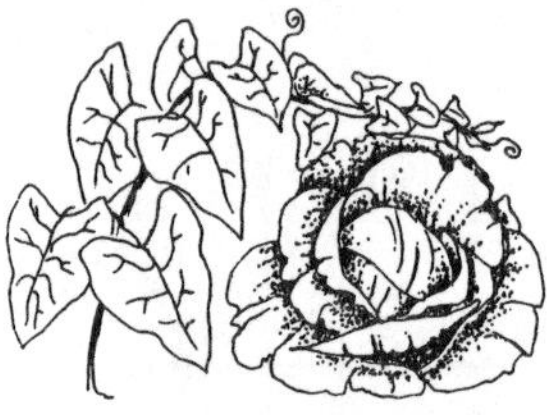

Figure 1019

Figure 1020

Spreading. Extending nearly to the horizontal; almost prostrate. Figure 1021.

Spumose. Frothy or foamy.

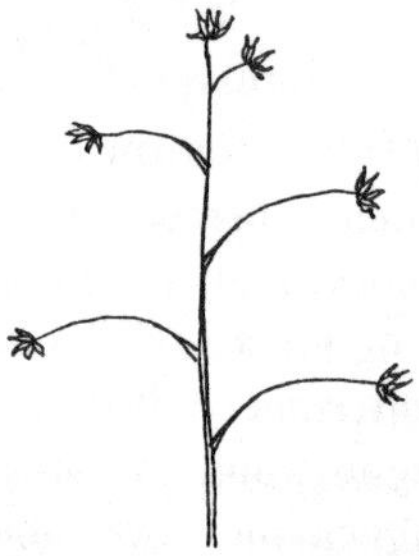

Figure 1021

Spur. A hollow, slender, saclike appendage of a petal or sepal, or of the calyx or corolla. Figure 1022; a short shoot bearing leaves or flowers and fruits. Figure 1023.

Spurred. Bearing a spur or spurs. Figures 1022 and 1023.

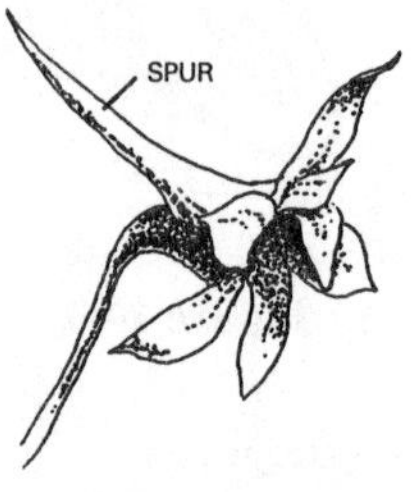

Figure 1022

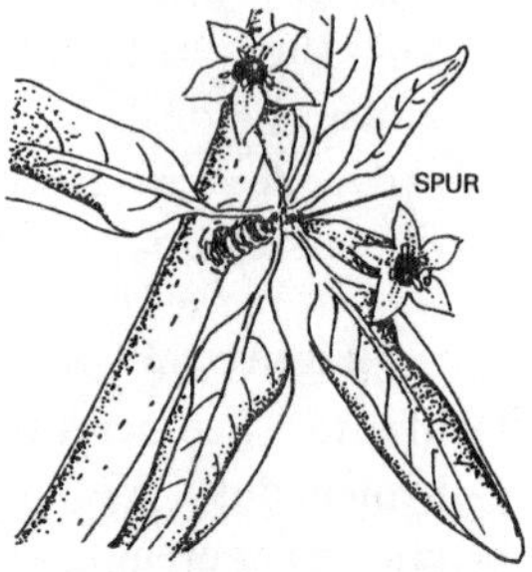

Figure 1023

Squama (pl. **squamae**). A scale, as in some types of pappus in the Compositae (Asteraceae). Figure 1024.

Squamaceous. See **squamate**.

Squamate. Covered with scales (squamae). Figure 1025.

Squamella (pl. **squamellae**). A small scale or squama. Figure 1025.

Squamellate. With squamellae. Figure 1025.

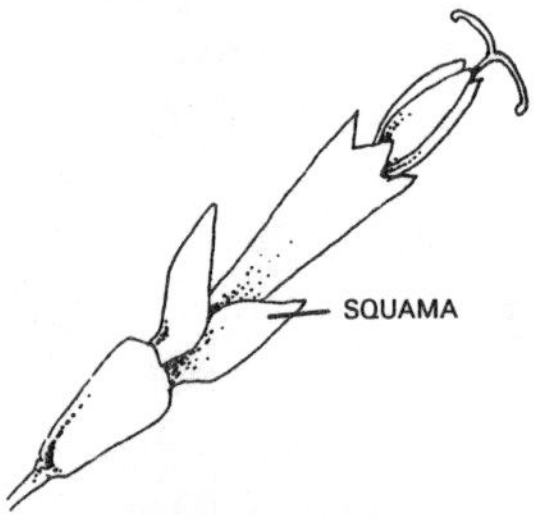

Figure 1024

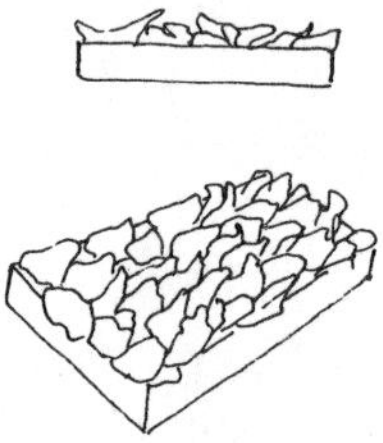

Figure 1025

Squamiform. Scalelike.

Squamose. See **squamate**.

Squamous. See **squamate**.

Squamule. The lodicule of a grass flower. Figure 1026.

Squamulose. With minute squamellae.

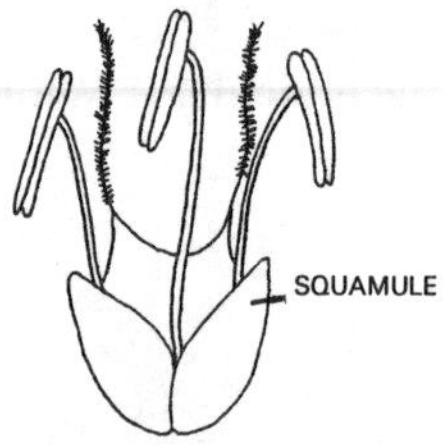

Figure 1026

Squarrose. Abruptly recurved or spreading above the base; rough or scurfy due to the presence of recurved or spreading processes. Figure 1027.

Squarrulose. With minute recurved processes.

Stalk. The supporting structure of an organ, usually narrower in diameter than the organ. Figure 1028.

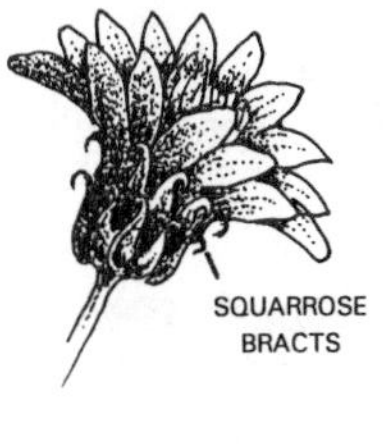

Figure 1027

Figure 1028

Stamen (pl. **stamens**, **stamina**). The male reproductive organ of a flower, consisting of an anther and filament. Figure 1029; the angiosperm microsporophyll.

Staminal. Of or pertaining to the stamens.

Staminate. Bearing stamens but not pistils, as a male flower which does not produce fruit or seeds. Figure 1030. (compare **pistillate**); bearing stamens.

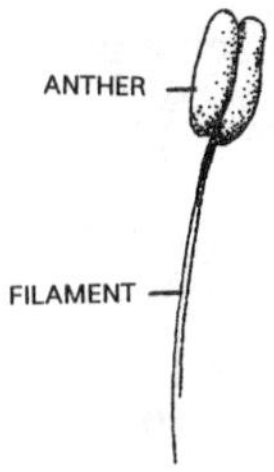

Figure 1029

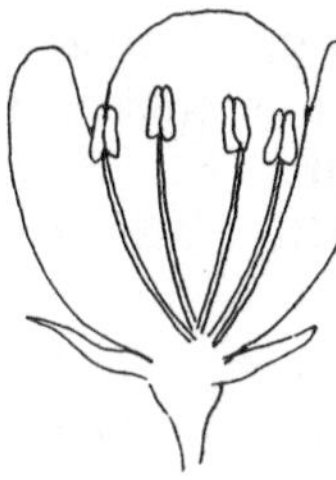

Figure 1030

Stamineal. See **staminal**.

Staminiferous. See **staminate**.

Staminode (pl. **staminodia**). A modified stamen which is sterile, producing no pollen. Figure 1031.

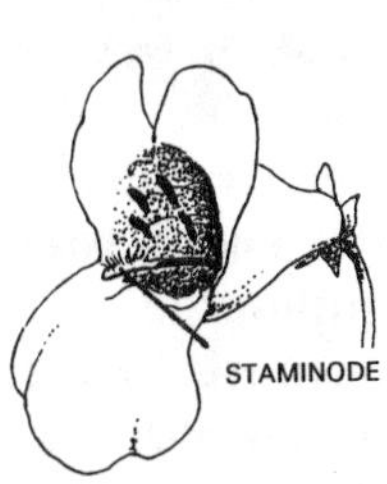

Figure 1031

Staminodium. See **staminode**.

Staminody. A condition in which other organs, such as petals or sepals, become stamens.

Standard. The upper and usually largest petal of a papilionaceous flower, as in peas and sweet peas. Figure 1032.

Stegium. Threadlike hairs on the styles of some members of the Asclepiadaceae.

Stele. The primary vascular structure of a stem or root, including the vascular tissues and all tissues internal to the vascular tissues. Figure 1033.

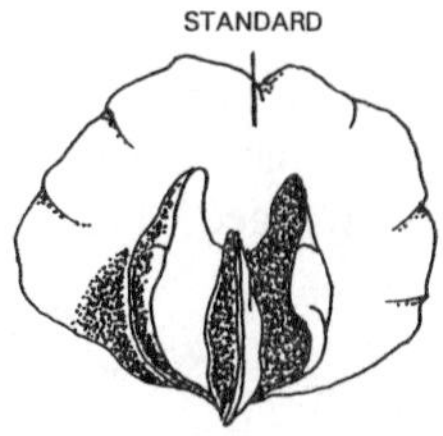

Figure 1032

Stelipilous. With stellate hairs. Figure 1034.

Stellate. Star-shaped, as in hairs with several to many branches radiating from the base. Figure 1034.

Stelliform. Star-shaped. Figure 1034.

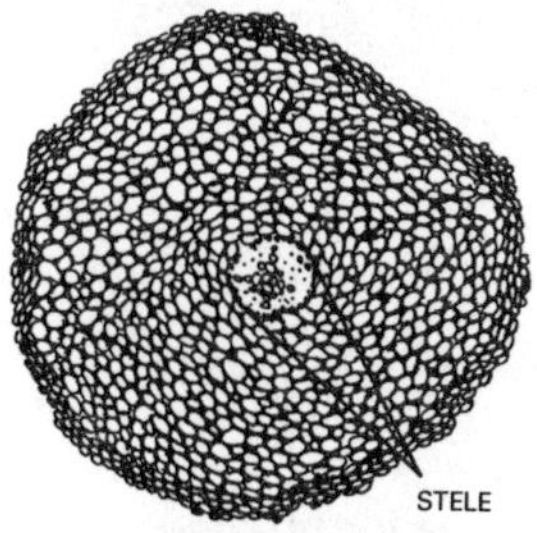

Figure 1033

Figure 1034

Stem. The portion of the plant axis bearing nodes, leaves, and buds and usually found above ground. Figure 1035.

Stenopetalous. With narrow petals. Figure 1036.

Figure 1035

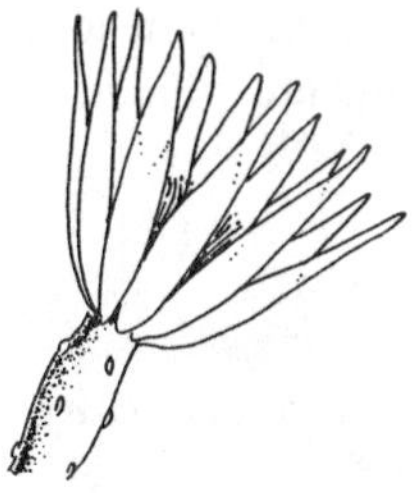

Figure 1036

Stenophyllous. With narrow leaves. Figure 1037.

Steppe. Grassland; plain; prairie.

Stereomorphic. Radially symmetrical, so that a line drawn through the middle of the structure along any plane will produce a mirror image on either side; essentially the same as **actinomorphic**. Figure 1038.

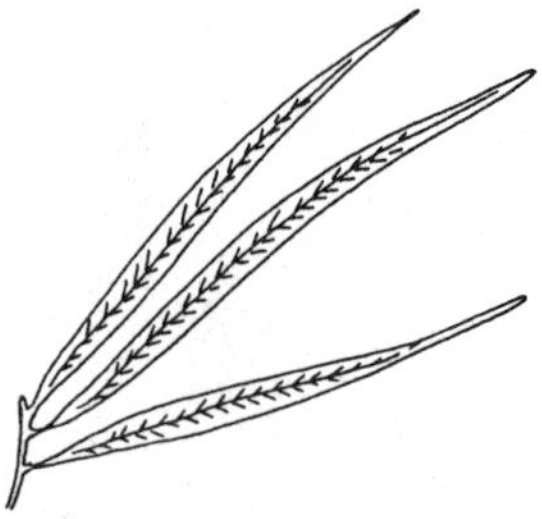

Figure 1037

Figure 1038

Sterigma. The persistent leaf base of the leaves of some coniferous trees. Figure 1039.

Sterile. Infertile, as a stamen that does not bear pollen, or a flower that does not bear seed. Figure 1040.

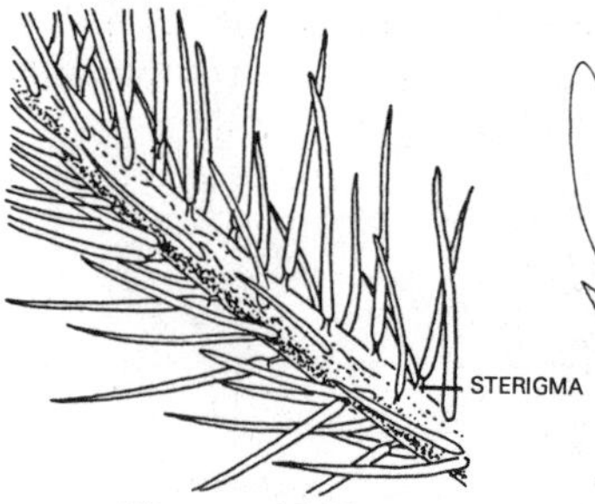

Figure 1039

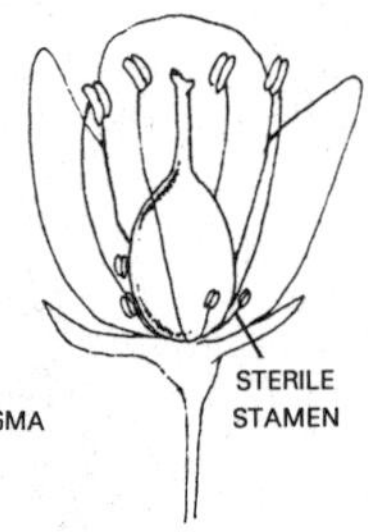

Figure 1040

Stigma. The portion of the pistil which is receptive to pollen. Figure 1041.

Stigmatic. Belonging to the stigma or having the characteristics of a stigma.

Stigmatiferous. Bearing a stigma. Figure 1041.

Figure 1041

Stipe. A stalk supporting a structure, as the stalk attaching the ovary to the receptacle in some flowers. Figure 1042.

Stipel. A small, stipule-like structure at the base of a leaflet. Figure 1043.

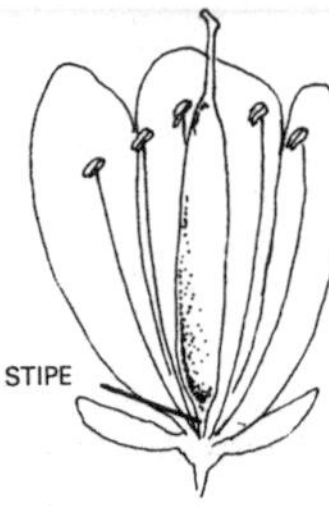

Figure 1042

Stipellate. Bearing stipels. Figure 1043.

Stipellule. See **stipel**.

Stipitate. Borne on a stipe or stalk. Figure 1042.

Stipitiform. With the form of a stipe.

Stipular. Of or pertaining to a stipule.

Stipulate. Bearing stipules. Figure 1044.

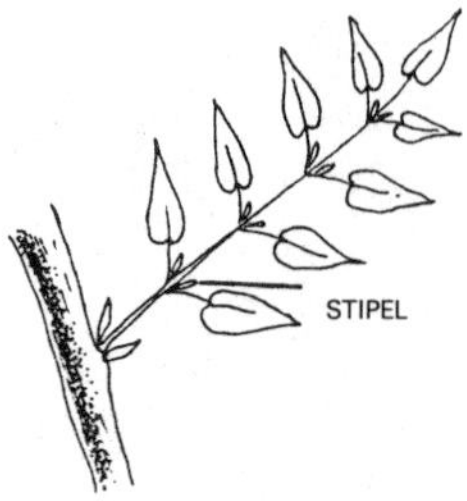

Figure 1043

Stipule. One of a pair of leaflike appendages found at the base of the petiole in some leaves. Figure 1044.

Stipuliform. Stipule-shaped.

Stipulose. See **stipulate**.

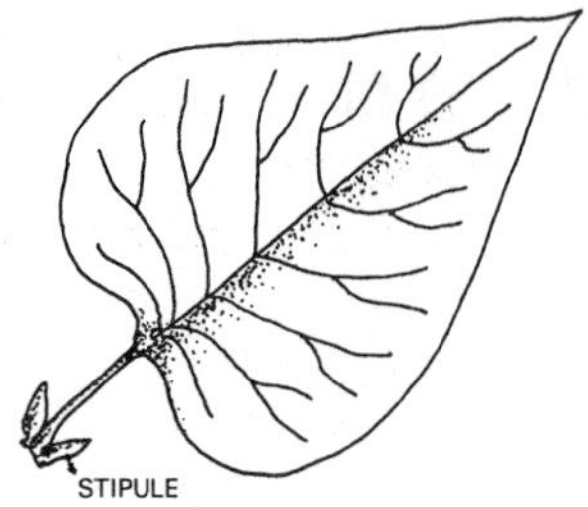

Figure 1044

Stolon. An elongate, horizontal stem creeping along the ground and rooting at the nodes or at the tip and giving rise to a new plant. Figure 1045.

Stoloniferous. Bearing stolons. Figure 1045.

Stoloniform. Stolonlike.

Stoma. See **stomate**.

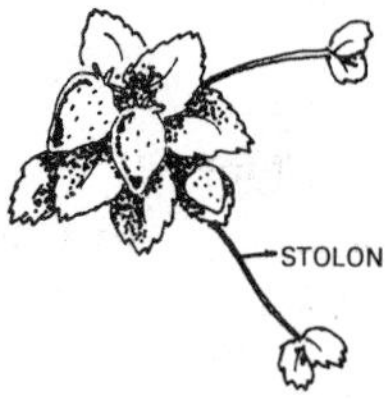

Figure 1045

Stomate (pl. **stomata**). A pore or aperature, surrounded by two guard cells, which allows gaseous exchange. Figure 1046.

Stomatiferous. Bearing stomata. Figure 1046.

Stone. The hard, woody endocarp enclosing the seed of a drupe. Figure 1047.

Stone fruit. A drupe; a fruit with a stony pit. Figure 1047.

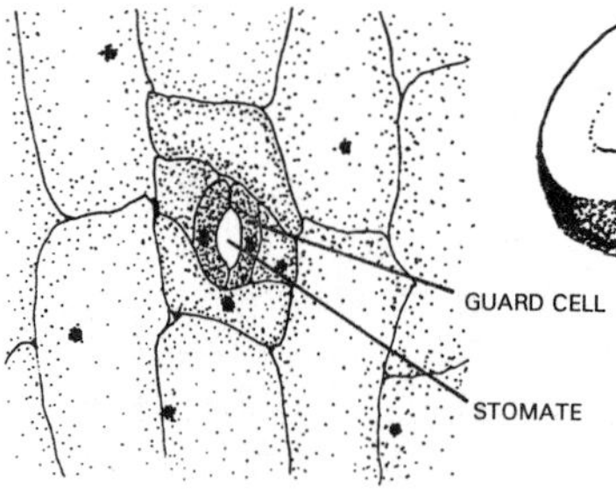

Figure 1046

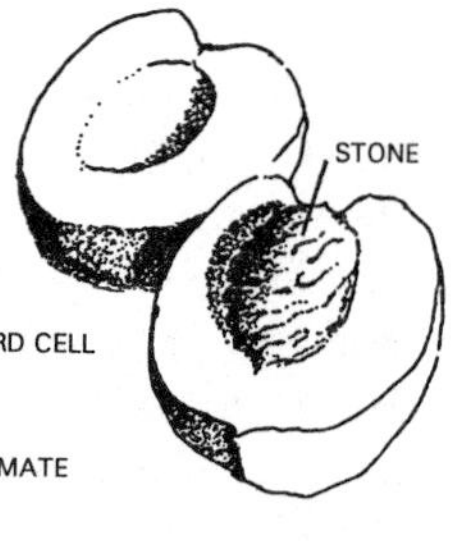

Figure 1047

Stool. The base of plants which produce new stems each year. Figure 1048; a group of stems arising from a single root.

Stramineous. Strawlike in color or texture.

Strap. The ligule of a ray flower in the Compositae (Asteraceae). Figure 1049.

Figure 1048

Strap-shaped. Elongated and flat. Figure 1049.

Streptocarpous. With twisted fruits. Figure 1050.

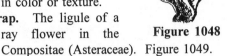

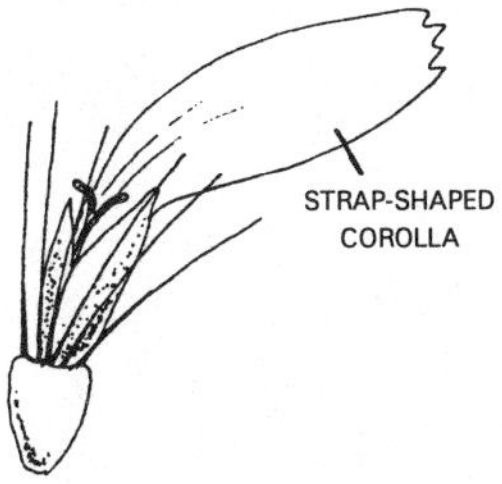

Figure 1049

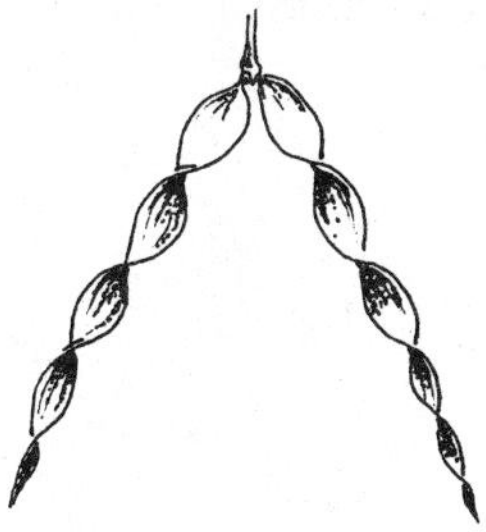

Figure 1050

Stria (pl. **striae**). A fine line or groove. Figure 1051.

Striate. Marked with fine, usually parallel lines or grooves. Figure 1051.

Strict. Very straight and upright, not at all spreading. Figure 1052.

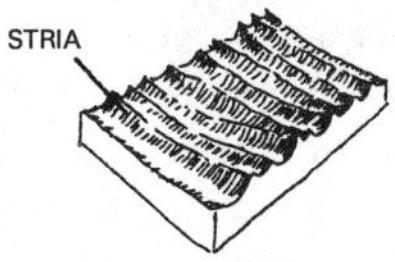

Figure 1051

Striga. A bristle; a straight, stiff, sharp, appressed hair. Figure 1053.

Figure 1052

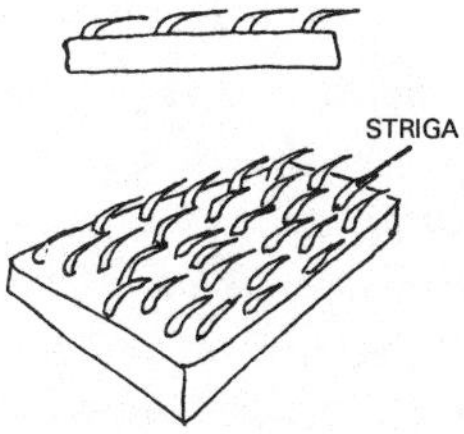

Figure 1053

Strigillose. Minutely strigose. Figure 1054.

Strigose. Bearing straight, stiff, sharp, appressed hairs. Figure 1053.

Strigulose. See **strigillose**.

Strobilaceous. Of or pertaining to a cone; conelike.

Strobile. A cone or an inflorescence resembling a cone. Figure 1055.

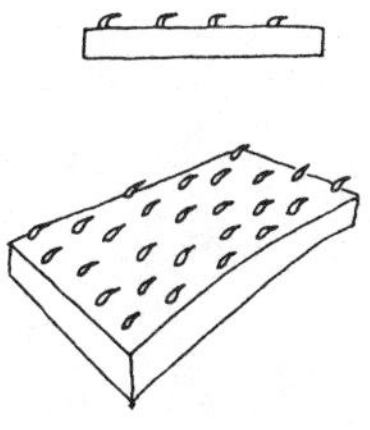

Figure 1054

Figure 1055

Strobilus (pl. **strobili**). A conelike cluster of sporophylls on an axis. Figure 1056; a cone. Figure 1055.

Strombus. A legume which is spirally coiled, as in *Medicago*. Figure 1057.

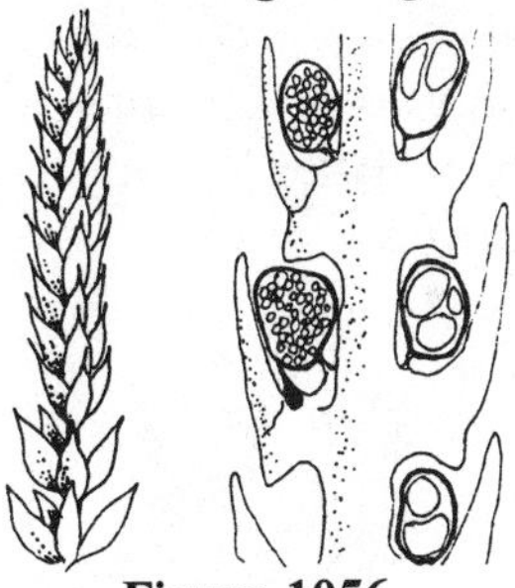

Figure 1056

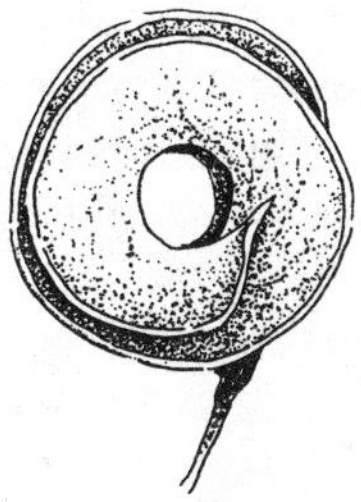

Figure 1057

Strophiole. An appendage at the hilum in some seeds. Figure 1058.

Struma (pl. **strumae**). A cushion-like swelling. Figure 1059.

Strumose. With a covering of cushion-like swellings; bullate. Figure 1059.

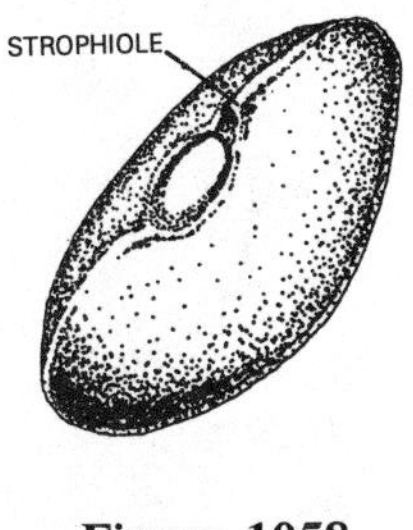

Figure 1058

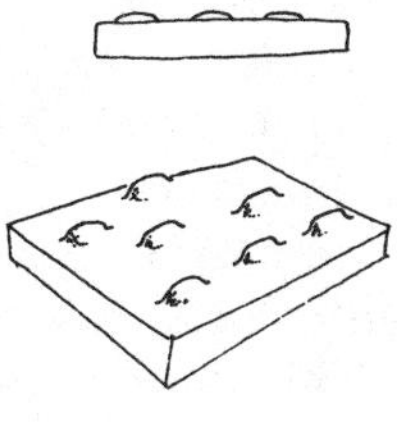

Figure 1059

Stylar. Of or pertaining to a style.

Style. The usually narrowed portion of the pistil connecting the stigma to the ovary. Figure 1060.

Stylocarpellous. With a style, but without a stipe. Figure 1061. (compare **astylocarpellous**, and see **stylocarpepodic**)

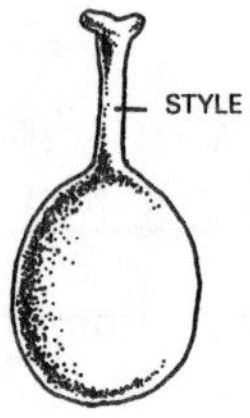

Figure 1060

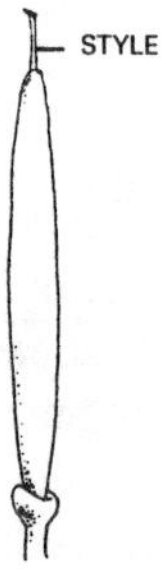

Figure 1061

Stylocarpepodic. With a style and a stipe. Figure 1062. (compare **astylocarpepodic**, and see **stylocarpellous**)

Stylodious. See **unicarpellous.**

Stylopod. See **stylopodium.**

Stylopodic. With a stylopodium. Figure 1063.

Stylopodium. A disklike expansion or enlargement at the base of the style in the Umbelliferae (Apiaceae). Figure 1063.

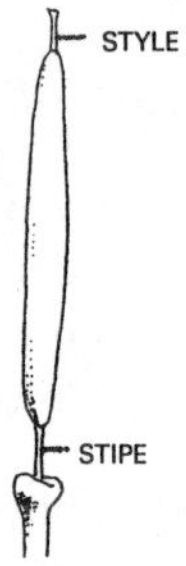

Figure 1062

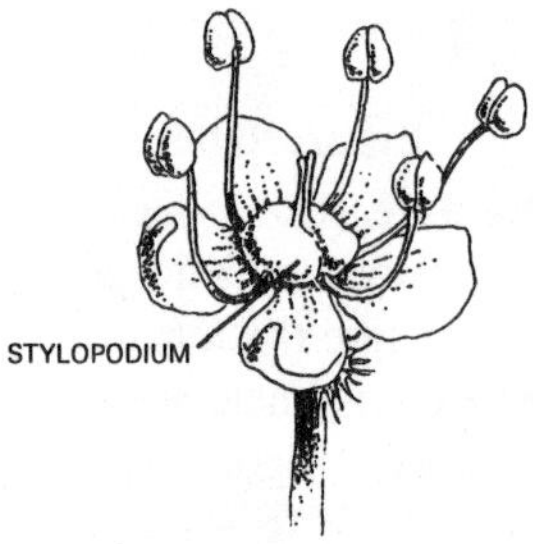

Figure 1063

Suaveolent. Fragrant.

Sub- (prefix). Meaning under, slightly, somewhat, or almost.

Subacute. Slightly acute. Figure 1064.

Subalpine. Growing in the mountains below the alpine zone and above the montane zone.

Figure 1064

Subapical. Near the apex. Figure 1065.

Subbasal. Near the base. Figure 1066.

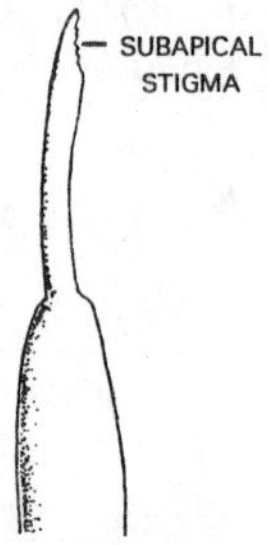

Figure 1065

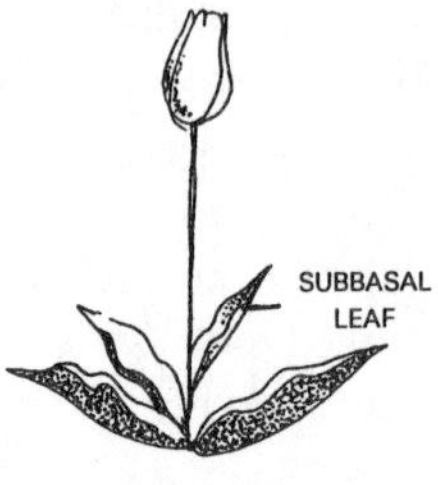

Figure 1066

Subcapitate. Almost capitate. Figure 1067.

Subcordate. Almost cordate. Figure 1068.

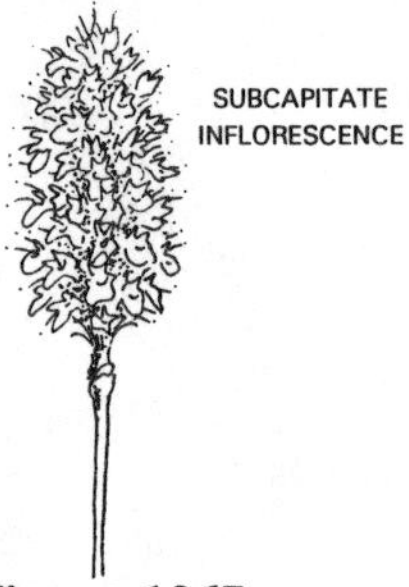

Figure 1067

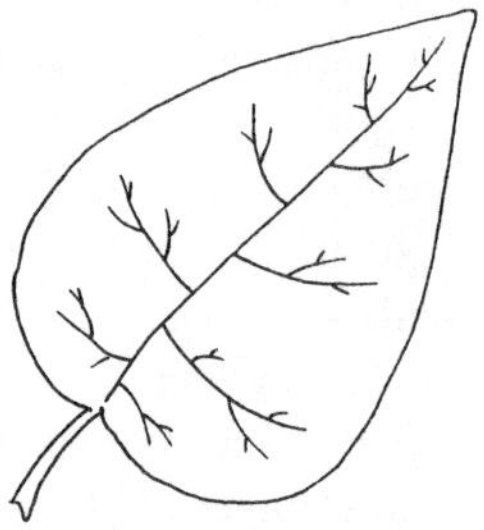

Figure 1068

Subcorymbose. Almost corymbose.

Subcylindric. Almost cylindric in shape. Figure 1069.

Subentire. Almost entire. Figure 1070.

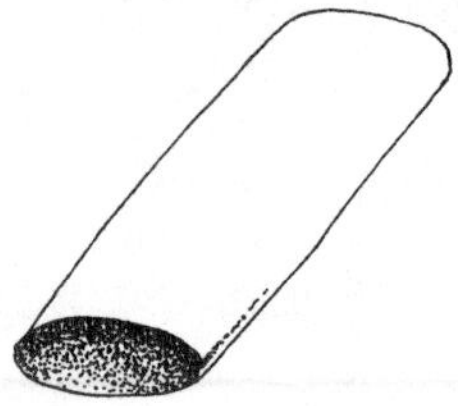

Figure 1069

Figure 1070

Suber. Cork.

Suberose. Corky in texture.

Suberous. See **suberose.**

Subfoliaceous. Almost foliaceous.

Subglabrate. Almost glabrous. Figure 1071.

Sublignous. Almost woody.

Submersed. Submerged.

Subrhizomatous. Almost rhizomatous.

Subscapose. Almost scapose. Figure 1066.

Subshrub. A suffrutescent perennial plant; a small shrub.

Subspicate. Almost spicate, but with short pedicels on some or all of the flowers or florets. Figure 1072.

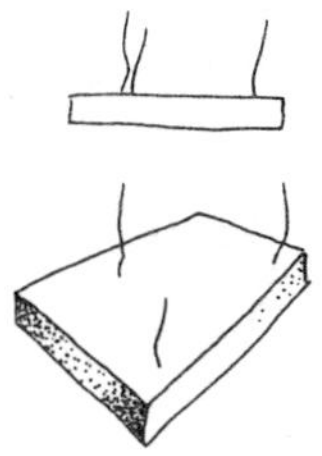

Figure 1071

Figure 1072

Subtend. To be below and close to, as a bract may subtend an inflorescence. Figure 1073.

Subterete. Almost terete. Figure 1074.

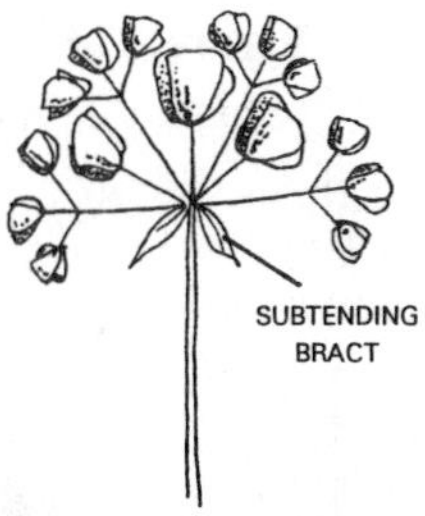

Figure 1073

Figure 1074

Subterranean. Below the surface of the ground.

Subterraneous. See **subterranean**.

Subtropical. Distributed in areas intermediate between tropical and temperate regions; nearly tropical.

Figure 1075

Subula. A fine, sharp point. Figure 1075.

Subulate. Awl-shaped. Figure 1075.

Succulent. Juicy and fleshy, as the stem of a cactus or the leaves of *Aloe*. Figures 1076 and 1077.

Sucker. A shoot originating from below ground. Figure 1078.

Suffrutescent. Somewhat shrubby; slightly woody at the base.

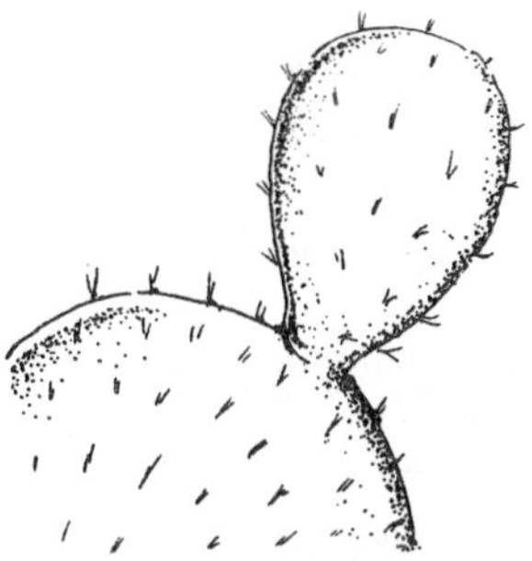

Figure 1076

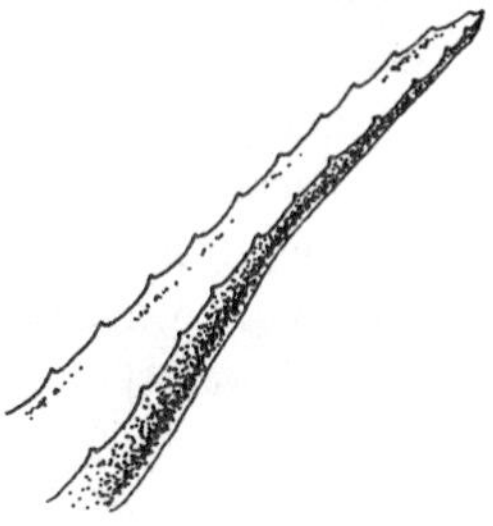

Figure 1077

Suffruticose. Somewhat woody.

Suffruticulose. See **suffruticose**.

Suffused. Tinted or tinged.

Sulcate. With longitudinal grooves or furrows. Figure 1079.

Sulcus (pl. **sulci**). A groove or furrow. Figure 1079.

Sulfureous. Sulfur-colored.

Summer annual. A plant with seeds germinating in spring or early summer and completing flowering and fruiting in late summer or early fall and then dying.

Superaxillary. Attached above the axil.

Superior. Attached above, as an ovary that is attached above the point of attachment of the other floral whorls. Figure 1080.

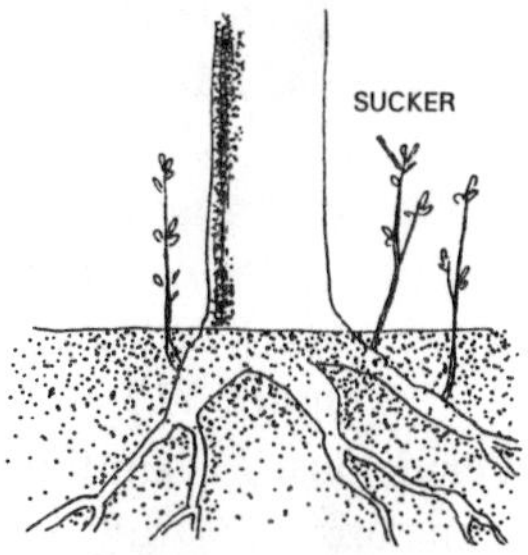

Figure 1078

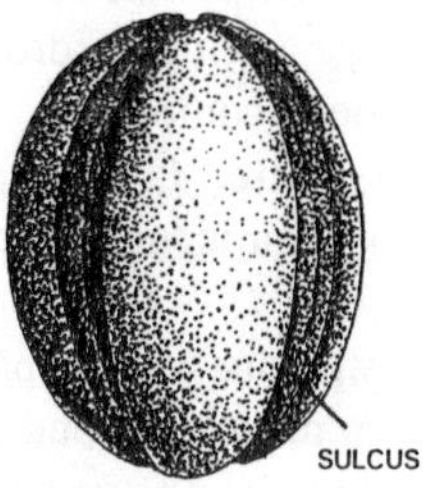

Figure 1079

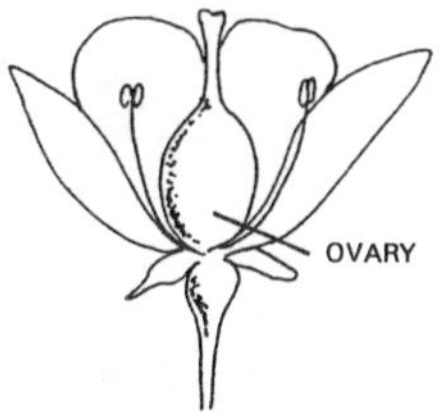

Figure 1080

Supra-axillary. See **superaxillary**.

Supraligular. Attached above the ligule.

Surculose. Producing suckers or runners from the base or from rootstocks. Figure 1081.

Surculose-proliferous. Reproducing by suckers or runners. Figure 1081.

Surculum. A fern rhizome. Figure 1082.

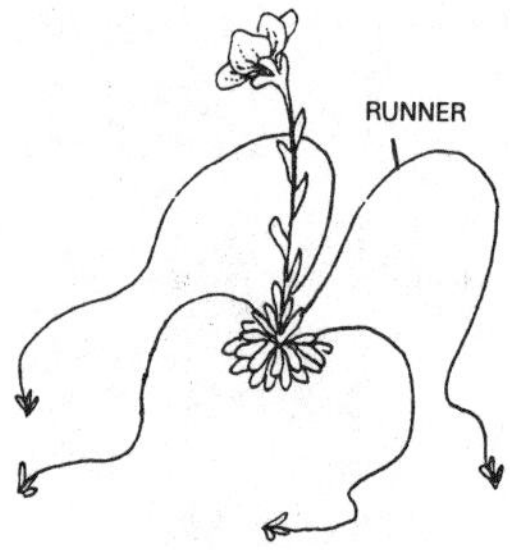

Figure 1081

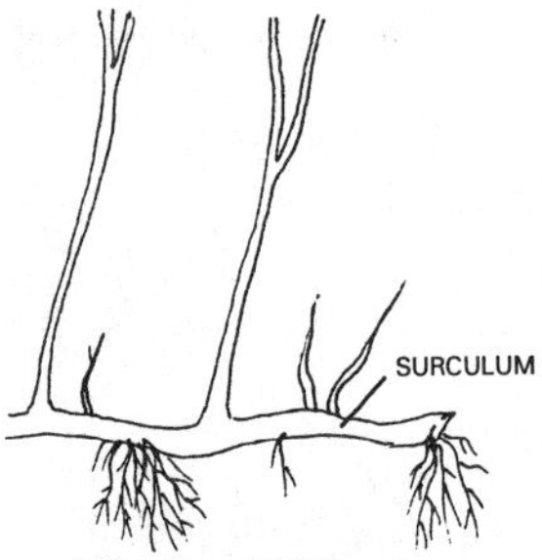

Figure 1082

Surcurrent. Extending upward from the point of insertion, as a leaf base that extends up along the stem. Figure 1083.

Surficial. Growing near the ground, or spread over the surface of the ground. Figure 1084.

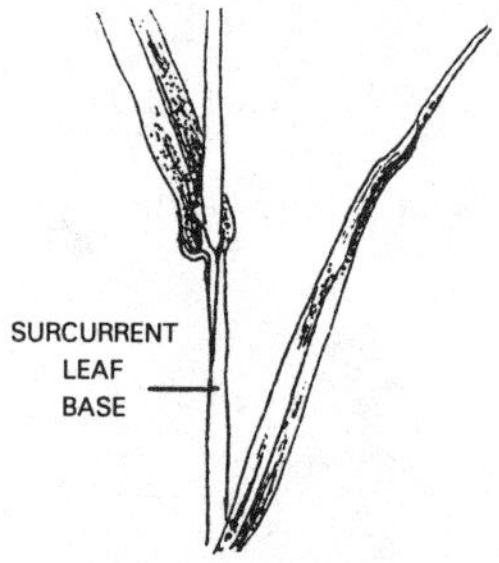

Figure 1083

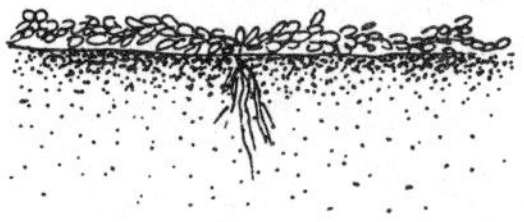

Figure 1084

Suture. A line of fusion; the line of dehiscence of a fruit or anther. Figure 1085.

Syconium (pl. **syconia**). The fruit of a fig, consisting of an entire ripened inflorescence with a hollow, inverted receptacle bearing flowers internally. Figure 1086.

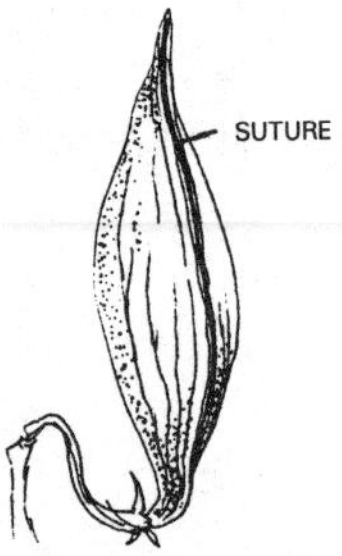

Figure 1085

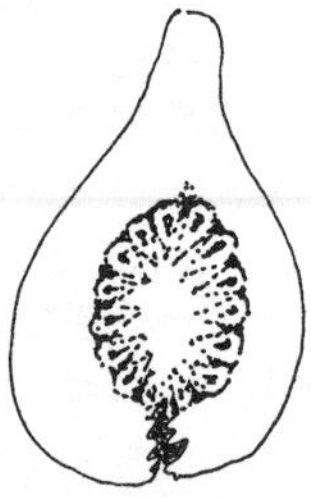

Figure 1086

Symmetric. Said of a flower having the same number of parts in each floral whorl. Figure 1087.

Sympatric. Occupying the same geographic region. (compare **allopatric**)

Sympetalous. With the petals united, at least near the base. Figure 1088. (same as **gamopetalous**; compare **apopetalous** and **polypetalous**)

Symphysis. The fusion or coalescence of like parts as in a sympetalous corolla. Figure 1088.

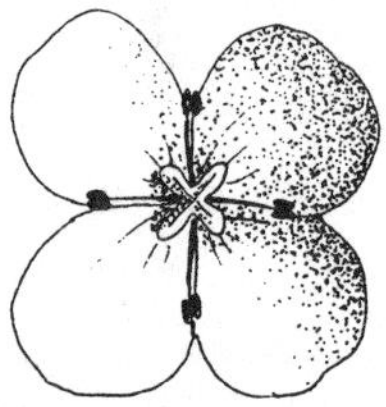

Figure 1087

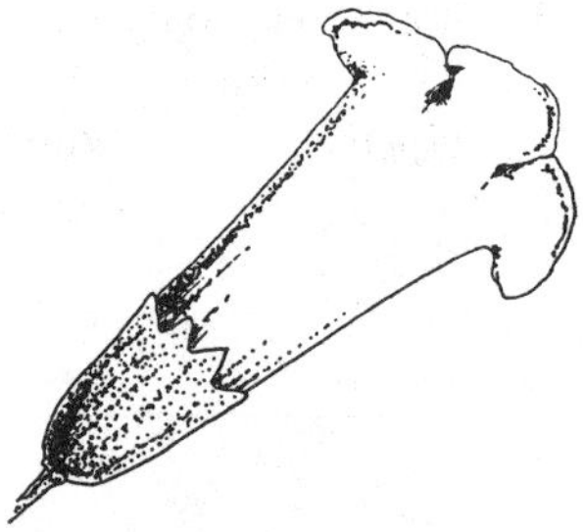

Figure 1088

Sympodial. Of or pertaining to a sympodium; in the form of a sympodium.

Sympodium. A main axis appearing to be simple, but actually consisting of a number of short axillary branches rather than a continuation of the main axis. Figure 1089. (compare **monopodium**)

Syn-, Sym- (prefix). Meaning united.

Synandrous. With united anthers. Figure 1090.

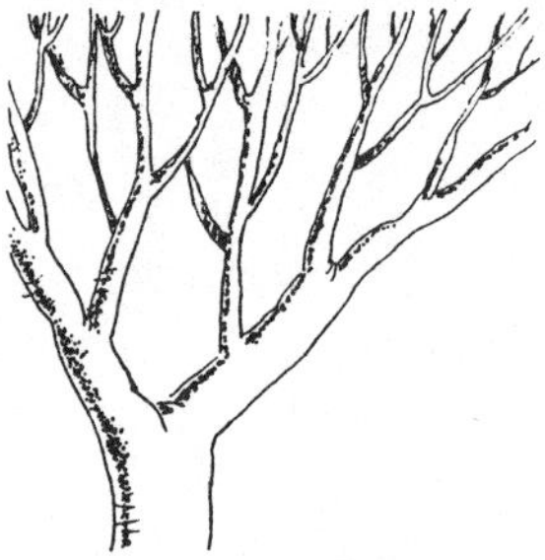

Figure 1089

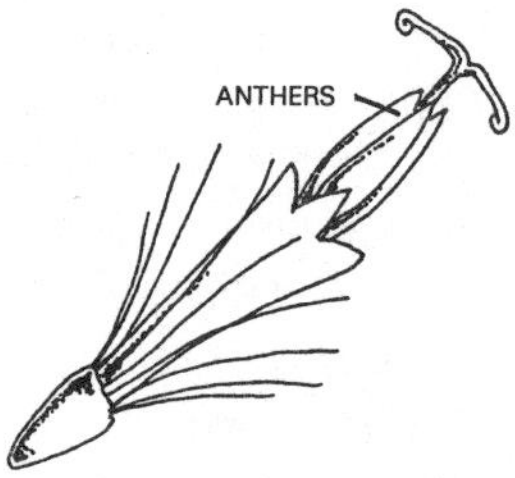

Figure 1090

Synantherous. See **synandrous**.

Syncarp. A multiple fruit. Figure 1091; an aggregate fruit. Figure 1092.

Figure 1091

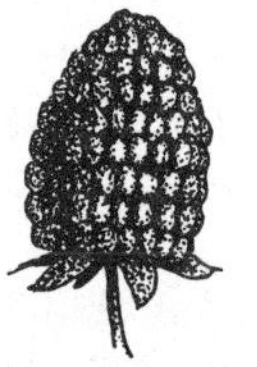

Figure 1092

Syncarpous. Of or pertaining to a syncarp; with united carpels. Figure 1093. (compare **apocarpous**)

Synema. The column composed of united filaments in a flower with monadelphous stamens. Figure 1094.

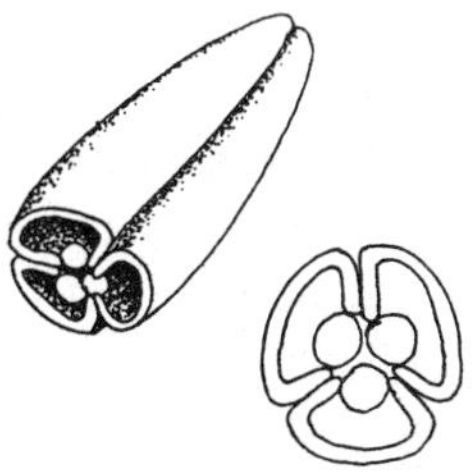

Figure 1093

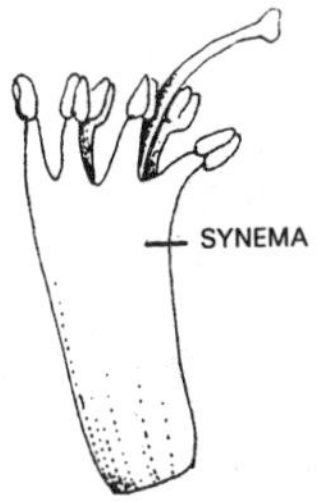

Figure 1094

Syngenesious. With stamens united by their anthers. Figure 1090.

Synobasic. With a united base.

Synoecious. With staminate and pistillate flowers together in the same head. Figure 1095.

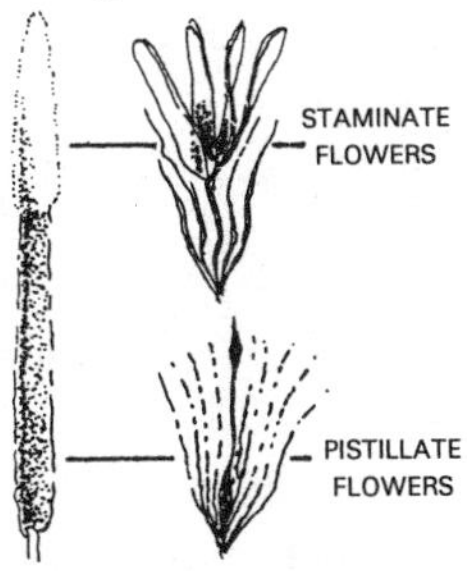

Figure 1095

Synsepalous. With united sepals. Figure 1088. (same as **gamosepalous**; compare **polysepalous**)

Tailed. With a tail-like appendage or appendages. Figure 1096.

Taproot. The main root axis from which smaller root branches arise; a root system with a main root axis and smaller branches, as in most dicots. Figure 1097. (compare **fibrous roots**)

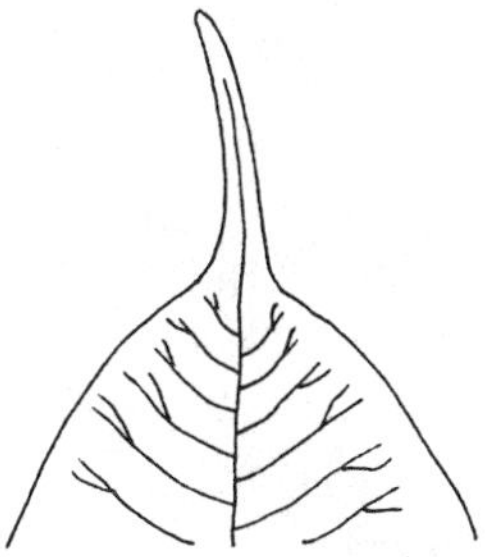

Figure 1096

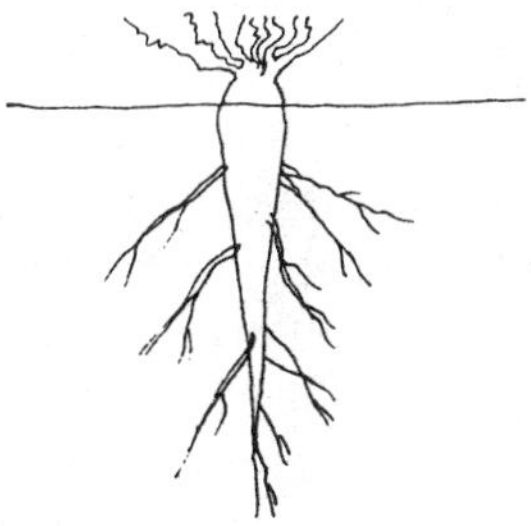

Figure 1097

Tassel. The staminate inflorescence in corn (*Zea*). Figure 1098.

Tawny. Tan in color.

Taxon (pl. **taxa**). A taxonomic entity of any rank, such as order, family, genus, or species.

Tectum. The outermost layer of a pollen grain. Figure 1099.

Figure 1098

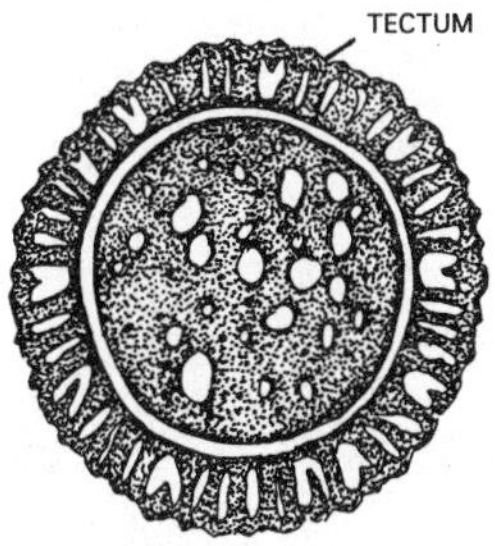

Figure 1099

Tegule. One of the bracts of the involucre in the Compositae (Asteraceae). Figure 1100.

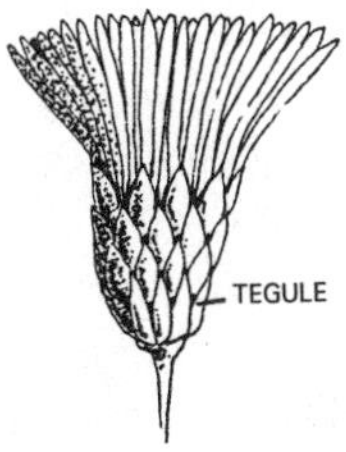

Figure 1100

Temperate. Distributed in those regions of the earth lying between the tropic of Cancer (23 1/2 degrees north latitude) and the Arctic Circle (66 2/3 degrees north latitude) or between the tropic of Capricorn (23 1/2 degrees south latitude) and the Antarctic Circle (66 2/3 degrees south latitude).

Tendril. A slender, twining organ used to grasp support for climbing. Figure 1101.

Tendril-pinnate. Pinnately compound, but ending in a tendril, as in the sweet pea. Figure 1101.

Tentacle. A sensitive filament, as the glandular hairs of *Drosera*. Figure 1102.

Tentacular. Bearing tentacles. Figure 1102.

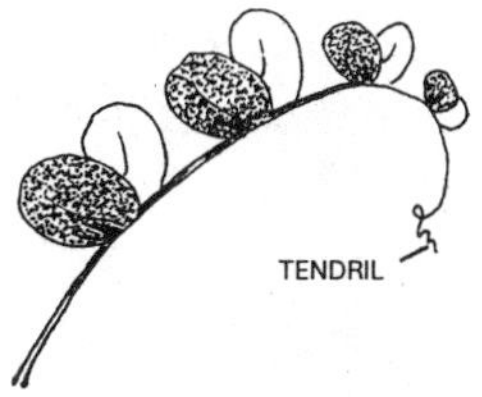

Figure 1101

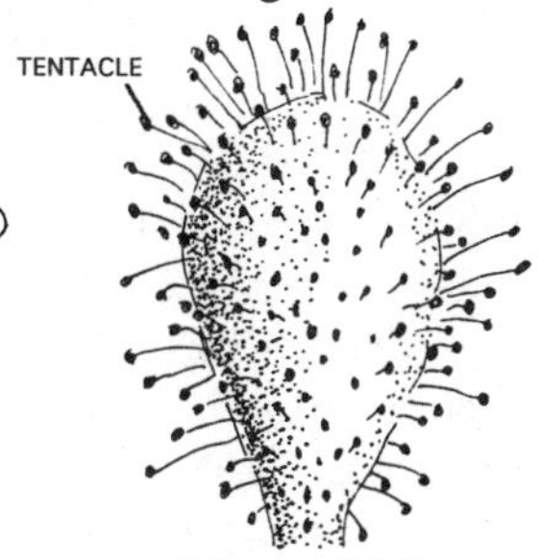

Figure 1102

Tenuous. Slender or thin.

Tepal. A segment of a perianth which is not differentiated into calyx and corolla; a sepal or petal. Figure 1103.

Terete. Round in cross section; cylindrical. Figure 1104.

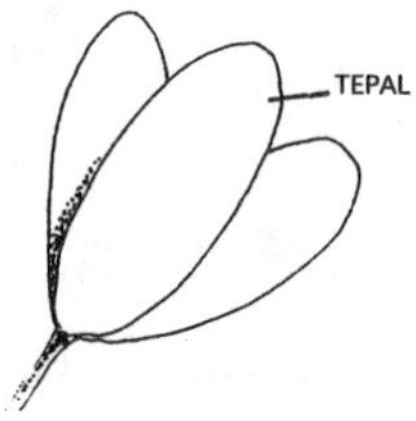

Figure 1103

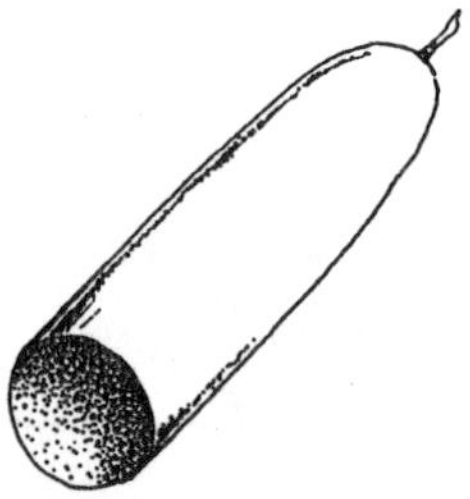

Figure 1104

Tergeminate. Thrice divided into equal pairs; paired leaflets ternately compound. Figure 1105.

Terminal. At the tip or apex.

Ternary. Consisting of threes or involving threes; triple.

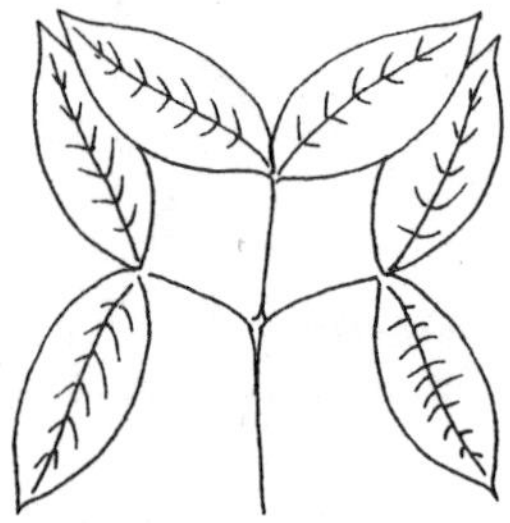

Figure 1105

Ternate. In threes, as a leaf which is divided into three leaflets. Figure 1106.

Terrestrial. Growing on ground; not aquatic.

Tesselate. With a checkered pattern. Figure 1107.

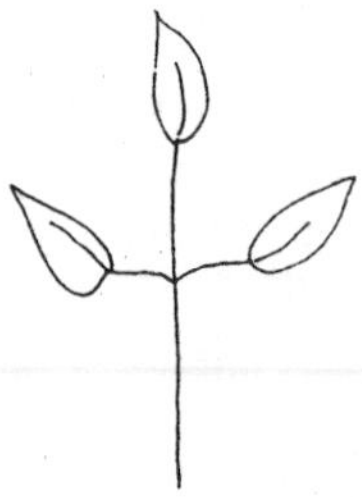

Figure 1106

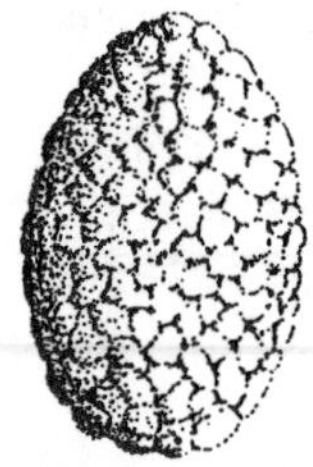

Figure 1107

Testa (pl. **testae**). The seed coat, from the integuments of the ovule. Figure 1108.

Testaceous. Brick-red or brownish-red in color.

Tetra- (prefix). Meaning four.

Tetracyclic. With four whorls. Figure 1109.

Tetrad. A group of four.

Tetradinous. Occurring in tetrads.

Tetradymous. With four cells.

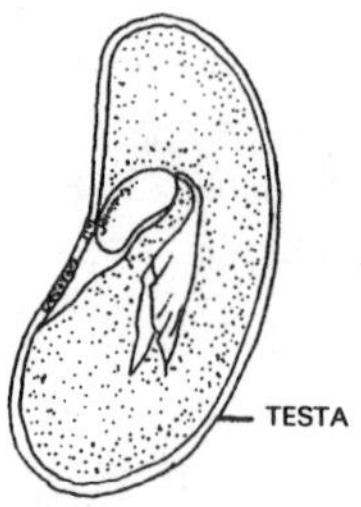

Figure 1108

Figure 1109

Tetradynamous. Having four long and two short stamens, as in most members of the Cruciferae (Brassicaceae). Figures 1109 and 1110.

Tetragonal. Four-angled. Figure 1111.

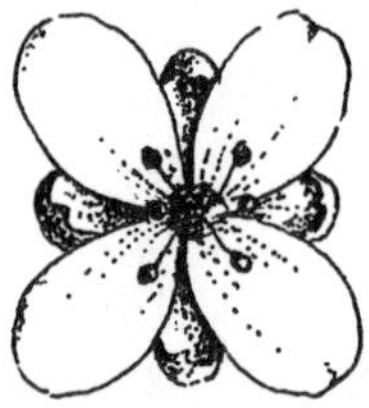

Figure 1110

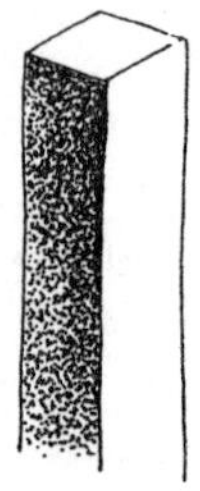

Figure 1111

Tetrahedral. Four-sided, each side triangular. Figure 1112.

Tetramerous. With parts arranged in sets or multiples of four. Figure 1113.

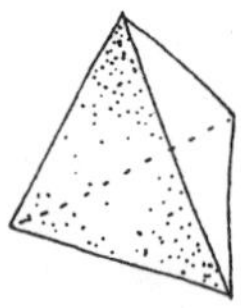

Figure 1112

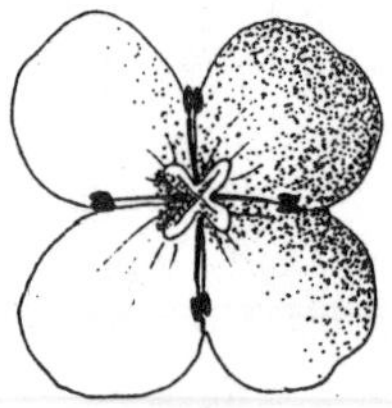

Figure 1113

Tetrandrous. With four stamens. Figure 1113.

Tetrangular. With four angles. Figure 1111.

Tetrapetalous. With four petals. Figure 1113.

Tetraploid. With four representatives of each type of chromosome, or four complete sets of chromosomes, in each cell; 4x. (compare **diploid** and **haploid**)

Tetrapterous. With four wings or winglike appendages. Figure 1114.

Tetrasepalous. With four sepals. Figure 1110.
Tetrastachyous. With four spikes.
Tetrastichous. In four vertical ranks or rows on an axis. Figure 1115.

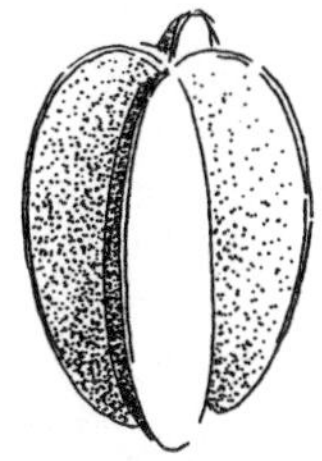

Figure 1114

Figure 1115

Thalamous. See **thalamus**.
Thalamus. The receptacle of a flower. Figure 1116.
Thalloid. Consisting of a thallus; resembling a thallus.

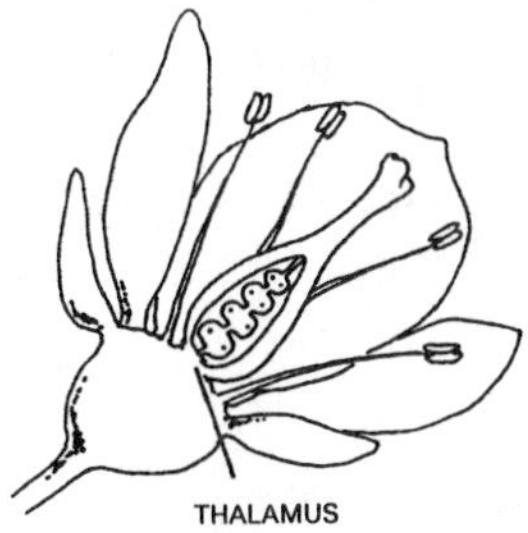

Figure 1116

Thallus (pl. **thalli**). A plant body which is not obviously differentiated into stems, roots, and leaves. Figure 1117.
Theca (pl. **thecae**). A pollen sac or cell of the anther. Figure 1118.
Thecate. With a theca. Figure 1118.

Figure 1117

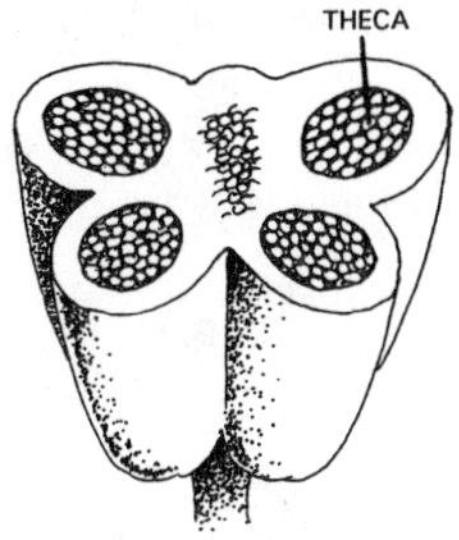

Figure 1118

Thelephorous. With nipplelike protuberances. Figure 1119.
Thorn. A stiff, woody, modified stem with a sharp point; sometimes applied to any structure resembling a true thorn. Figure 1120. (compare **spine** and **prickle**)
Three-ranked. In three vertical ranks or rows around an axis. Figure 1121.
Throat. The orifice of a gamopetalous corolla or gamosepalous calyx. Figure 1122; the expanded portion of the corolla between the limb and the tube. Figure 1123; the upper margin of the leaf sheath in grasses. Figure 1124.

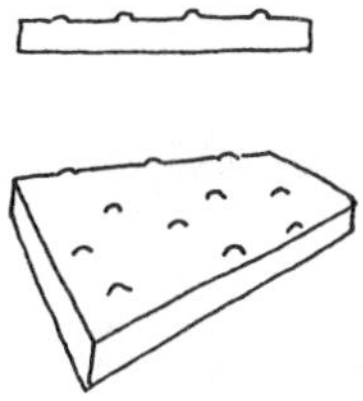

Figure 1119

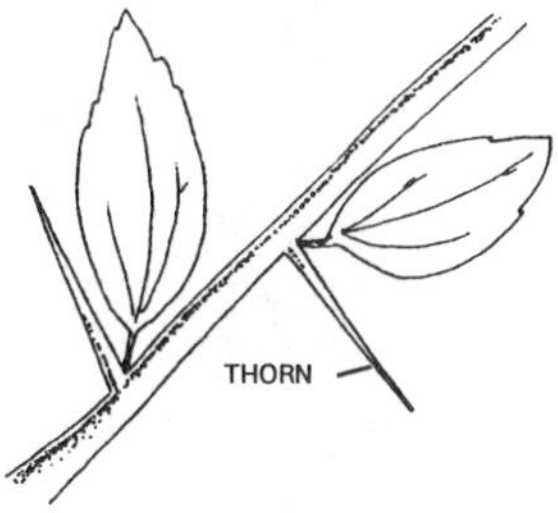

Figure 1120

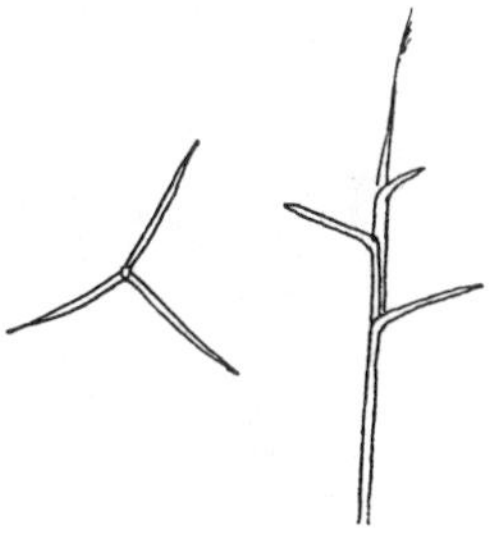

Figure 1121

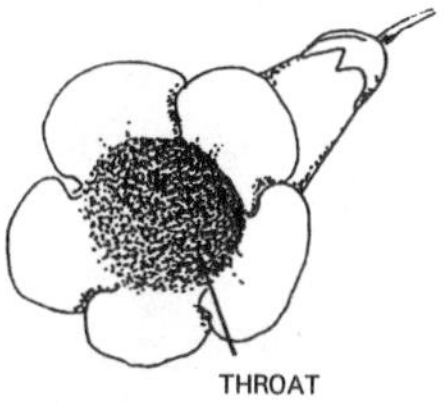

Figure 1122

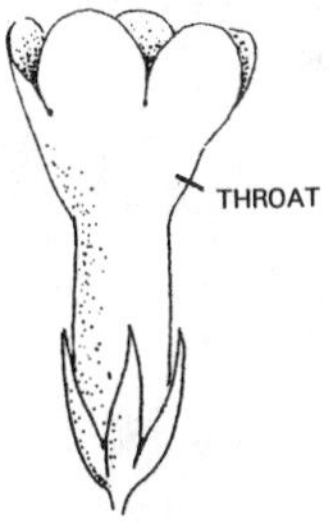

Figure 1123

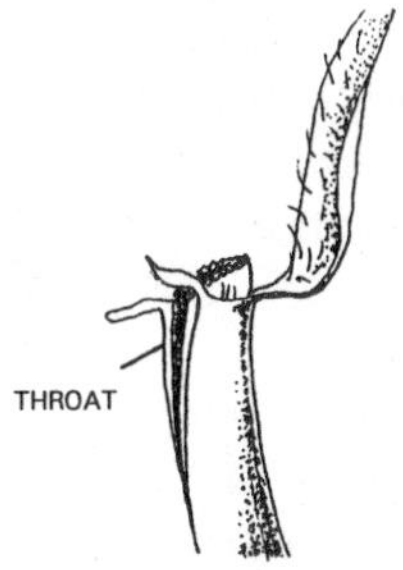

Figure 1124

Thrum. A heterostylic flower with a fairly short style and long stamens. Figure 1125. (compare **pin**)
Thyrse. A compact, cylindrical, or ovate panicle with an indeterminate main axis and cymose sub-axes. Figure 1126.
Thyrsoid. Thyrse-like.

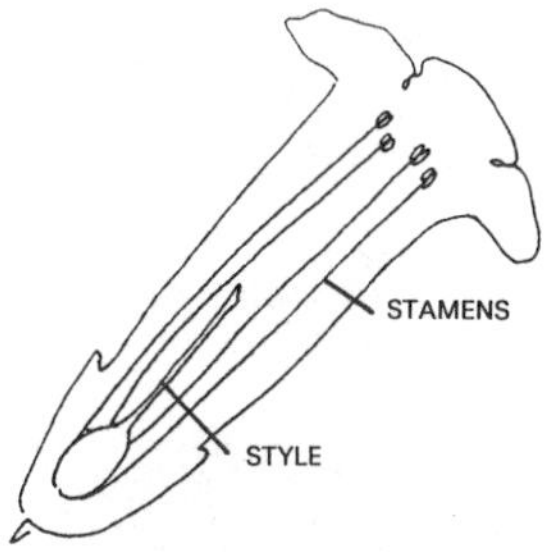

Figure 1125

Thyrsula. A small cyme borne in the leaf axil, as in many members of the Labiatae (Lamiaceae). Figure 1127.

Figure 1126

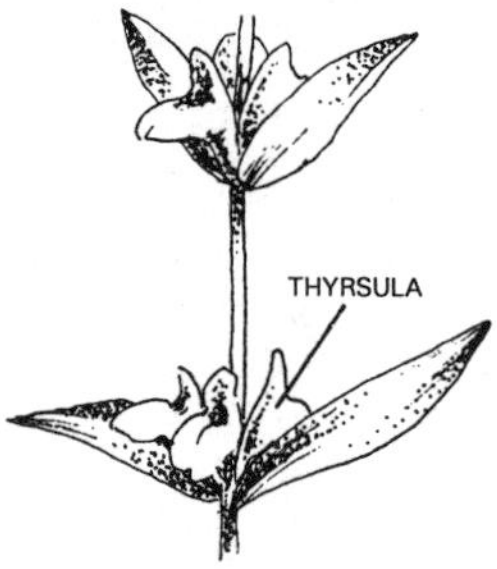

Figure 1127

Thyrsus. See **thyrse**.

Tiller. A basal or subterranean shoot which is more or less erect. Figure 1128. (compare **stolon** and **rhizome**)

Tillering. A type of vegetative reproduction accomplished by tiller production. Figure 1128.

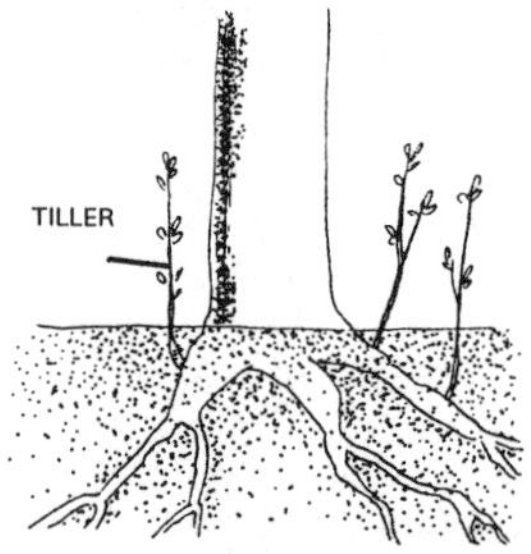

Figure 1128

Tissue. A group of cells organized to perform a specific function, as epidermal tissue or vascular tissue. Figure 1129.

Tolerant. Capable of growing in the shade.

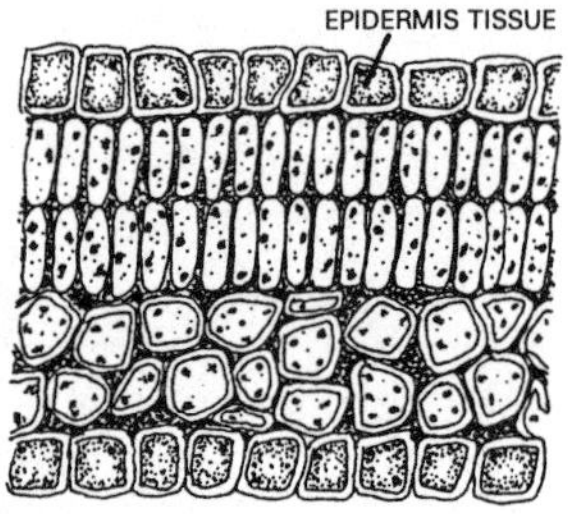

Figure 1129

Tomentellous. See **tomentulose**.

Tomentose. With a covering of short, matted or tangled, soft, wooly hairs; with tomentum. Figure 1130. (compare **lanate** and **canescent**)

Tomentulose. Slightly tomentose. Figure 1131.

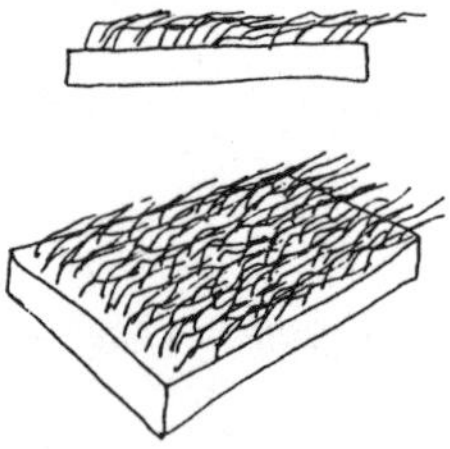

Figure 1130

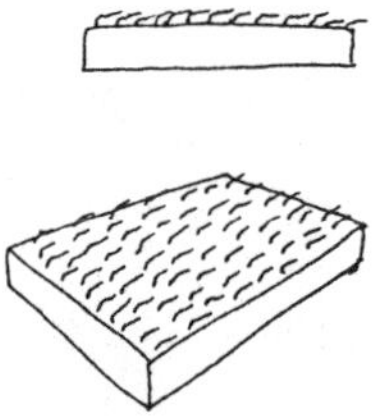

Figure 1131

Tomentum (pl. **tomenta**). A covering of short, soft, matted, wooly hairs. Figure 1130.

Tongue. Ligule. Figure 1132.

Tooth. Any small lobe or point along a margin. Figure 1133.

Toothed. Dentate. Figure 1133.

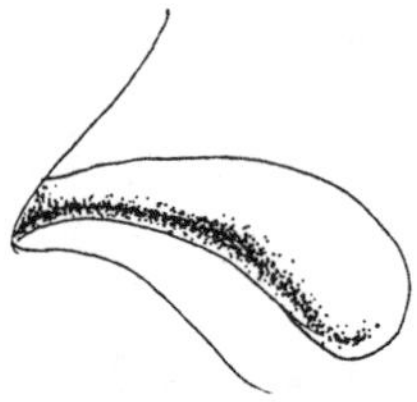

Figure 1132

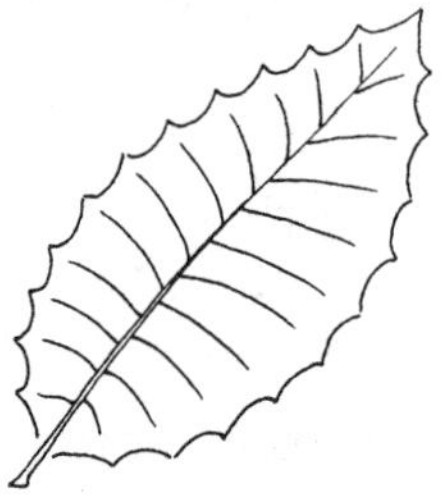

Figure 1133

Torose. Cylindrical with alternate swellings and contractions. Figure 1134.

Tortuous. Twisted or bent. Figure 1135.

Figure 1134

Figure 1135

Torulose. Slightly torose, as in a small fruit which is constricted between the seeds. Figure 1136.

Torus (pl. **tori**). The receptacle of a flower. Figure 1137.

Figure 1136

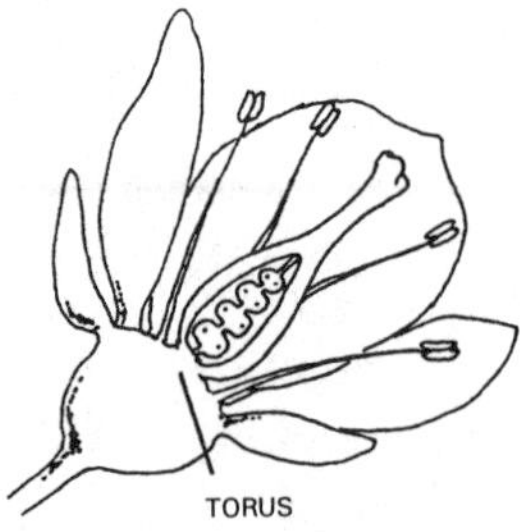

Figure 1137

Trabecula (pl. **trabeculae**). A structure resembling a beam or crossbar. Figure 1138.

Trabecular. Of or pertaining to trabeculae.

Trabeculate. With a crossbar. Figure 1138.

Tracheid. A xylem cell which is long, slender, and tapered at the ends. Figure 1139.

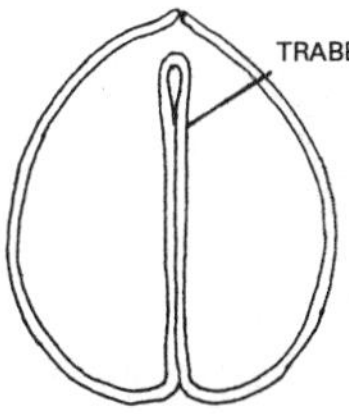

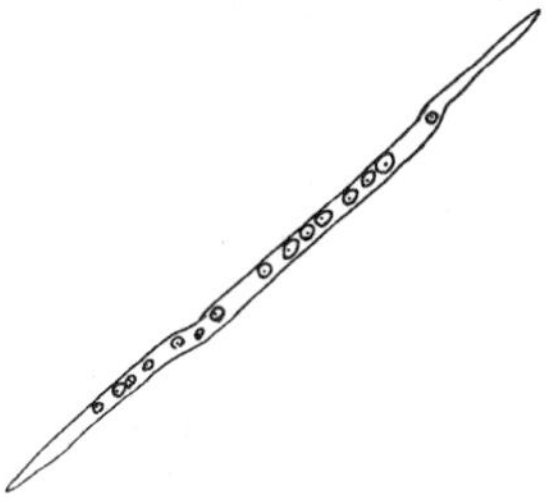

Figure 1138 **Figure 1139**

Trailing. Prostrate and creeping but not rooting. Figure 1140.

Transcorrugated. Corrugated transversely to the axis. Figure 1141.

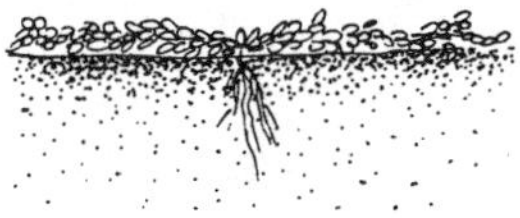

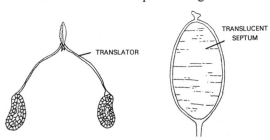

Figure 1140 **Figure 1141**

Translator. The connecting structure between the pollinia of adjacent anthers in the Asclepiadaceae. Figure 1142.

Translucent. Almost transparent. Figure 1143.

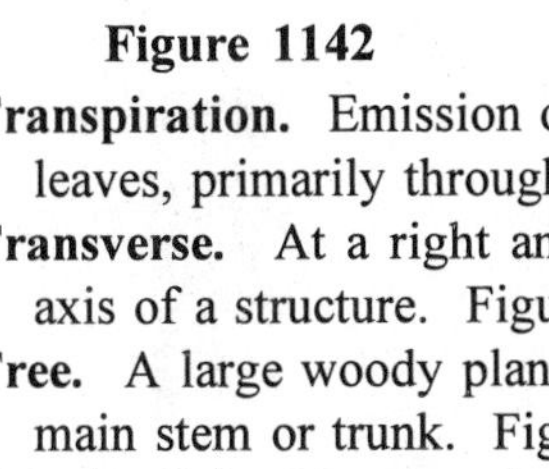

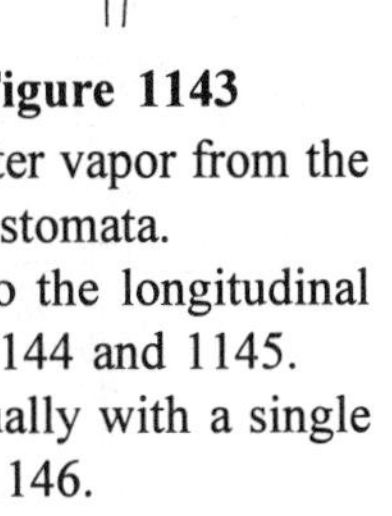

Figure 1142 **Figure 1143**

Transpiration. Emission of water vapor from the leaves, primarily through the stomata.

Transverse. At a right angle to the longitudinal axis of a structure. Figures 1144 and 1145.

Tree. A large woody plant, usually with a single main stem or trunk. Figure 1146.

Tri- (prefix). Meaning three.

Triadelphous. With stamens arranged into three groups. Figure 1147.

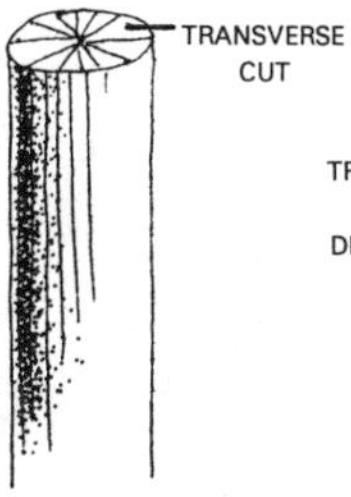

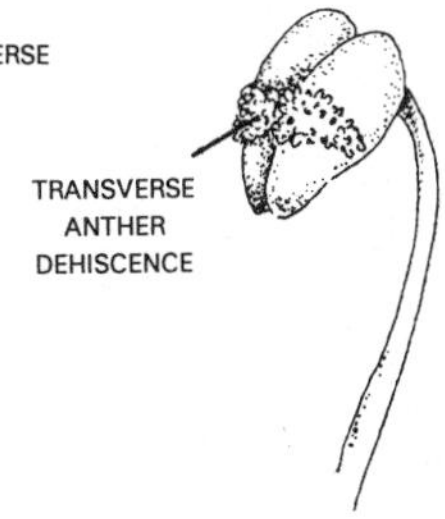

Figure 1144 **Figure 1145**

Figure 1146 **Figure 1147**

Triandrous. With three stamens. Figure 1148.

Triangulate. Three-angled. Figure 1149.

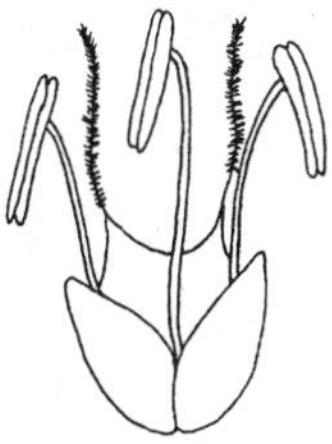
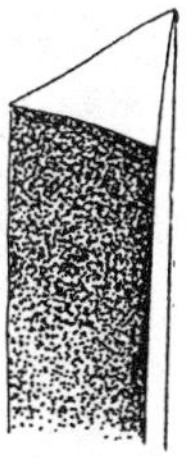

Figure 1148 **Figure 1149**

Triaristate. Three-awned. Figure 1150.

Tricamarous. With three locules. Figure 1151.

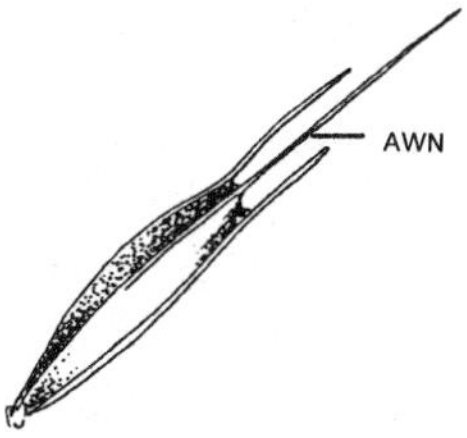

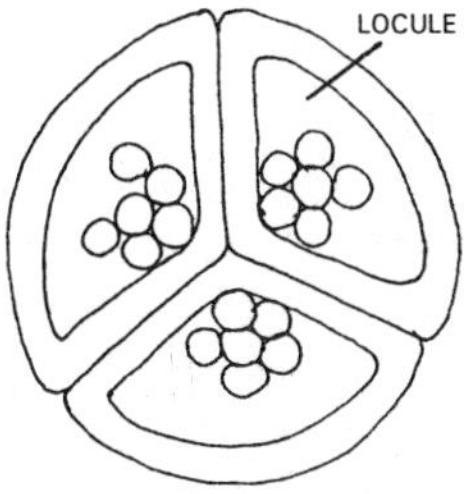

Figure 1150 **Figure 1151**

Tricarinate. With three ridges or keels. Figure

1152.

Tricarpellary. With three carpels. Figure 1153.

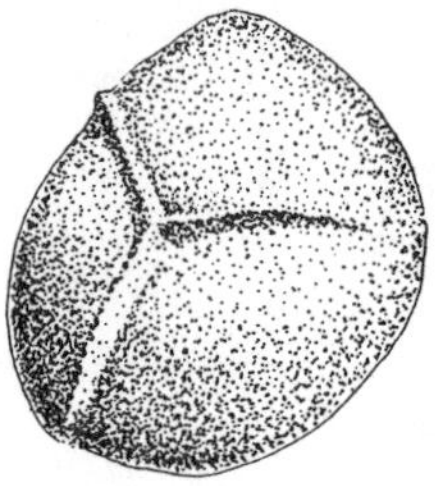
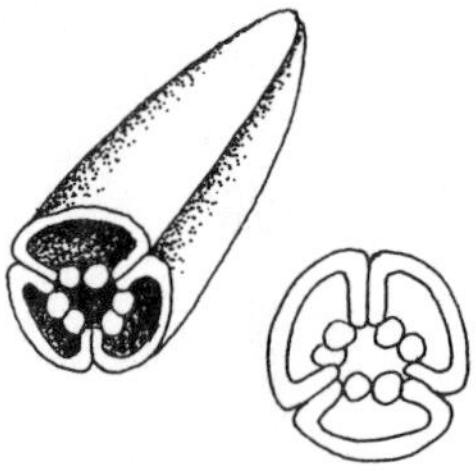

Figure 1152 **Figure 1153**

Trichasium. A cymose inflorescence with three branches. Figure 1154.

Trichocarpous. With hairy fruit. Figure 1155.

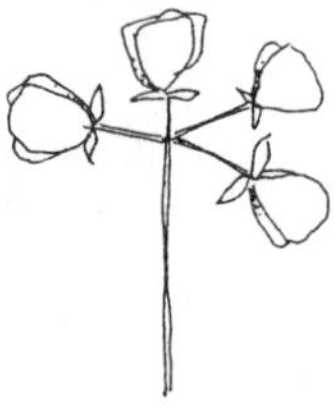

Figure 1154 **Figure 1155**

Trichome. A hair or hairlike outgrowth of the epidermis. Figure 1156.

Trichotomous. Three-forked. Figure 1157.

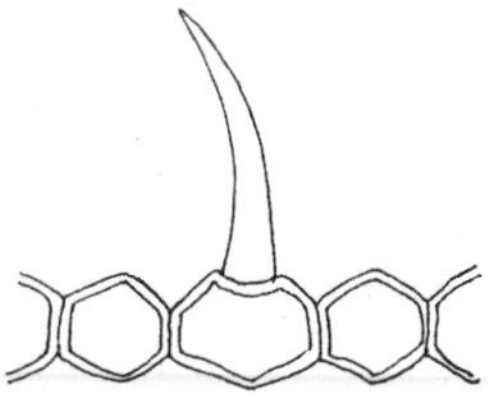

Figure 1156 **Figure 1157**

Tricolor. With three colors.

Tricyclic. With three whorls. Figure 1158.

Tridentate. Three-toothed. Figure 1159.

Tridigitate. Divided into three fingerlike lobes or divisions. Figure 1160.

Tridynamous. With stamens arranged in two groups of three, one group often longer than the other. Figure 1161.

Triecious. See **trioecious**.

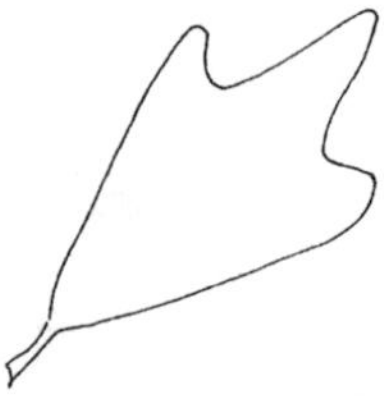

Figure 1158 **Figure 1159**

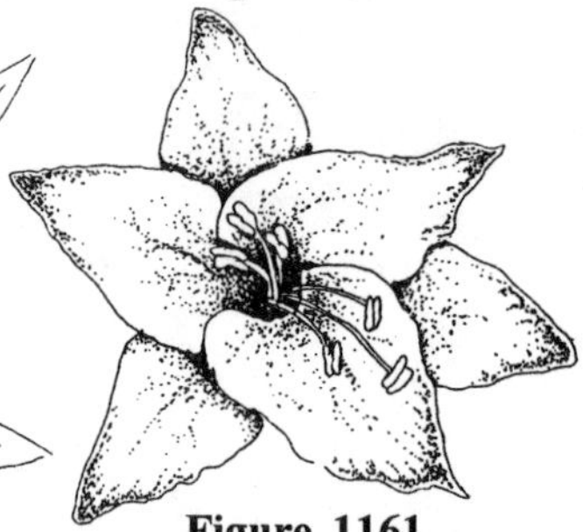

Figure 1160 **Figure 1161**

Trifid. Three-cleft. Figure 1162.

Trifoliate. With three leaves or three leaflets. Figure 1163.

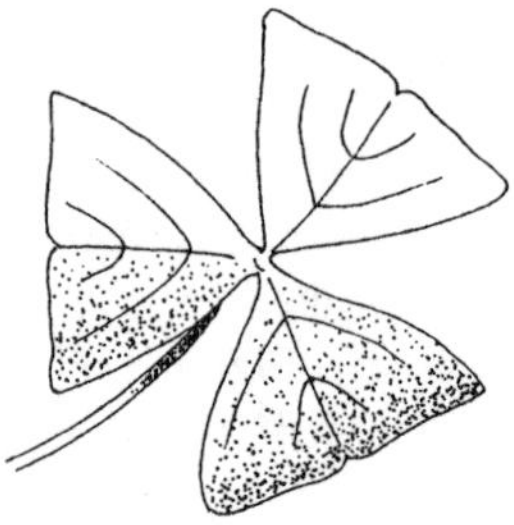

Figure 1162 **Figure 1163**

Trifoliolate. See **trifoliate**.

Trifurcate. Three-forked; divided into three branches. Figure 1164.

Trigonal. See **trigonous**.

Trigonous. Three-angled. Figure 1165.

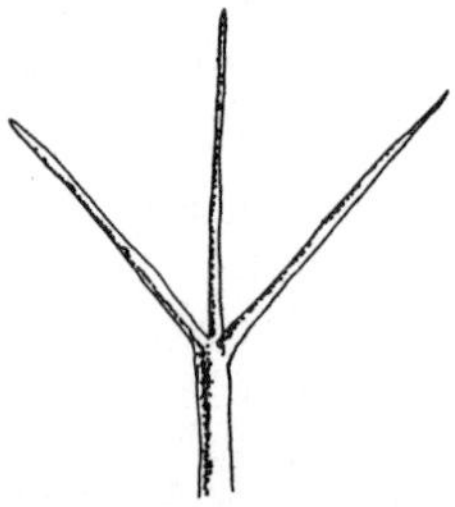
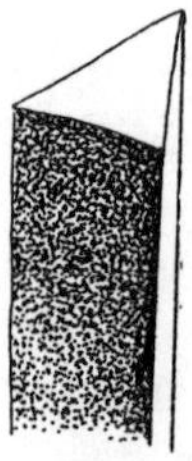

Figure 1164 **Figure 1165**

Trilobate. With three lobes. Figure 1162.
Trilocular. With three locules. Figure 1166.
Trimerous. With parts arranged in sets or multiples of three. Figure 1167.

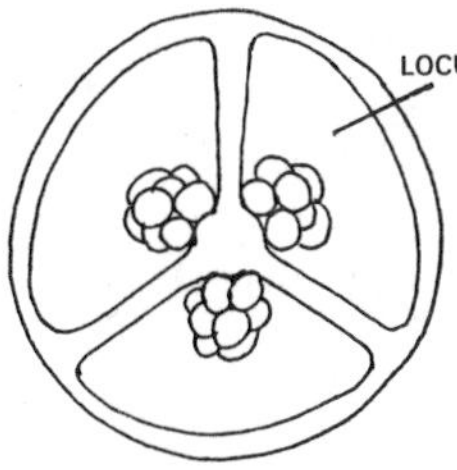

Figure 1166 Figure 1167

Trimonoecious. With male, female, and bisexual flowers on the same plant.
Trimorphic. With three forms.
Trinervate. See **trinerved**.
Trinerved. Three-nerved, with the nerves all arising from near the base. Figure 1168. (compare **triplinerved**)

Figure 1168

Trioecious. With male, female, and bisexual flowers on different plants.
Tripalmate. Palmately compound three times. Figure 1169.
Tripartite. Three-parted. Figure 1170.

Figure 1169 Figure 1170

Tripetalous. With three petals. Figure 1161.
Triphyllous. With three leaves.
Tripinnate. Pinnately compound three times, with pinnate pinnules. Figure 1171.
Tripinnatifid. Thrice pinnately cleft. Figure 1172.

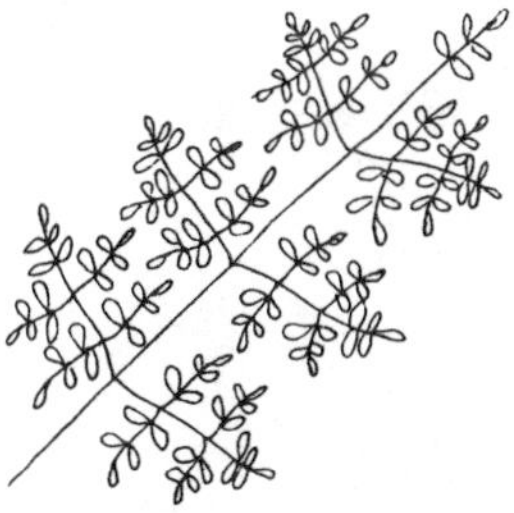

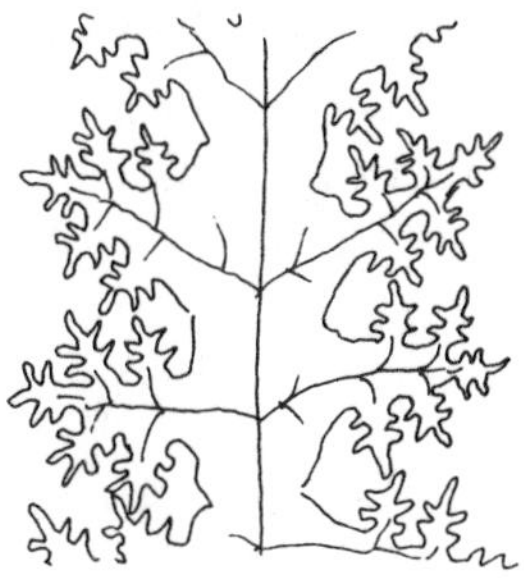

Figure 1171 Figure 1172

Triple-nerved. See **triplinerved**.
Triplinerved. Three-nerved, with the two lateral nerves arising from the midnerve above the base. Figure 1173. (compare **trinerved**)
Tripterous. With three wings or winglike appendages. Figure 1174.

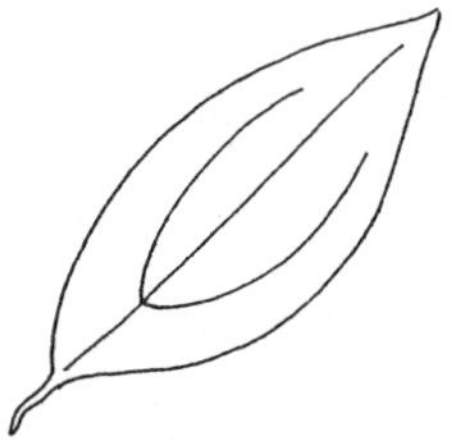

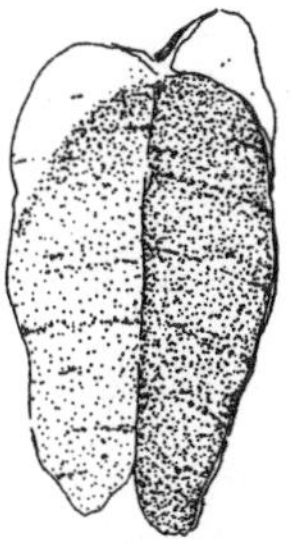

Figure 1173 Figure 1174

Triquetrous. Three-edged; with three protruding angles. Figure 1175.
Trispermous. Three-seeded.
Tristichous. In three vertical ranks or rows; three-ranked. Figure 1176.

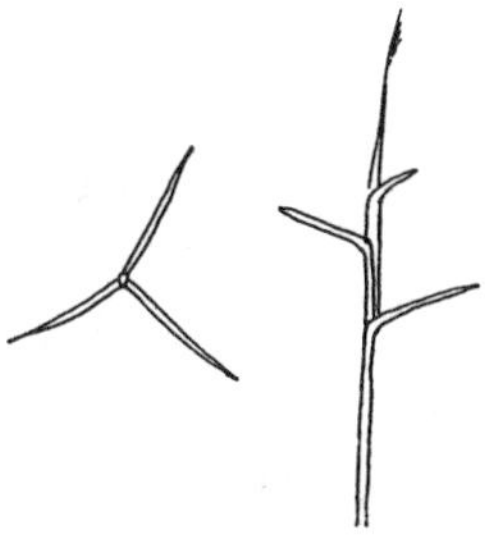

Figure 1175 Figure 1176

Tristylous. With three styles. Figure 1177.
Trisulcate. With three furrows or grooves. Figure 1178.
Triternate. Triply ternate. Figure 1179.
Tropical. Distributed in the tropics (i.e. between the tropic of Cancer and the tropic of Capricorn,

or between 23 ½ degrees north latitude and 23 ½ degrees south latitude).

Truncate. With the apex or base squared at the end as if cut off. Figure 1180.

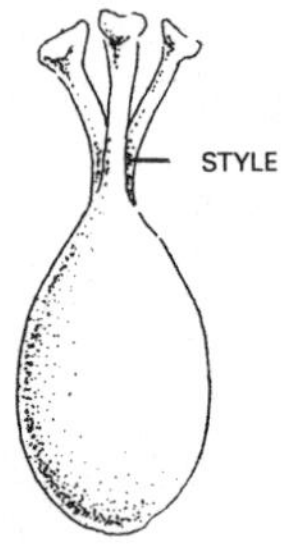

Figure 1177

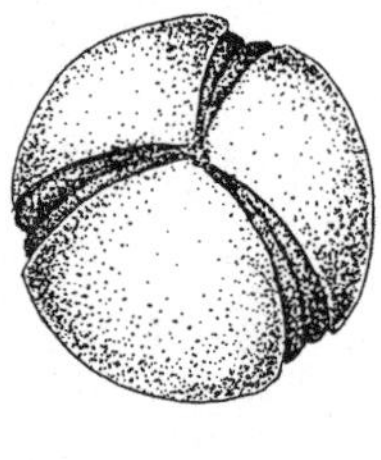

Figure 1178

Figure 1179

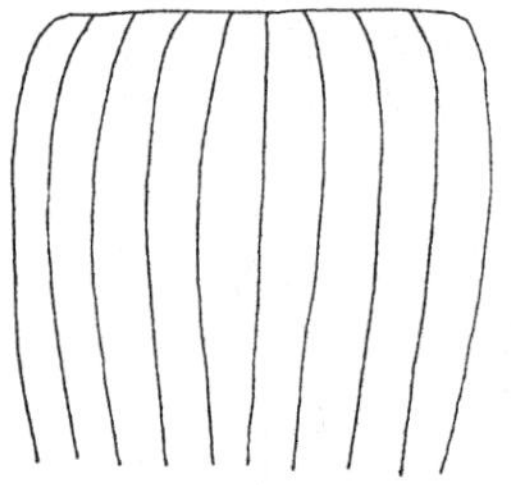

Figure 1180

Trunk. The main stem of a tree below the branches. Figure 1181.

Tryma. A drupelike nut with a fleshy, dehiscent exocarp, as a walnut or hickory nut. Figure 1182.

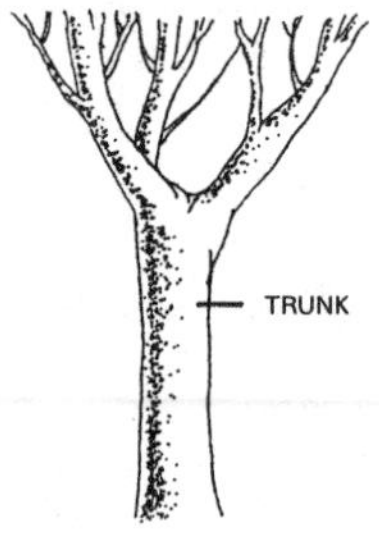

Figure 1181

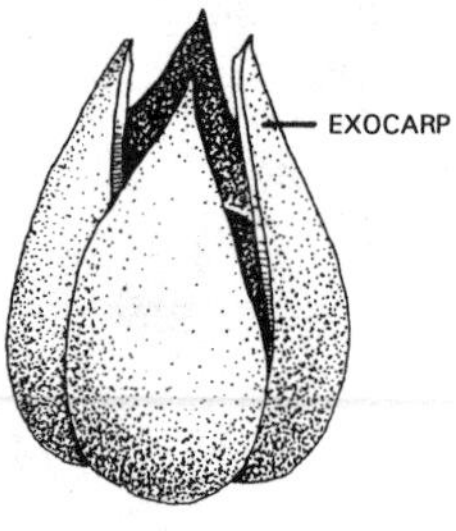

Figure 1182

Tube. A hollow, cylindrical structure, as the constricted basal portion of some gamopetalous corollas. Figure 1183.

Tuber. The thickened portion of a rhizome bearing nodes and buds; underground stem modified for food storage. Figure 1184.

Tubercle. A small tuberlike swelling or projection. Figures 1185 and 1186; the base of the

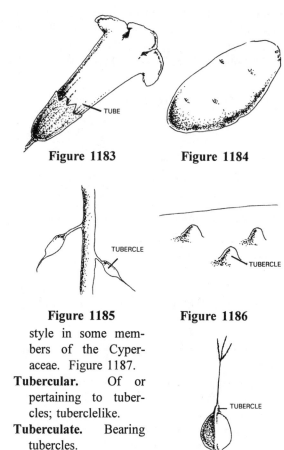

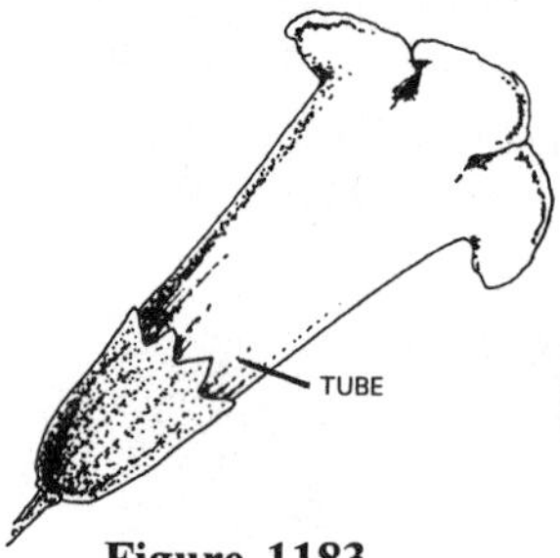

Figure 1183

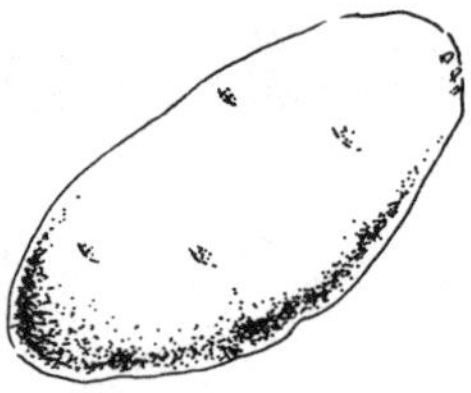

Figure 1184

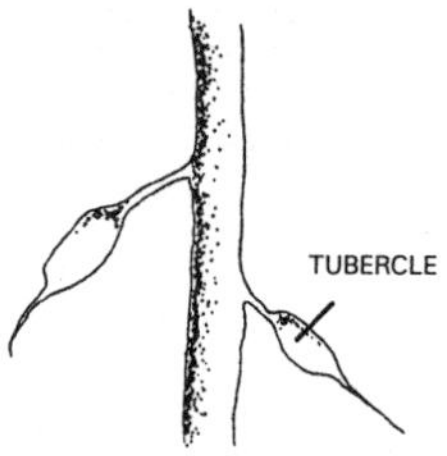

Figure 1185

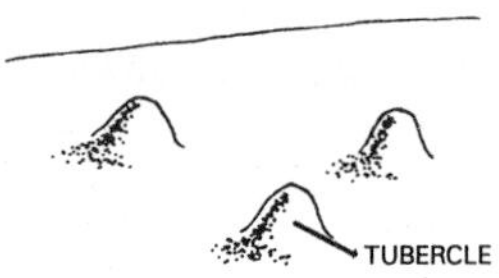

Figure 1186

style in some members of the Cyperaceae. Figure 1187.

Tubercular. Of or pertaining to tubercles; tuberclelike.

Tuberculate. Bearing tubercles.

Tuberculation. See **tubercle**.

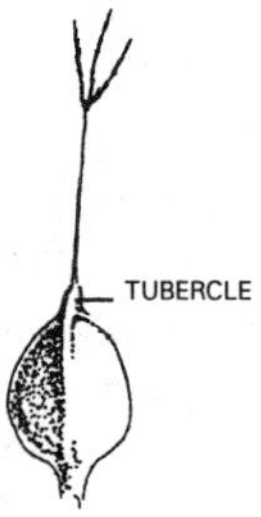

Figure 1187

Tubercule. A nodule, as on the roots of some legumes. Figure 1188.

Tuberiferous. See **tuberculate**.

Tuberoid. A thickened root which resembles a tuber.

Tuberous. Resembling a tuber; producing tubers.

Tubular. With the form of a tube or cylinder. Figure 1189.

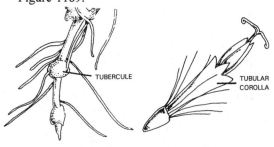

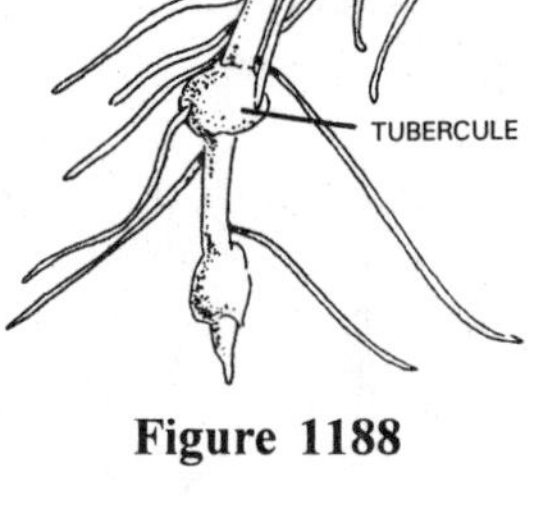

Figure 1188

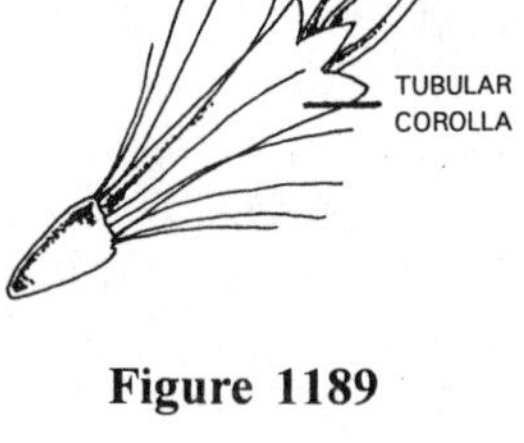

Figure 1189

Tubuliflorous. Having tubular corollas in the perfect flowers of a head, as in some members of the Compositae (Asteraceae).

Tubulous. With tubular flowers. Figure 1190.

Tufted. Arranged in a dense cluster.

Figure 1190

Tumescent. Somewhat tumid; swelling. Figure 1191.

Tumid. Swollen. Figure 1192.

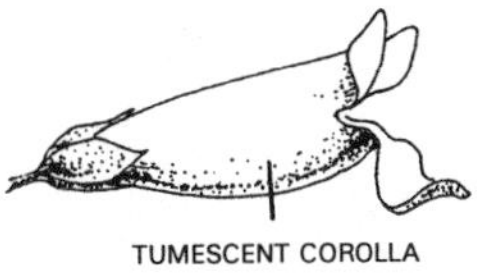

Figure 1191

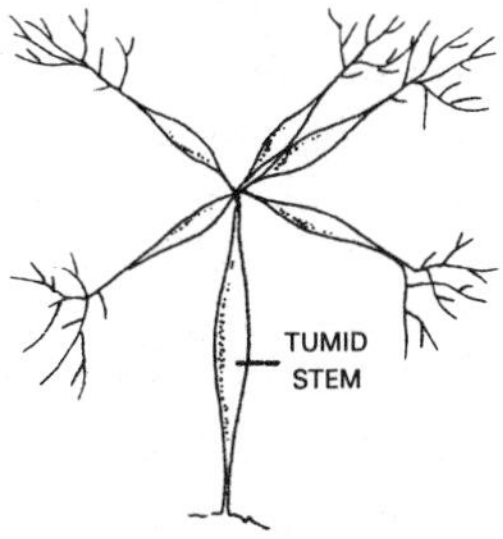

Figure 1192

Tunic. An integument; the outer coating of a seed or bulb. Figure 1193.

Tunicate. Arranged in sheathing, concentric layers, as the leaves of an onion bulb. Figure 1194.

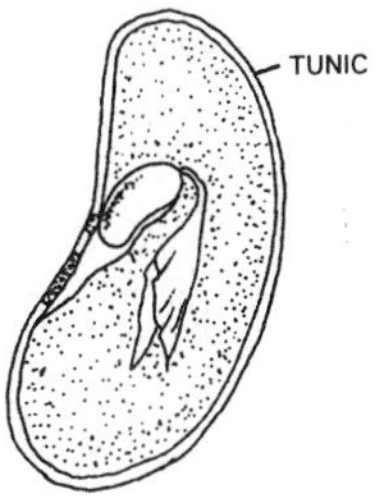

Figure 1193

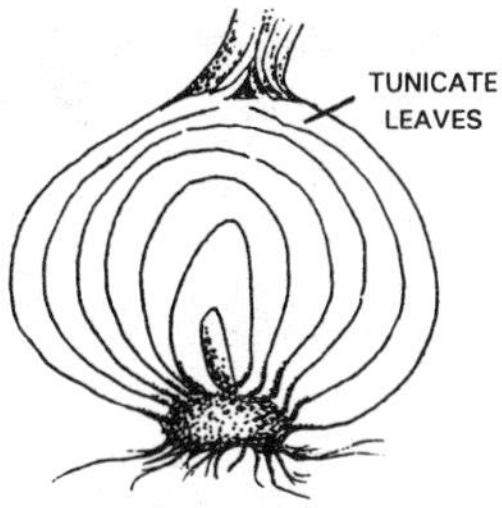

Figure 1194

Turbinate. Top-shaped. Figure 1195.

Turgid. Swollen; expanded or inflated. Figure 1196.

Turion. A small shoot which often over winters, as in some species of *Epilobium*.

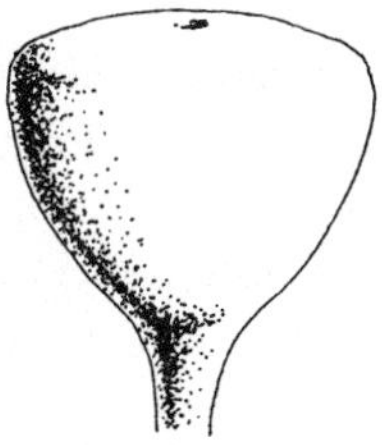

Figure 1195

Tussock. A tuft or clump of grasses or sedges. Figure 1197.

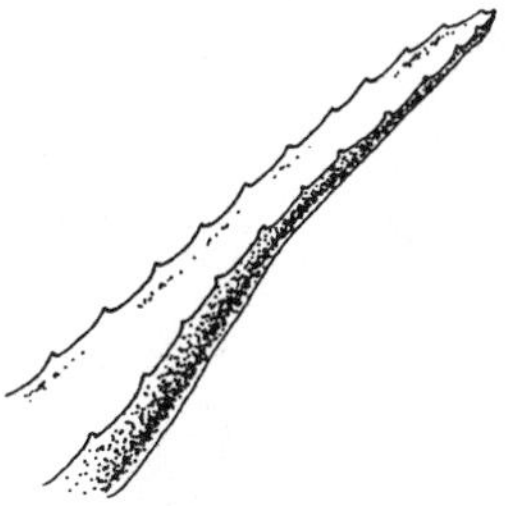

Figure 1196

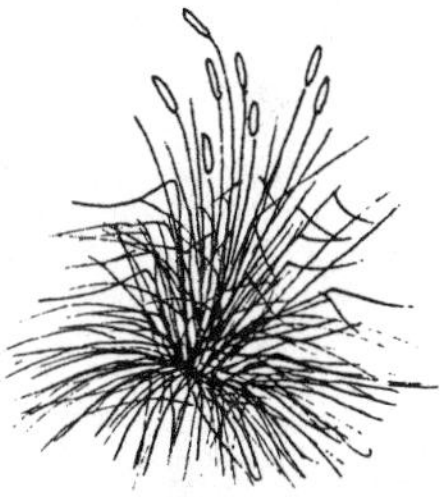

Figure 1197

Twig. A small shoot or branch from a tree. Figure 1198.

Twining. Coiling or spiraling around a support (usually another stem) for climbing. Figure 1199.

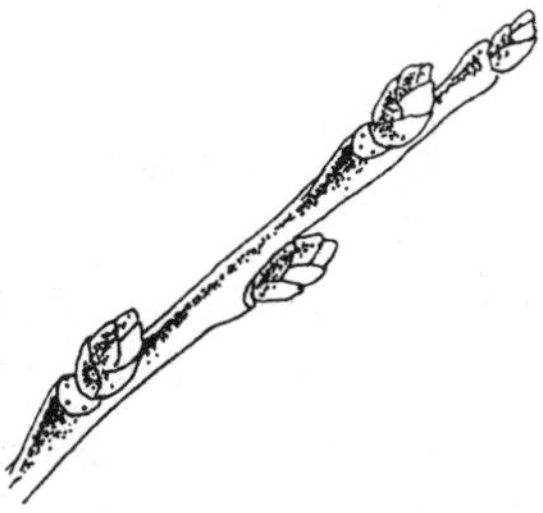

Figure 1198

Figure 1199

Two-ranked. In two vertical ranks or rows on opposite sides of an axis; distichous. Figure 1200.

Type. The specimen that serves as the basis for a plant name.

Figure 1200

Ubiquitous. Widespread; occurring in a wide variety of habitats.

Ultimate. The final section or division of a structure. Figure 1201.

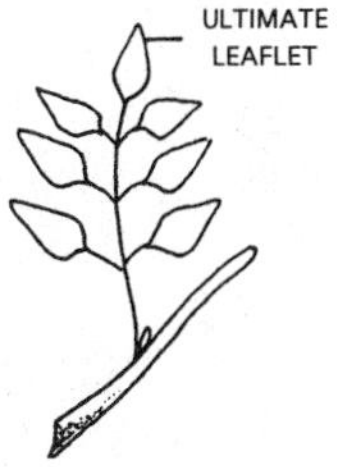

Figure 1201

Umbel. A flat-topped or convex inflorescence with the pedicels arising more or less from a common point, like the struts of an umbrella. Figures 1202 and 1203; a highly condensed raceme.

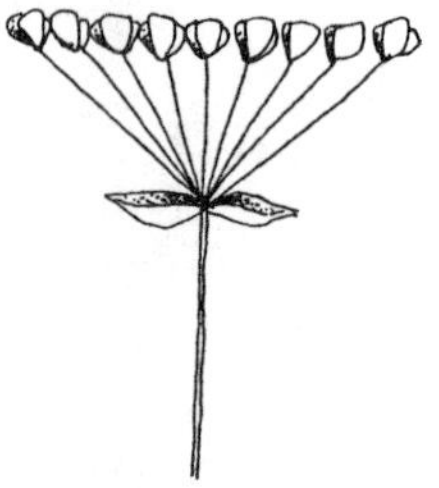

Figure 1202

Figure 1203

Umbellate. In umbels; umbel-like.

Umbellet. An ultimate umbellate cluster of a compound umbel. Figure 1204.

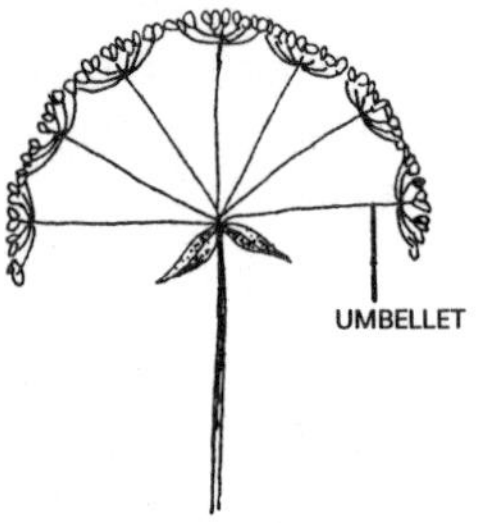

Figure 1204

Umbelliferous. Bearing umbels; pertaining or belonging to the Umbelliferae (Apiaceae) family.

Umbelliform. An inflorescence with the general appearance, but not necessarily the structure, of a true umbel. The term is often applied to inflorescences which are condensed cymes rather than condensed racemes.

Umbellule. See **umbellet**.

Umbilicate. With a depression in the middle, like a navel. Figure 1205.

Umbilicus. A navellike structure, as the hilum of a seed. Figure 1206.

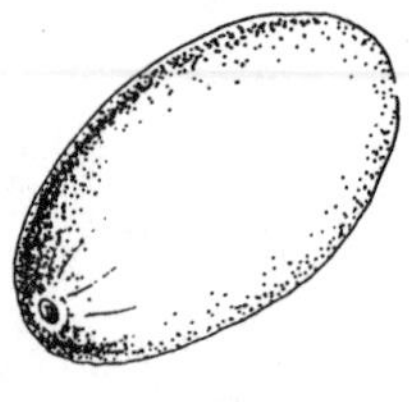

Figure 1205

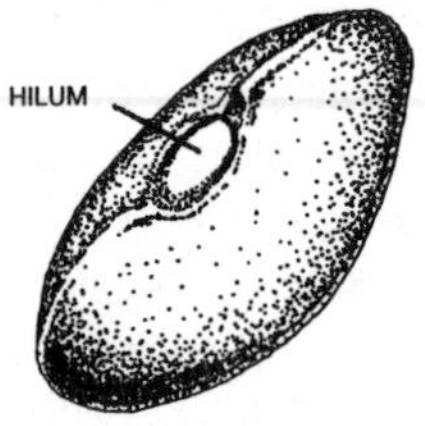

Figure 1206

Umbo. A blunt or rounded protuberance, as on the ends of the scales of some pine cones. Figure 1207.

Umbonate. Possessing an umbo. Figure 1207.

Umbonulate. With a very small umbo.

Umbraculate. Umbrella-shaped. Figure 1208.

Umbraculiferous. Bearing umbrella-shaped structures. Figure 1208.

Figure 1207

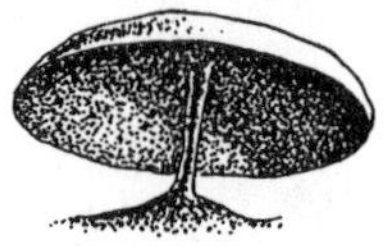

Figure 1208

Umbraculiform. See **umbraculate**.

Unarmed. Lacking spines, prickles, or thorns.

Uncate. See **uncinate**.

Uncinate. Hooked at the tip. Figure 1209.

Figure 1209

Unctuous. Greasy; oily.

Undate. See **undulate**.

Undershrub. See **subshrub**.

Undulate. Wavy, but not so deeply or as pronounced as sinuate. Figure 1210. (Same as **repand**.)

Unequally pinnate. See **odd-pinnate**.

Unguiculate. Clawed. Figure 1211.

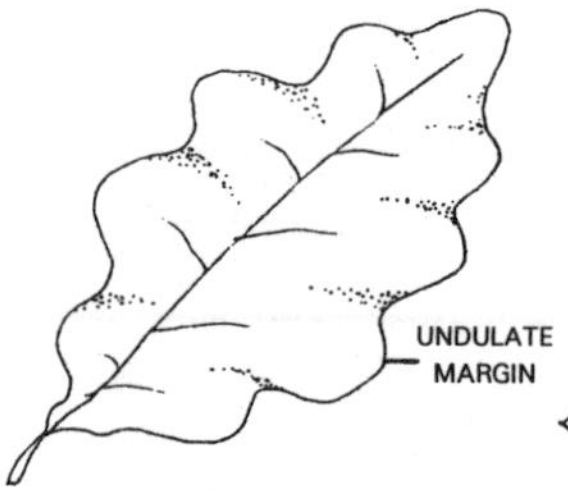

Figure 1210

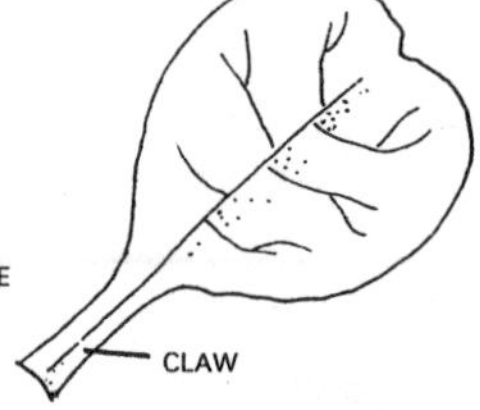

Figure 1211

Ungulate. See **unguiculate**.

Uni- (prefix). Meaning one.

Uniaristate. One-awned. Figure 1212.

Uniaxial. With a single unbranched stem terminating in a flower. Figure 1213.

Unicarpellate. See **unicarpellous**.

Unicarpellous. With a single, free carpel. Figure

1214.

Unicostate. With a single obvious rib, as in some leaves. Figure 1215.

Figure 1212

Figure 1213

Figure 1214

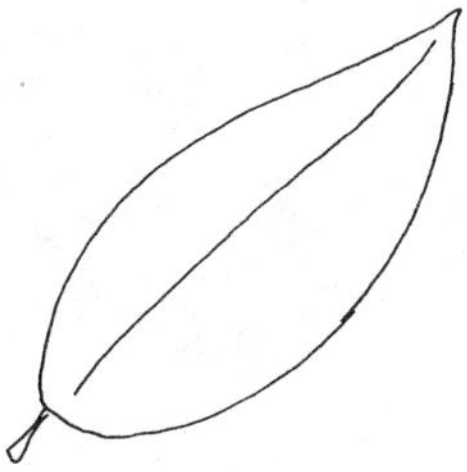

Figure 1215

Uniflorous. With a single flower. Figure 1213.

Unifoliate. With a single leaf; unifoliolate. Figure 1216.

Unifoliolate. A leaf theoretically compound, though only expressing a single leaflet and appearing simple, as in *Cercis*. Figure 1217.

Figure 1216

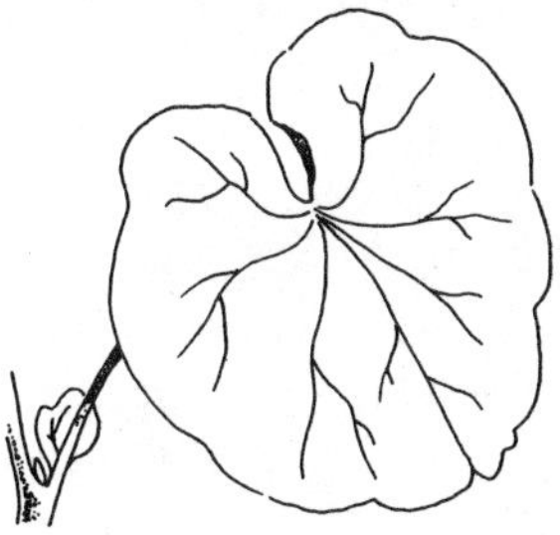

Figure 1217

Unijugate. A leaf pinnately compound, but consisting of only two leaflets. Figure 1218.

Unilateral. One-sided, as in an inflorescence with the flowers all on one side of the axis. Figure 1219.

Unilocular. With a single locule or compartment, as in some ovaries. Figure 1220.

Uniparous. With only a single axis produced at each branching, as in some cymes. Figure 1221.

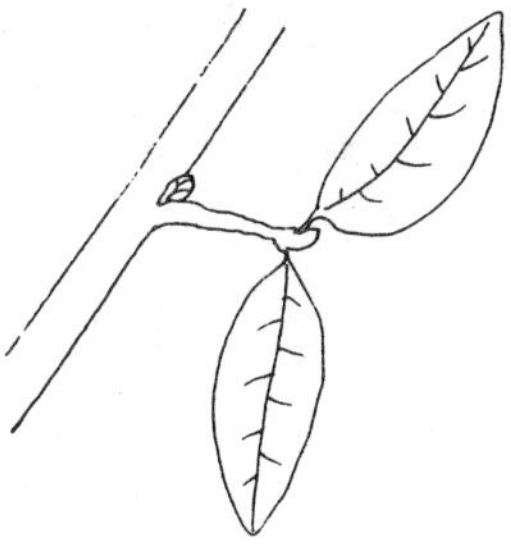

Figure 1218

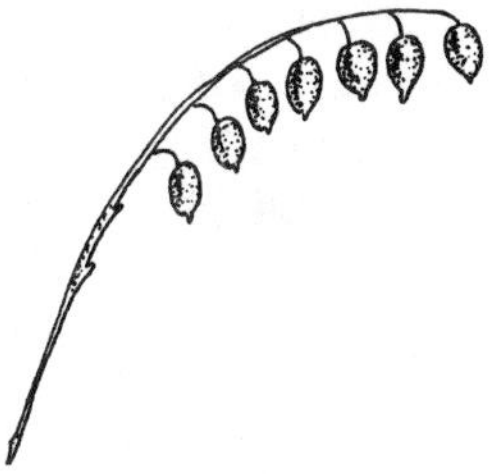

Figure 1219

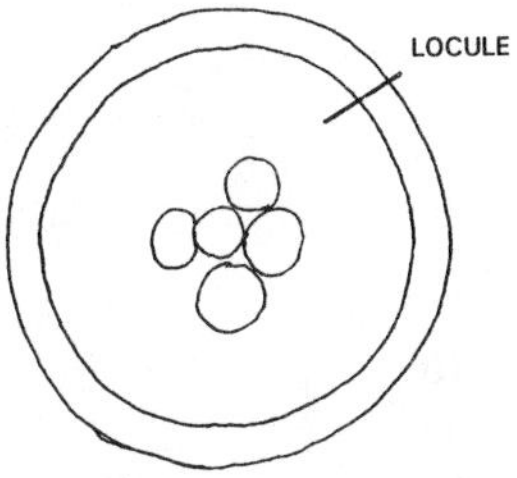

Figure 1220

Figure 1221

Unipetalous. With only a single petal.

Uniseptate. With only one septum, as in a silicle or silique. Figure 1222.

Uniseriate. Arranged in a single row or series. Figure 1223.

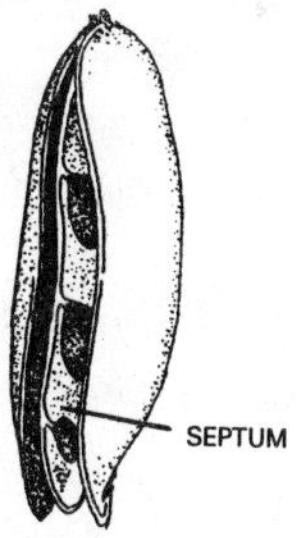

Figure 1222

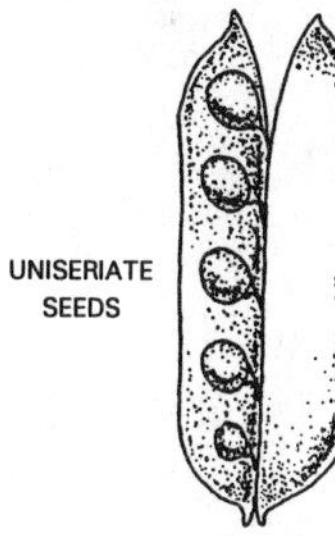

Figure 1223

Unisexual. A flower with either male or female reproductive parts, but not both. The term is also applied to plants possessing such flowers. Figure 1224. (compare **bisexual** and **perfect**)

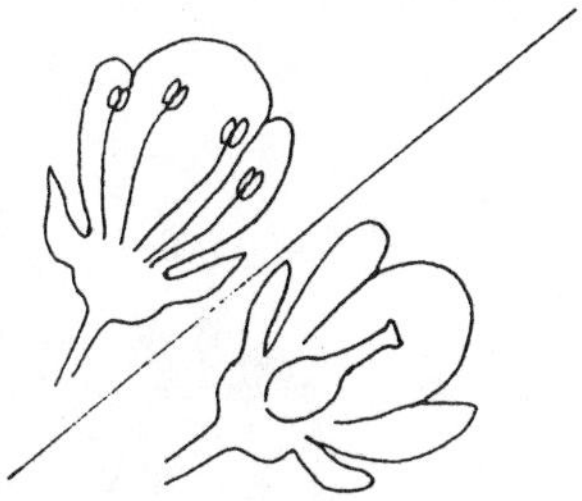

Figure 1224

Urceolate. Pitcherlike; hollow and contracted near

the mouth like a pitcher or urn. Figure 1225.

Urceolus. Any structure resembling a small pitcher.

Urceus. Any structure resembling a pitcher. Figure 1225.

Urent. Stinging. Figure 1226.

Figure 1225

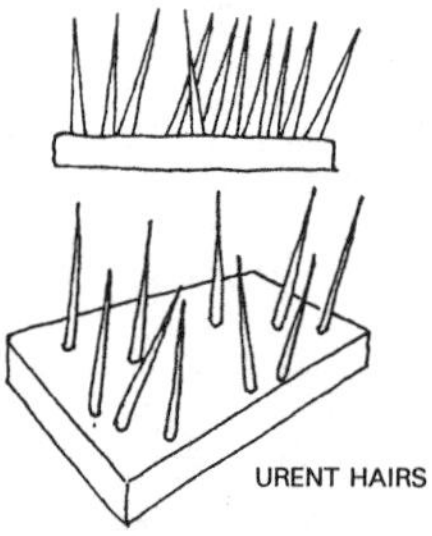

Figure 1226

Urn. The basal portion of a pyxis. Figure 1227.

Urn-shaped. See **urceolate**.

Utricle. A small, thin-walled, one-seeded, more or less bladdery-inflated fruit. Figure 1228.

Utricular. Of or pertaining to a utricle; bladder-like. Figure 1228.

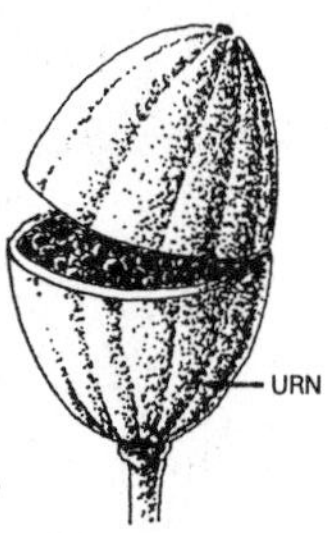

Figure 1227

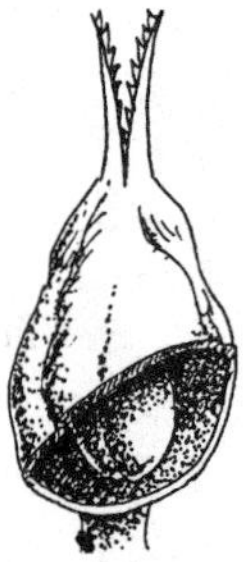

Figure 1228

Vagina. A sheath, as the sheathing petiole in grasses. Figure 1229.

Vaginate. Sheathed. Figure 1229.

Vaginiferous. Bearing sheaths.

Vallecula (pl. **valleculae**). A furrow, groove or depression. Figure 1230.

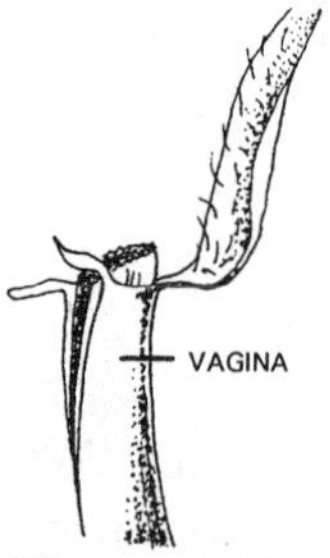

Figure 1229

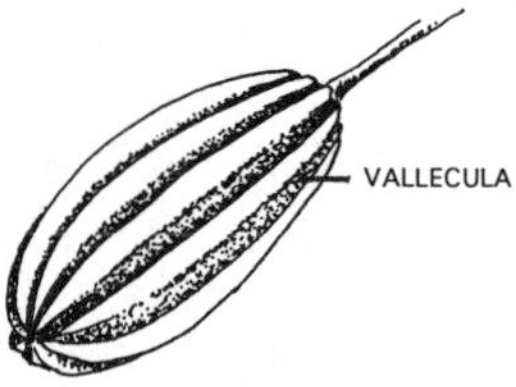

Figure 1230

Vallecular. Of or pertaining to the valleculae.

Valleculate. Having valleculae. Figure 1230.

Valvate. Opening by valves, as in many dehiscent fruits. Figure 1231; a flower with the petals or sepals edge to edge along their entire length, but not overlapping. Figure 1232.

Valve. One of the segments of a dehiscent fruit, separating from other such segments at maturity. Figure 1231.

Varicose. Swollen or enlarged in places. Figure 1233.

Variegated. Marked with patches or spots of different colors. Figure 1234.

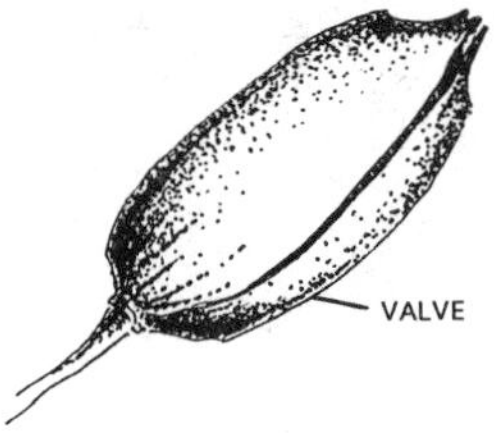

Figure 1231

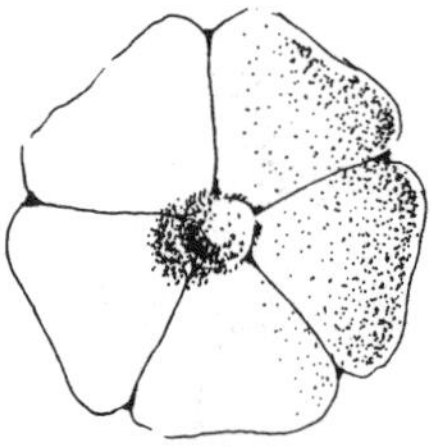

Figure 1232

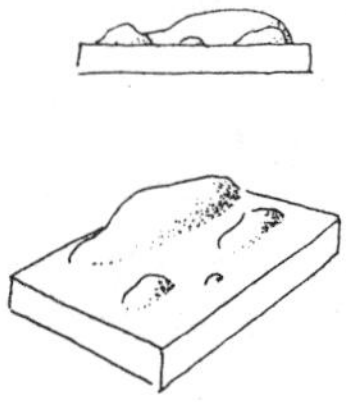

Figure 1233

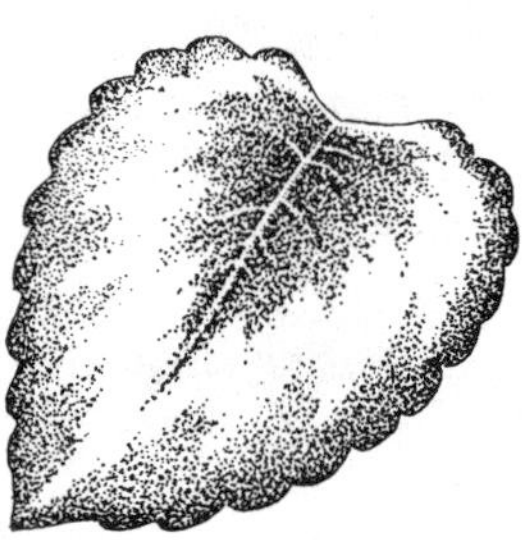

Figure 1234

Variety. A category in the taxonomic hier-archy below the species and subspecies level.

Vascular. Conductive tissue (i.e., xylem and phloem); plants possessing such conductive tissue.

Vascular bundle. A cluster or group of vascular tissues. Figure 1235.

Vegetative. The non-floral parts of a plant.

Vein. A vascular bundle, usually visible externally, as in leaves.

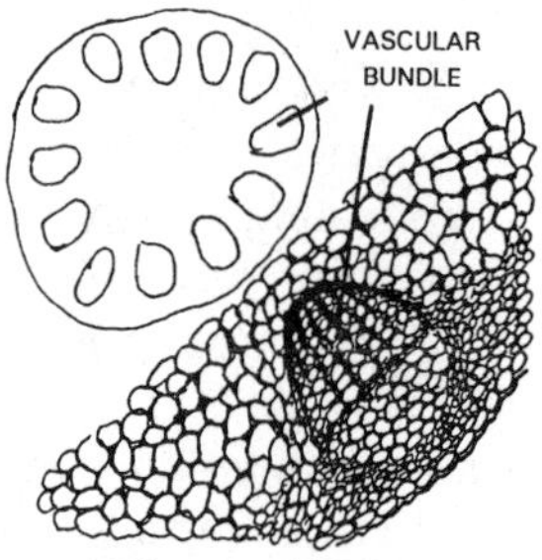

Figure 1235

Figure 1236.

Veinlet. A small vein. Figure 1236.

Velamen (pl. **velamina**). The thick, spongy integument layer on the roots of some epiphytic orchids.

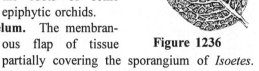

Velum. The membranous flap of tissue partially covering the sporangium of *Isoetes*. Figure 1237.

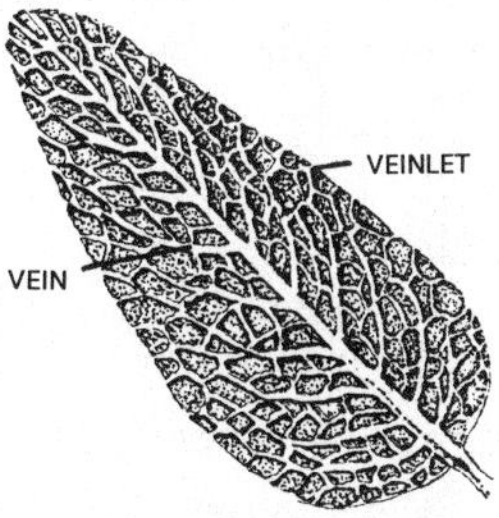

Figure 1236

Velumen. A covering of short, soft hairs. Figure 1238.

Velutinous. Velvety; covered with short, soft, spreading hairs. Figure 1238.

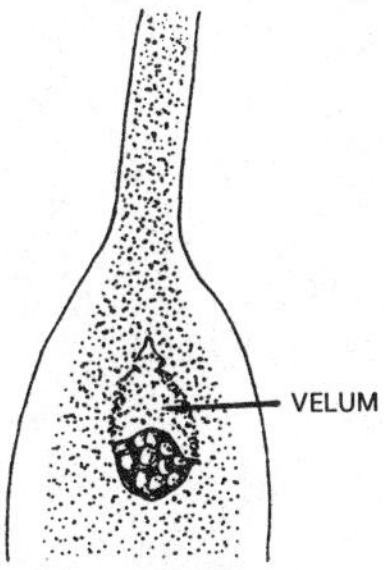

Figure 1237

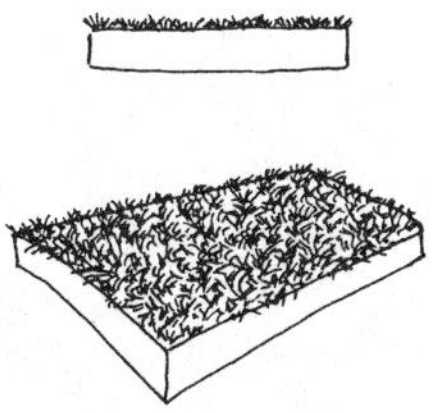

Figure 1238

Venation. The pattern of veining on a leaf. Figure 1236.

Venenose. See **venomous.**

Venomous. Poisonous.

Venose. Veiny; venous. Figure 1236.

Venous. Of or pertaining to veins; veinlike.

Ventral. Pertaining to the front or inward surface of an organ in relation to the axis, as in the upper surface of a leaf; adaxial. Figure 1239. (compare **dorsal**)

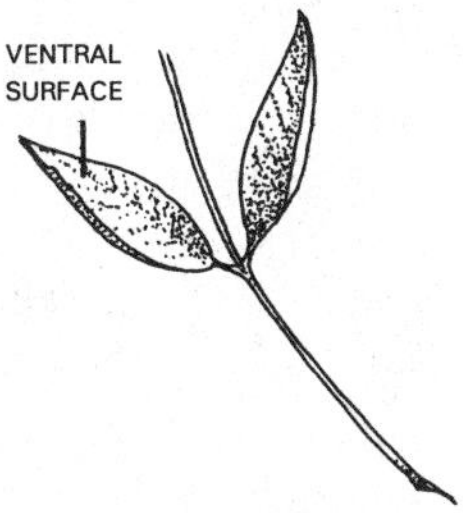

Figure 1239

Ventricose. Inflated or swollen on one side only, as in some corollas, especially in the genus *Penstemon*. Figure 1240.

Venulose. With veinlets. Figure 1236.

Venulous. See **venulose.**

Vermicular. See **vermiform.**

Vermiform. Worm-shaped. Figure 1241.

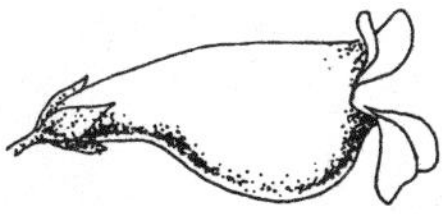

Figure 1240

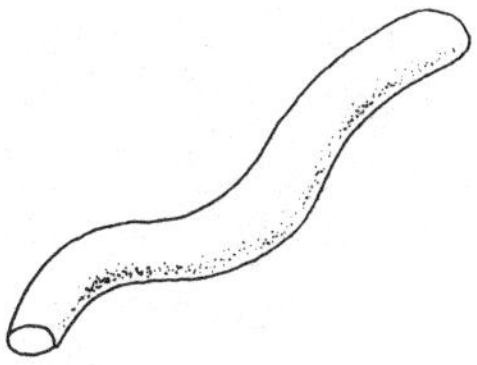

Figure 1241

Vernal. Flowering or appearing in the spring.

Vernation. The arrangement of leaves within the bud.

Verrucose. Warty; covered with wartlike elevations. Figure 1242.

Versatile. Attached near the middle rather that at one end, as some anthers. Figure 1243. (compare **basifixed** and **dorsifixed**)

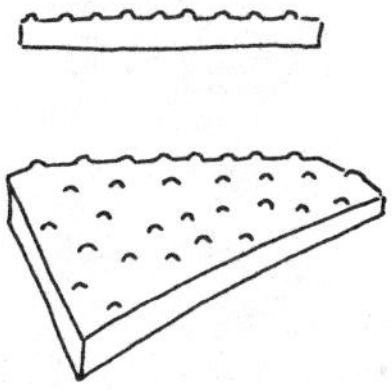

Figure 1242

Figure 1243

Versicolor. Of various colors; changeable in color.

Vertical. Positioned lengthwise, in the same direction as the axis; leaves positioned with the blade perpendicular, so that neither surface is obviously the upper or lower. Figure 1244.

Verticil. An arrangement of similar parts around a central axis or point of attachment; a whorl. Figure 1245.

Figure 1244

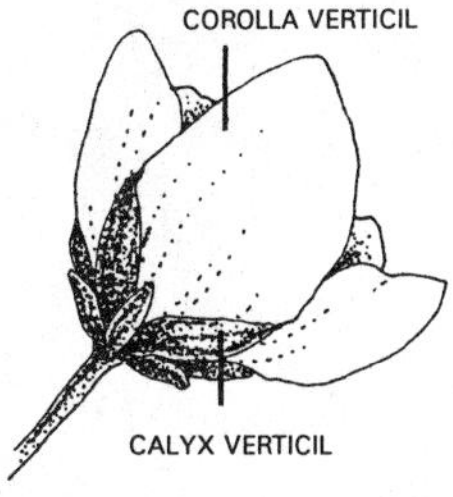

Figure 1245

Verticillaster. A pair of axillary cymes arising from opposite leaves or bracts and forming a false whorl. Figure 1246.

Verticillate. Arranged in verticils; whorled. Figure 1247.

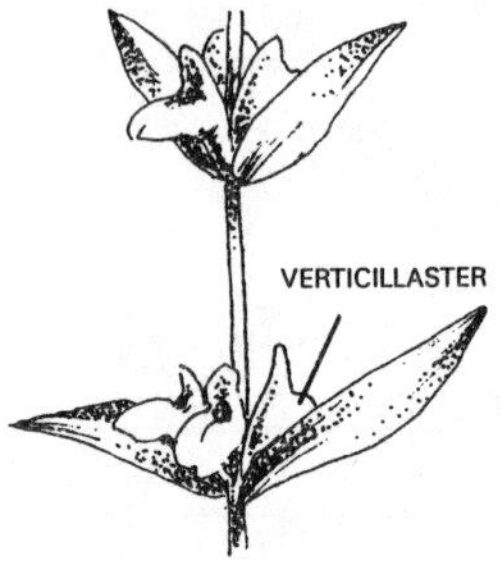

Figure 1246

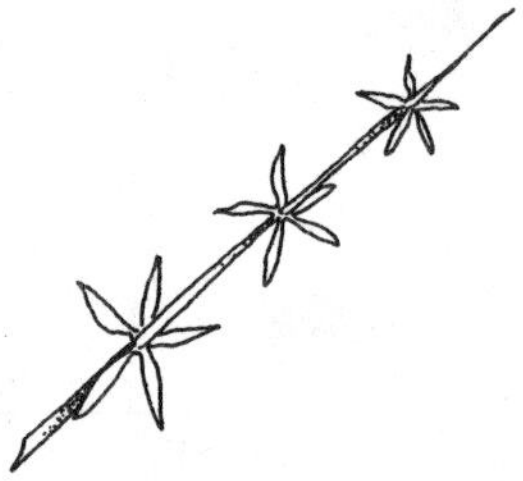

Figure 1247

Vesicle. A small bladderlike structure. Figure 1248.

Vesicular. Of or pertaining to vesicles.

Vespertine. Opening, or functioning, in the evening.

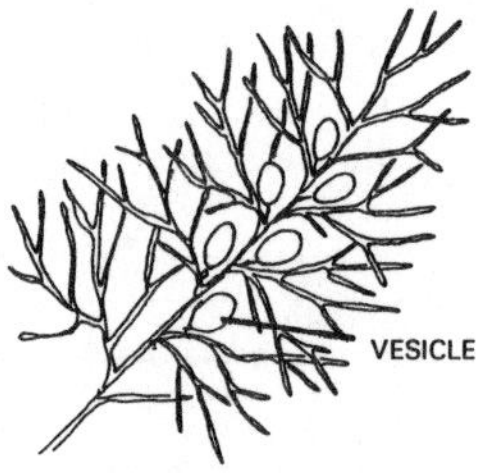

Figure 1248

Vessel. A tubelike xylem structure composed of vessel elements attached end to end.

Vessel element. A short, thick xylem cell with blunt ends. Figure 1249.

Vessel member. See **vessel element**.

Vestigial. An organ or structure which is much reduced and likely nonfunctional, though believed at one time to have been more perfectly formed. Figure 1250. (see **rudimentary** and **obsolete**)

Figure 1249

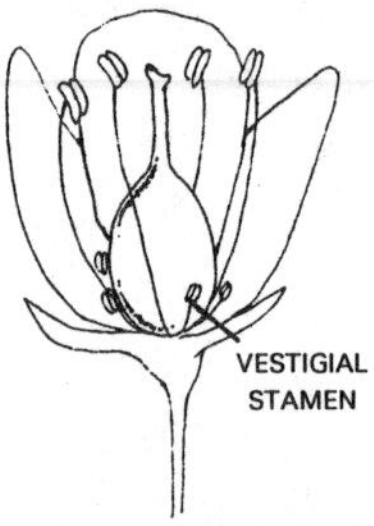

Figure 1250

Vestiture. The epidermal coverings of a plant, collectively.

Vexillum. The upper and usually largest petal of a papilionaceous flower, as in peas and sweet peas. Figure 1251.

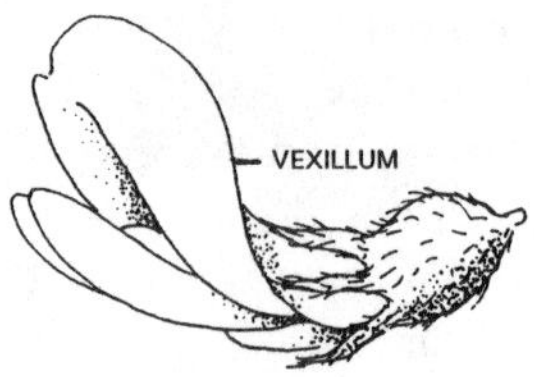

Figure 1251

Villose. See **villous**.

Villosulous. Diminutive if **villous**.

Villous. Bearing long, soft, shaggy, but unmatted, hairs. Figure 1252.

Villus (pl. **villi**). A long, soft, shaggy hair. Figure 1252.

Vimineous. With long, flexible twigs; composed of twigs; twiglike.

Vinaceous. Wine-colored.

Vine. A plant with the stem not self-supporting, but climbing or trailing on some support. Figure 1253.

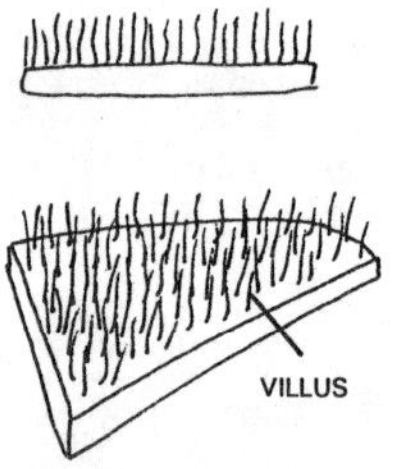

Figure 1252

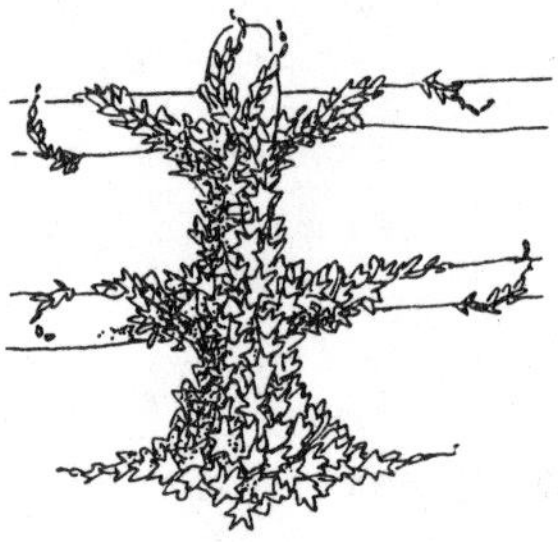

Figure 1253

Vinicolor. See **vinaceous**.

Violaceous. Violet-colored; of or pertaining to the Violaceae.

Virescence. The condition of becoming green.

Virescent. Becoming green; greenish.

Virgate. Wandlike; straight, slender, and erect. Figure 1254.

Viridescent. See **virescent**.

Viscid. Sticky or gummy.

Viscidulous. Slightly sticky.

Vitreous. Transparent.

Vitta (pl. **vittae**). An oil tube in the carpel walls of the fruits of the Umbelliferae (Apiaceae). Figure

Figure 1254

1255.

Vittate. Having vittae. Figure 1255.

Viviparous. Sprouting on the parent plant, as the bulblets forming in some inflorescences. Figure 1256.

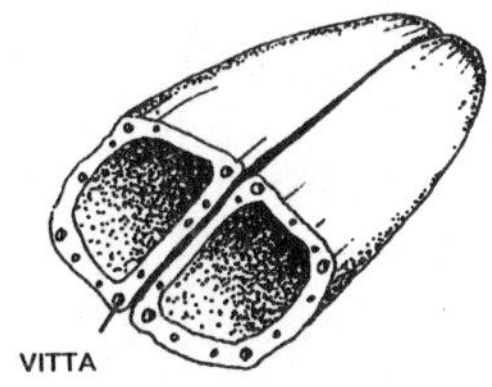

Figure 1255

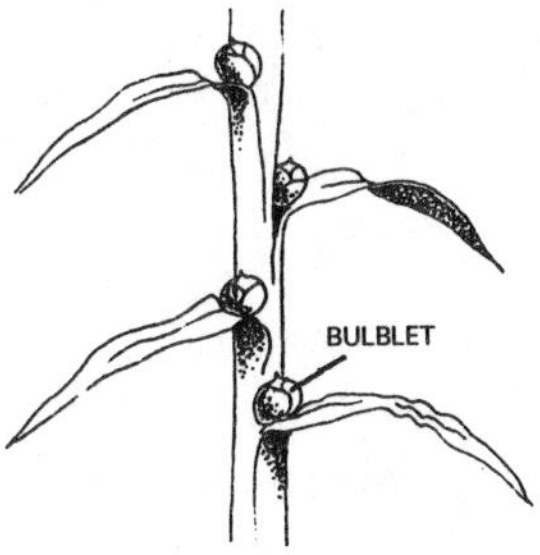

Figure 1256

Volute. Rolled up. Figure 1257.

Wart. A firm protuberance. Figure 1258.

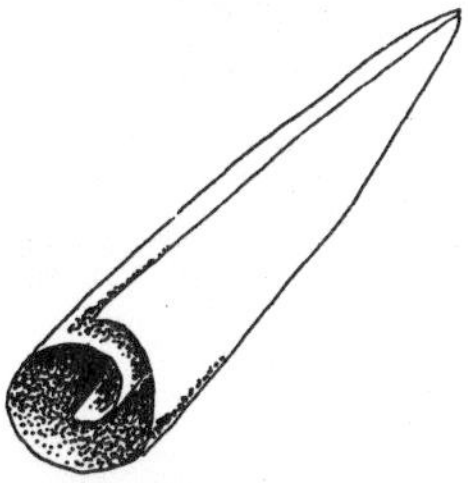

Figure 1257

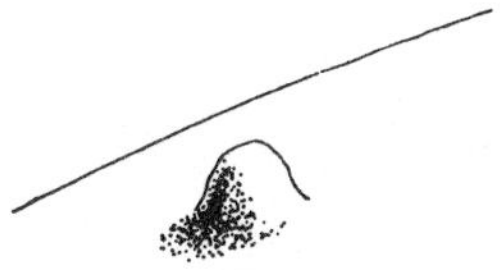

Figure 1258

Webbed. With an interlacing network of filaments, fibers, hairs, or veins. Figure 1259.

Weed. An aggressive plant which colonizes disturbed habitats and cultivated lands.

Whorl. A ringlike arrangement of similar parts arising from a common point or node; a verticil. Figure 1260.

Figure 1259

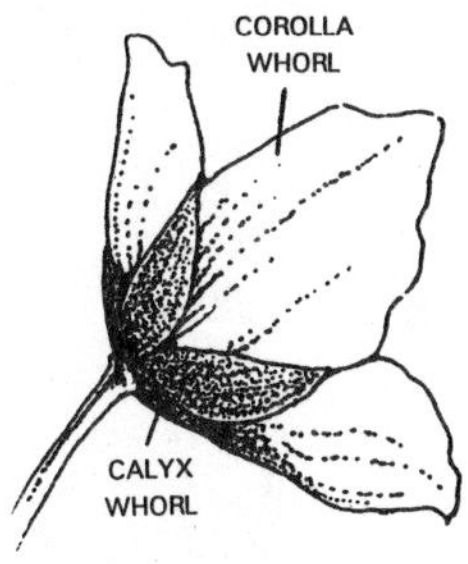

Figure 1260

Whorled. With parts arranged in whorls, as in a leaf arrangement with three or more leaves arising from a node. Figure 1261. (Same as **verticillate.**)

Wing. A thin, flat margin bordering a structure. Figure 1262; one of the two lateral petals of a papilionaceouscorolla. Figure 1263.

Winged. Possessing wings. Figures 1262 and 1263.

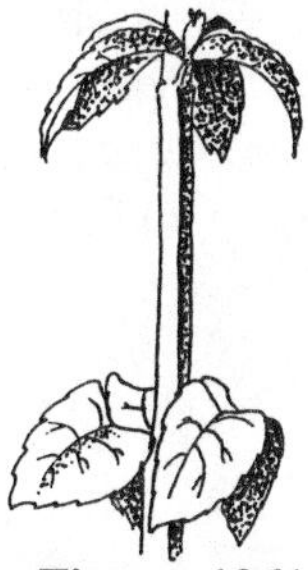

Figure 1261

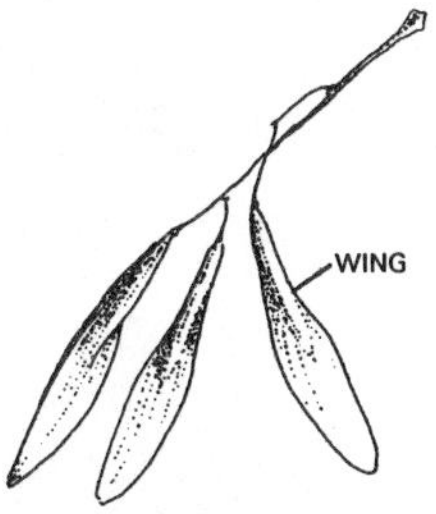

Figure 1262

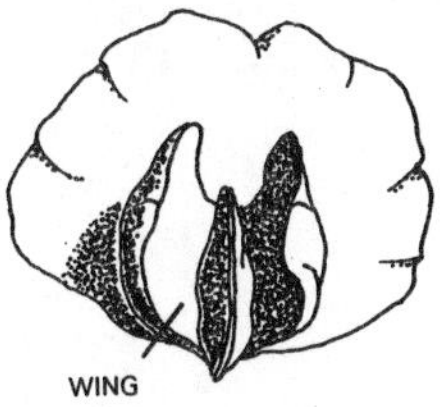

Figure 1263

Winter annual. A plant with seeds germinating in late summer or fall and completing flowering and fruiting in spring or summer of the following year and then dying.

Winter bud. A hibernating vegetative shoot.

Woolly. With long, soft, entangled hairs; lanate. Figure 1264.

X. When placed before a specific epithet, indicates the taxon is of known hybrid origin.

Xanthic. Yellowish.

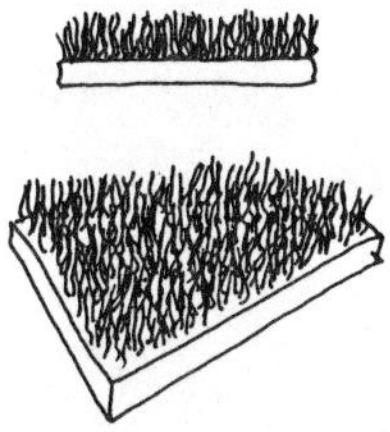

Figure 1264

Xanthophyll. Yellow, orange, or red fat-soluble pigments.

Xenogamy. Pollination between flowers of separate plants.

Xeric. Of dry areas.

Xero- (prefix). Meaning dry.

Xeromorphic. Possessing obvious physical adaptations for a dry environment, such as the succulent, water storing stem of a cactus.

Xerophilous. See **xeric**.

Xerophyte (adj. **xerophytic**). A plant adapted to life in dry environments.

Xylem. The water conducting tissue of vascular plants. Figure 1265.

Zonate. Marked or colored in circular rings or zones. Figure 1266.

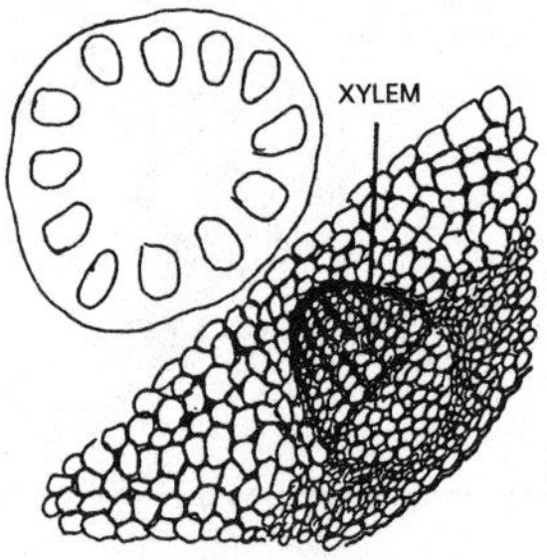

Figure 1265

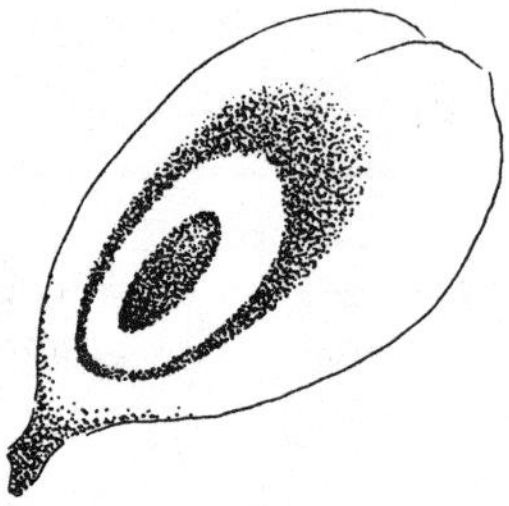

Figure 1266

Zoophilous. Animal-pollinated.

Zygomorphic. Bilaterally symmetrical, so that a line drawn through the middle of the structure along only one plane will produce a mirror image on either side. Figure 1267. (compare **actinomorphic**, and see **irregular**)

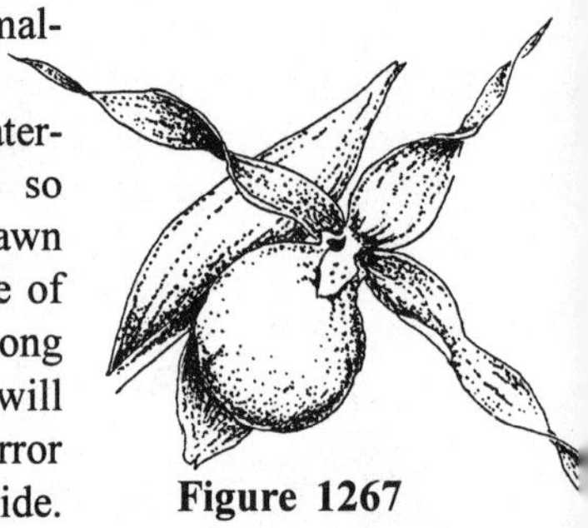

Figure 1267

Zygomorphous. See **zygomorphic**.

PART TWO

SPECIFIC TERMINOLOGY

SPECIFIC TERMINOLOGY

ROOTS

That portion of the plant axis lacking nodes and leaves and usually found below ground.

ROOT PARTS

Cortex. Root tissue between the epidermis and the stele. Figure 1268.

Meristem. Undifferentiated, actively dividing tissues at the growing tips of shoots and roots. Figure 1269.

Pith. The spongy, parenchymatous central tissue in some roots. Figure 1270.

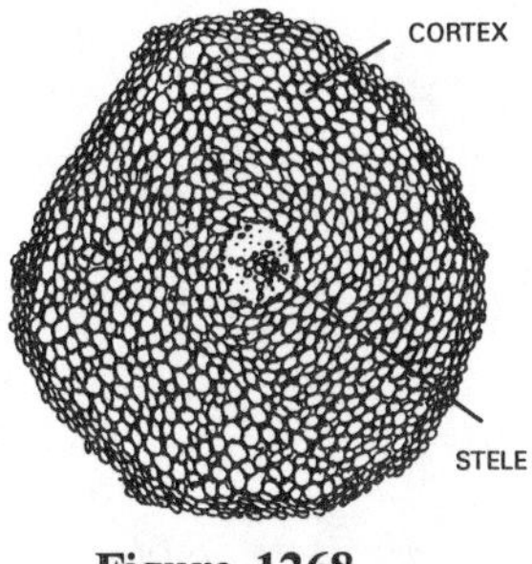

Figure 1268

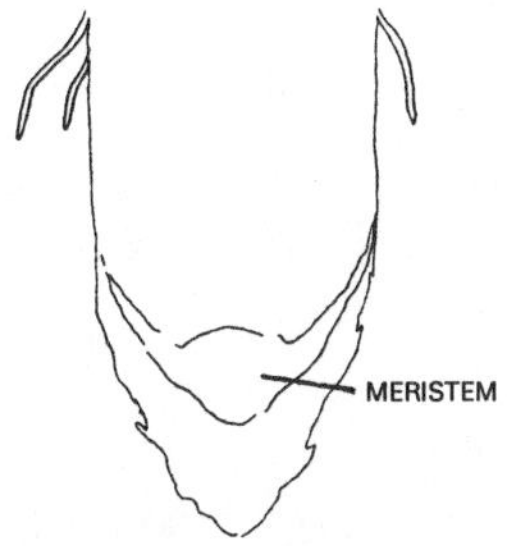

Figure 1269

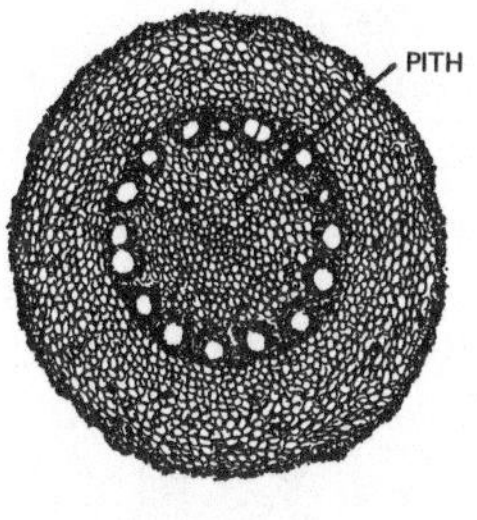

Figure 1270

Rootlet. A small root. Figure 1271.

Stele. The primary vascular structure of a root. Figure 1268.

Tubercule. A nodule, as on the roots of some legumes. Figure 1272.

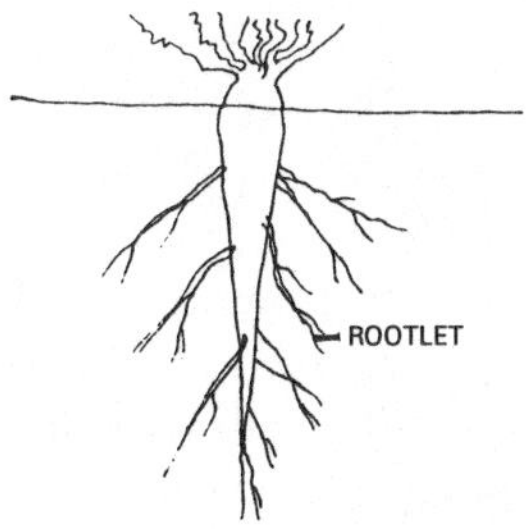

Figure 1271

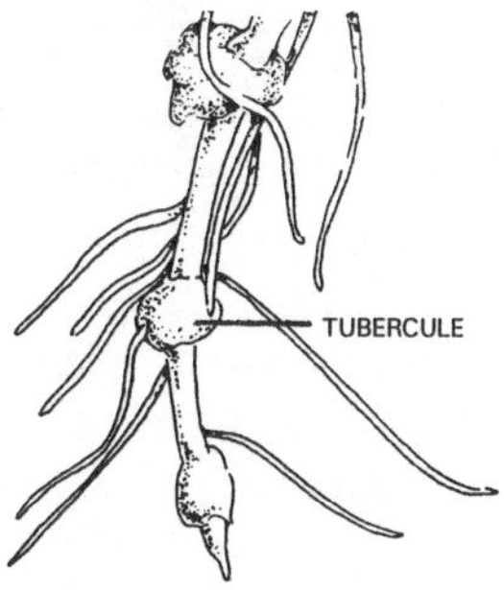

Figure 1272

ROOT SHAPES

Fusiform. Spindle-shaped; broadest near the middle and tapering toward both ends. Figure 1273.

Napiform. Turnip-shaped. Figure 1274.

Rapiformis. See **napiform**.

Spindle-shaped. See **fusiform**.

Turbinate. Top-shaped. Figure 1275.

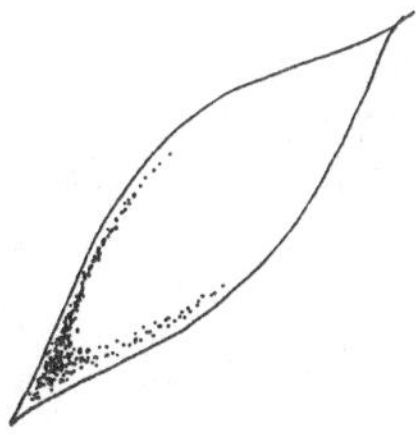

Figure 1273

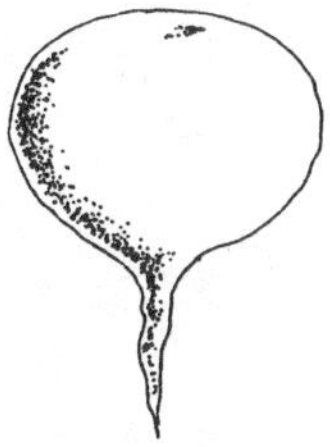

Figure 1274

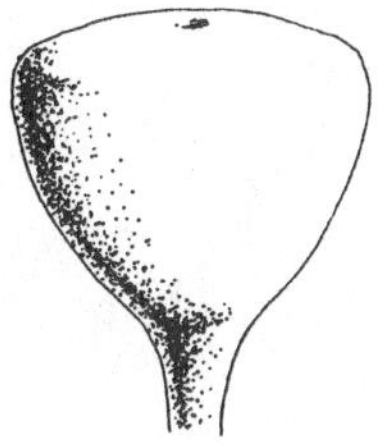

Figure 1275

ROOT TYPES

Adventitious. Structures or organs developing in an unusual position, as roots originating on the stem. Figure 1276.

Aerial. Occurring above ground or water.

Buttressed. With props or supports, as in the flared trunks of some trees. Figure 1277.

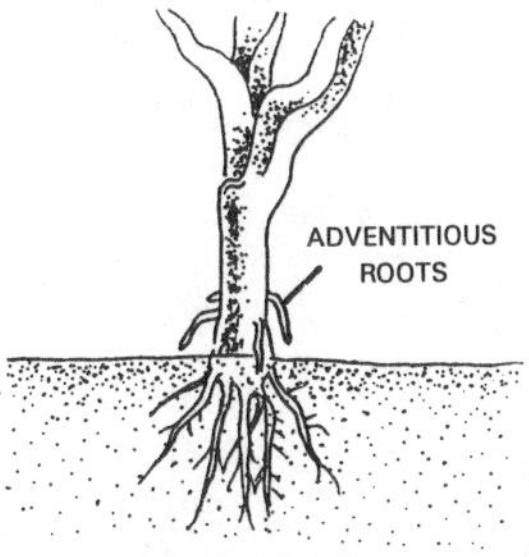

Figure 1276

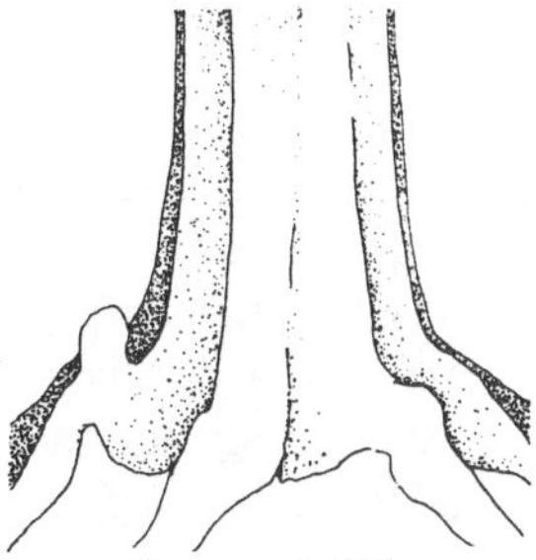

Figure 1277

Fibrous roots. A root system with all of the branches of approximately equal thickness, as in the grasses and other monocots. Figure 1278.

Haustorium (pl. **haustoria**). A specialized root-like organ used by parasitic plants to draw nourishment from host plants.

Figure 1278

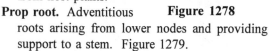

Prop root. Adventitious roots arising from lower nodes and providing support to a stem. Figure 1279.

Radicant. A root arising from the node of a prostrate stem. Figure 1280.

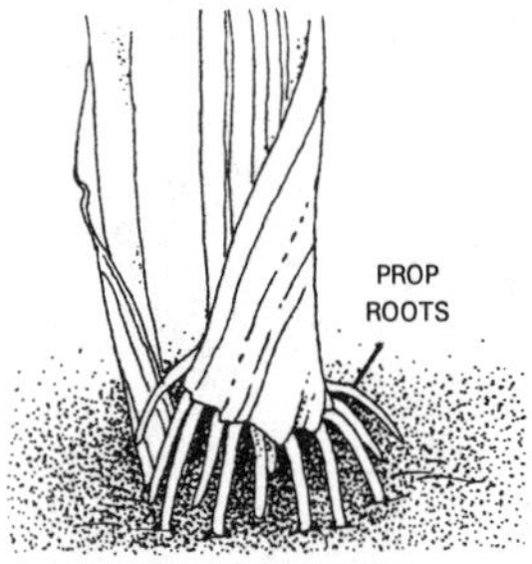

Figure 1279

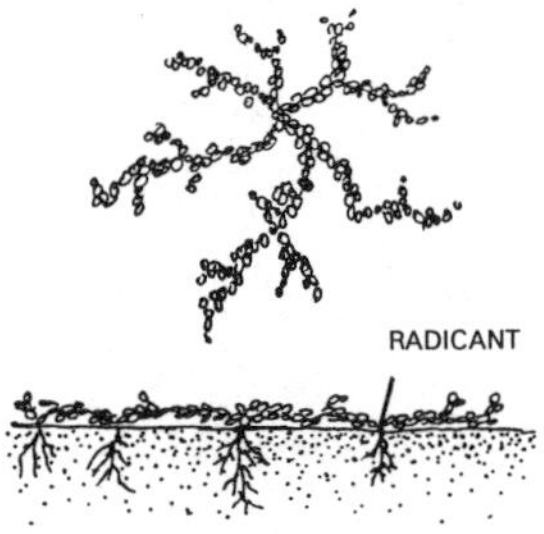

Figure 1280

Rhizoid. A rootlike structure lacking conductive tissues (xylem and phloem). Figure 1281.

Rootlet. A small root. Figure 1282.

Figure 1281

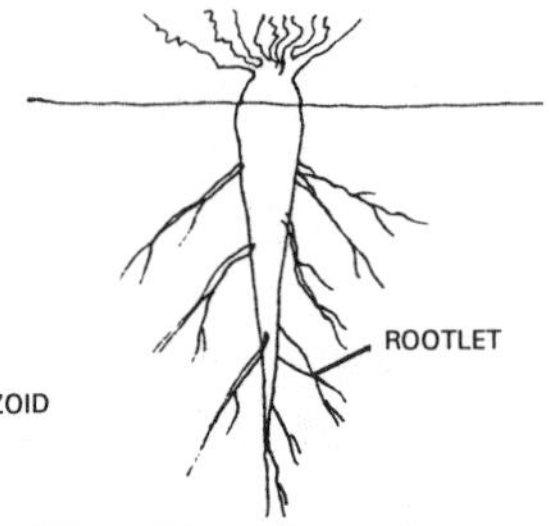

Figure 1282

Surculum. A fern rhizome. Figure 1283.

Taproot. The main root axis from which smaller root branches arise; a root system with a main root axis and smaller branches, as in most dicots. Figure 1282.

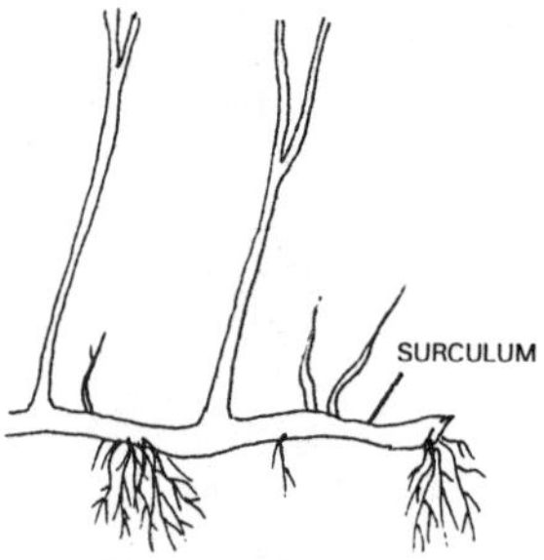

Figure 1283

Tuberoid. A thickened root which resembles a tuber. Figure 1284.

Tubercule. A nodule, as on the roots of some legumes. Figure 1285.

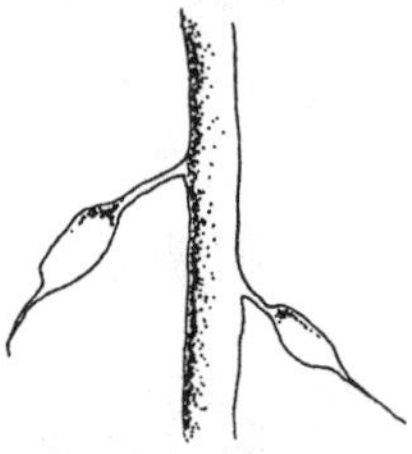

Figure 1284

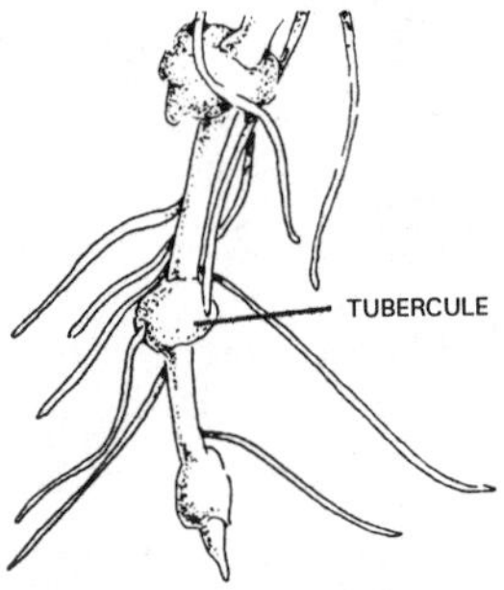

Figure 1285

STEMS

The portion of the plant axis bearing nodes, leaves, and buds and usually found above ground.

STEM PARTS

Bark. The outermost layers of a woody stem including all of the living and nonliving tissues external to the cambium. Figure 1286.

Bole. See **trunk**.

Branchlet. A small branch. Figure 1287.

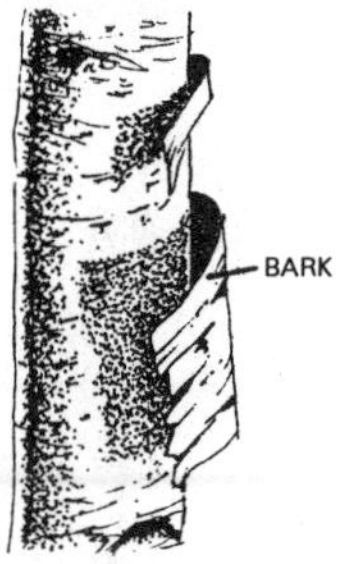

Figure 1286

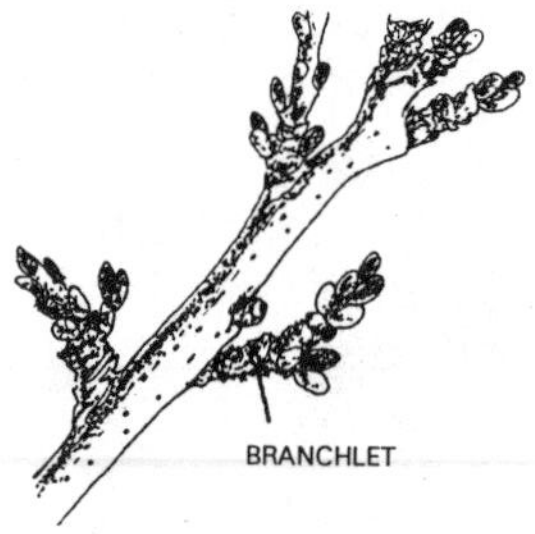

Figure 1287

Bud. An undeveloped shoot or flower. Figure 1288.

Bundle scar. Scar left on a twig by the vascular bundles when a leaf falls. Figure 1288.

Cambium. A tissue composed of cells capable of active cell division, producing xylem to the inside of the plant and phloem to the outside; a lateral meristem. Figure 1289.

Caudex (pl. **caudices**, **caudexes**). The persistent

and often woody base of a herbaceous perennial. Figure 1290.

Cortex. Bark or rind; root tissue between the epidermis and the stele. Figure 1289.

Crown. The persistent base of a herbaceous perennial, a caudex. Figure 1290; the top part of a tree. Figure 1291.

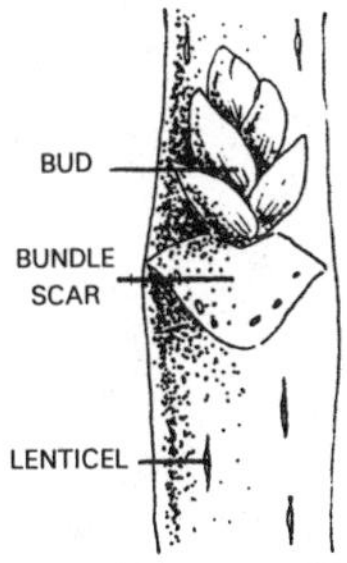

Figure 1288

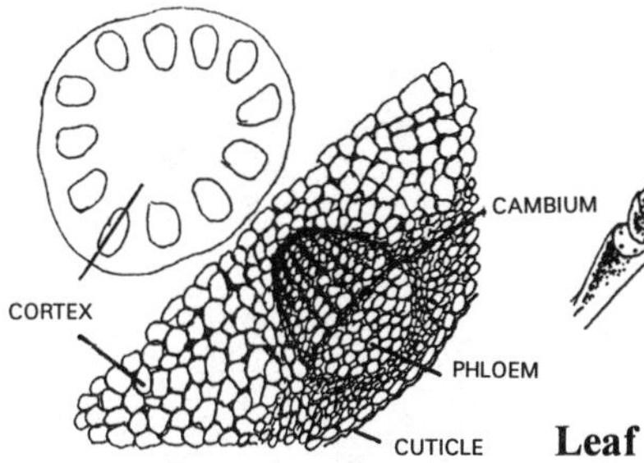

Figure 1289

Figure 1290

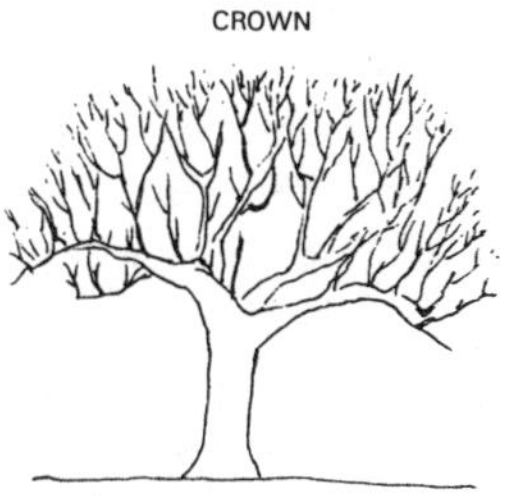

Figure 1291

Cuticle. The waxy layer on the surface of a stem. Figure 1289.

Epicotyl. That portion of the embryonic stem above the cotyledons. Figure 1292.

Epidermis. The outermost cellular layer of a stem. Figure 1289.

Heartwood. The innermost, usually somewhat darker wood of a woody stem. Figure 1293.

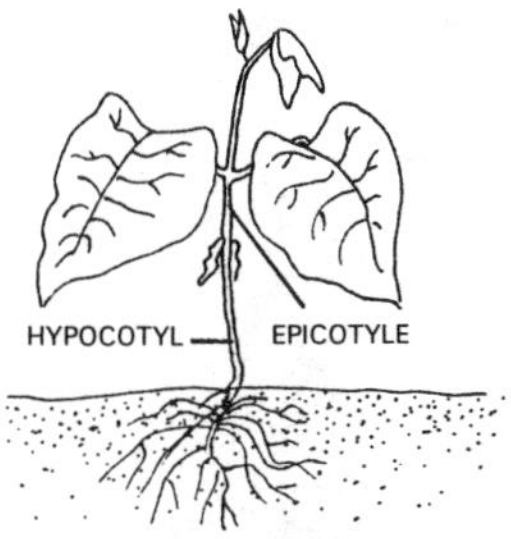

Figure 1292

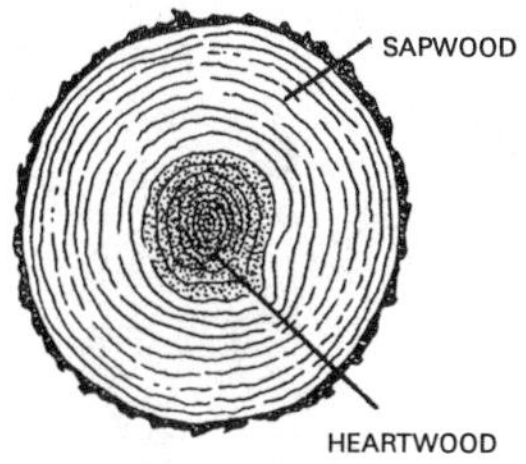

Figure 1293

Hypocotyl. That portion of the embryonic stem below the cotyledons. Figure 1292.

Internode. The portion of a stem between two nodes. Figure 1294.

Joint. The section of a stem from which a leaf or branch arises; a node, especially on a grass stem. Figure 1295.

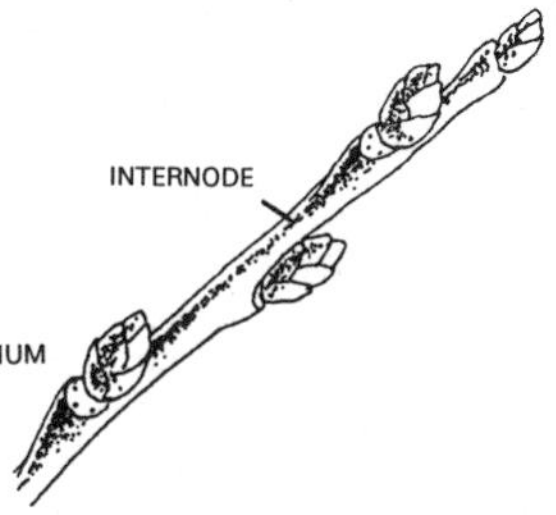

Figure 1294

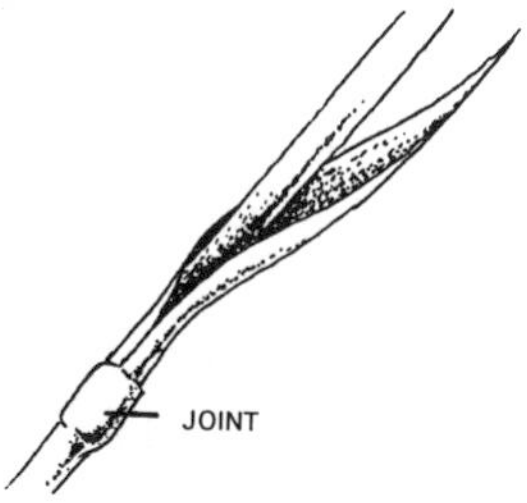

Figure 1295

Leaf scar. The scar remaining on a twig after a leaf falls. Figure 1288.

Lenticel. A slightly raised, somewhat corky, often lens-shaped area on the surface of a young stem. Figure 1288.

Meristem. Undifferentiated, actively dividing tissues at the growing tip of a shoot.

Node. The position on the stem where leaves or branches originate. Figure 1296.

Phloem. The food conducting tissue of vascular plants; bark. Figure 1289.

Pith. The spongy, parenchymatous central tissue in some stems and roots. Figure 1297.

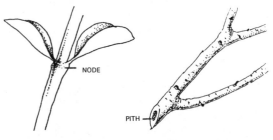

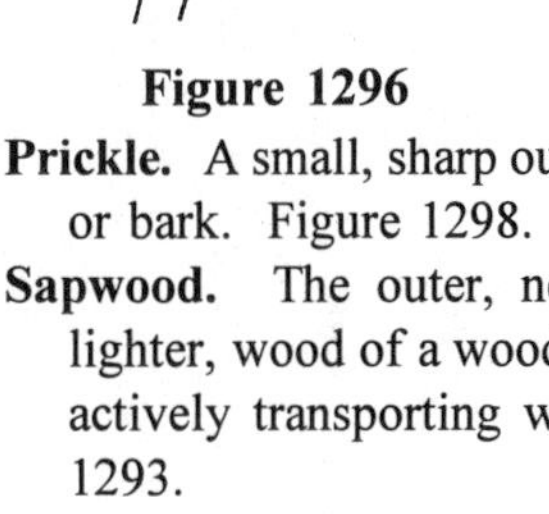

Figure 1296

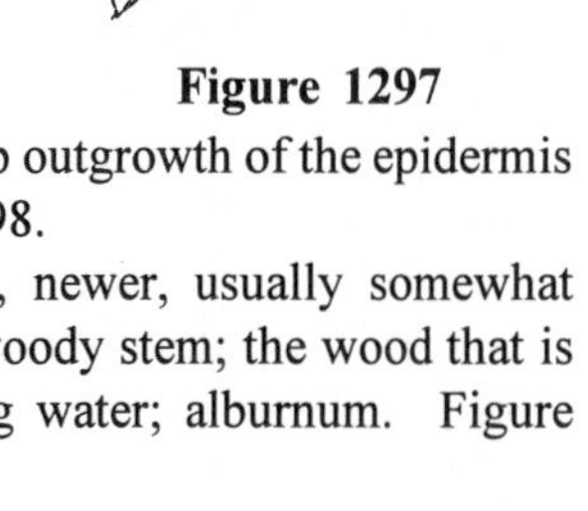

Figure 1297

Prickle. A small, sharp outgrowth of the epidermis or bark. Figure 1298.

Sapwood. The outer, newer, usually somewhat lighter, wood of a woody stem; the wood that is actively transporting water; alburnum. Figure 1293.

Stele. The primary vascular structure of a stem, including the vascular tissues and all tissues internal to the vascular tissues. Figure 1289.

Trunk. The main stem of a tree below the branches. Figure 1299.

Twig. A small shoot or branch from a tree. Figure 1300.

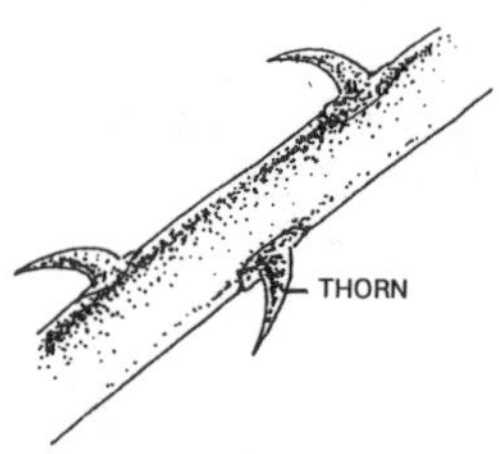

Figure 1298

Caulis. The main stem of a herbaceous plant. Figure 1304.

Cladode. See **cladophyll**.

Cladophyll. A stem with the form and function of a leaf. Figures 1305 and 1306.

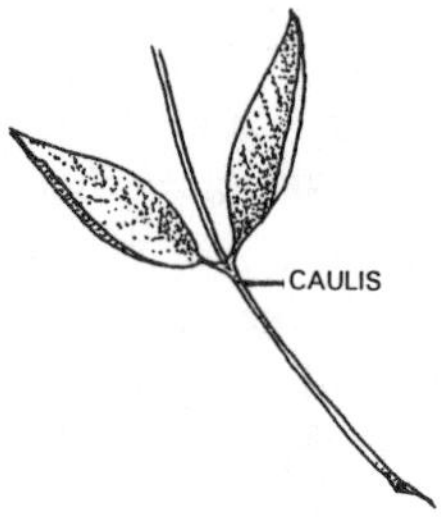

Figure 1304

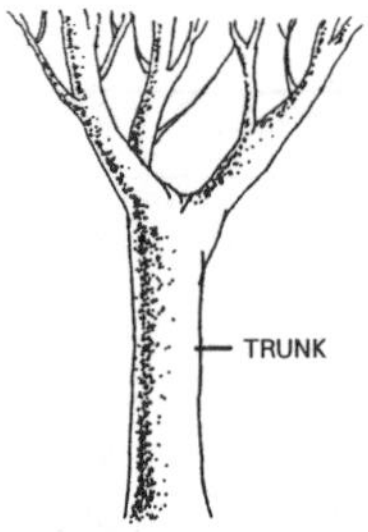

Figure 1299

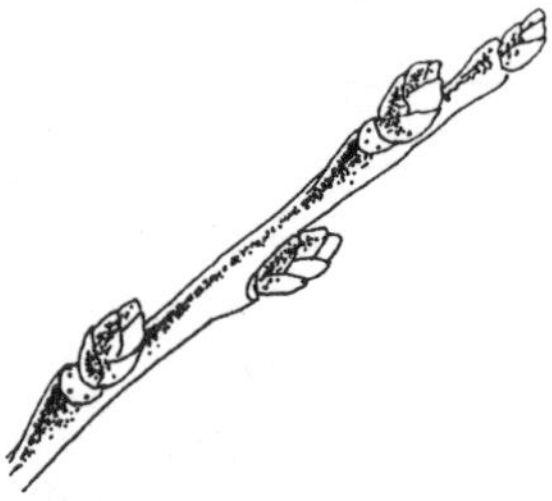

Figure 1300

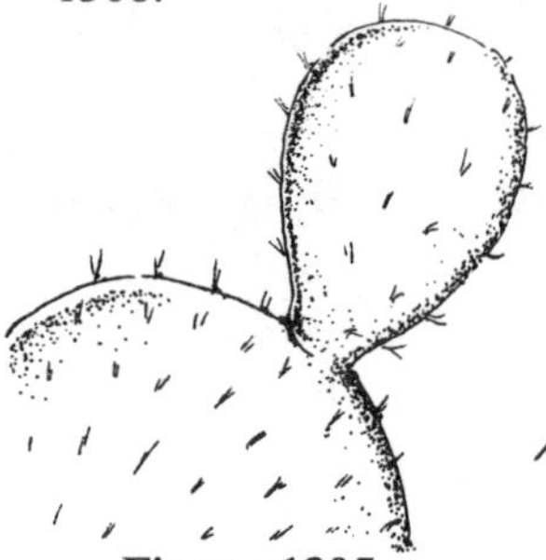

Figure 1305

Figure 1306

STEM TYPES

Bulb. An underground bud with thickened fleshy scales, as in the onion. Figure 1301.

Bulbel. A small bulb arising from the base of a larger bulb. Figure 1302.

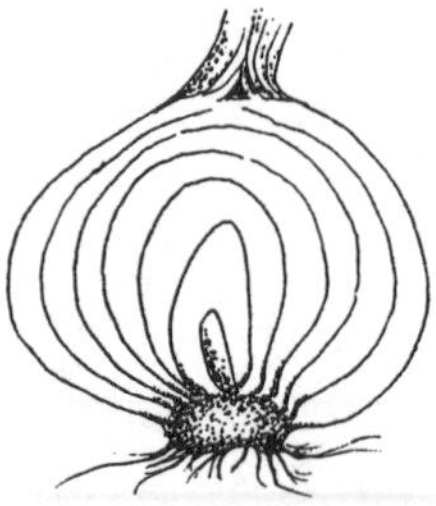

Figure 1301

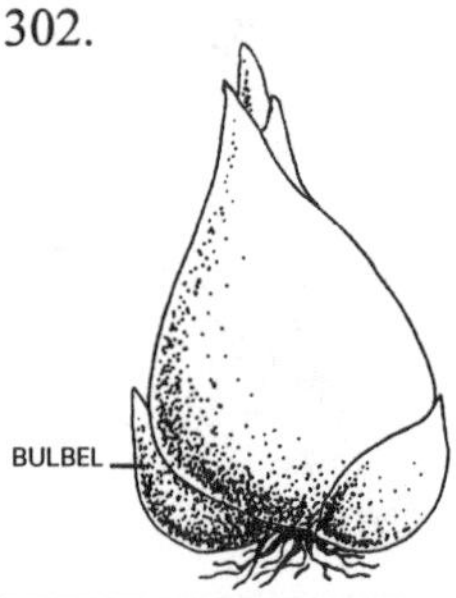

Figure 1302

Bulbil. See **bulblet**.

Bulblet. A small bulb.

Caudex (pl. **caudices**, **caudexes**). The persistent and often woody base of a herbaceous perennial. Figure 1303.

Caulicle. A small stem; a rudimentary stem.

Figure 1303

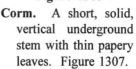

Corm. A short, solid, vertical underground stem with thin papery leaves. Figure 1307.

Cormel. A small corm arising at the base of a larger corm.

Culm. A hollow or pithy stalk or stem, as in the grasses, sedges, and rushes. Figure 1308.

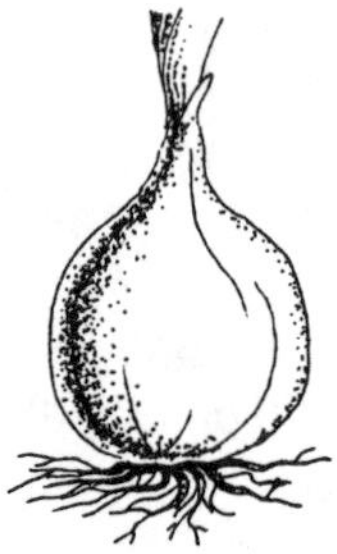

Figure 1307

Floricane. The second-year flowering and fruiting cane (shoot) of *Rubus*. (compare **primocane**)

Liana. A woody, climbing vine.

Monopodium (pl. **monopodia**). A single main axis giving rise to lateral branches. Figure 1309.

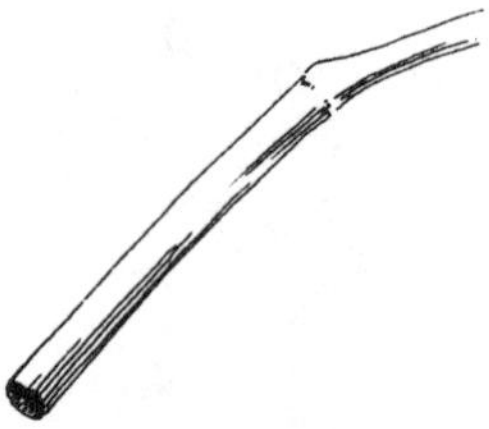

Figure 1308

Figure 1309

Offset. A short, often prostrate, shoot originating near the ground at the base of another shoot. Figure 1310.

Offshoot. A shoot or branch arising from a main stem. Figure 1311.

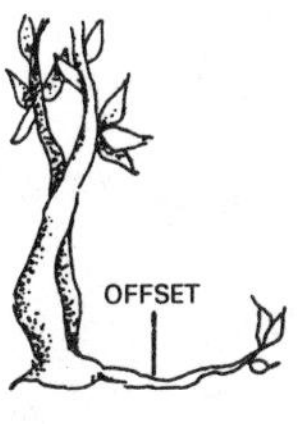

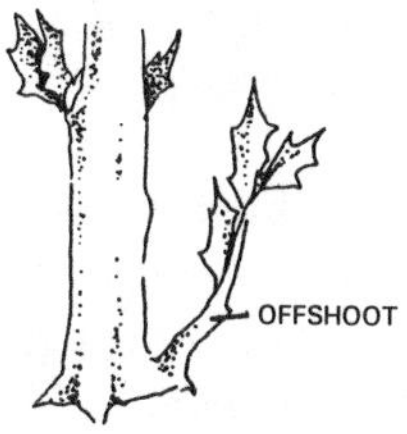

Figure 1310 **Figure 1311**

Phylloclade. See **cladophyll**.

Primocane. The first-year, usually flowerless, cane (shoot) of *Rubus*. (compare **floricane**)

Pseudoscape. A false scape, where not all of the leaves are truly basal in origin though, superficially, they appear to be so. Figure 1312.

Ratoon. A shoot arising from the root of a plant that has been cut down. Figure 1313.

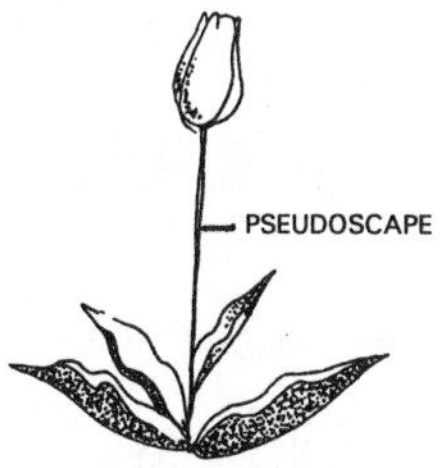

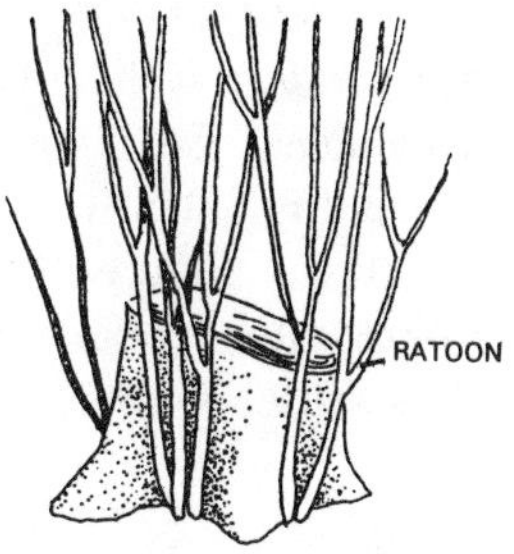

Figure 1312 **Figure 1313**

Rhizome. A horizontal underground stem; rootstock. Figure 1314.

Runner. A slender stolon or prostrate stem rooting at the nodes or at the tip. Figure 1315.

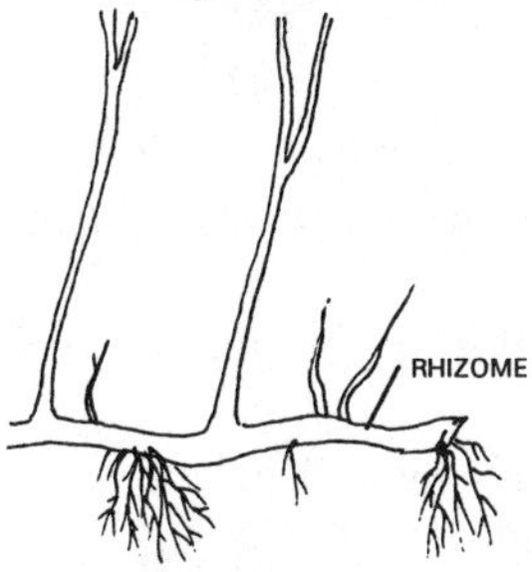

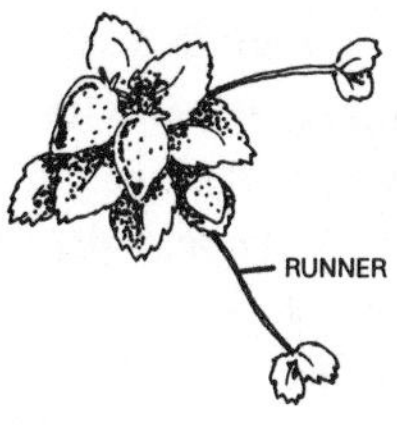

Figure 1314 **Figure 1315**

Sarment. A long, slender runner. See **runner**.

Shoot. A young stem or branch. Figure 1316.

Sobol. Elongated caudex branches; a shoot arising from the base of a stem or from the rhizome. Figure 1314.

Sobole. See **sobol**.

Spray. A slender shoot or branch with its leaves, flowers, or fruits. Figure 1317.

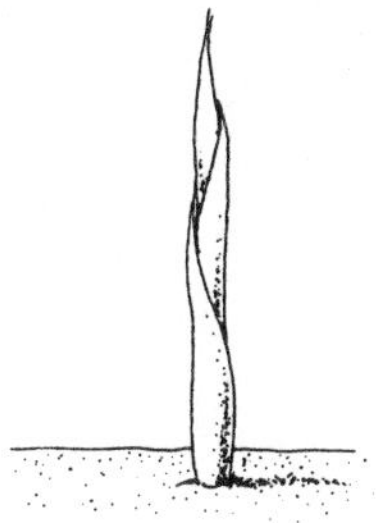

Figure 1316 **Figure 1317**

Stolon. An elongate, horizontal stem creeping along the ground and rooting at the nodes or at the tip and giving rise to a new plant. See **runner**.

Stool. The base of plants which produce new stems each year. See **caudex**; a group of stems arising from a single root.

Sucker. A shoot originating from below ground. Figure 1318.

Sympodium. A main axis appearing to be simple, but actually consisting of a number of short axillary branches rather than a continuation of the main axis. Figure 1319.

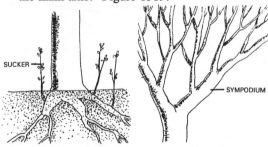

Figure 1318 **Figure 1319**

Thorn. A stiff, woody, modified stem with a sharp point; sometimes applied to any structure resembling a true thorn. Figure 1320.

Tiller. A basal or subterranean shoot which is more or less erect. See **sucker**.

Trunk. The main stem of a tree below the

branches. Figure 1321.

Tuber. The thickened portion of a rhizome bearing nodes and buds; underground stem modified for food storage. Figure 1322.

Turion. A small shoot which often over winters, as in some species of *Epilobium*.

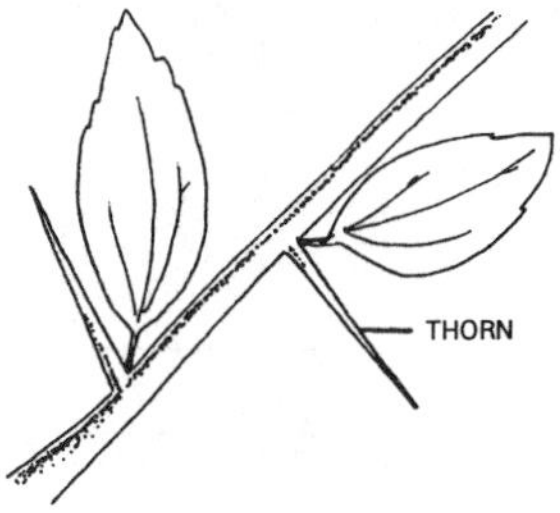

Figure 1320

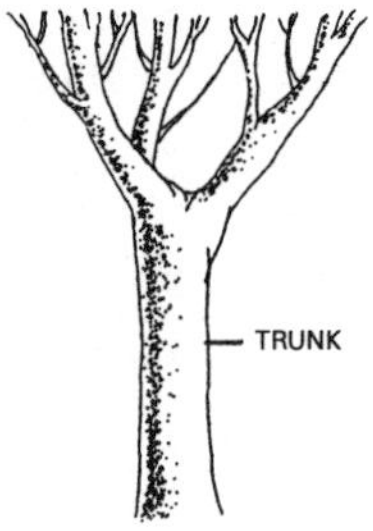

Figure 1321

Figure 1322

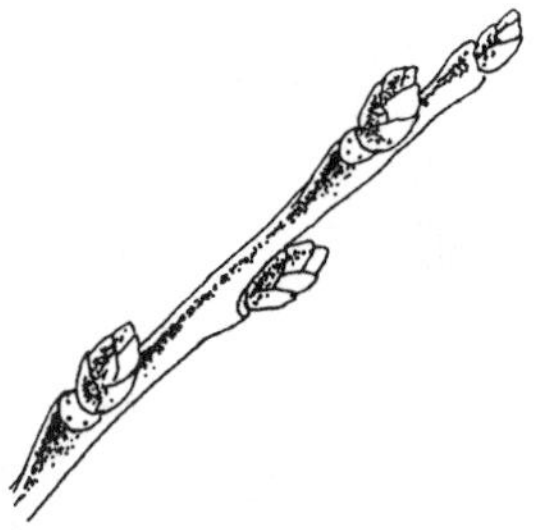

Figure 1323

Twig. A small shoot or branch from a tree. Figure 1323.

Vine. A plant with the stem not self-supporting, but climbing or trailing on some support. Figure 1324.

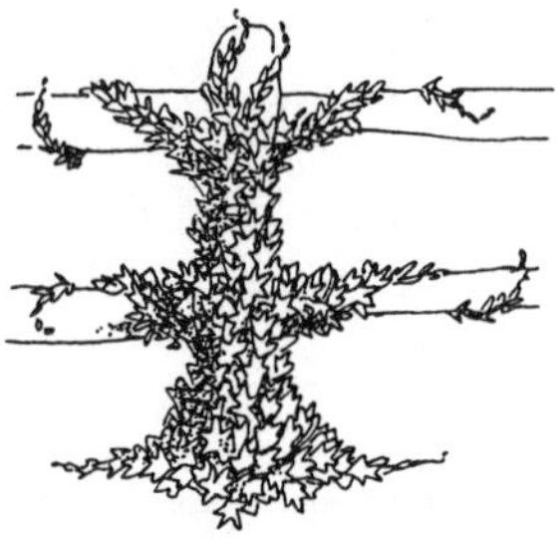

Figure 1324

STEM FORMS

Acaulescent. Without a stem, or the stem so short that the leaves are apparently all basal, as in the dandelion. Note: the peduncle should not be confused with the stem. Figure 1325.

Adscendent. See **ascending**.

Adsurgent. See **ascending**.

Alate. Winged. Figure 1326.

Ancipital. Two-edged, as the winged stem of

Figure 1325

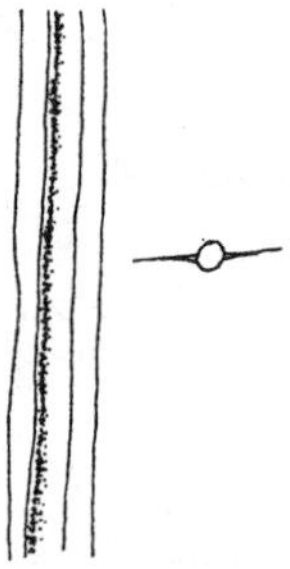

Figure 1326

Sisyrinchium. Figure 1326.

Angulate. Angled. Figure 1327.

Aphyllopodic. Having the lowermost leaves reduced to small scales. Figure 1328.

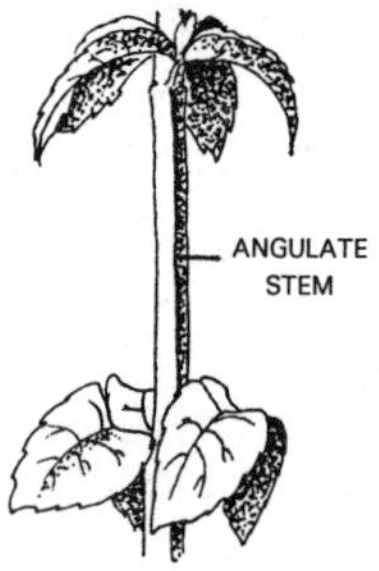

Figure 1327

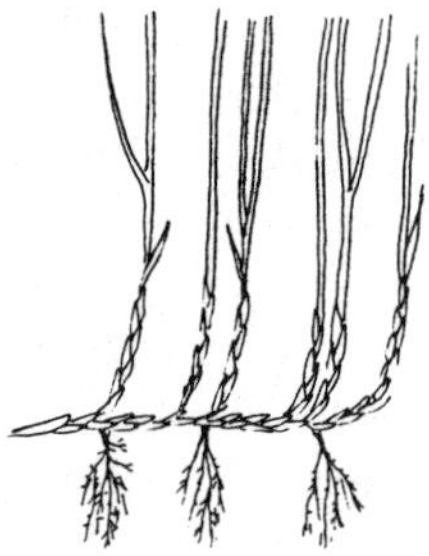

Figure 1328

Aphyllous. Without leaves.

Ascendent. See **ascending**.

Ascending. Growing obliquely upward, usually curved. Figure 1329.

Assurgent. See **ascending**.

Caespitose. Growing in dense tufts. Figure 1330.

Figure 1329

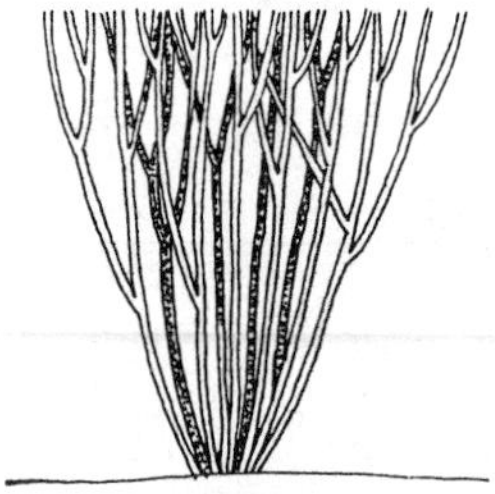

Figure 1330

Caulescent. With an obvious leafy stem rising above the ground.

Cauliflorous. Bearing flowers on the stem or trunk.

Cauline. Of, on, or pertaining to the stem.

Cespitose. See **caespitose**.

Clambering. Weakly climbing on other plants or

surrounding objects. Figure 1331.

Climbing. Growing more or less erect by leaning or twining on another structure for support. Figure 1332.

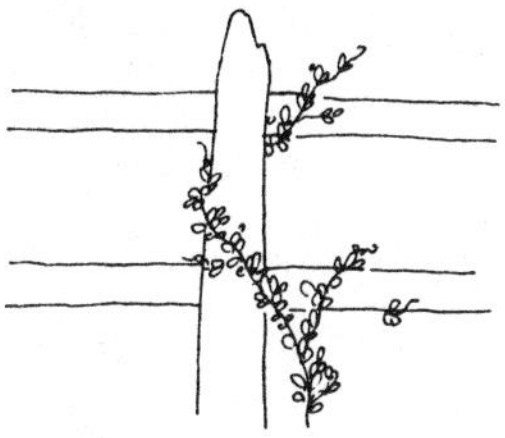

Figure 1331

Figure 1332

Creeping. Growing along the surface of the ground, or just beneath the surface, and producing roots, usually at the nodes. Figure 1333.

Decumbent. Reclining on the ground but with the tip ascending. Figure 1334.

Figure 1333

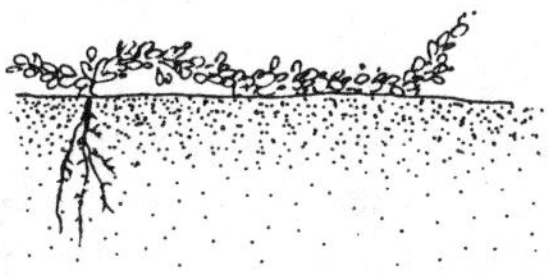

Figure 1334

Deliquescent. An irregular pattern of branching without a well defined central axis from bottom to top. Figure 1335.

Dichotomous. Branched or forked into two more or less equal divisions. Figure 1336.

Figure 1335

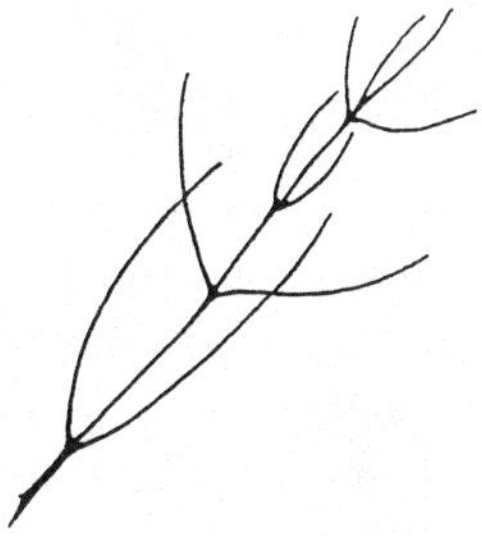

Figure 1336

Divaricate. Widely diverging or spreading apart. Figure 1337.

Divergent. Diverging or spreading. Figure 1338.

Figure 1337

Figure 1338

Eramous. With unbranched stems. Figure 1339.

Erect. Vertical, not declining or spreading. Figure 1340.

Figure 1339

Figure 1340

Fasciated. Compressed into a bundle or band; grown closely together; with the stems malformed and flattened as if several separate stems had been fused together. Figure 1341.

Fasciculate. Arranged in fascicles. Figure 1342.

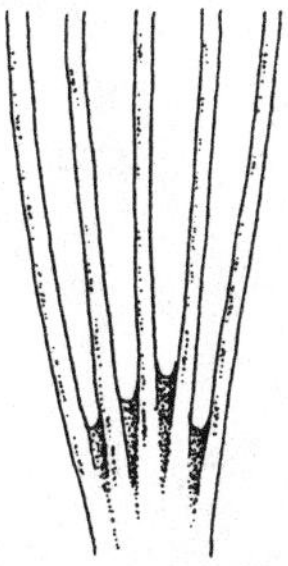

Figure 1341

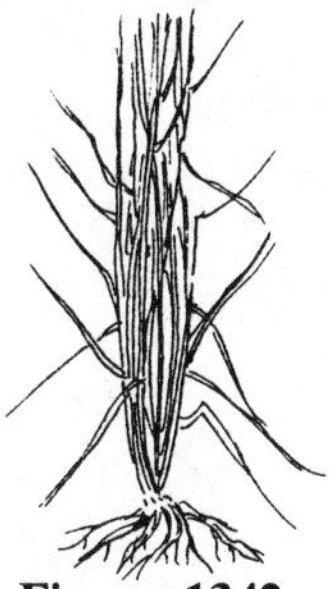

Figure 1342

Fastigiate. Clustered, parallel, and erect, giving a broom-like appearance. Figure 1330.

Fleshy. Thick and pulpy; succulent. Figure 1343.

Frutescent. Shrubby or shrublike.

Fruticose. See **frutescent**.

Fruticulose. Somewhat shrubby; small and shrubby.

Herbaceous. Not woody.

Jointed. Having nodes or points of articulation, as in the stems of *Opuntia*. Figure 1344.

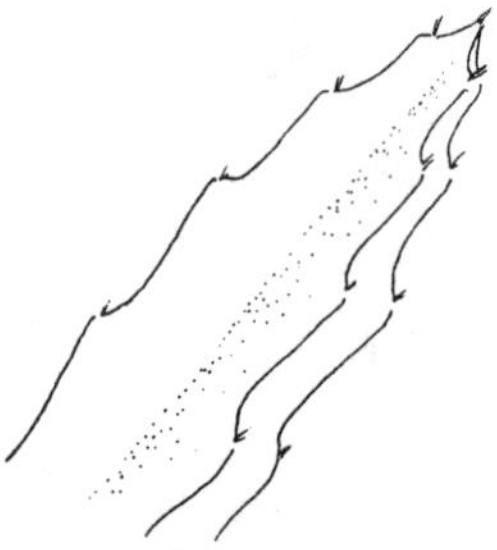

Figure 1343

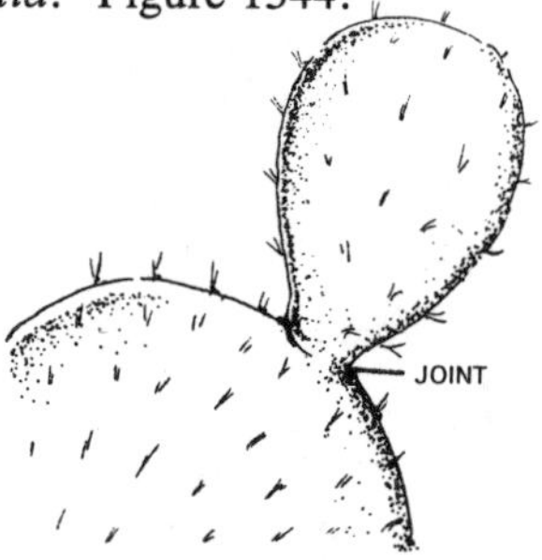

Figure 1344

Macrocladous. With long branches.

Multicipital. With many heads, as in a highly branched caudex. Figure 1345.

Nodiferous. See **nodose.**

Nodose. With nodes. Figure 1346.

Figure 1345

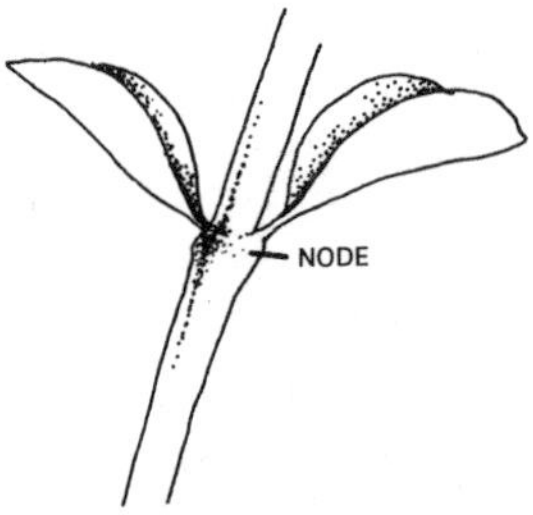

Figure 1346

Nudicaul. With leafless stems.

Orthocladous. With straight branches. Figure 1347.

Orthotropic. Of, pertaining to, or exhibiting an essentially vertical growth habit. Figure 1348.

Figure 1347

Figure 1348

Pachycladous. With thick branches.

Pluricipital. See **multicipital**.

Procumbent. Lying or trailing on the ground, but not rooting at the nodes. Figure 1349.

Prostrate. Lying flat on the ground. Figure 1350.

Pterocaulous. With winged stems. Figure 1351.

Pulvinate. Cushion-like or mat-like. Figure 1352.

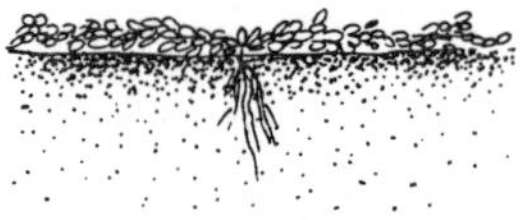

Figure 1349

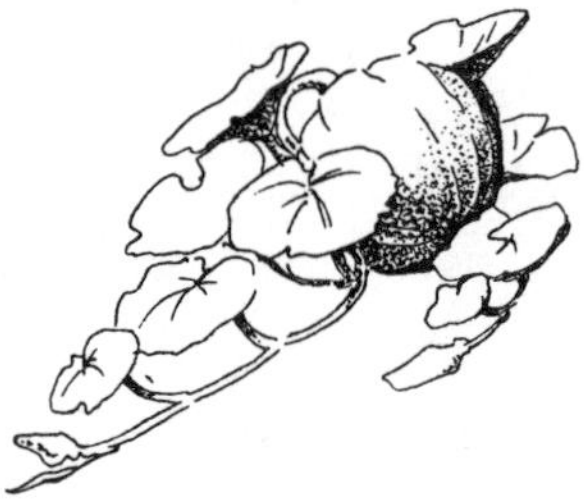

Figure 1350

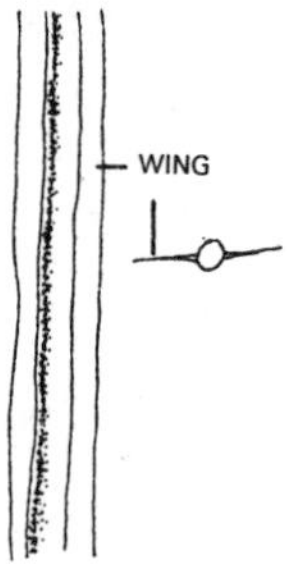

Figure 1351

Figure 1352

Pulviniform. See **pulvinate**.

Quadrangular. Four-angled. Figure 1353.

Quadrangulate. See **quadrangular**.

Ramiform. Branchlike in form; branched.

Ramose. With many branches; branching. Figure 1354.

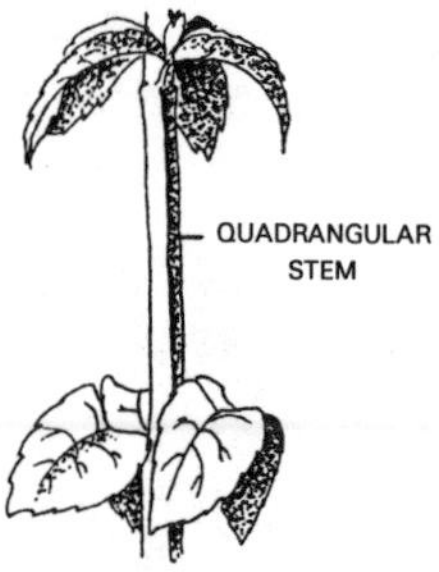

Figure 1353

Figure 1354

Ramous. See **ramose**.

Ramulose. See **ramose**.

Reclining. Bending or curving downward; lying upon something and being supported by it. Figure 1355.

Recumbent. Leaning or resting on the ground; prostrate. Figure 1356.

Repent. Prostrate; creeping. Figure 1357.

Rhizomatous. With rhizomes. Figure 1358.

Figure 1355

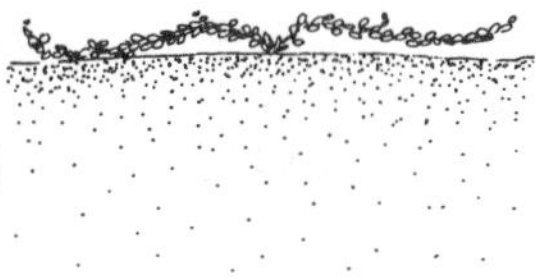

Figure 1356

Figure 1357

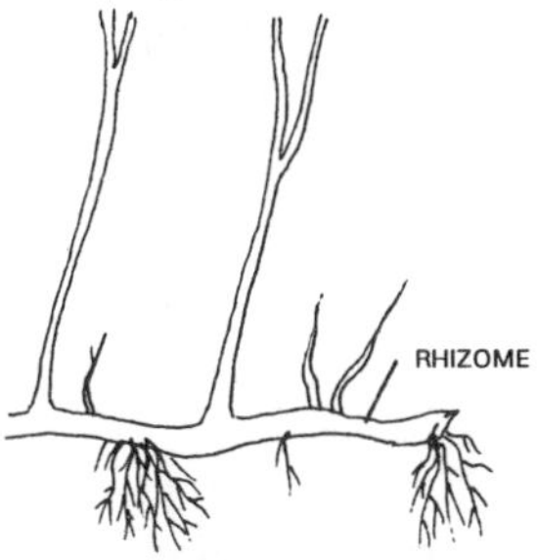

Figure 1358

Rosulate. With the leaves arranged in basal rosettes, the stem very short or lacking. Figure 1359.

Rushlike. Grasslike in appearance, with inconspicuous flowers. Figure 1360.

Figure 1359

Figure 1360

Sarcocaulous. With fleshy stems. Figure 1361.

Sarcous. Fleshy. See **sarcocaulous**.

Sarmentose. With long, slender runners. Figure 1362.

Scandent. Climbing. Figure 1363.

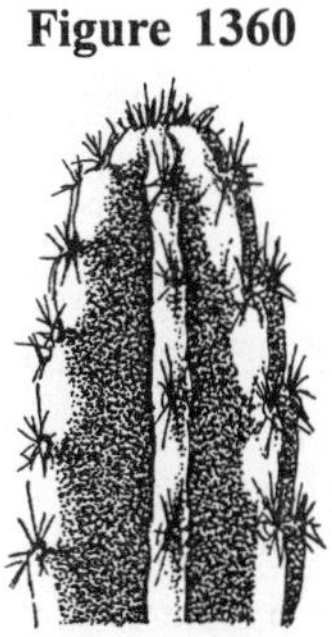

Figure 1361

Scapiform. Scapelike but not entirely leafless. Figure 1364.

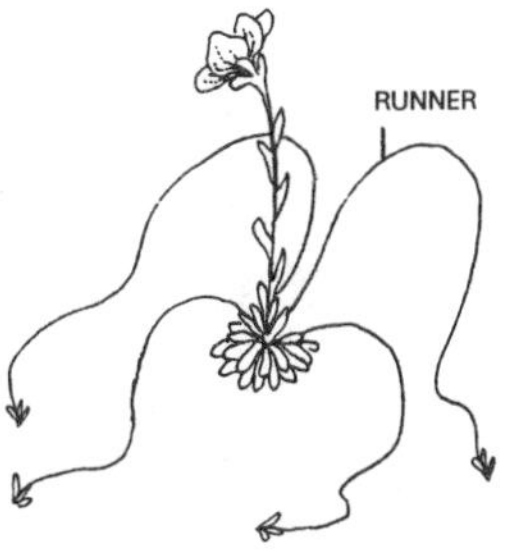

Figure 1362

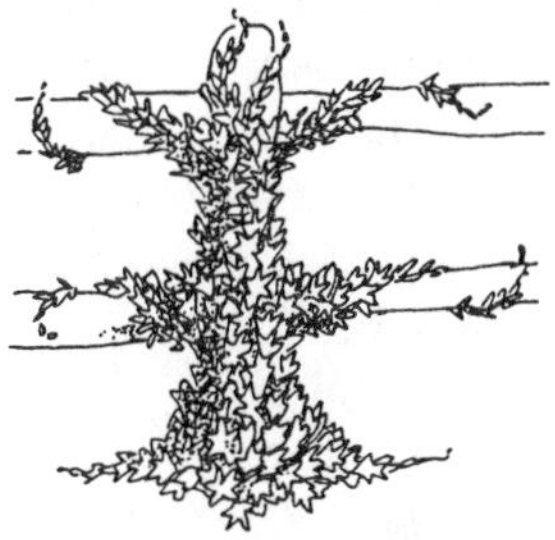

Figure 1363

Simple. Unbranched.

Soboliferous. Of or pertaining to sobols; bearing sobols. Figure 1358.

Sprawling. Bending or curving downward; lying upon something and being supported by it. See **reclining**.

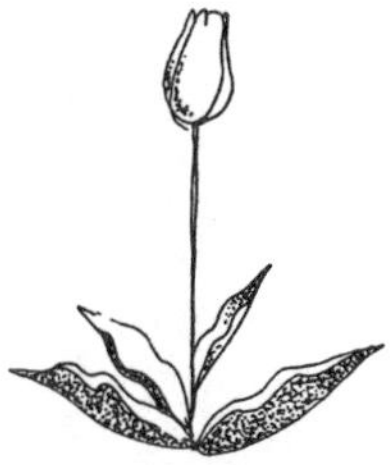

Figure 1364

Spreading. Extending nearly to the horizontal; almost prostrate. Figure 1365.

Stoloniferous. Bearing stolons. Figure 1366.

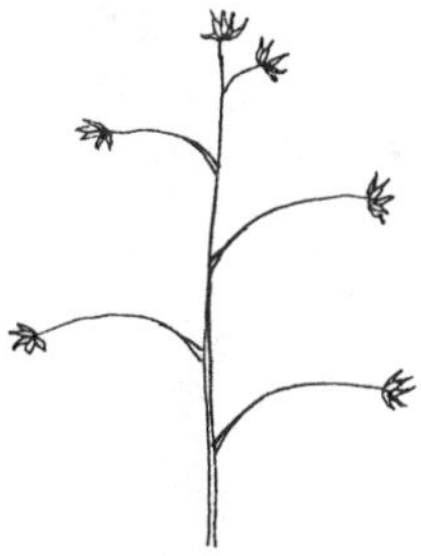

Figure 1365

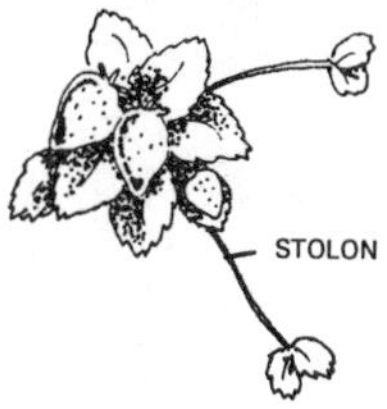

Figure 1366

Stoloniform. Stolonlike.

Strict. Very straight and upright, not at all spreading. Figure 1367.

Subscapose. Almost scapose. Figure 1364.

Subterranean. Below the surface of the ground.

Figure 1367

Subterraneous. See **subterranean**.

Succulent. Juicy and fleshy, as the stem of a

cactus. Figure 1368.

Suffrutescent. Somewhat shrubby; slightly woody at the base.

Suffruticose. Somewhat woody.

Suffruticulose. See **suffruticose**.

Surculose. Producing suckers or runners from the base or from rootstocks. Figure 1369.

Figure 1368

Surficial. Growing near the ground, or spread over the surface of the ground. Figure 1370.

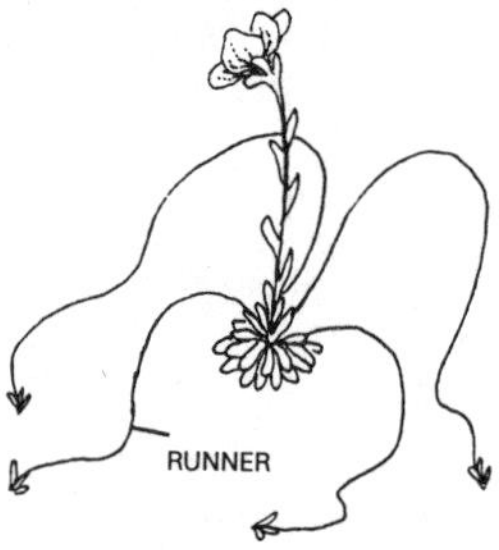

Figure 1369

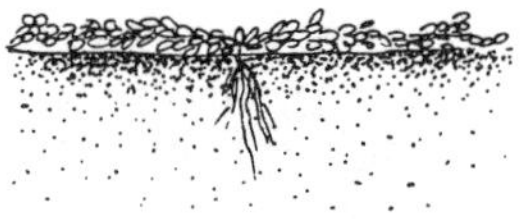

Figure 1370

Tetragonal. Four-angled. Figure 1371.

Trailing. Prostrate and creeping but not rooting. Figure 1370.

Triangulate. Three-angled. Figure 1372.

Figure 1371

Figure 1372

Trigonal. See **trigonous**.

Trigonous. Three-angled. Figure 1372.

Tufted. Arranged in a dense cluster.

Twining. Coiling or spiraling around a support (usually another stem) for climbing. Figure 1373.

Unarmed. Lacking spines, prickles, or thorns.

Uniaxial. With a single unbranched stem terminating in a flower. Figure 1374.

Vimineous. With long, flexible twigs; composed of twigs; twiglike.

Figure 1373

Figure 1374

Virgate. Wandlike; straight, slender, and erect. Figure 1375.

Winged. Possessing wings. Figure 1376.

Figure 1375

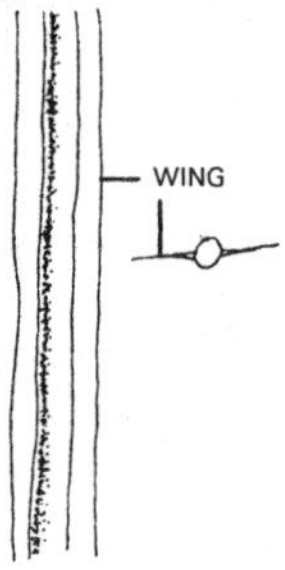

Figure 1376

LEAVES

The usually expanded, photosynthetic organs of a plant.

LEAF PARTS

Apex. The tip; the point farthest from the point of attachment. Figure 1377.

Base. The end of the leaf blade nearest to the point of attachment. Figure 1377.

Blade. The broad part of a leaf. Figure 1377.

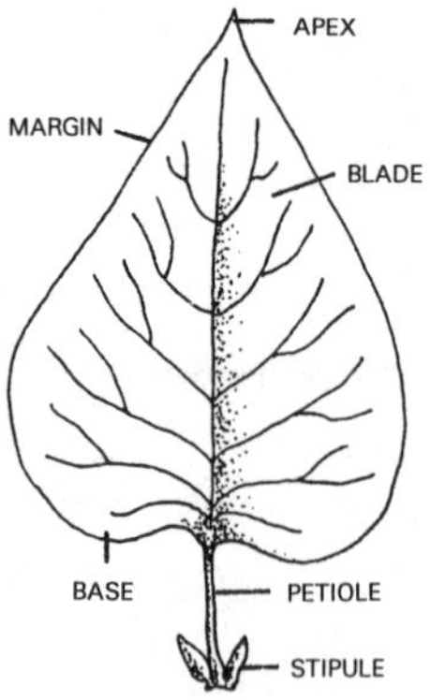

Figure 1377

Collar. The area on the outside of a grass leaf at the juncture of the blade and sheath. Figure 1378.

Costa (pl. **costae**). A rib or prominent midvein of

a leaf. See **midvein**.

Denticle. A small tooth or toothlike projection. Figure 1379.

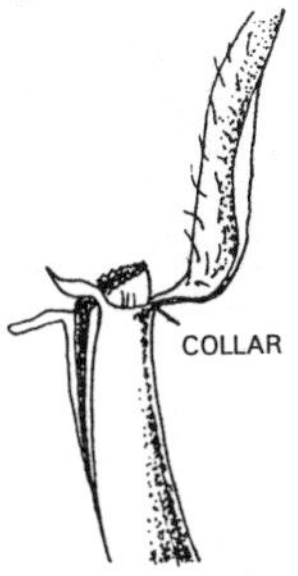

Figure 1378

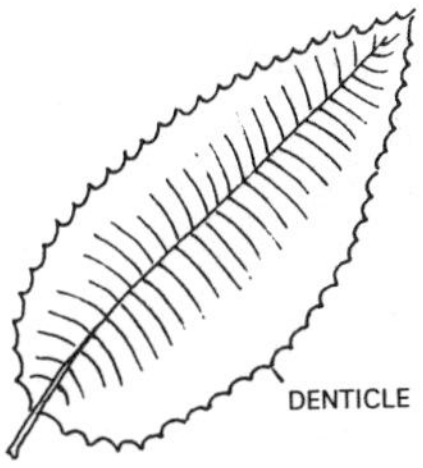

Figure 1379

Leaflet. A division of a compound leaf. Figure 1380.

Limb. The expanded part of a leaf. See **blade**.

Lobe. A rounded division or segment of an organ, as of a leaf. Figure 1381.

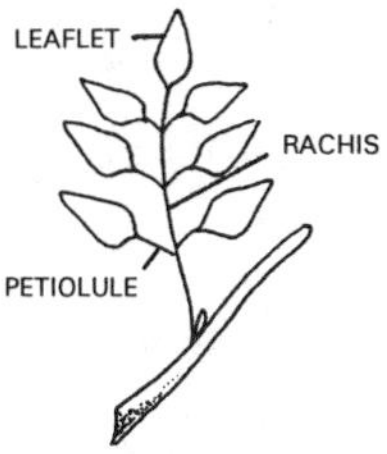

Figure 1380

Figure 1381

Lobule. A small lobe; a lobelike subdivision of a lobe. Figure 1382.

Margin. The edge of a leaf blade. Figure 1377.

Midlobe. The central lobe of a leaf. Figure 1381.

Midnerve. The central nerve of a leaf. See **midvein**.

Midrib. The central rib or vein of a leaf. See **midvein**.

Midvein. The central vein of a leaf. Figure 1383.

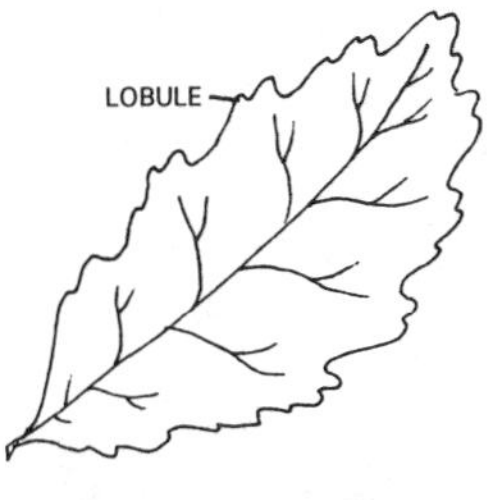

Figure 1382

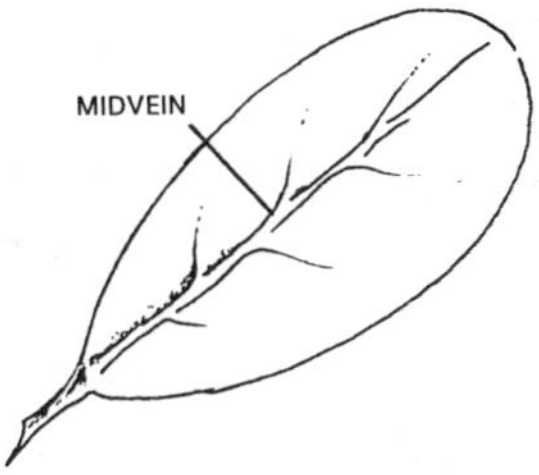

Figure 1383

Mucro. A short, sharp, abrupt point, usually at the tip of a leaf. Figure 1384.

Nerve. A prominent, simple vein or rib of a leaf. Figure. See illustration for **midvein**.

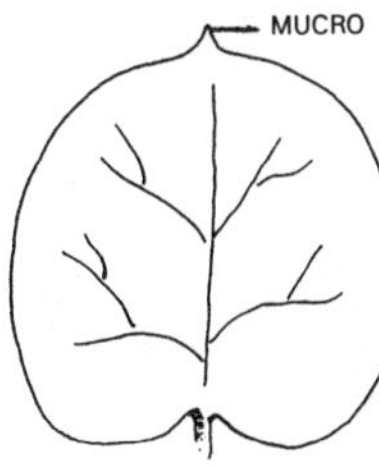

Figure 1384

Petiole. A leaf stalk. Figure 1377.

Petiolule. The stalk of a leaflet of a compound leaf. Figure 1380.

Pinna (pl. **pinnae**). One of the primary divisions or leaflets of a pinnate leaf. See illustration for **leaflet**.

Pinnule. The pinnate division of a pinna in a bipinnately compound leaf, or the ultimate divisions of a leaf which is more than twice pinnately compound. Figure 1385.

Rachis. The main axis of a compound leaf. Figure 1380.

Rhachis. See **rachis**.

Rib. A main longitudinal vein in a leaf. See **vein**.

Segment. A section or division of a leaf. Figure 1386.

Sinus. The cleft, depression, or recess between two lobes of a leaf. Figure 1386.

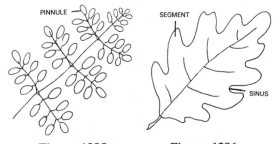

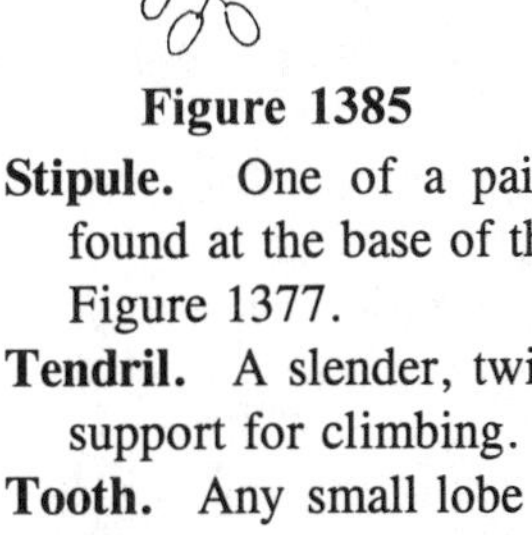

Figure 1385

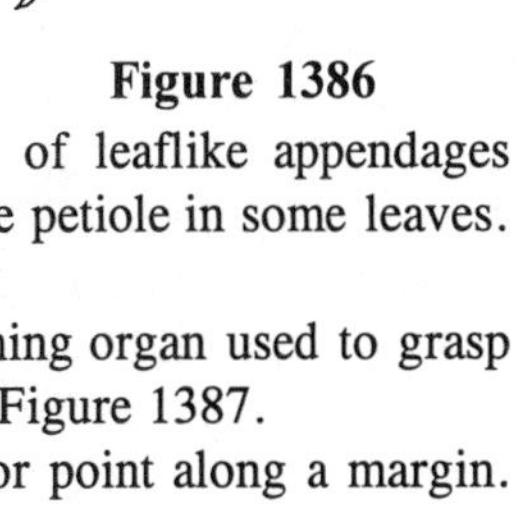

Figure 1386

Stipule. One of a pair of leaflike appendages found at the base of the petiole in some leaves. Figure 1377.

Tendril. A slender, twining organ used to grasp support for climbing. Figure 1387.

Tooth. Any small lobe or point along a margin. See **denticle** illustration.

Vein. A vascular bundle, usually visible externally, as in leaves. Figure 1388.

Veinlet. A small vein. Figure 1388.

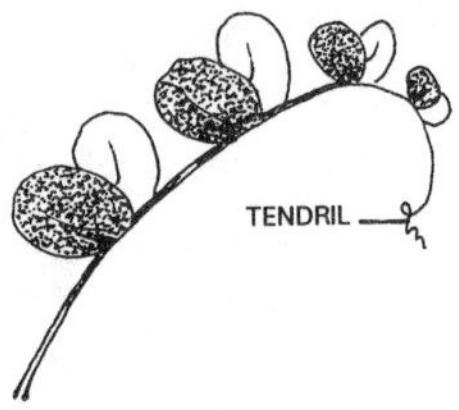

Figure 1387

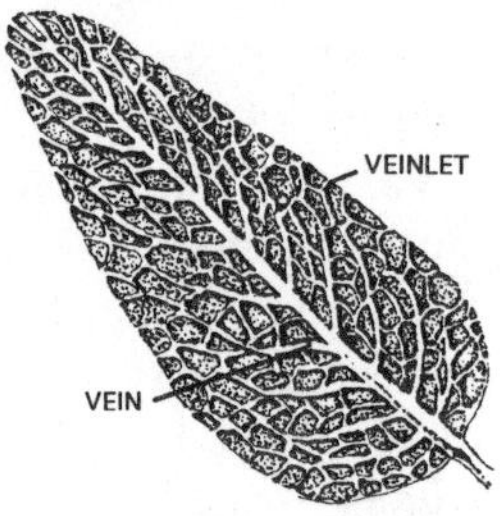

Figure 1388

LEAF SHAPES (Figure 1389.)

Acerose. Needle-shaped, as the leaves of pine or spruce.

Awl-shaped. Short, narrowly triangular, and sharply pointed like an awl.

Cordate. Heart-shaped, with the notch at the base.

Deltoid. With the shape of the Greek letter delta; shaped like an equilateral triangle.

Elliptic. In the shape of an ellipse, or a narrow oval; broadest at the middle and narrower at the two equal ends.

Elliptical. See **elliptic**.

Ensiform. Sword-shaped, as an *Iris* leaf.

Falcate. Sickle-shaped; hooked; shaped like the beak of a falcon.

Flabellate. Fan-shaped.

Flabelliform. See **flabellate**.

Gladiate. Sword-shaped.

Halberd-shaped. See **hastate**.

Hastate. Arrowhead-shaped, but with the basal lobes turned outward rather than downward; halberd-shaped. (compare **sagittate**)

Lanceolate. Lance-shaped; much longer than wide, with the widest point below the middle.

Linear. Resembling a line; long and narrow with more or less parallel sides.

Lyrate. Lyre-shaped; pinnatifid, with the terminal lobe large and rounded and the lower lobes much smaller.

Obcordate. Inversely cordate, with the attachment at the narrower end; sometimes refers to any leaf with a deeply notched apex.

Obdeltoid. Deltoid, with the attachment at the pointed end.

Obelliptic or **obelliptical.** Almost elliptic, but with the distal end somewhat larger than the proximal end.

Oblanceolate. Inversely lanceolate, with the attachment at the narrower end.

Oblong. Two to four times longer than broad with nearly parallel sides.

Obovate. Inversely ovate, with the attachment at the narrower end.

Orbicular. Approximately circular in outline.

Orbiculate. See **orbicular**.

Oval. Broadly elliptic, the width over one-half the length.

Ovate. Egg-shaped in outline and attached at the broad end (applied to plane surfaces).

Pandurate. Fiddle-shaped.

Panduriform. See **pandurate**.

Peltate. Shield-shaped; borne on a stalk attached to the lower surface rather than to the base or margin.

Perfoliate. A leaf with the margins entirely surrounding the stem, so that the stem appears to pass through the leaf.

Quadrate. Square; rectangular.

Reniform. Kidney-shaped.

Rhombic. Diamond-shaped.

Rhomboid. See **rhomboidal**.

Rhomboidal. Quadrangular, nearly rhombic, with obtuse lateral angles.

Rotund. Round or rounded in outline.

Sagittate. Arrowhead-shaped, with the basal lobes directed downward. (compare **hastate**)

Spathulate. See **spatulate**.

Spatulate. Like a spatula in shape, with a rounded blade above gradually tapering to the base.

Subulate. Awl-shaped.

LEAF BASES (Figure 1390.)

Aequilateral. Equal-sided, as opposed to oblique.

Attenuate. Tapering gradually to a narrow base.

Auriculate. With ear-shaped appendages.

Cordate. Heart-shaped, with the notch at the base.

Cuneate. Wedge-shaped, triangular and tapering to a point at the base.

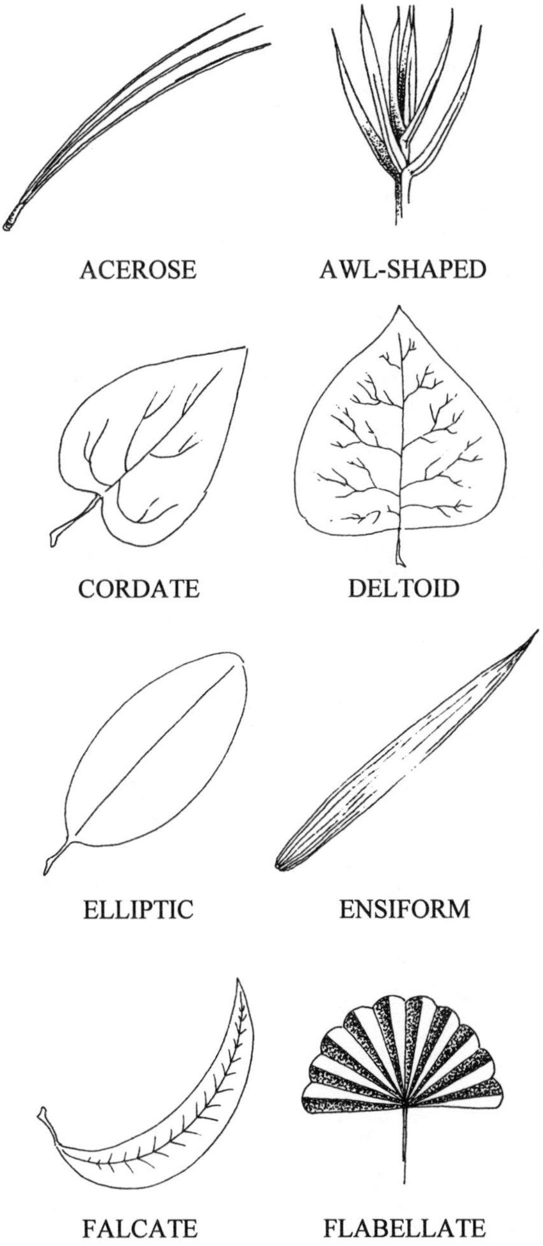

Figure 1389a.

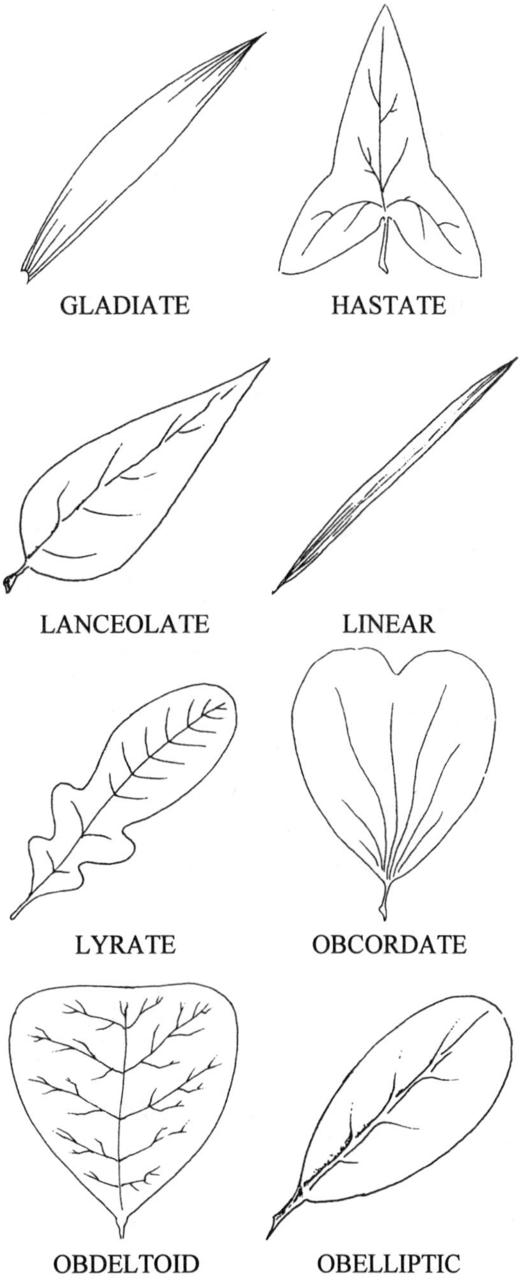

Figure 1389b.

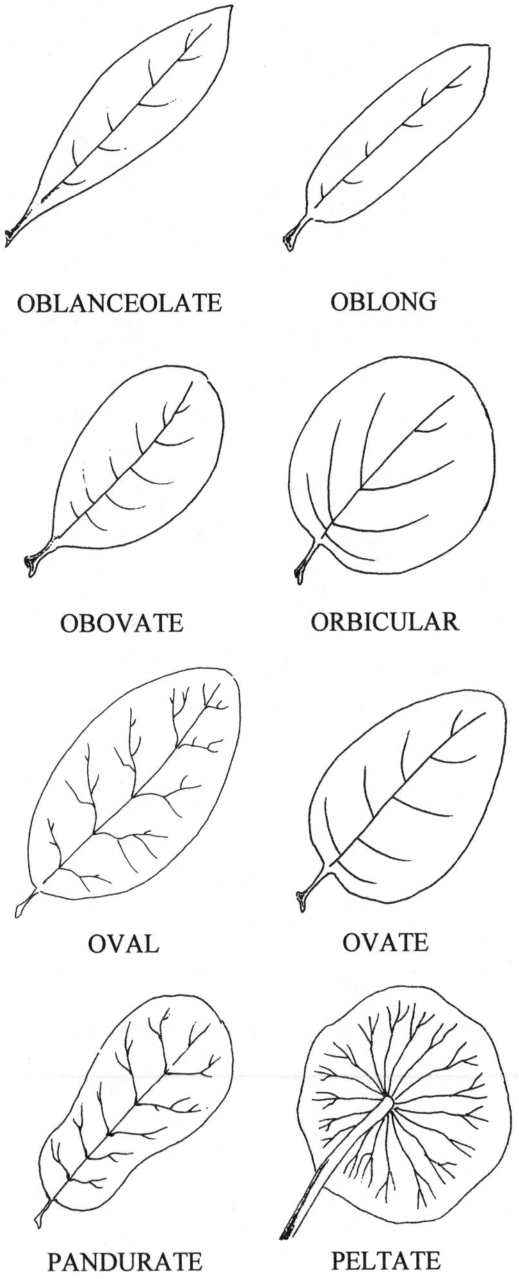

Figure 1389c.

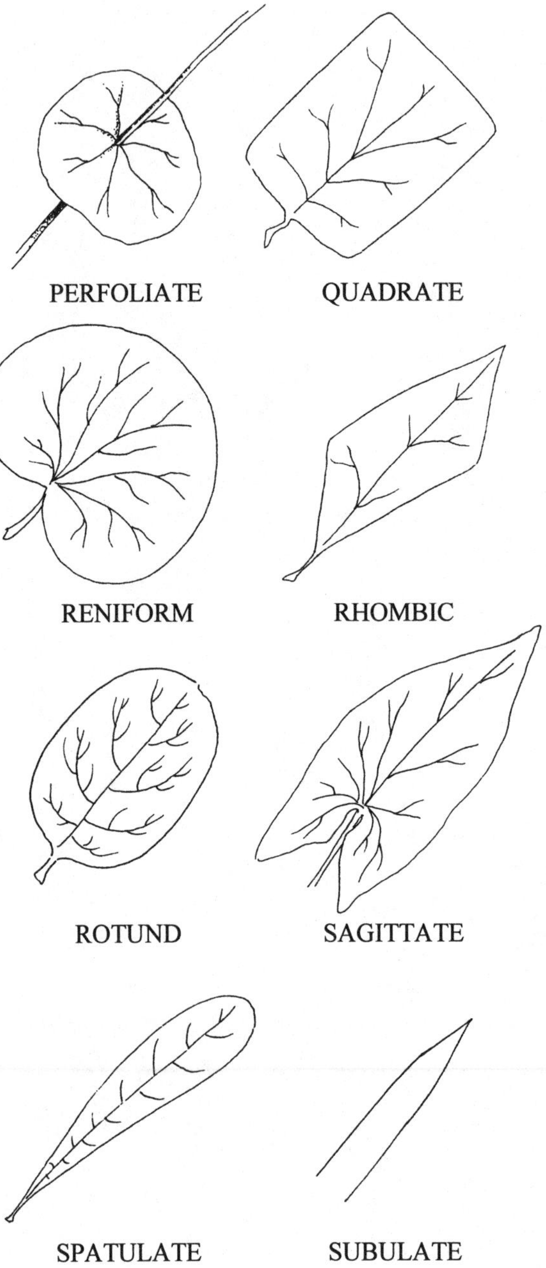

Figure 1389d.

Eared. See **auriculate**.

Halberd-shaped. See **hastate**.

Hastate. Arrowhead-shaped, but with the basal lobes turned outward rather than downward; halberd-shaped. (compare **sagittate**)

Inequilateral. See **oblique**.

Oblique. With unequal sides; slanting.

Rounded. With a rounded base.

Sagittate. Arrowhead-shaped, with the basal lobes directed downward. (compare **hastate**)

Truncate. With the base squared at the end as if cut off.

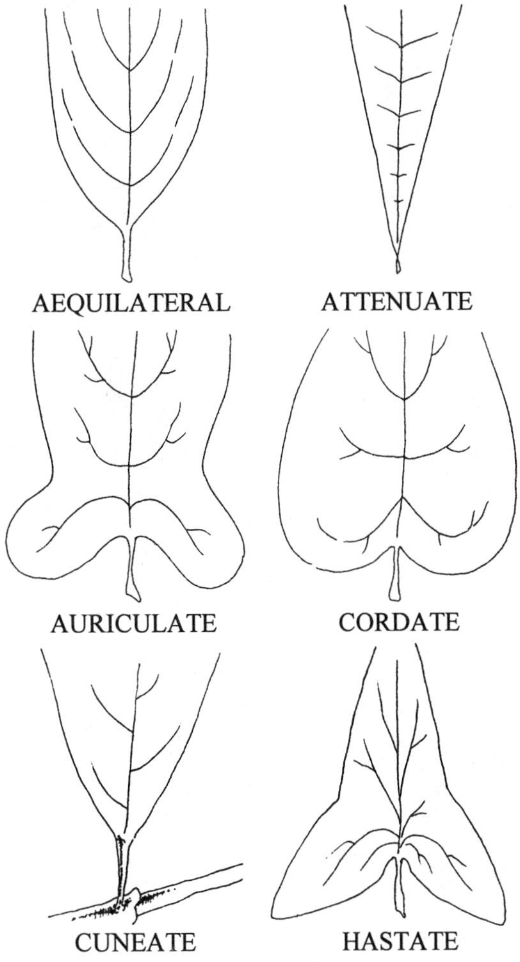

Figure 1390a.

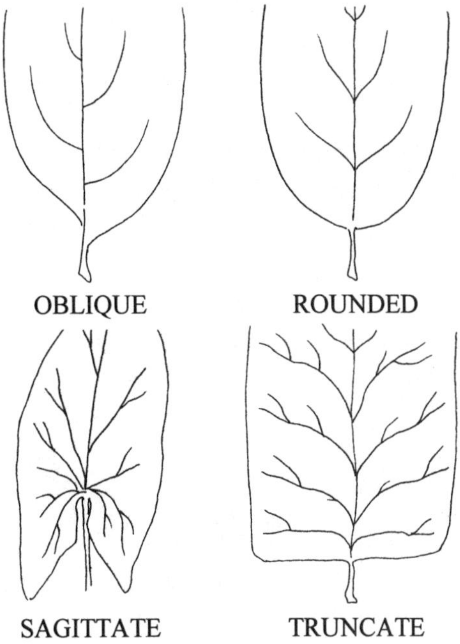

Figure 1390b.

LEAF APICES (Figure 1391.)

Abrupt. Terminating suddenly. See **truncate**.

Acuminate. Gradually tapering to a sharp point and forming concave sides along the tip.

Acute. Tapering to a pointed apex with more or less straight sides.

Apiculate. Ending abruptly in a small, slender point.

Aristate. Bearing an awn or bristle at the tip.

Aristulate. Bearing a minute awn or bristle at the tip.

Caudate. With a taillike appendage.

Cirrhous. See **cirrose**.

Cirrose. With a cirrus (tendril).

Cuspidate. Tipped with a short, sharp, abrupt point (cusp).

Emarginate. With a notch at the apex.

Mucronate. Tipped with a short, sharp, abrupt point (mucro).

Mucronulate. Tipped with a very small mucro.

Muticous. Blunt, without a point or spine.

Obcordate. With a deeply notched apex.

Obtuse. Blunt or rounded at the apex; with the

sides coming together at the apex at an angle greater than 90 degrees.

Praemorse. Terminating abruptly, as if bitten off. See **truncate**.

Pungent. Tipped with a sharp, rigid point; with a sharp, acrid odor or taste.

Retuse. With a shallow notch in a round or blunt apex.

Rounded. With a rounded apex.

Subacute. Slightly acute.

Truncate. With the apex squared at the end as if cut off.

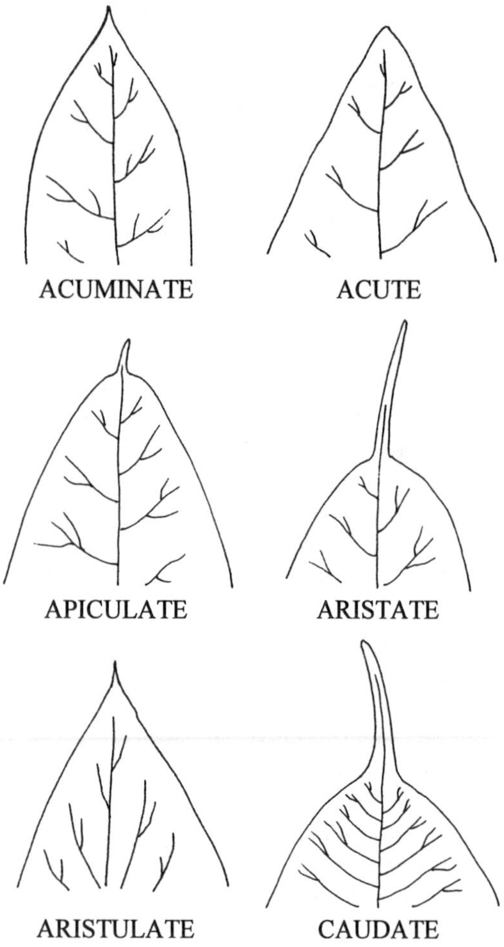

Figure 1391a.

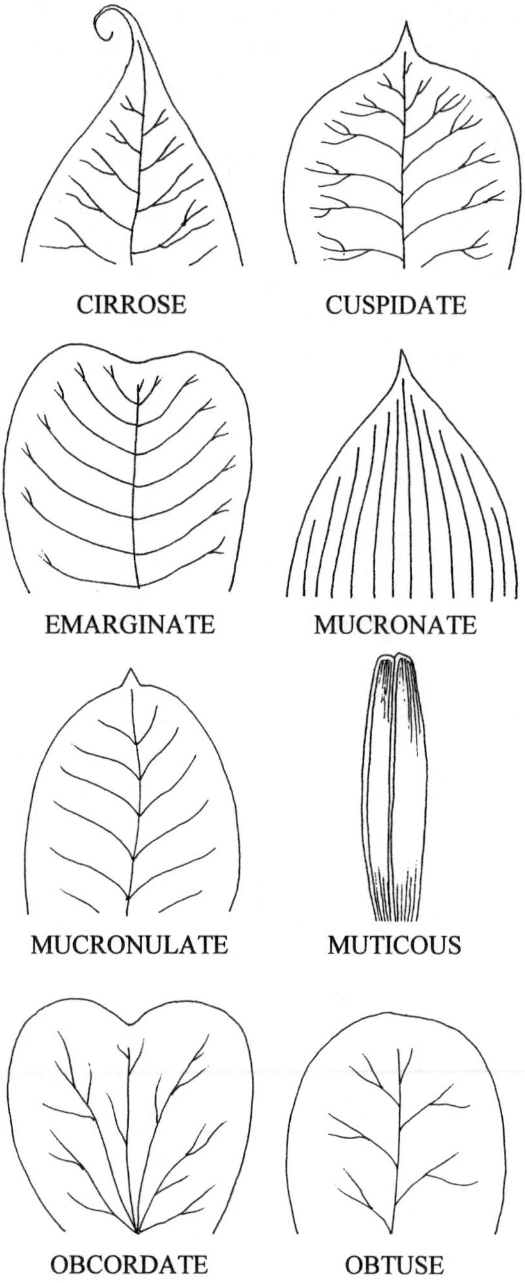

Figure 1391b.

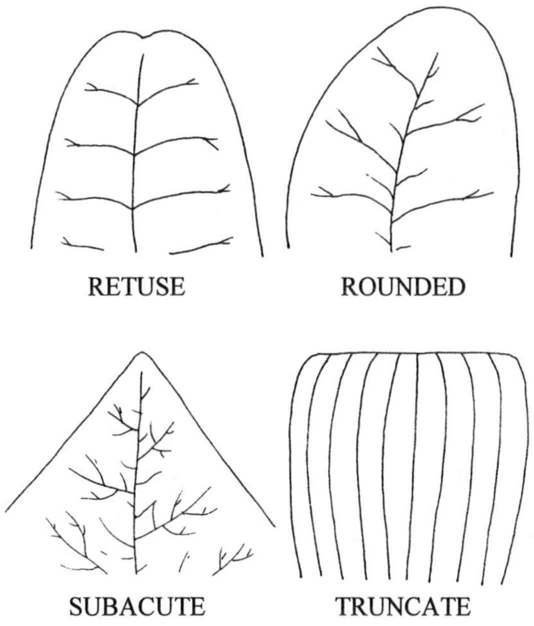

Figure 1391c.

LEAF DIVISION (Figure 1392.)

Abruptly pinnate. Pinnate without an odd leaflet at the tip. Same as **even-pinnate**.

Bifoliate. With two leaves or two leaflets.

Bipinnate. Twice pinnate; with the divisions again pinnately divided.

Biternate. Doubly ternate with the ternate divisions again ternately divided.

Compound leaf. A leaf separated into two or more distinct leaflets.

Decompound. More than once-compound, the leaflets again divided.

Even-pinnate. Pinnately compound with a terminal pair of leaflets or a tendril rather than a single terminal leaflet, so that there is an even number of leaflets.

Foliolate. Pertaining to or having leaflets; usually used in compounds, such as **bifoliolate** or **trifoliolate**.

Imparipinnate. Odd-pinnate; unequally pinnate.

Interruptedly pinnate. Pinnate with leaflets of various sizes intermixed.

Key to Common Types of Leaf Divisions

1 Leaf blade not divided into separate leaflets. . . . . **Simple**
1 Leaf blade divided into separate leaflets. (**Compound**)
  2 Leaflets arising from a common point, like the fingers of a hand. (**Palmate**)
    3 Leaflets simple.
      4 Leaflets three. . . . . **Ternate, Trifoliate**
      4 Leaflets more than three. . . . . **Palmate**
    3 Leaflets divided into secondary leaflets. (**Decompound**)
      5 Leaves twice divided. . . . . **Biternate**
      5 Leaves thrice divided. . . . . **Triternate**
  2 Leaflets arising from opposite sides of an elongated axis. (**Pinnate**)
    6 Leaflets simple.
      7 Leaflets even in number.
        8 Leaf ending in a tendril. . . . . **Tendril-pinnate**
        8 Leaf not ending in a tendril. . . . . **Even-pinnate, Abruptly pinnate**
      7 Leaflets odd in number.
        9 Leaflets three. . . . . **Ternate, Trifoliate**
        9 Leaflets more than three. . . . . . . **Odd-pinnate, Unequally pinnate, Imparipinnate**
    6 Leaflets divided into secondary leaflets. (**Decompound**)
        10 Leaflets twice divided. . . . . **Bipinnate**
        10 Leaflets thrice divided. . . . . **Tripinnate**

Odd-pinnate. Pinnately compound with a terminal leaflet rather than a pair of leaflets or a tendril, so that there is an odd number of leaflets.

Palmate. Lobed, veined, or divided from a common point, like the fingers of a hand. (compare **pinnate**)

Pinnate. A compound leaf with leaflets arranged on opposite sides of an elongated axis. (compare **palmate**)

Simple. Undivided, as a leaf blade which is not separated into leaflets (though the blade may be deeply lobed or cleft).

Tendril-pinnate. Pinnately compound, but ending in a tendril, as in the sweet pea.

Ternate. In threes, as a leaf which is divided into three leaflets.

Trifoliate. With three leaves or three leaflets.

Trifoliolate. See **trifoliate**.

Tripinnate. Pinnately compound three times, with pinnate pinnules.

Triternate. Triply ternate.

Unequally pinnate. See **odd-pinnate**.

Unifoliate. A leaf theoretically compound, though only expressing a single leaflet and appearing simple, as in *Cercis*.

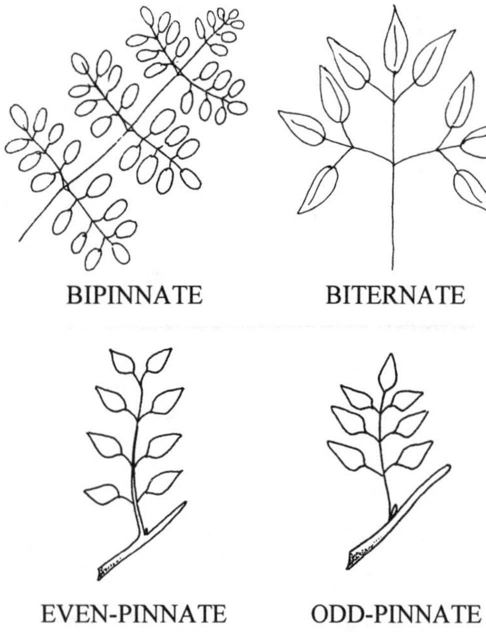

Figure 1392a.

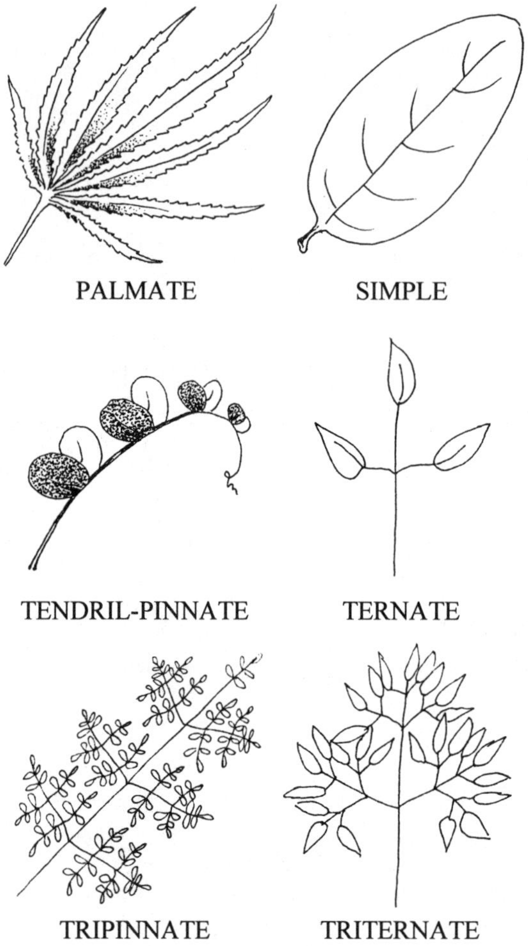

Figure 1392b.

LEAF VENATION

The pattern of veining on a leaf.

Costate. Ribbed. Figure 1393.

Net-veined. In the form of a network; reticulate. Figure 1394.

Parallel-veined. With the main veins parallel to the leaf axis or to each other. Figure 1395. (compare **net-**

Figure 1393

Figure 1394

Figure 1395

veined)

Pinnipalmate. Intermediate between pinnate and palmate, as in a leaf with the first pair of veins larger and more distinctive than the others. Figure 1396.

Figure 1396

Reticulate. In the form of a network; net-veined. Figure 1394.

Ribbed. With prominent ribs or veins. See illustration for **costate**.

Trinerved. Three-nerved, with the nerves all arising from near the base. Figure 1397. (compare **triplinerved**)

Triplinerved. Three-nerved, with the two lateral nerves arising from the midnerve above the base. Figure 1398. (compare **trinerved**)

Venose. Veiny. Figure 1394.

Figure 1397

Figure 1398

LEAF MARGINS (Figure 1399.)

Bidentate. With two teeth.

Bifid. Deeply two-cleft or two-lobed, usually from the tip.

Bilobed. Divided into two lobes.

Bipartite. Divided almost to the base into two divisions.

Bipinnatifid. Twice pinnately cleft.

Bisected. Split into two parts.

Biserrate. Doubly serrate, as when the teeth of a serrate leaf are also serrate.

Cleft. Cut or split about half-way to the middle or base.

Crenate. With rounded teeth along the margin.

Crenulate. With very small rounded teeth along the margin.

Crisped. Curled, wavy or crinkled.

Dentate. Toothed along the margin, the teeth directed outward rather than forward.

Denticulate. Dentate with very small teeth.

Digitate. Lobed, veined, or divided from a common point, like the fingers of a hand. (same as **palmate**)

Dissected. Deeply divided into many narrow segments.

Divided. Cut or lobed to the base or to the midrib.

Edentate. Without teeth.

Entire. Not toothed, notched, or divided, as the continuous margins of some leaves.

Erose. With the margin irregularly toothed, as if gnawed.

Erosulate. More or less erose.

Incised. Cut sharply, deeply, and usually irregularly.

Inrolled. Curled or rolled inward; involute.

Involute. With the margins rolled inward toward the upper side. (compare **revolute**)

Lacerate. Cut or cleft irregularly, as if torn.

Laciniate. Cut into narrow, irregular lobes or segments.

Lobed. Bearing lobes which are cut less than half way to the base or midvein.

Lobulate. With lobules.

Multifid. Cleft into many narrow segments or lobes.

Palmate. Lobed, veined, or divided from a common point, like the fingers of a hand. (compare **pinnate**)

Palmatifid. Palmately cleft or lobed.

Palmatisect. Palmately divided.

Parted. Deeply cleft, usually more than half the

distance to the base or midvein.

Pedate. Palmately divided, with the lateral lobes 2-cleft.

Pinnatifid. Pinnately cleft or lobed half the distance or more to the midrib, but not reaching the midrib.

Pinnatilobate. With pinnately arranged lobes.

Pinnatisect. Pinnately cleft to the midrib.

Quadripinnatifid. Four times pinnately cleft.

Repand. With a slightly wavy or weakly sinuate margin. Same as **undulate**.

Revolute. With the margins rolled backward toward the underside. (compare **involute**)

Runcinate. Sharply pinnatifid or cleft, the segments directed downward.

Serrate. Toothed along the margin, the sharp teeth pointing forward.

Serrulate. Toothed along the margin with minute, sharp, forward-pointing teeth.

Sinuate. With a strongly wavy margin.

Key to Common Leaf Margin Types

1 Margin continuous, not toothed, notched, lobed, or divided. . . . . **Entire**
1 Margin toothed, notched, lobed, or divided.
 2 Leaf toothed, notched, lobed, or incised less than half the distance to the base or midvein.
  3 Margin not distinctly toothed or lobed, merely wavy.
   4 Margin tightly wavy, producing a crinkled appearance. . . . . **Crisped**
   4 Margin loosely wavy, producing a smoother appearance. . . . . **Repand, Sinuate**
   5 Leaf toothed or lobed only at the apex.
    6 Apex with two teeth or lobes. **Bidentate**
    6 Apex with three teeth or lobes. . . . . **Tridentate**
   5 Leaf toothed or lobed below the apex.
    7 Margin coarsely lobed or cleft.
     8 Margin irregularly cleft. . . . . **Cleft**
     8 Margin regularly lobed. . . . . **Lobed, Pinnatilobate**
    7 Margin finely toothed or lobed.
     9 Margin not sharply toothed. **Crenate**
     9 Margin sharply toothed.
      10 Margin irregularly toothed. . . . . **Erose**
      10 Margin regularly toothed.
       11 Teeth directed forward. . . . . **Serrate**
       11 Teeth directed outward. . . . . **Dentate**
 2 Leaf lobed, divided, or incised half or more the distance to the base or midvein.
  12 Leaf parted or divided only at the apex.
   13 Leaf divided at the apex into two sections. . . . . **Bipartite**
   13 Leaf divided at the apex into three sections. . . . . **Trifid**
  12 Leaf parted or divided below the apex.
   14 Leaf palmately lobed or divided.
    15 Leaf divisions resembling the fingers of a hand. . . . . **Digitate**
    15 Leaf divisions not particularly resembling the fingers of a hand. . . . . . **Palmatifid**
   14 Leaf pinnately lobed or divided.
    16 Leaf irregularly cleft or divided.
     17 Leaf divided into many very narrow segments. . . . . **Dissected**
     17 Leaf divided more broadly.
      18 Leaf segments directed outward or forward. . . . . **Incised, Lacerate**
      18 Leaf segments directed backward. . . . . **Runcinate**
    16 Leaf regularly cleft or divided.
     19 Leaf cleft to the midvein. **Pinnatisect**
     19 Leaf not cleft to the midvein. . . . . **Pinnatifid**

Sinuous. Of a wavy or serpentine form. See illustration for **sinuate**.
Tridentate. Three-toothed.
Trifid. Three-cleft.
Tripartite. Three-parted.
Tripinnatifid. Thrice pinnately cleft.
Undulate. Wavy, but not so deeply or as pronounced as sinuate. See illustration for **repand**.

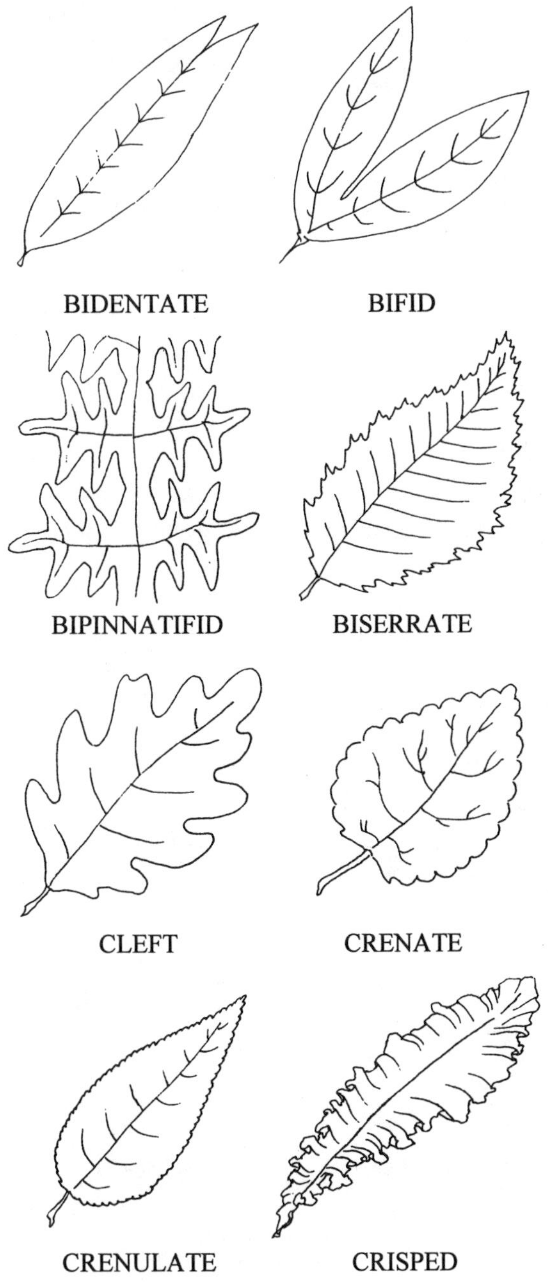

Figure 1399a.

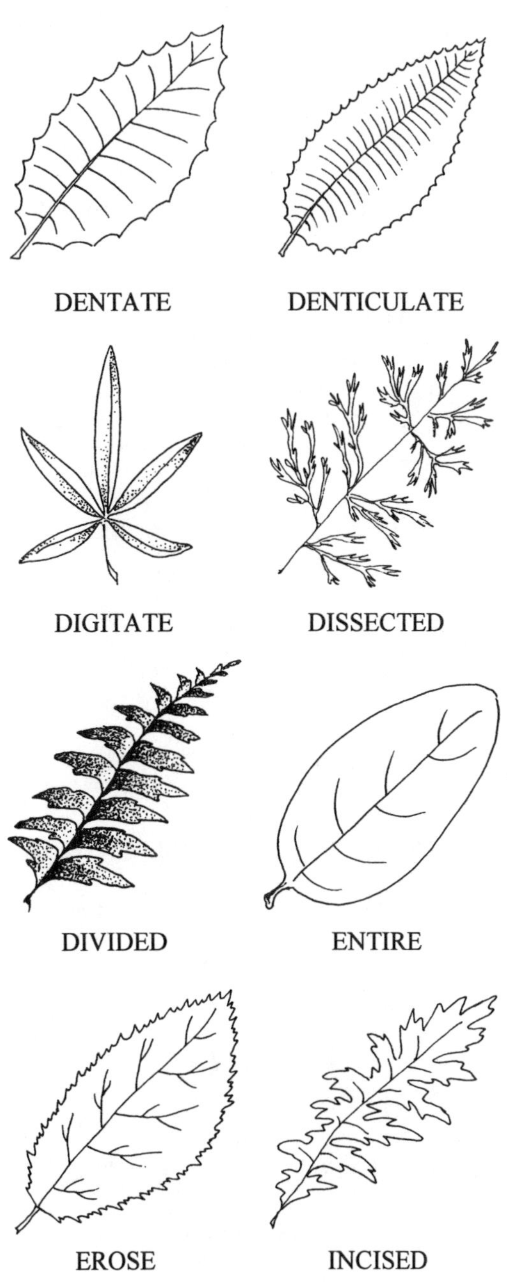

Figure 1399b.

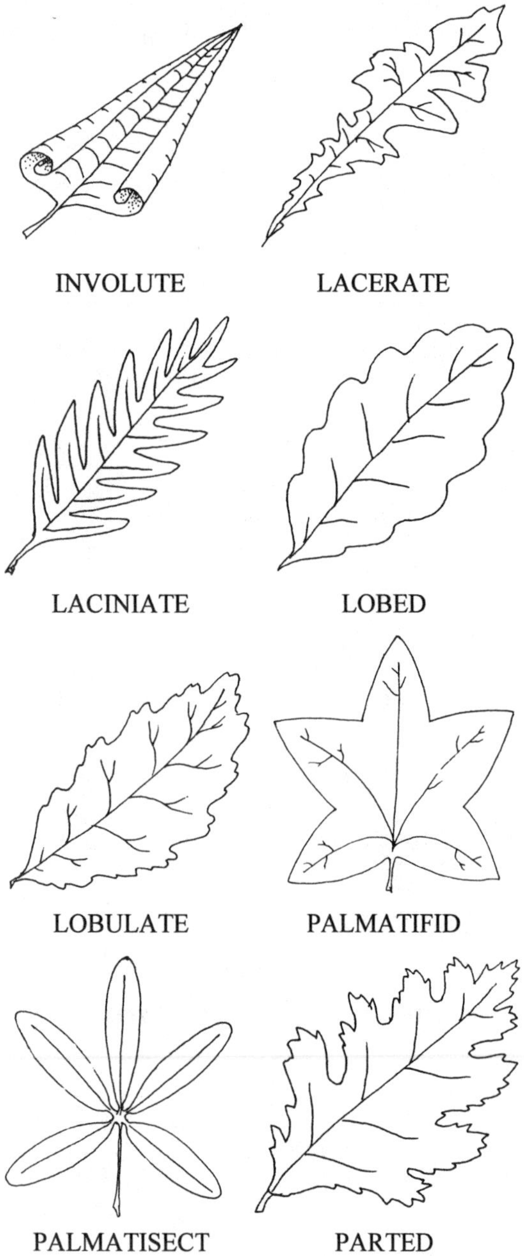

Figure 1399c.

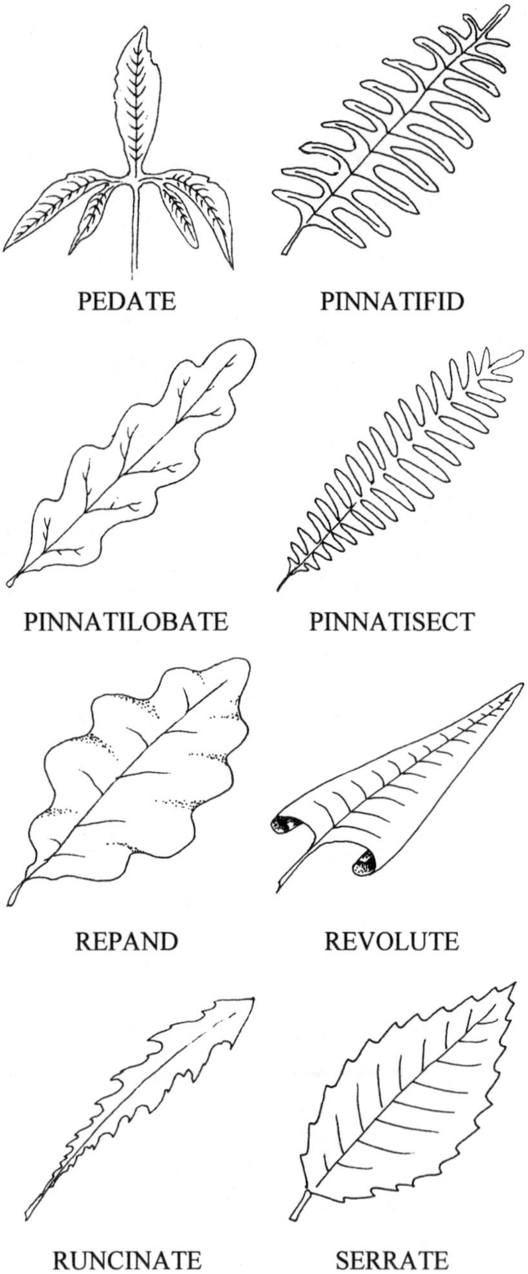

Figure 1399d.

SERRULATE

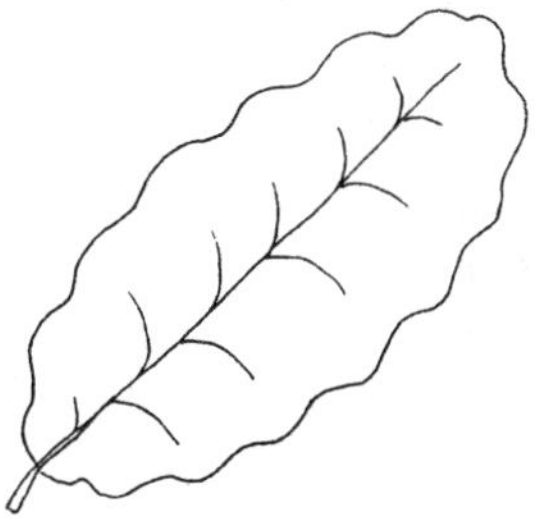

SINUATE

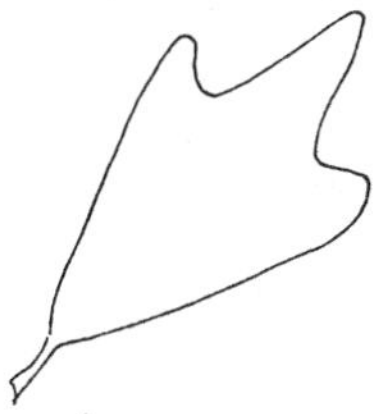

TRIDENTATE

TRIFID

TRIPARTITE

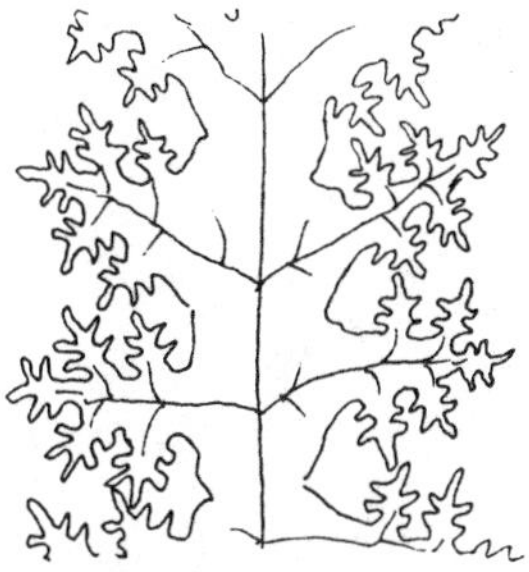

TRIPINNATIFID

Figure 1399e.

LEAF ATTACHMENT

Amplexicaul. Clasping the stem, as the base or stipules of some leaves. Figure 1400.

Auriculate-clasping. Earlike lobes at the base of a leaf, encircling the stem. Figure 1401.

Clasping. Wholly or partly surrounding the stem. Figure 1400.

Connate-perfoliate. With the bases of opposite leaves fused around the stem. Figure 1402.

Decurrent. Extending downward from the point of insertion, as a leaf base that extends down along the stem. Figure 1403.

Excurrent. Extending beyond what is typical, as in a leaf base which extends down the stem. See illustration for **decurrent**.

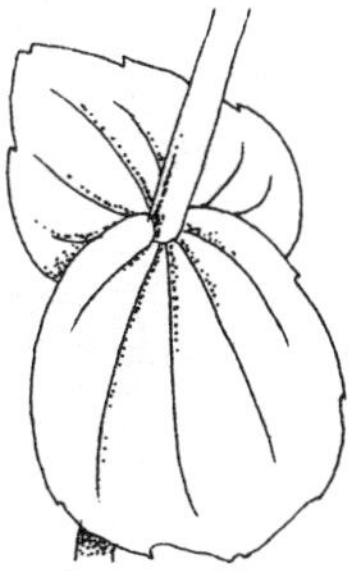

Figure 1400

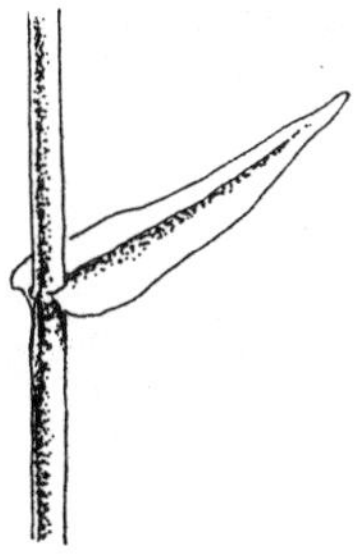

Figure 1401

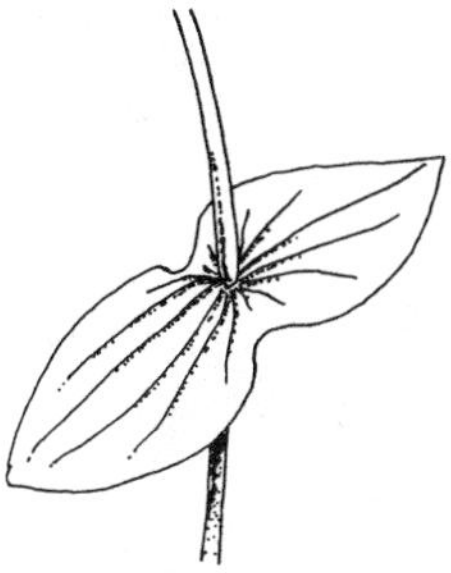

Figure 1402

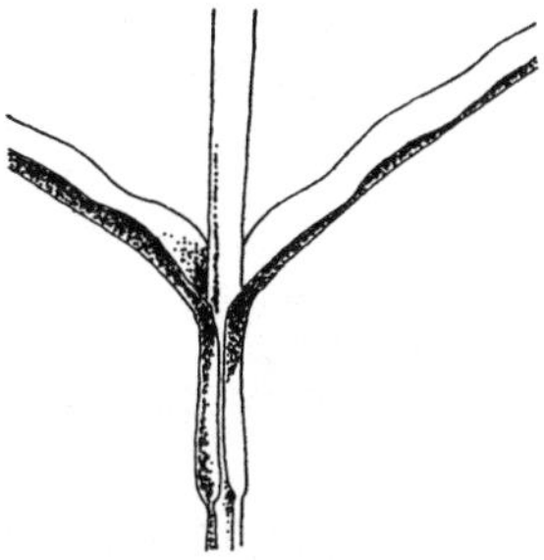

Figure 1403

Ocreate. With sheathing stipules. Figure 1404.

Perfoliate. A leaf with the margins entirely surrounding the stem, so that the stem appears to pass through the leaf. Figure 1405.

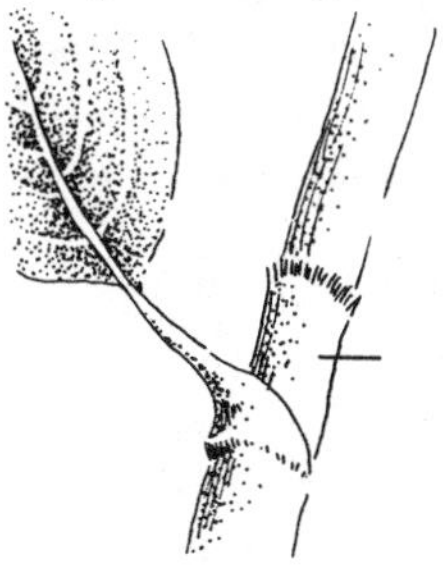

Figure 1404

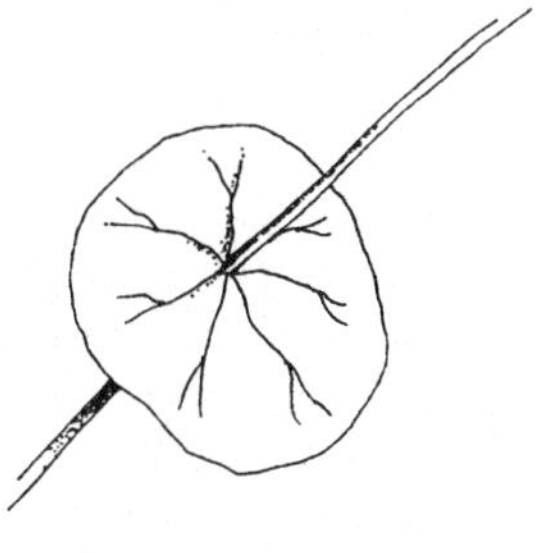

Figure 1405

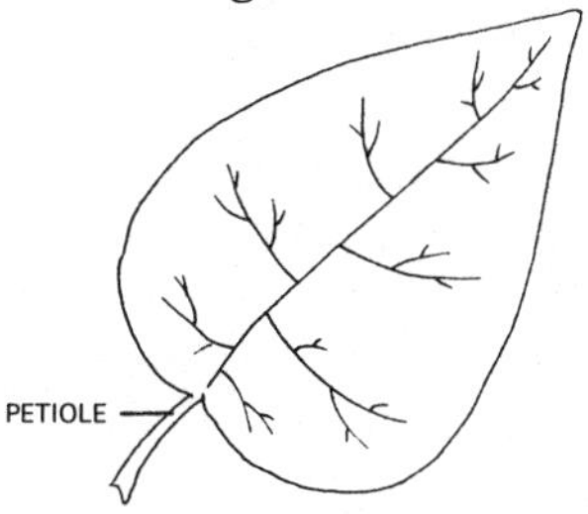

Figure 1406

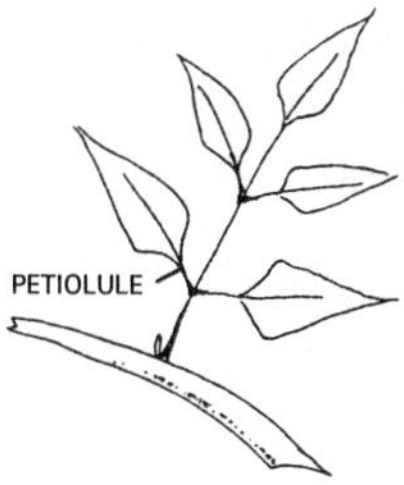

Figure 1407

Petiolate. With a petiole. Figure 1406.

Petioled. See **petiolate.**

Petiolulate. With a petiolule. Figure 1407.

Sessile. Attached directly, without a supporting stalk, as a leaf without a petiole. Figure 1408.

Sheathing. Forming a sheath, as the leaf base of a grass forms a sheath as it surrounds the stem. Figure 1409.

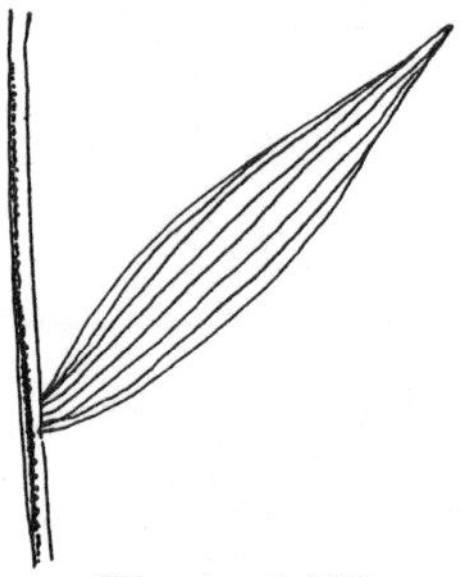

Figure 1408

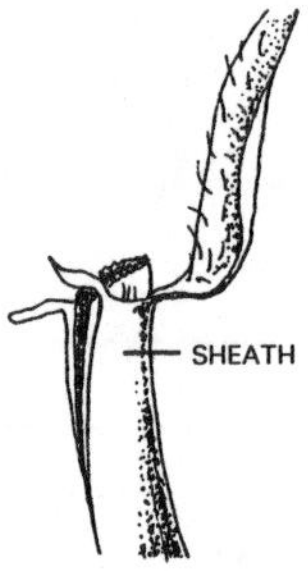

Figure 1409

LEAF ARRANGEMENT (Figure 1410.)

Key to Common Leaf Arrangments

1 Leaves positioned at the base of the stem. **Basal**
1 Leaves positioned on an elongated stem. **(Cauline)**
  2 Leaves one per node. . . . . **Alternate**
  2 Leaves two or more per node.
    3 Leaves two per node. . . . . . . **Opposite**
    3 Leaves three or more per node. **Whorled**

Alternate. Borne singly at each node, as leaves on a stem. (compare **opposite**)

Basal. Positioned at or arising from the base, as leaves arising from the base of the stem.

Bilateral. Arranged on two sides, as leaves on a stem.

Cauline. Leaves arising from the stem above ground level.

Decussate. Arranged along the stem in pairs, with each pair at right angles to the pair above or below.

Dextrorse. Turned to the right or spirally arranged to the right, as in the leaves on some stems.

Distichous. In two vertical ranks or rows on opposite sides of an axis.

Equitant. Overlapping or straddling in two ranks, as the leaves of *Iris*.

Opposite. Borne across from one another at the same node, as in a stem with two leaves per node. (compare **alternate**)

Ranked. Arranged into vertical rows.

Rosette. A dense radiating cluster of leaves usually at or near ground level.

Rosulate. With the leaves arranged in basal rosettes, the stem very short or lacking.

Sinistrorse. Turned to the left or spirally arranged to the left, as in the leaves on some stems.

Verticillate. Arranged in verticils; whorled.

Whorled. With parts arranged in whorls, as in a leaf arrangement with three or more leaves arising from a node. Same as **verticillate.**

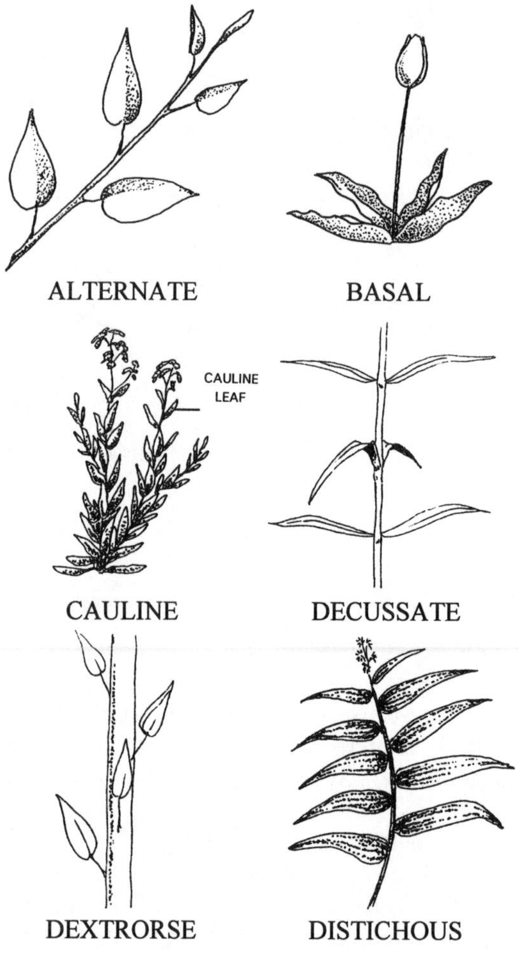

Figure 1410a.

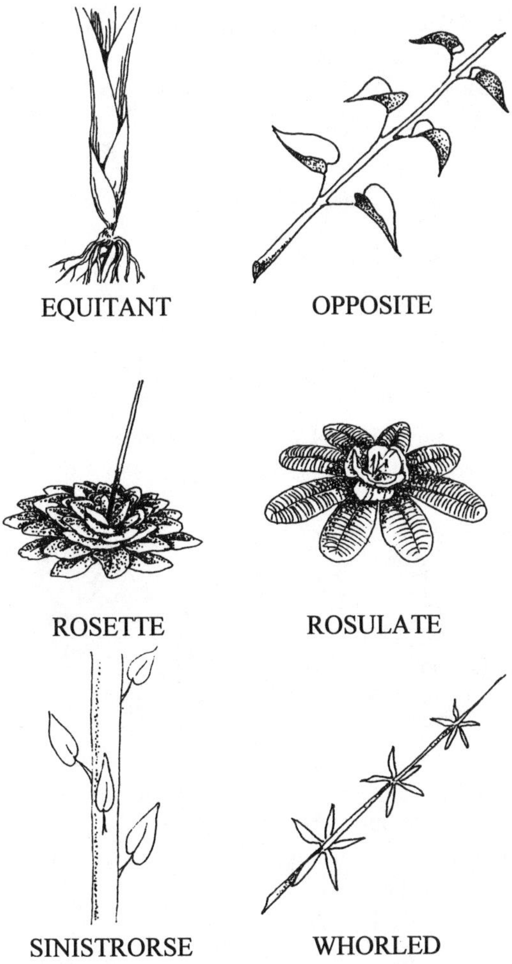

Figure 1410b.

MISCELLANEOUS LEAF TERMS

Circinate. Coiled from the tip downward, as in the young leaves of a fern. Figure 1411.

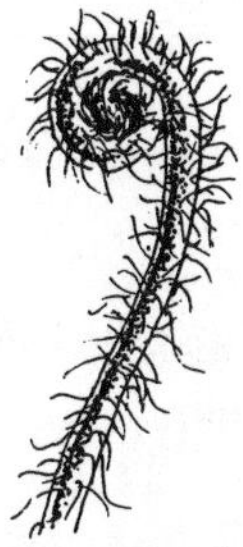

Figure 1411

Cirriferous. Bearing a tendril. Figure 1412.

Complicate. Folded together. Figure 1413.

Conduplicate. Folded together lengthwise with the upper surface within, as the leaves of many grasses. Figure 1414.

Deciduous. Falling off, as leaves from a tree; not evergreen; not persistent.

Estipulate. Without stipules. Figure 1415.

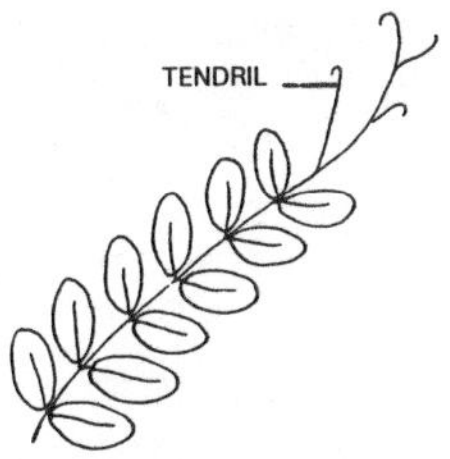

Figure 1412

Figure 1413

Figure 1414

Figure 1415

Evergreen. Having green leaves through the winter; not deciduous.

Exstipulate. Same as **estipulate**.

Frond. A large, divided leaf; a fern or palm leaf. Figure 1416.

Gamophyllous. With the leaves united, usually by the margins.

Heterophyllous. With different kinds of leaves on the same plant. Figure 1417.

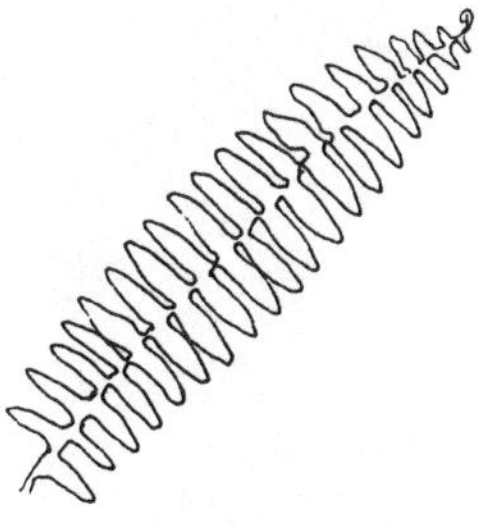

Figure 1416

Figure 1417

Limbate. Bordered, as in a leaf or flower in which one color forms an edging or margin around another. Figure 1418.

Macrophyll. The relatively large, expanded leaf

of higher vascular plants. Figure 1419.

Marcescent. Withering but persistent, as the sepals and petals in some flowers or the leaves at the base of some plants. Figure 1420.

Marginate. With a distinct margin.

Phyllopodic. With the lowest leaves well developed, not reduced to scales. Figure 1421.

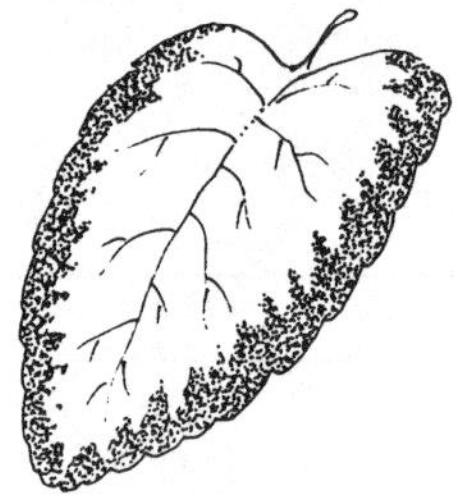

Figure 1418

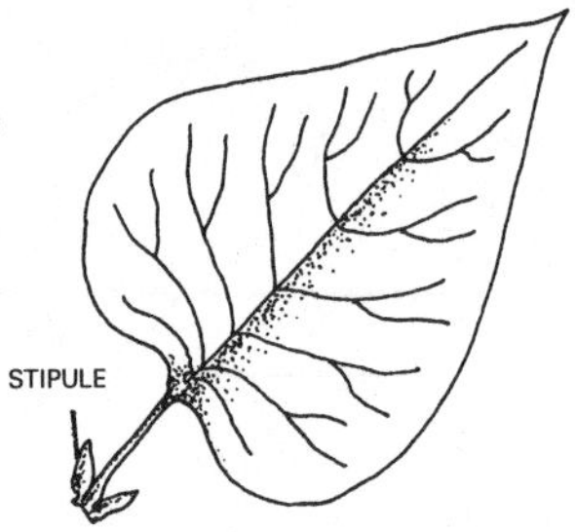

Figure 1419

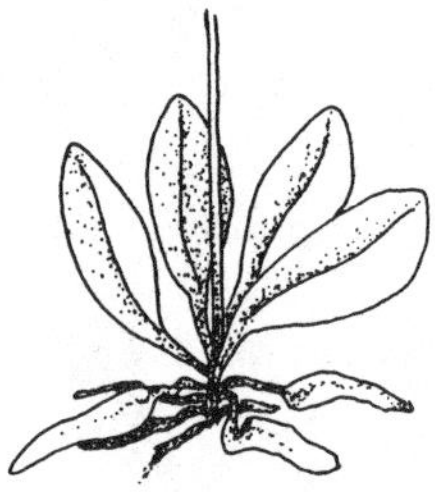

Figure 1420

Figure 1421

Stipulate. Bearing stipules. Figure 1419.

Succulent. Juicy and fleshy, as the leaves of *Aloe*. Figure 1422.

Venation. The pattern of veining on a leaf. Figure 1423.

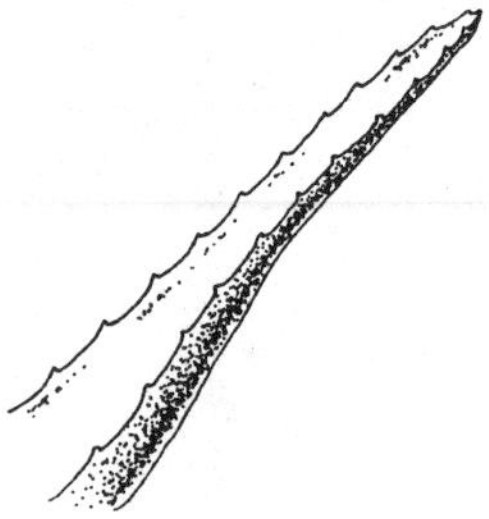

Figure 1422

Figure 1423

SURFACES

Surfaces of leaves, stems, fruits, and other organs.

Aculeate. Prickly; covered with prickles. Figure 1424.

Aculeolate. Minutely prickly; covered with tiny prickles. Figure 1425.

Figure 1424

Figure 1425

Alveolar. See **alveolate**.

Alveolate. Honeycombed, with pits separated by thin, ridged partitions. Figure 1426.

Aperturate. With one or more openings or apertures. In pollen grains, these apertures may be only thin spots rather than actual perforations. Figure 1427.

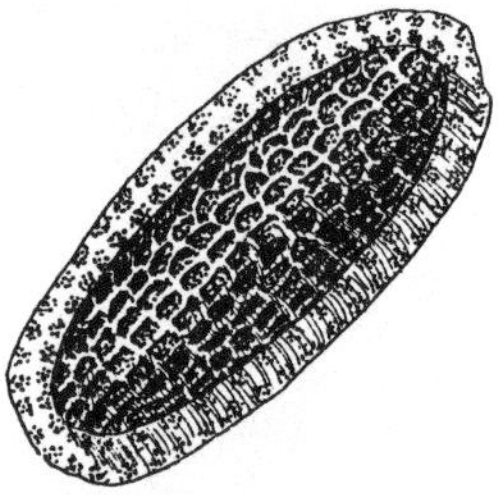

Figure 1426

Arachnoid. Bearing long, cobwebby, entangled hairs. Figure 1428.

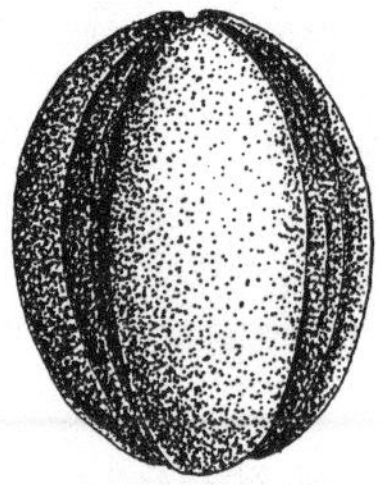

Figure 1427

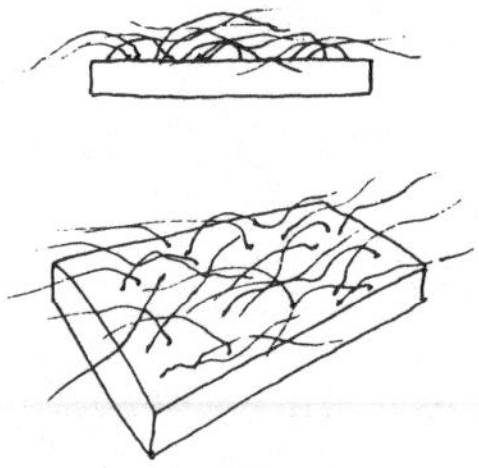

Figure 1428

Argenteous. Silvery.

Armed. Bearing thorns, spines, barbs, or prickles.

Asperous. Rough to the touch.

Barbellate. With short, stiff hairs or barbs. Figure 1429.

Barbellulate. With very tiny short, stiff hairs or barbs. Figure 1430.

Bullate. With rounded, blistery projections covering the surface. Figure 1431.

Canaliculate. With longitudinal channels or grooves. Figure 1432.

Figure 1429

Figure 1430

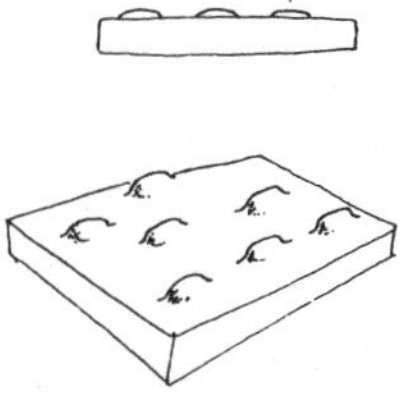
Figure 1431

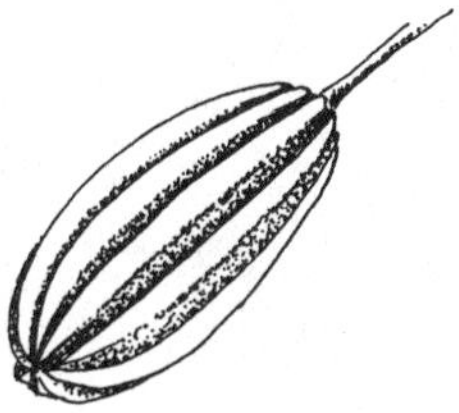
Figure 1432

Cancellate. Latticed with a fine, regular, reticulate pattern. Figure 1433.

Canescent. Gray or white in color due to a covering of short, fine gray or white hairs. Figure 1434.

Figure 1433

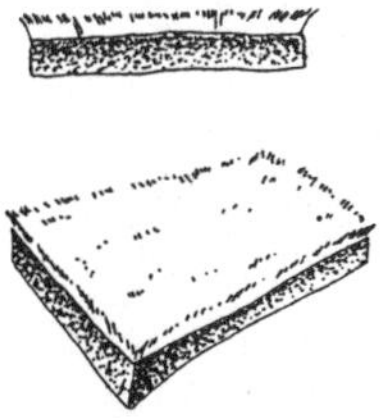
Figure 1434

Channeled. With one or more deep longitudinal grooves. Figure 1435.

Ciliate. With a marginal fringe of hairs. Figure 1436.

Ciliolate. With a marginal fringe of minute hairs. Figure 1437.

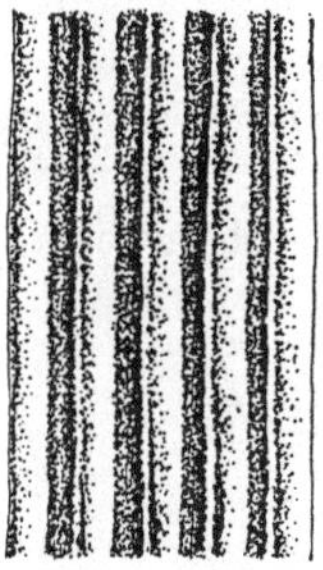
Figure 1435

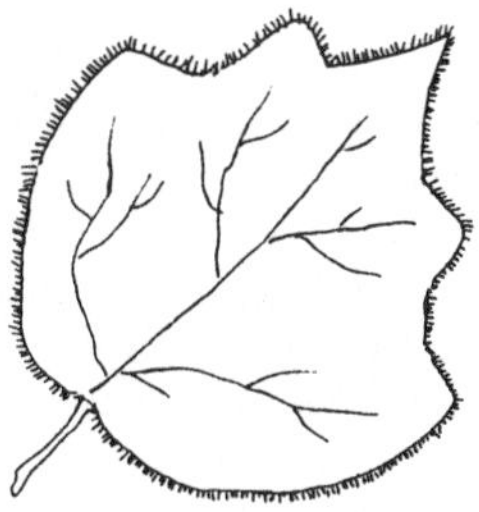
Figure 1436

Figure 1437

Cinereous. Ash-colored; grayish due to a covering of short hairs.

Coriaceous. With a leathery texture.

Corrugated. Wrinkled or folded into alternating furrows and ridges. Figure 1438.

Crinite. With tufts of long, soft hairs. Figure 1439.

Figure 1438

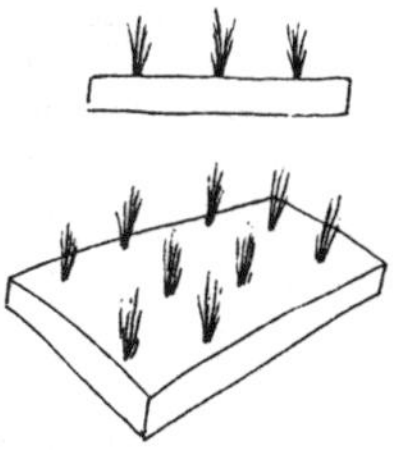
Figure 1439

Echinate. With prickles or spines. Figure 1440.

Echinulate. With very small prickles or spines. Figure 1441.

Figure 1440

Figure 1441

Faceted. With many plane surfaces, like a cut gem, as in some seeds. Figure 1442.

Farinose. Covered with a mealy, powdery substance. Figure 1443.

Faveolate. Honeycombed or pitted; alveolate.

Favose. See **faveolate**.

Fenestrate. With window-like perforations,

Key to Common Surfaces

1 Surface lacking hairs, grooves, scales, glands, pits, perforations, or projections; smooth.
  2 Surface sticky to the touch. . . . . **Viscid, Glutinous**
  2 Surface not obviously sticky to the touch.
    3 Surface covered with a whitish or bluish waxy coating. . . . . **Glaucous, Pruinose**
    3 Surface lacking a whitish or bluish waxy coating.
      4 Surface covered with a mealy, powdery substance. . . . . **Farinose**
      4 Surface not covered with a mealy, powdery substance. . . . . **Glabrous**
1 Surface with hairs, grooves, scales, glands, pits, perforations, or projections.
  5 Surface with hairs.
    6 Hairs limited to the margins of the surface. **Ciliate, Fimbriate, Fringed**
    6 Hairs not limited to the margins of the surface.
      7 Hairs not evenly spread over the surface, occurring in tufts. . . . . **Floccose, Crinite**
      7 Hairs evenly spread over the surface.
        8 Hairs with hooks or barbs.
          9 Hairs hooked at the tip. . . . . **Uncinate**
          9 Hairs barbed.
            10 Hairs barbed from apex to base. . . . . **Barbellate**
            10 Hairs barbed only near the apex. . . . . **Glochidiate**
        8 Hairs lacking hooks or barbs.
          11 Hairs bearing glands. . . . . **Glandular**
          11 Hairs lacking glands.
            12 Hairs with several branches radiating from a central point. . . . . **Stellate**
            12 Hairs simple or forked, but not with several branches radiating from a central point.
              13 Hairs straight, not interwoven or entangled.
                14 Hairs stiff and sharp.
                  15 Hairs borne on swollen, nipplelike bases. . . . . **Papillose-hispid**
                  15 Hairs not borne on swollen, nipplelike bases.
                    16 Hairs appressed. . . . . **Strigose**
                    16 Hairs not appressed.
                      17 Hairs very short. . . . . **Scabrous**
                      17 Hairs longer.
                        18 Hairs stiff enough to break the skin. . . . . **Hirsute**
                        18 Hairs not stiff enough to break the skin. . . . . **Hispid**
                14 Hairs soft and flexible.
                  19 Hairs long.
                    20 Hairs appressed. . . . . **Sericeous**
                    20 Hairs not appressed. . . . . **Pilose**
                  19 Hairs short.
                    21 Hairs very short and dense, producing a whitish appearance. . . . . . **Canescent**
                    21 Hairs somewhat longer, not producing a whitish appearance. . . . . . **Pubescent**
              13 Hairs curly or wavy and usually interwoven or entangled.
                22 Hairs long.
                  23 Hairs dense, so that the leaf surface is obscured. . . . . **Lanate, Woolly**
                  23 Hairs less dense, the leaf surface not obscured. . . . . **Villous, Holosericeous**
                22 Hairs short.
                  24 Hairs matted. . . . . **Tomentose**
                  24 Hairs not matted. . . . . **Velutinous**

5 Surface with grooves, scales, glands, pits, perforations, or projections.
  25 Surface with projections.
    26 Surface with prickles or spines. . . . . **Aculeate, Echinate**
    26 Surface lacking prickles or spines.
      27 Projections rounded and blistery. . . . . **Bullate, Strumose**
      27 Projections nipplelike. . . . . **Papillate, Mammillate**
  25 Surface with grooves, scales, glands, pits, or perforations.
    28 Surface with glands. . . . . **Punctate**
    28 Surface with grooves, scales, pits, or perforations.
      29 Surface with scales. . . . . **Lepidote, Squamate, Scurfy**
      29 Surface with grooves, pits, or perforations.
        30 Surface with perforations. . . . . **Fenestrate, Perforate**
        30 Surface with grooves or pits.
          31 Surface pitted.
            32 Pits separated by thin ridges. . . . . **Alveolate, Faveolate**
            32 Pits not separated by thin ridges. . . . . **Foveate**
          31 Surface grooved. . . . . **Canaliculate, Channeled**

openings, or translucent areas. Figure 1444.

Fimbriate. Fringed, usually with hairs or hairlike structures (fimbrillae) along the margin. Figure 1445.

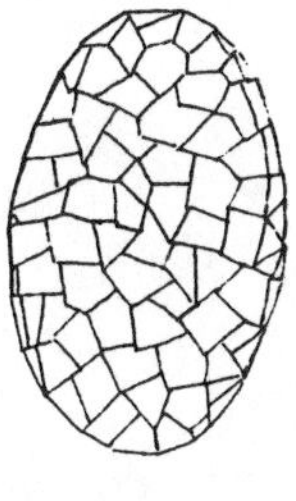

Figure 1442

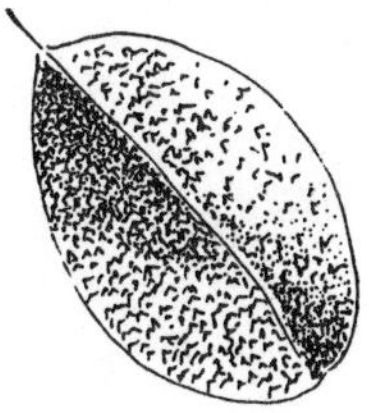

Figure 1443

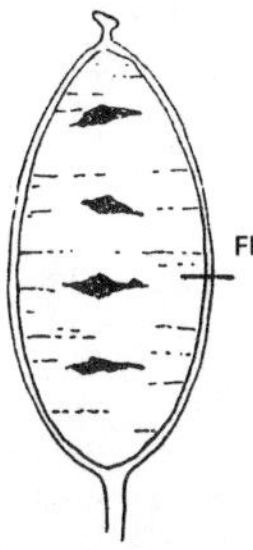

Figure 1444

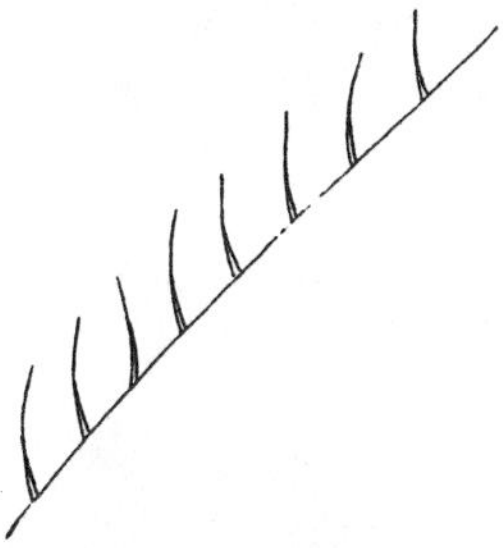

Figure 1445

Floccose. Bearing tufts of long, soft, tangled hairs. Figure 1446.

Flocculent. Bearing tufts of very fine woolly hairs; floccose.

Fluted. With furrows or grooves. Figure 1447.

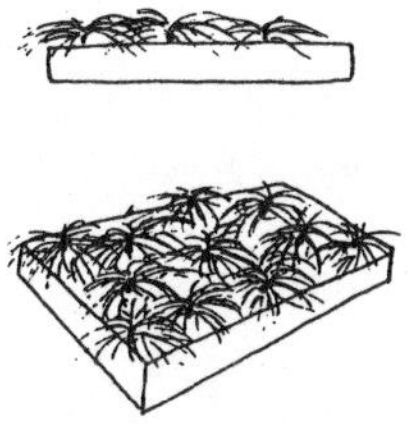

Figure 1446

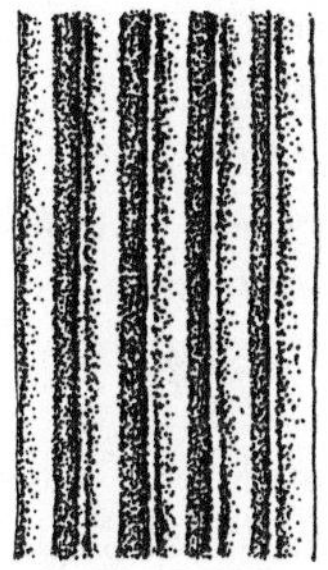

Figure 1447

Foveate. With foveae; pitted. Figure 1448.

Foveolate. With foveolae; minutely pitted. Figure 1449.

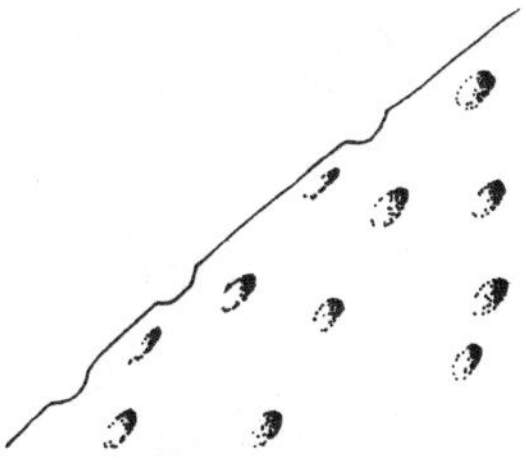

Figure 1448

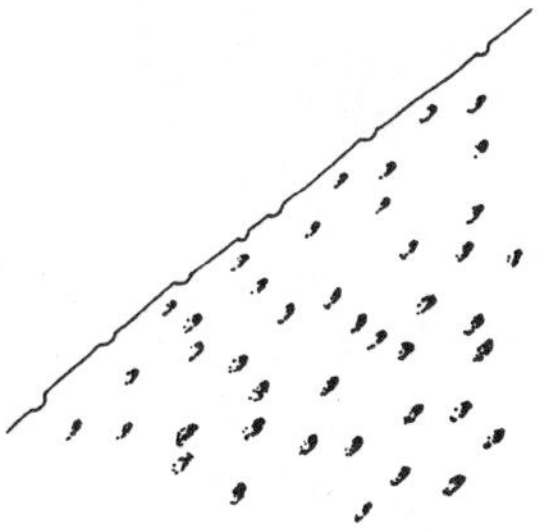

Figure 1449

Fringed. With hairs or bristles along the margin.

Furfuraceous. Scurfy; branlike; flaky. Figure 1450.

Glabrate. Becoming glabrous; almost glabrous.

Glabrescent. See **glabrate**.

Glabrous. Smooth; hairless.

Glandular. Bearing glands. Figure 1451.

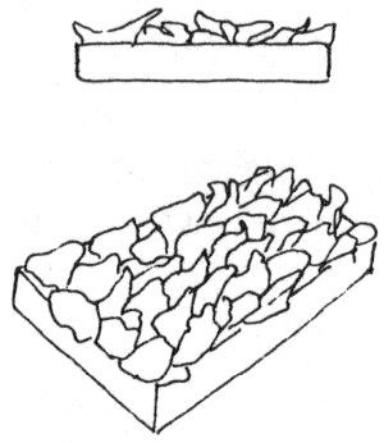

Figure 1450

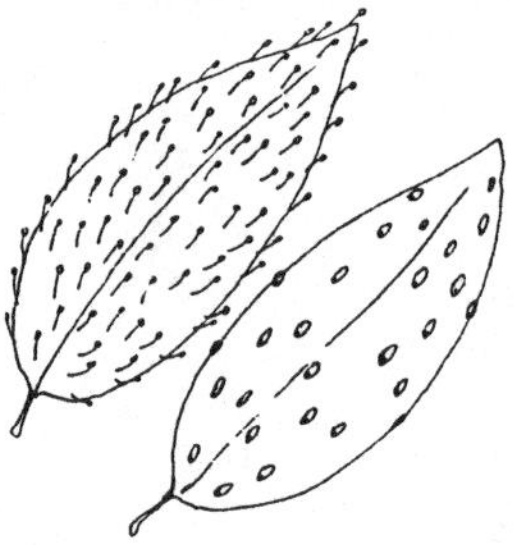

Figure 1451

Glaucescent. Somewhat glaucous; becoming glaucous.

Glaucous. Covered with a whitish or bluish waxy coating (bloom), as on the surface of a plum.

Glochidiate. Barbed at the tip. Figure 1452.

Figure 1452

Glutinous. Gluey; sticky; gummy; covered with a sticky exudation.

Hirsute. Pubescent with coarse, stiff hairs. Figure 1453.

Hirsutulous. Pubescent with very small, coarse, stiff hairs. Figure 1454.

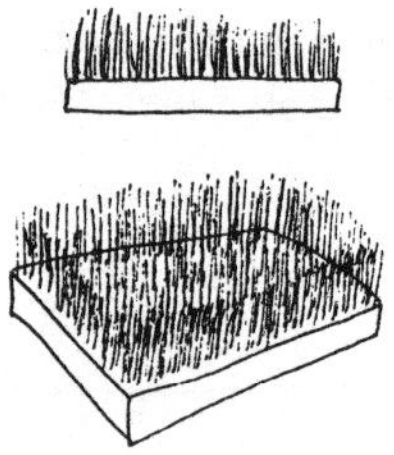

Figure 1453

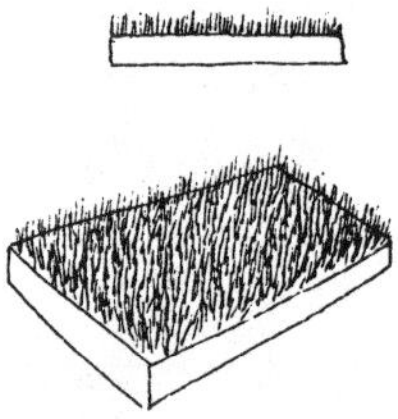

Figure 1454

Hirtellate. Same as **hirsutulous**.

Hirtellous. Same as **hirsutulous**.

Hispid. Rough with firm, stiff hairs. Figure 1455.

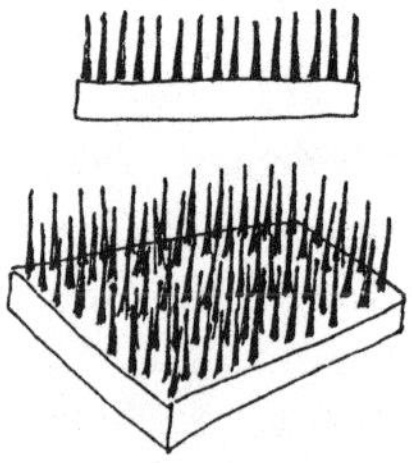

Figure 1455

Hispidulous. Minutely hispid. Figure 1456.

Hoary. With gray or white short, fine hairs. Figure 1457.

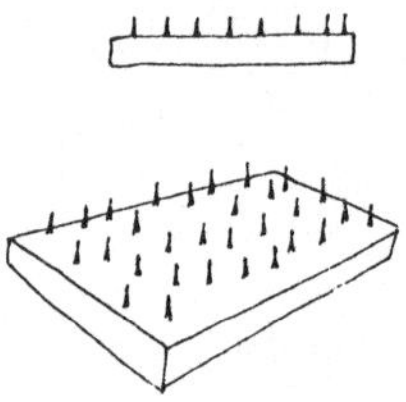

Figure 1456

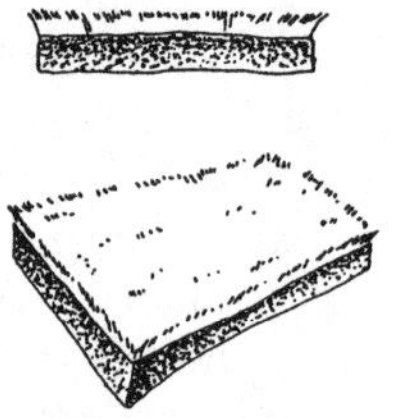

Figure 1457

Holosericeous. Covered with fine, silky hairs.

Incanous. With a whitish pubescence.

Inermous. See **unarmed**.

Innocuous. Harmless; lacking thorns or spines.

Laevigate. Lustrous; shining.

Lanate. Woolly; densely covered with long tangled hairs. Figure 1458.

Lanuginose. See **lanuginous**.

Lanuginous. Downy or woolly; with soft downy hairs. Figure 1459.

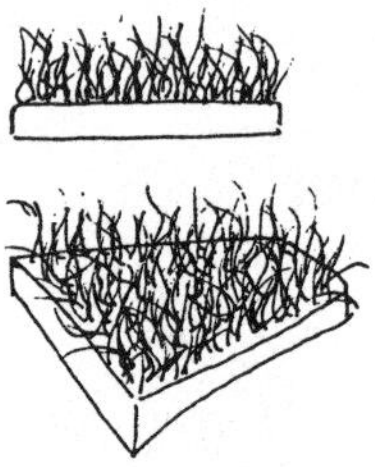

Figure 1458

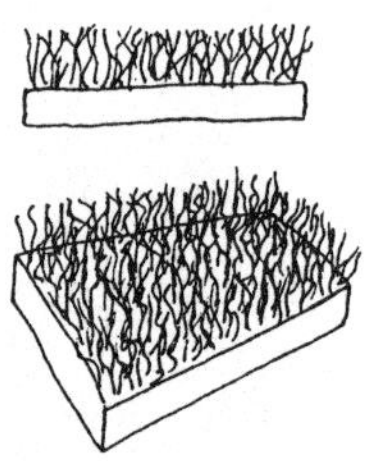

Figure 1459

Lanulose. Diminutive of lanate; minutely woolly. Figure 1460.

Lepidote. Covered with small, scurfy scales. Figure 1461.

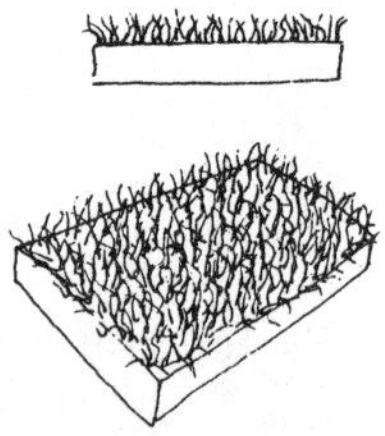

Figure 1460

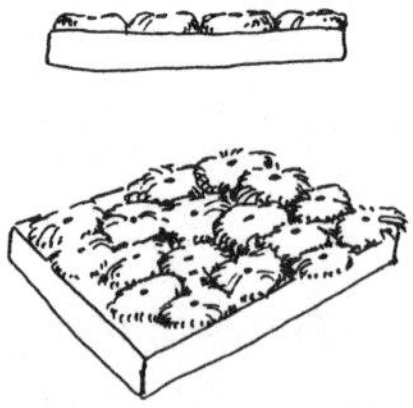

Figure 1461

Lucid. Luminous; shining.

Lustrous. Shiny or glossy.

Mammillate. With nipple-like protuberances. Figure 1462.

Manicate. With a thick, interwoven pubescence. Figure 1463.

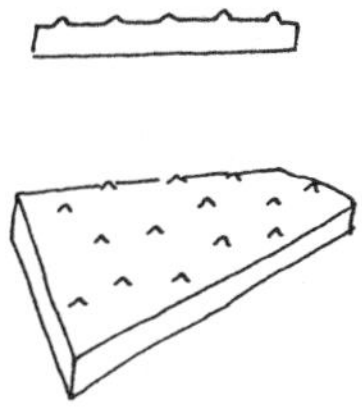

Figure 1462

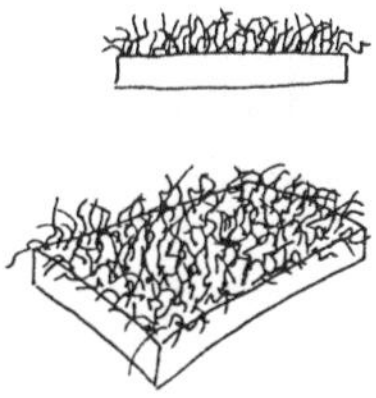

Figure 1463

Mealy. With the consistency of meal; powdery, dry, and crumbly. Figure 1464.

Muricate. Rough with small, sharp projections or points. Figure 1465.

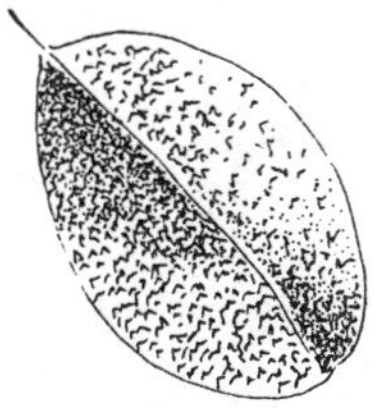

Figure 1464

Figure 1465

Muriculate. Very finely muricate. Figure 1466.

Nacreous. With a pearly luster; pearlescent.

Nitid. Lustrous; shining.

Notate. Marked with lines or spots. Figure 1467.

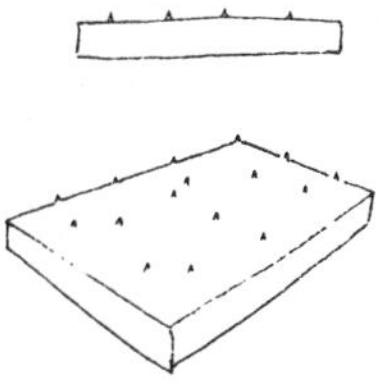

Figure 1466

Figure 1467

Paleaceous. Chaffy; with chaffy scales. Figure 1468.

Pannose. Covered with a short, dense, felt-like tomentum. Figure 1469.

Papillate. Having papillae. Figure 1470.

Papillose. Having minute papillae. Figure 1471.

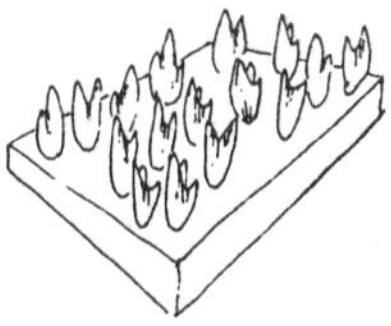

Figure 1468

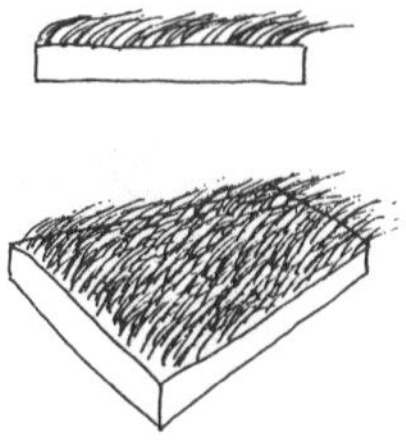

Figure 1469

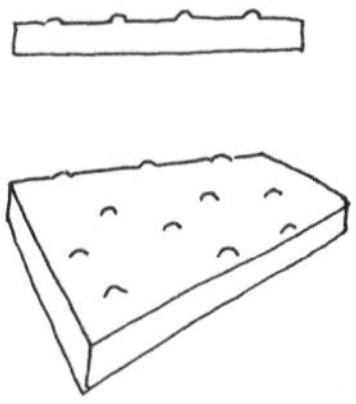

Figure 1470

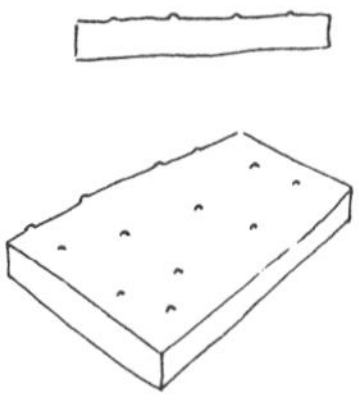

Figure 1471

Papillose-hispid. With stiff hairs borne on swollen, nipplelike bases. Figure 1472.

Papyraceous. Papery in texture and usually color.

Pellucid. Transparent or translucent.

Perforate. With holes or perforations. Figure 1473.

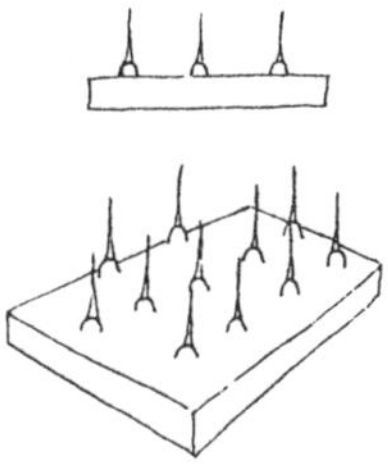

Figure 1472

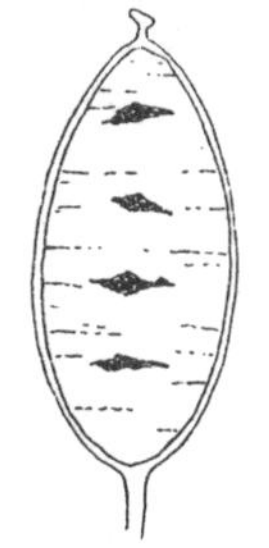

Figure 1473

Pilose. Bearing long, soft, straight hairs. Figure 1474.

Pilosulose. Bearing minute, long, soft, straight hairs. Figure 1475.

Pilosulous. See **pilosulose**.

Pitted. With small pits

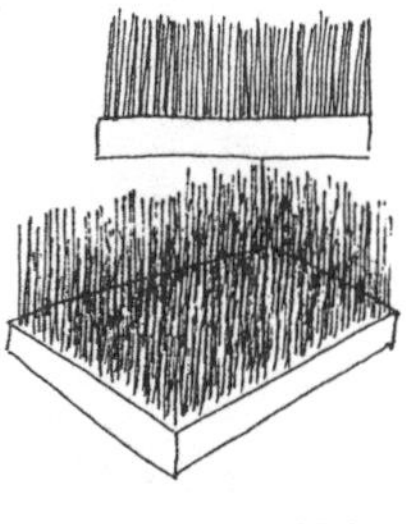

Figure 1474

or depressions. Figure 1476.

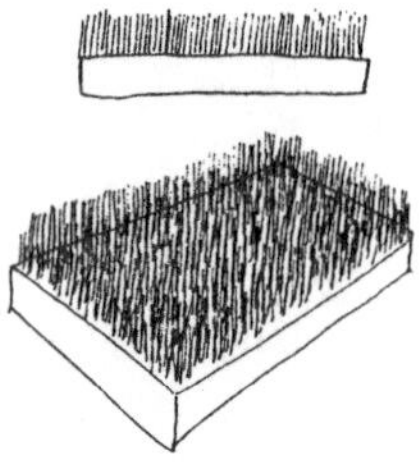

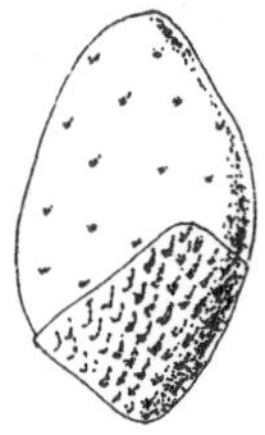

Figure 1475 **Figure 1476**

Plumbeous. Lead-colored.

Pruinate. See **pruinose**.

Pruinose. With a waxy, powdery, usually whitish coating (bloom) on the surface; conspicuously glaucous, like a prune.

Puberulent. Minutely pubescent; with fine, short hairs. Figure 1477.

Puberulous. See **puberulent**.

Pubescent. Covered with short, soft hairs; bearing any kind of hairs. Figure 1478.

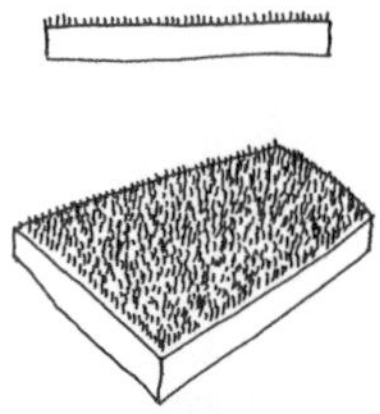

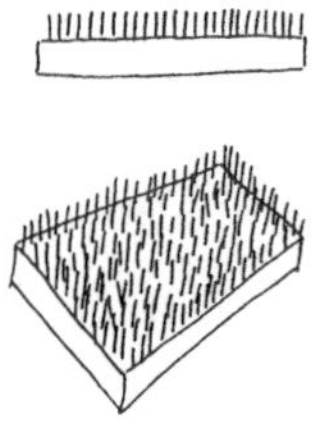

Figure 1477 **Figure 1478**

Pulverulent. Appearing dusty or powdery.

Punctate. Dotted with pits or with translucent, sunken glands or with colored dots. Figure 1479.

Puncticulate. Minutely punctate. Figure 1480.

Figure 1479 **Figure 1480**

Pustular. See **pustulose**.

Pustulate. See **pustulose**.

Pustuliferous. See **pustulose**.

Pustulose. With small blisters or pustules, often at the base of a hair. Figure 1481.

Ramentaceous. With flattened, scaly outgrowths, as on the epidermis of the stem and leaves of some ferns. Figure 1482.

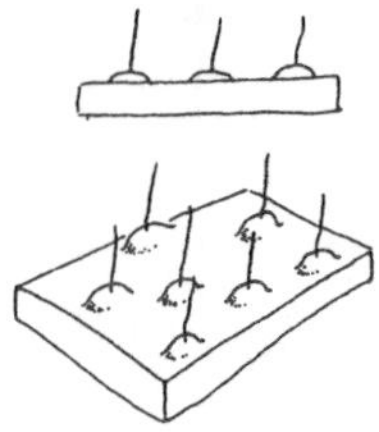

Figure 1481 **Figure 1482**

Roridulate. With a covering of waxy platelets, appearing moist.

Rugate. See **rugose**.

Rugose. Wrinkled. Figure 1483.

Rugulose. Slightly wrinkled. Figure 1484.

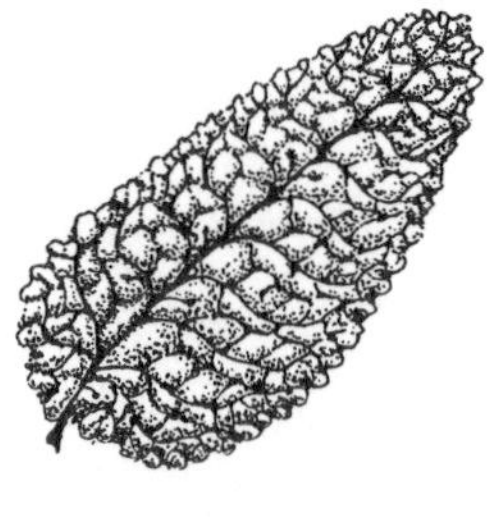

Figure 1483

Ruminate. Roughly wrinkled, as if chewed. Figure 1485.

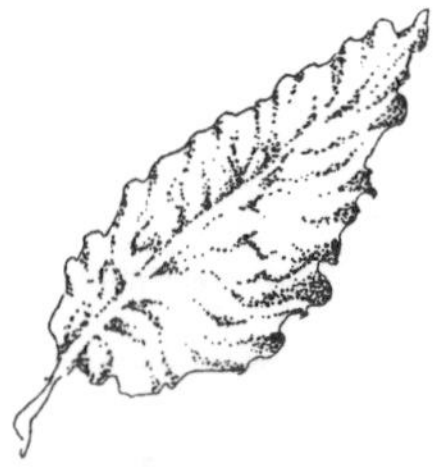

Figure 1484 **Figure 1485**

Scaberulent. See **scaberulose**.

Scaberulose. Slightly rough to the touch, due to the structure of the epidermal cells, or to the presence of short stiff hairs. Figure 1486.

Scaberulous. See **scaberulose**.

Scabrellate. Same as **scaberulose**.

Scabrid. Roughened.

Scabridulous. Minutely roughened.

Scabrous. Rough to the touch, due to the structure of the epidermal cells, or to the presence of short stiff hairs. Figure 1487.

Figure 1486

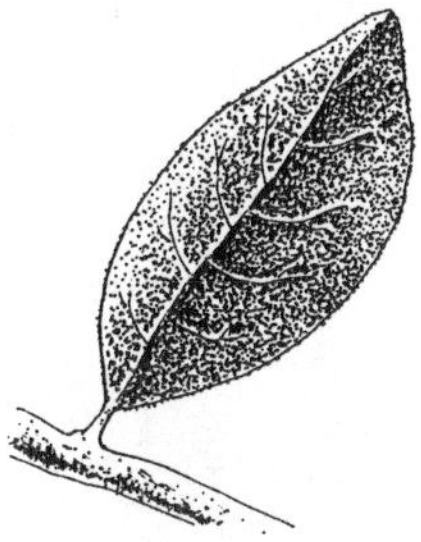
Figure 1487

Scurfy. Covered with small, branlike scales. Figure 1488.

Sericeous. Silky, with long, soft, slender, somewhat appressed hairs. Figure 1489.

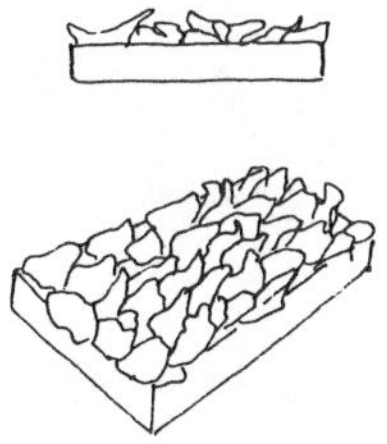
Figure 1488

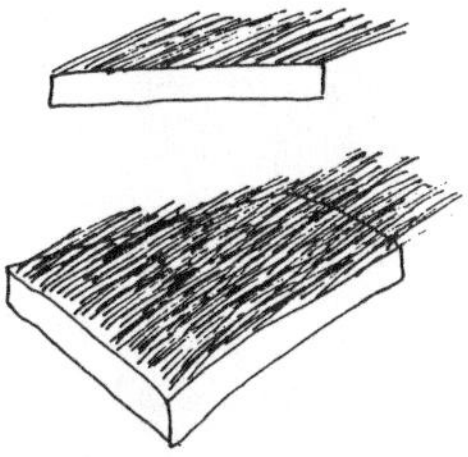
Figure 1489

Setose. Covered with bristles. Figure 1490.

Setulose. Covered with minute bristles. Figure 1491.

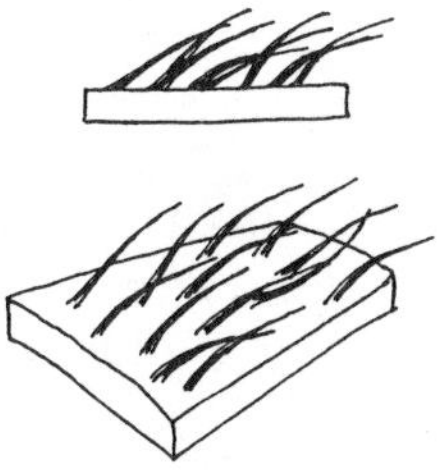
Figure 1490

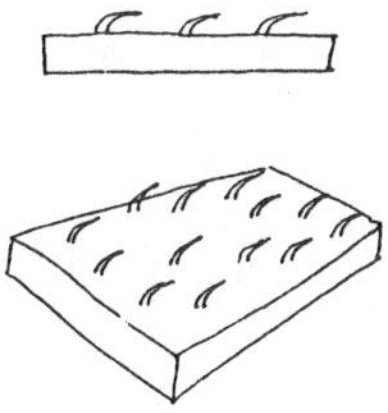
Figure 1491

Silky. Silk-like in appearance or texture; sericeous.

Smooth. With an even surface; not rough to the touch.

Spiniferous. See **spinose**.

Spinose. Bearing spines.

Spinous. See **spinose**.

Spinulose. Bearing spinules.

Spiny. With spines.

Squamate. Covered with scales (squamae). See illustration for **scurfy**.

Squamulose. With minute squamellae.

Stellate. Star-shaped, as in hairs with several to many branches radiating from the base. Figure 1492.

Striate. Marked with fine, usually parallel lines or grooves. Figure 1493.

Figure 1492

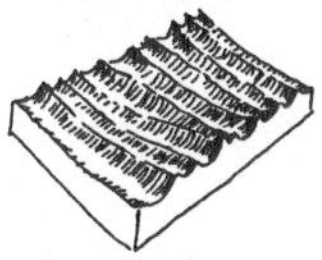
Figure 1493

Strigillose. Minutely strigose. Figure 1494.

Strigose. Bearing straight, stiff, sharp, appressed hairs. Figure 1495.

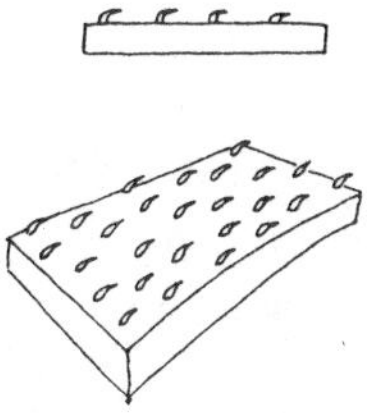
Figure 1494

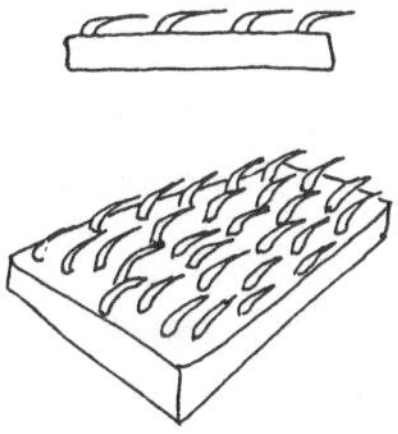
Figure 1495

Strigulose. See **strigillose**.

Strumose. With a covering of cushion-like swellings; bullate. Figure 1496.

Sulcate. With longitudinal grooves or furrows. Figure 1497.

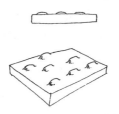

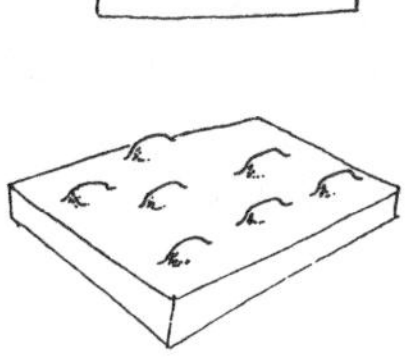
Figure 1496

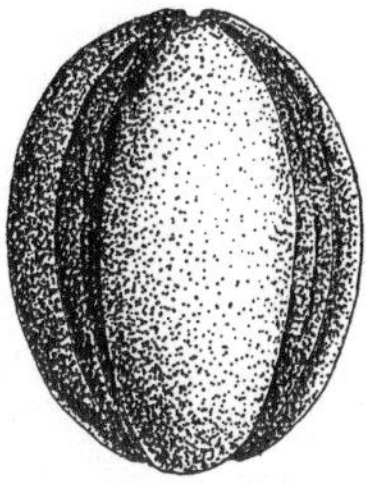
Figure 1497

Tesselate. With a checkered pattern. Figure 1498.
Tomentellous. See **tomentulose**.
Tomentose. With a covering of short, matted or tangled, soft, wooly hairs; with tomentum. Figure 1499.

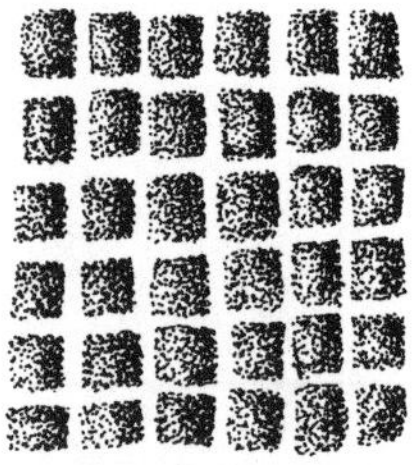

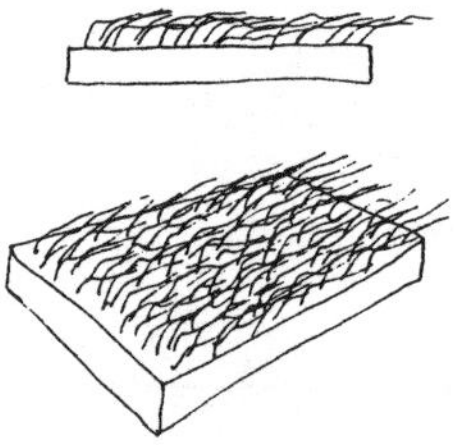

Figure 1498 **Figure 1499**

Tomentulose. Slightly tomentose. Figure 1500.
Translucent. Almost transparent. Figure 1501.

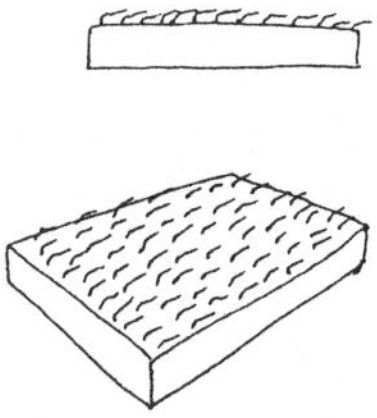

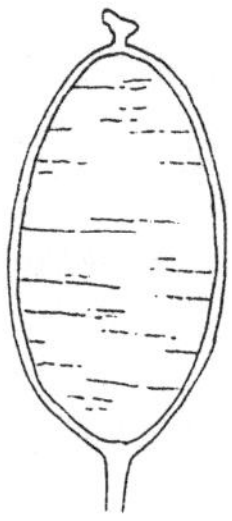

Figure 1500 **Figure 1501**

Unarmed. Lacking spines, prickles, or thorns.
Uncinate. Hooked at the tip. Figure 1502.
Urent. Stinging. Figure 1503.

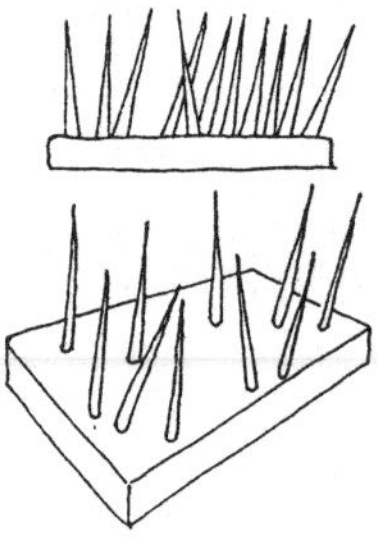

Figure 1502 **Figure 1503**

Velutinous. Velvety; covered with short, soft, spreading hairs. Figure 1504.
Verrucose. Warty; covered with wartlike elevations. Figure 1505.
Villose. Same as **villous**.
Villosulous. Diminutive if **villous**.
Villous. Bearing long, soft, shaggy, but unmatted, hairs. Figure 1506.
Viscid. Sticky or gummy.
Viscidulous. Slightly sticky.
Woolly. With long, soft, entangled hairs; lanate. Figure 1507.

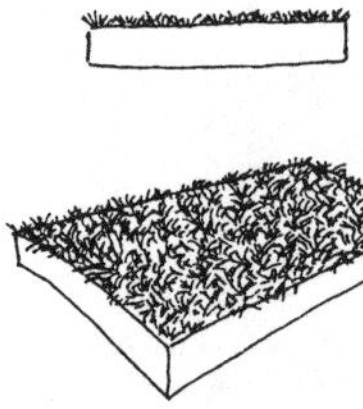

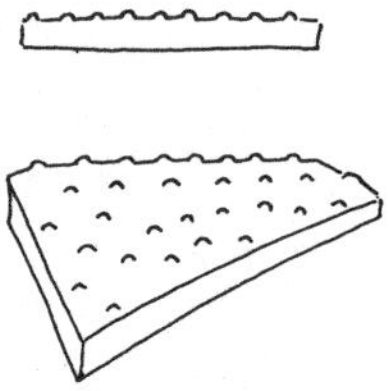

Figure 1504 **Figure 1505**

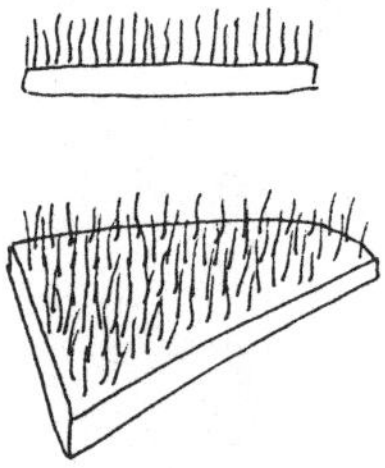

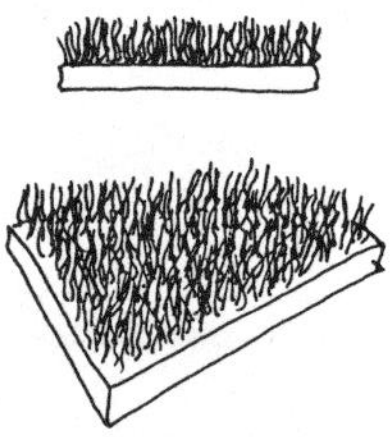

Figure 1506 **Figure 1507**

INFLORESCENCES

The flowering part of a plant; a flower cluster; the arrangement of the flowers on the flowering axis.

INFLORESCENCE PARTS

Bract. A reduced leaf or leaflike structure at the base of a flower or inflorescence. Figure 1508.
Bracteole. A small bract borne on a peduncle. Figure 1509.

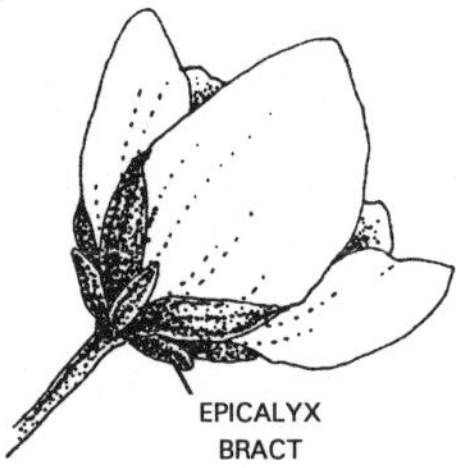

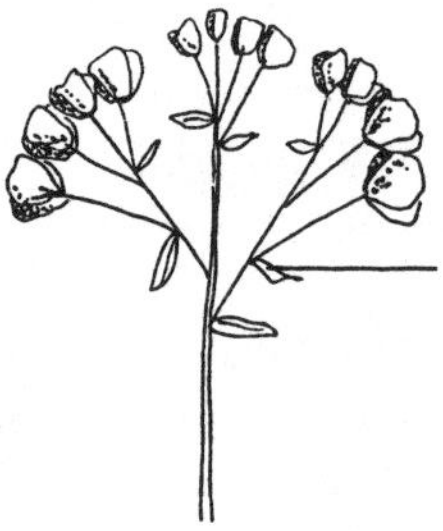

Figure 1508 **Figure 1509**

Bractlet. See **bracteole.**

Cupule. A cup-shaped involucre, as in an acorn. Figure 1510.

Disk. In the Compositae (Asteraceae), the central portion of the involucrate head bearing tubular or disk flowers. Figure 1511.

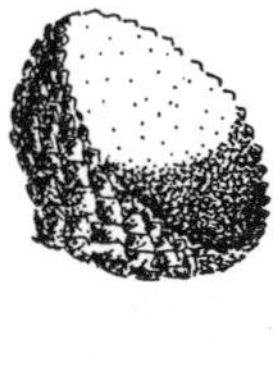

Figure 1510

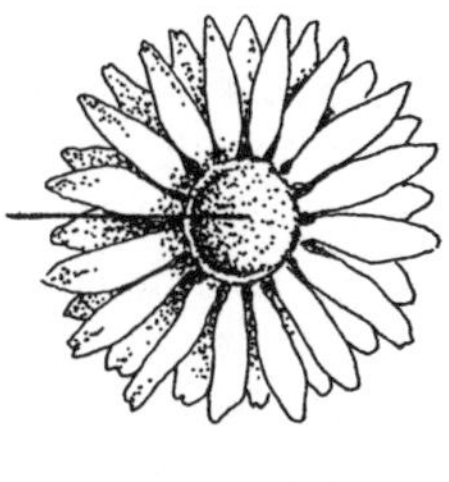

Figure 1511

Epicalyx. An involucre which resembles an outer calyx, as in *Malva*. Figure 1508.

Floret. A small flower; an individual flower within a dense cluster, as a grass flower in a spikelet, or a flower of the Compositae (Asteraceae) in an involucrate head. Figures 1512 and 1513.

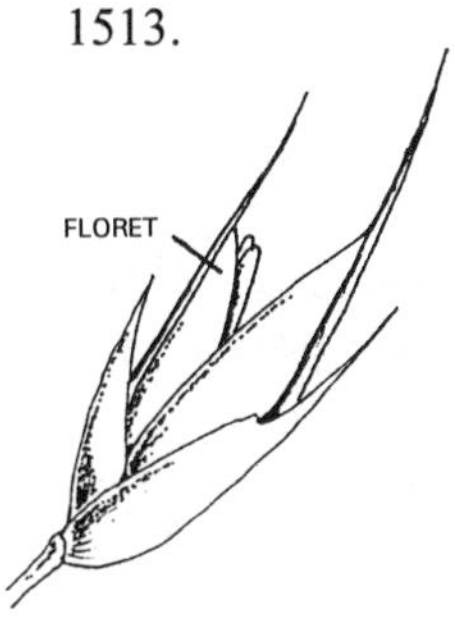

Figure 1512

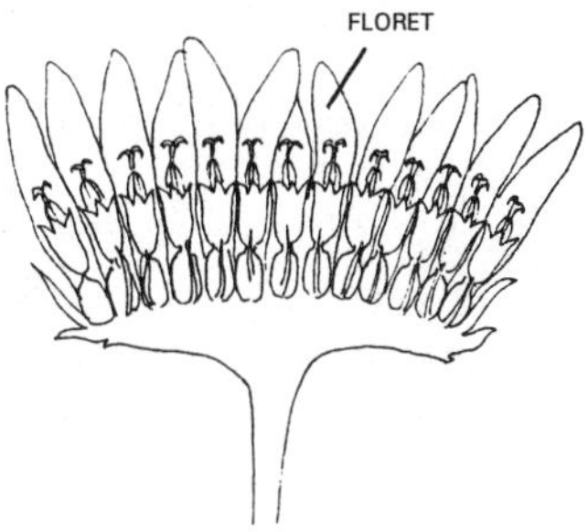

Figure 1513

Flower. The reproductive portion of the plant, consisting of stamens, pistils, or both, and usually including a perianth of sepals or both sepals and petals. Figure 1514.

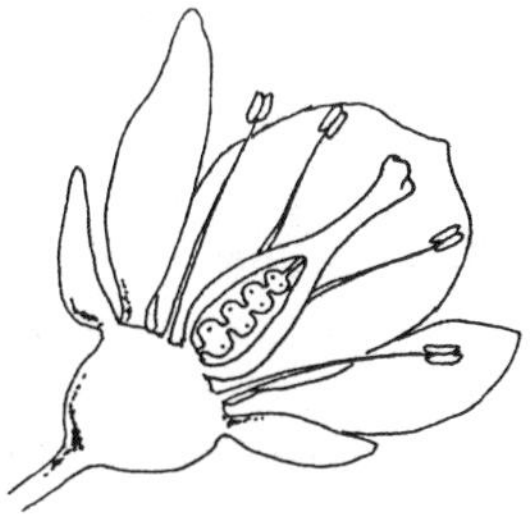

Figure 1514

Involucel. A small involucre; a secondary involucre, as in the bracts of the secondary umbels in the Umbelliferae (Apiaceae). Figure 1515.

Involucre. A whorl of bracts subtending a flower or flower cluster. Figure 1516.

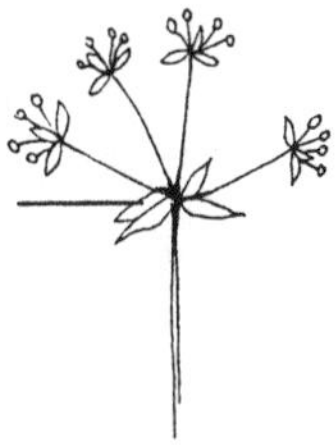

Figure 1515

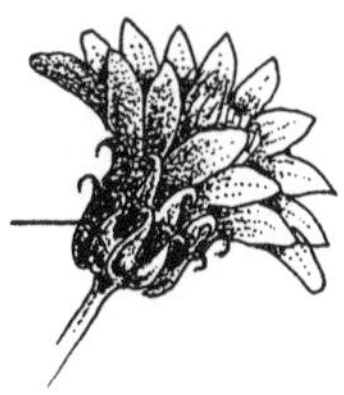

Figure 1516

Involucrum (pl. **involucra**). See **involucre.**

Ocreola (pl. **ocreolae**). A minute stipular sheath around the secondary divisions of the inflorescence in some members of the Polygonaceae.

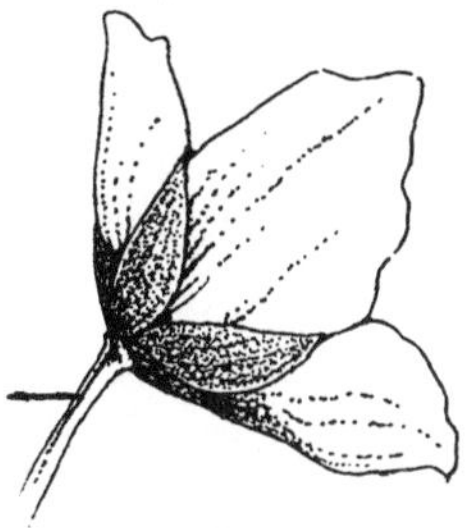

Figure 1517

Pedicel. The stalk of a single flower in an inflorescence, or of a grass spikelet. Figure 1517.

Peduncle. The stalk of a solitary flower or of an inflorescence. Figures 1518 and 1519.

Figure 1518

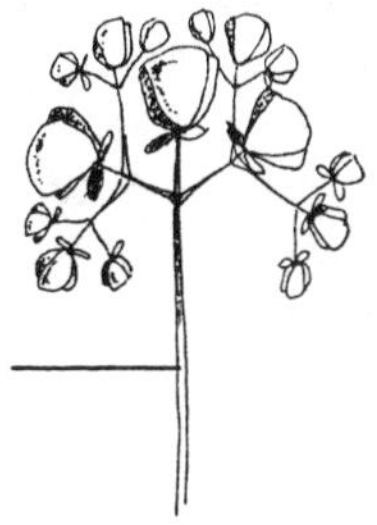

Figure 1519

Perigynium (pl. **perigynia**). A scalelike bract enclosing the pistil in *Carex*. Figure 1520.

Phyllary. An involucral bract of the Compositae (Asteraceae). Figure 1521.

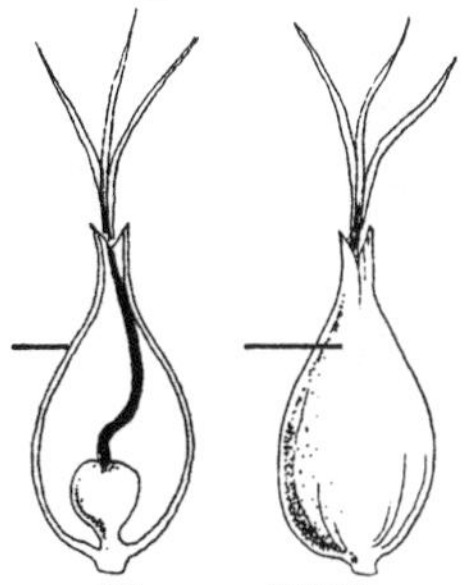

Figure 1520

Rachilla. The axis of a grass or sedge spikelet; a small rachis. Figure 1522.

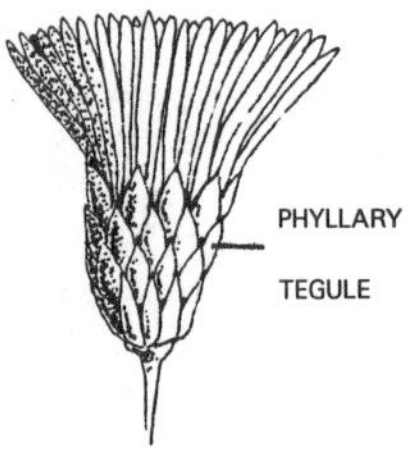

Figure 1521

Figure 1522

Rachis. The main axis of an inflorescence. Figure 1523.

Ray. An inflorescence branch in an umbel. Figure 1524.

Figure 1523

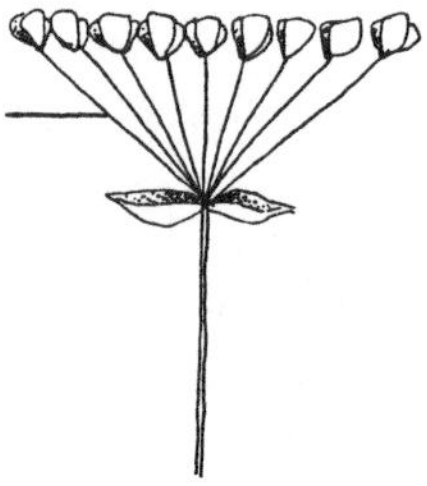

Figure 1524

Scape. A leafless peduncle arising from ground level (usually from a basal rosette) in acaulescent plants. Figure 1525.

Secondary peduncle. An inflorescence branch. Figure 1526.

Figure 1525

Spathe. A large bract or pair of bracts subtending and often enclosing an inflorescence. Figure 1527.

Tegule. One of the bracts of the involucre in the Compositae (Asteraceae). Figure 1521.

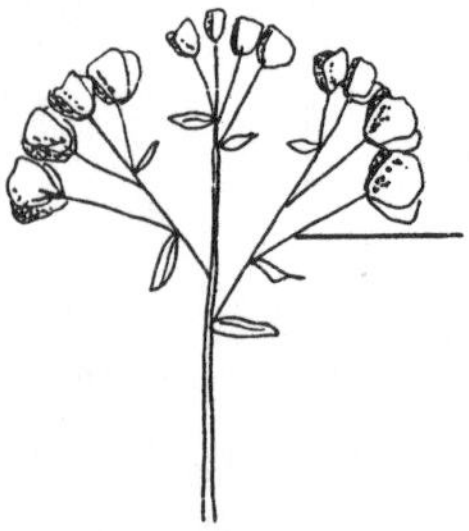

Figure 1526

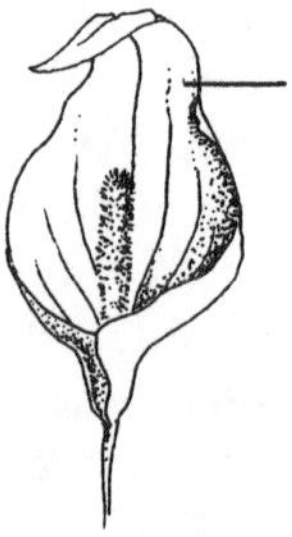

Figure 1527

INFLORESCENCE TYPES (Figure 1528.)

Ament. See **catkin**.

Capitulum. A small flower head.

Catkin. An inflorescence consisting of a dense spike or raceme of apetalous, unisexual flowers as in Salicaceae and Betulaceae; an ament.

Cincinnus. A dense helicoid cyme with the pedicels short on the developed side.

Corymb. A flat-topped or round-topped inflorescence, racemose, but with the lower pedicels longer than the upper.

Cyathium (pl. **cyathia**). The inflorescence in the genus *Euphorbia*, consisting of a cup-like involucre containing a single pistil and male flowers with a single stamen.

Cyme. A flat-topped or round-topped determinate inflorescence, paniculate, in which the terminal flower blooms first.

Cymule. A small cyme or a small section of a compound cyme.

Dichasium. A cymose inflorescence in which each axis produces two opposite or subopposite lateral axes.

Glomerule. A dense cluster; a dense, headlike cyme.

Head. A dense cluster of sessile or subsessile flowers; the involucrate inflorescence of the Compositae (Asteraceae).

Helicoid cyme. A one-sided cymose inflorescence coiled like a spiral or helix.

Hypanthodium. An inflorescence with flowers borne on the walls of a capitulum, as in *Ficus*.

Intercalary inflorescence. An inflorescence type in which the main vegetative axis of the plant continues to elongate after the flowers are produced.

Mixed inflorescence. An inflorescence with both racemose and cymose portions.

Monochasium. A type of cymose inflorescence with only a single main axis.

Panicle. A branched, racemose inflorescence with flowers maturing from the bottom upwards.

Pleiochasium. A cymose inflorescence with more than two branches from the main axis.

Polychasium. A cymose inflorescence in which each axis produces more than two lateral axes.

Key to Common Inflorescence Types

1 Flowers sessile.
  2 Inflorescence elongate.
    3 Flowers very small and densely clustered, obscuring the inflorescence axis.
      4 Perianth reduced to paired bracts. . . . . **Spikelet**
      4 Perianth not reduced to paired bracts.
        5 Inflorescence usually erect, bisexual, with a thickened axis. **Spadix**
        5 Inflorescence usually pendulous, unisexual, lacking a thickened axis. . . . . **Catkin, Ament**
    3 Flowers not very small and densely clustered, not obscuring the inflorescence axis.
      6 Perianth reduced to paired bracts. . . . . **Spikelet**
      6 Perianth not reduced to paired bracts, often showy. . . . . **Spike**
  2 Inflorescence not elongate.
    7 Flowers enclosed within the walls of a concave capitulum. . . . . **Hypanthodium**
    7 Flowers borne on a flat or convex receptacle. . . . . **Head, Capitulum**
1 Flowers pedicellate.
  8 Inflorescence unbranched.
    9 Pedicels arising from a common point, like the struts of an umbrella. . . . . **Umbel**
    9 Pedicels not arising from a common point.
      10 Inflorescence determinate, the central or terminal flower developing first. . . . . **Cyme**
      10 Inflorescence indeterminate, the lateral or basal flowers developing first.
        11 Inflorescence flat-topped or rounded. . . . . **Corymb**
        11 Inflorescence elongate. . . . . **Raceme**
  8 Inflorescence branched.
    12 Inflorescence flat-topped or rounded.
      13 Inflorescence branches arising from a common point, like the struts of an umbrella. . . . .
      . . . . **Compound umbel**
      13 Inflorescence branches not arising from a common point.
        14 Inflorescence determinate, the central or terminal flower developing first. . . **Compound cyme**
        14 Inflorescence indeterminate, the lateral or basal flowers developing first. . **Compound corymb**
    12 Inflorescence elongate.
      15 Flowers densely clustered in a compact, cylindrical or ovate inflorescence. . . . . **Thyrse**
      15 Flowers less densely clustered in a more open inflorescence. . . . . **Panicle**

Pseudanthium. A compact inflorescence of many small flowers which simulates a single flower.

Raceme. An unbranched, elongated inflorescence with pedicellate flowers maturing from the bottom upwards.

Scorpioid cyme. A determinate cymose inflorescence with a zigzag rachis.

Solitary. Flowers occurring singly and not borne in a cluster or group.

Spadix. A spike with small flowers crowded on a thickened axis.

Spike. An unbranched, elongated inflorescence with sessile or subsessile flowers or spikelets maturing from the bottom upwards.

Spikelet. A small spike or secondary spike; the ultimate flower cluster of grasses and sedges, consisting of one to many flowers subtended by two bracts (glumes).

Strobile. A cone or an inflorescence resembling a cone.

Thyrse. A compact, cylindrical, or ovate panicle with an indeterminate main axis and cymose sub-axes.

Thyrsus. See **thyrse**.

Umbel. A flat-topped or convex inflorescence with the pedicels arising more or less from a common point, like the struts of an umbrella; a highly condensed raceme.

Umbellet. An ultimate umbellate cluster of a compound umbel.

Verticillaster. A pair of axillary cymes arising from opposite leaves or bracts and forming a false whorl.

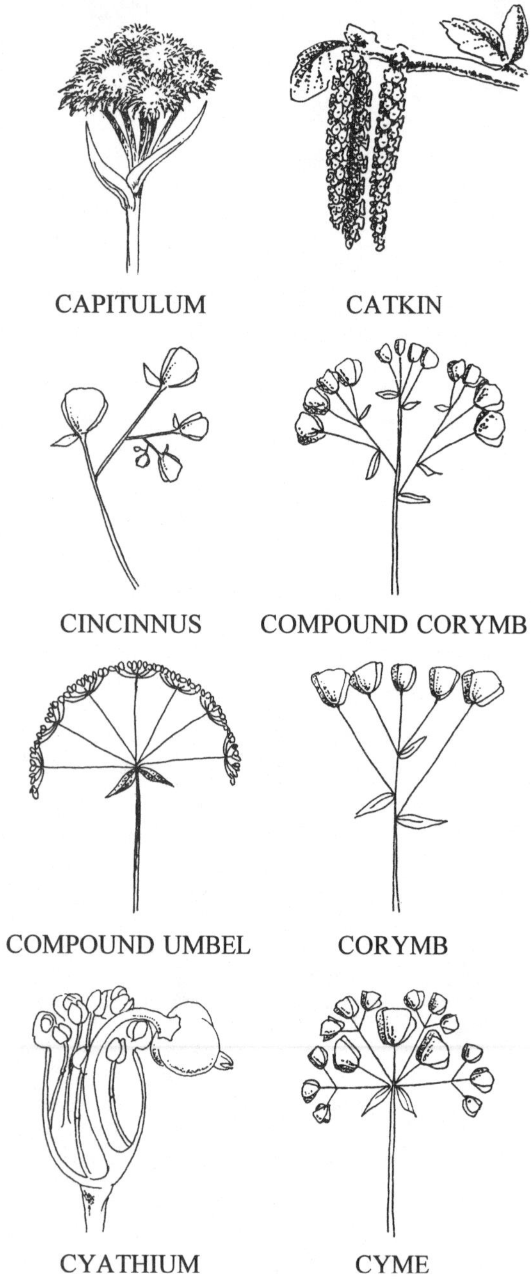

Figure 1528a.

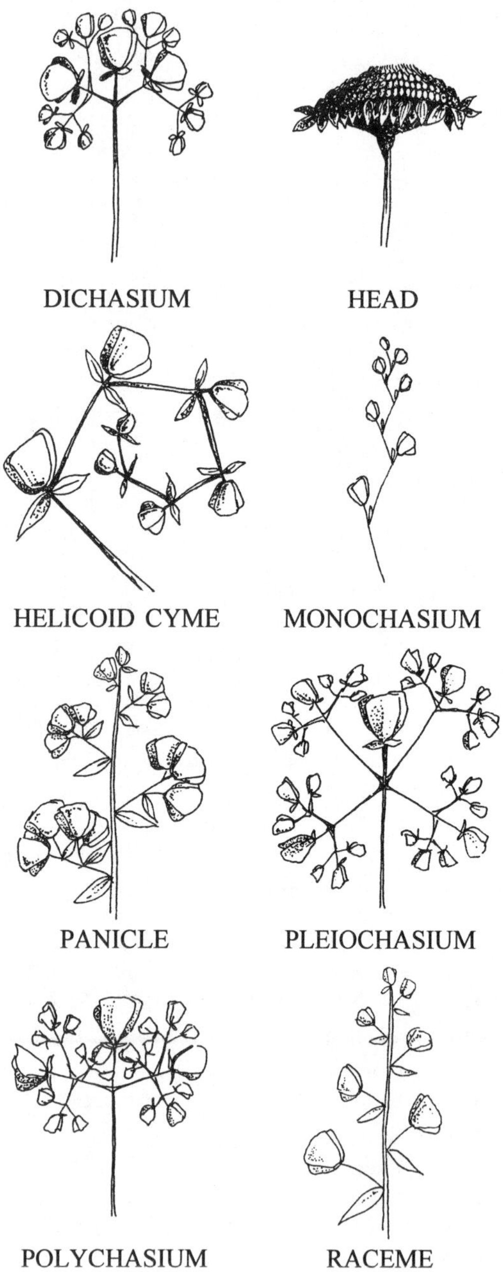

Figure 1528b.

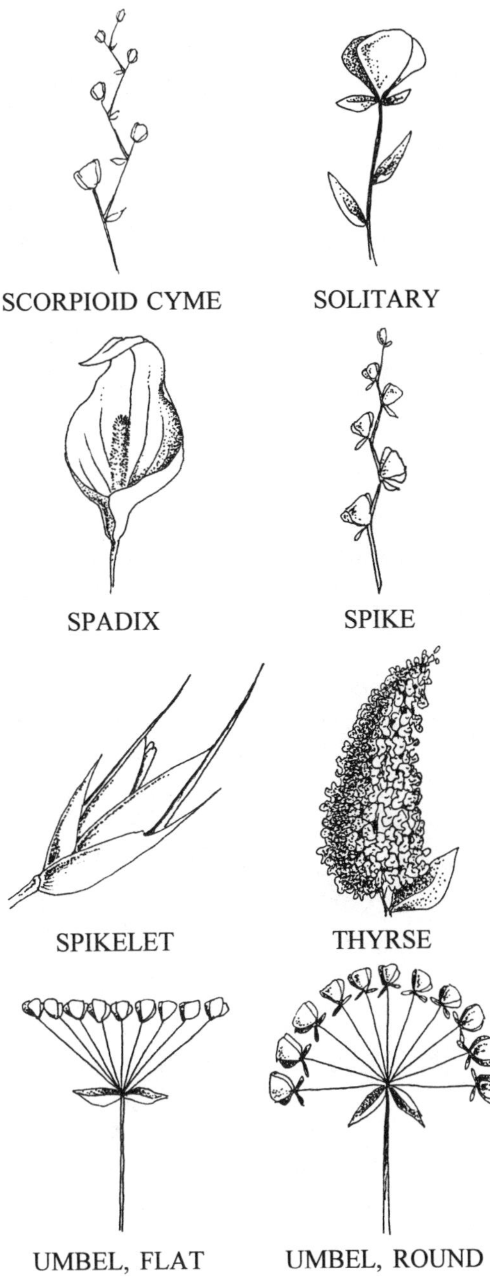

Figure 1528c.

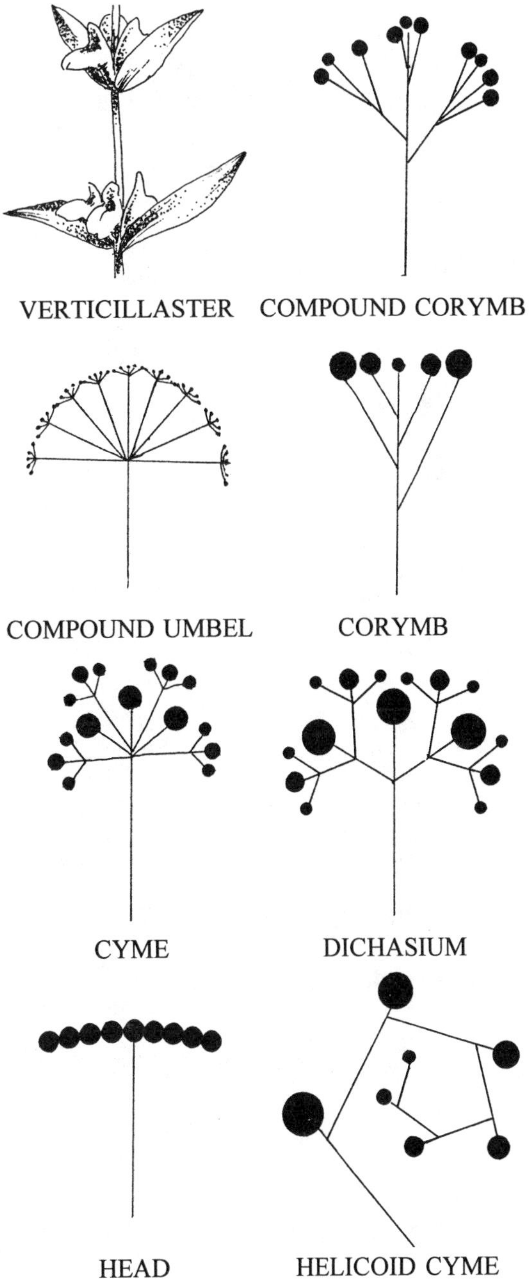

Figure 1528d.

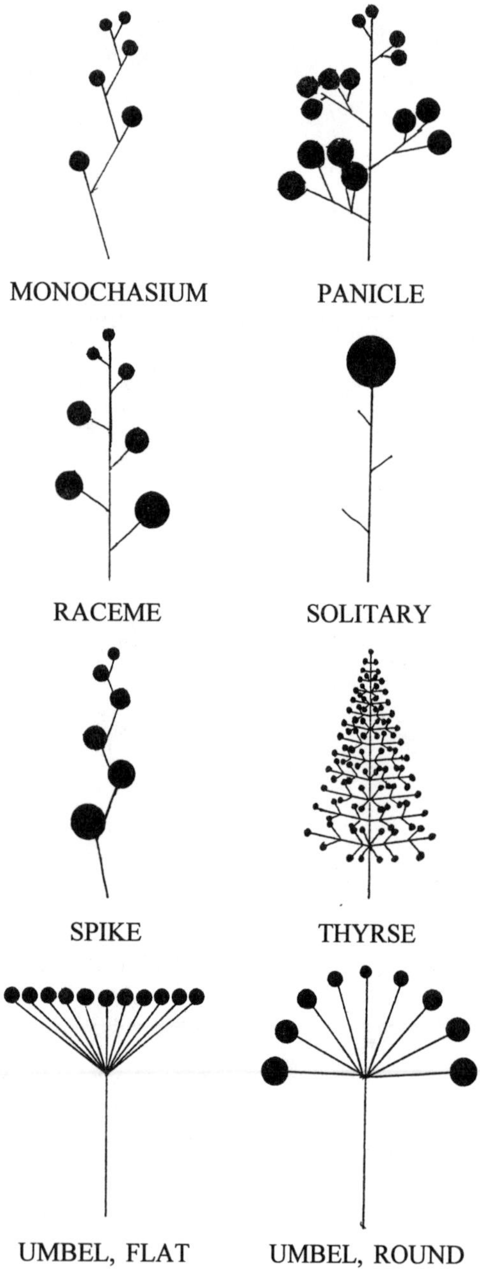

Figure 1528e.

INFLORESCENCE FORMS

Androgynous. An inflorescence with both staminate and pistillate flowers, the staminate flowers borne above the pistillate, as in some *Carex* species. (compare **gynaecandrous**)

Axillary. Positioned in or arising in an axil. Figure 1529.

Capitate. Head-like, or in a head-shaped cluster, as the flowers in the Compositae (Asteraceae). Figure 1530.

Figure 1529

Figure 1530

Capitellate. With small head-like structures, or with parts in very small head-shaped clusters. Figure 1531.

Centrifugal inflorescence. A flower cluster developing from the center outward, as in a cyme. Figure 1532.

Figure 1531

Figure 1532

Centripetal inflorescence. A flower cluster developing from the edge toward the center, as in a corymb. Figure 1533.

Corymbiform. An inflorescence with the general appearance, but not necessarily the

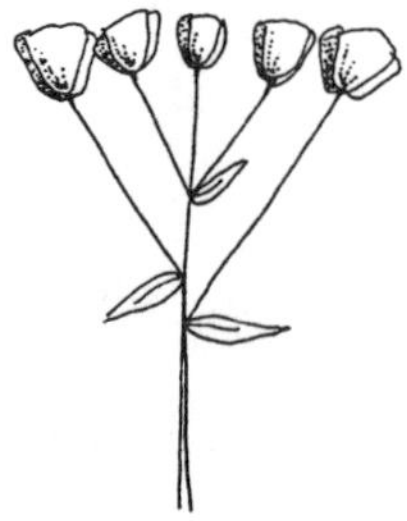

Figure 1533

structure, of a true corymb.

Corymbose. Having flowers in corymbs. The term is sometimes used in the same sense as **corymbiform**. Figure 1533.

Cyathiform. With the form of a cyathium; cup-shaped.

Cymose. With flowers in a cyme. Figure 1532.

Determinate. Describes an inflorescence in which the terminal flower blooms first, halting further elongation of the main axis. Figure 1532.

Glomerate. Densely clustered. Figure 1534.

Glomerulate. Arranged in very small, dense clusters. Figure 1535.

Figure 1534

Figure 1535

Gynaecandrous. An inflorescence with the pistillate flowers borne above the staminate, as in some *Carex* species. (compare **androgynous**)

Helicoid. Coiled like a spiral or helix, as in some one-sided cymose inflorescences in the Boraginaceae. Figure 1536.

Indeterminate. Describes an inflorescence in which the lower or outer flowers bloom first, allowing indefinite elongation of the main axis. Figure 1533.

Monochasial. With the form of a monochasium. Figure 1537.

Figure 1536

Figure 1537

Paniculate. Having flowers in panicles. Figure 1538.

Paniculiform. An inflorescence with the general appearance, but not necessarily the structure, of a true panicle.

Racemiform. An inflorescence with the general appearance, but not necessarily the structure, of a true raceme.

Racemose. Having flowers in racemes. The term is sometimes used in the same sense as **racemiform**. Figure 1539.

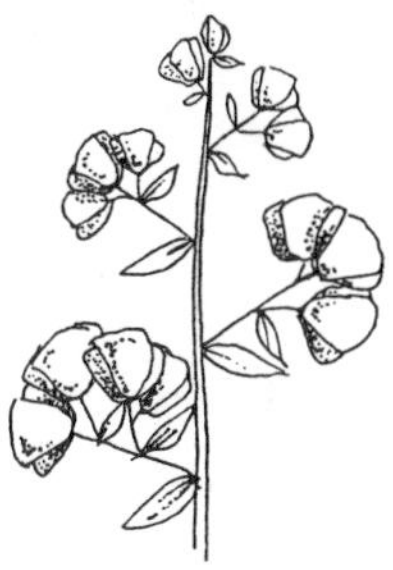

Figure 1538

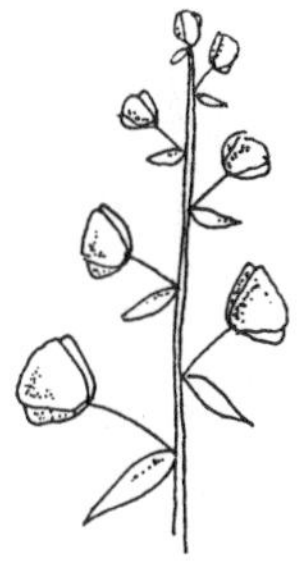

Figure 1539

Scorpioid. A determinate cymose inflorescence with a zigzag rachis. Figure 1540; often used in the same sense as **helicoid**.

Secund. Arranged on one side of the axis only. Figure 1541.

Figure 1540

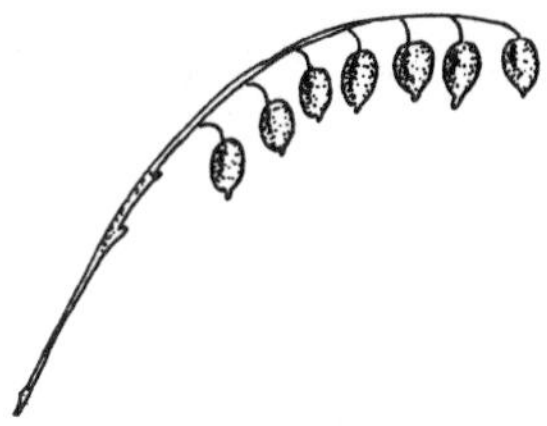

Figure 1541

Spathaceous. Spathe bearing; spathelike. Figure 1542.

Spicate. Arranged in a spike. Figure 1543.

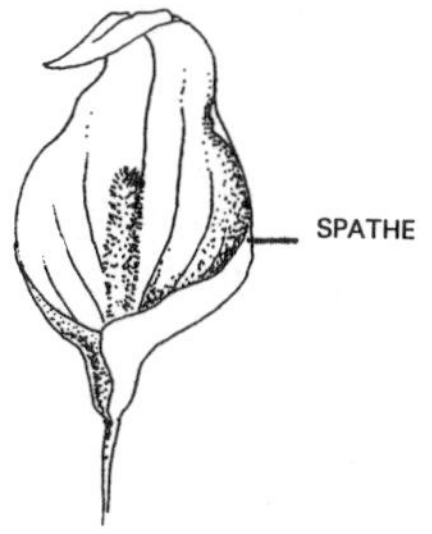

Figure 1542

Figure 1543

Spiciform. An inflorescence with the general appearance, but not necessarily the structure, of a true spike.

Terminal. At the tip or apex. Figure 1544.

Thyrsoid. Thyrse-like.

Umbellate. In umbels; umbel-like. Figure 1545.

Figure 1544

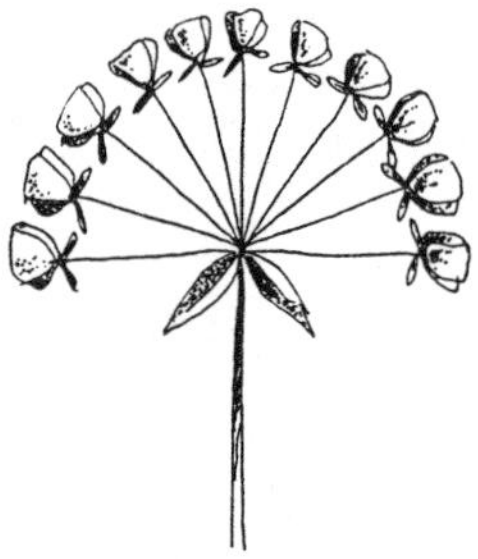

Figure 1545

Umbelliform. An inflorescence with the general appearance, but not necessarily the structure, of a true umbel. The term is often applied to inflorescences which are condensed cymes rather than condensed racemes.

FLOWERS

The reproductive portion of the plant, consisting of stamens, pistils, or both, and usually including a perianth of sepals or both sepals and petals.

FLOWER PARTS

Androecium. All of the stamens in a flower, collectively. Figure 1546.

Calyx (pl. **calyces**, **calyxes**). The outer perianth whorl; collective term for all of the sepals of a flower. Figure 1546.

Carpel. A simple pistil formed from one modified leaf, or that part of a compound pistil formed from one modified leaf; megasporophyll. Figure 1547.

Corolla. The collective name for all of the petals of a flower; the inner perianth whorl. Figure 1546.

Floral envelope. A collective term for the calyx and corolla. (same as **perianth**)

Gynecium. See **gynoecium**.

Gynoecium. All of the carpels or pistils of a flower, collectively. Figure 1546.

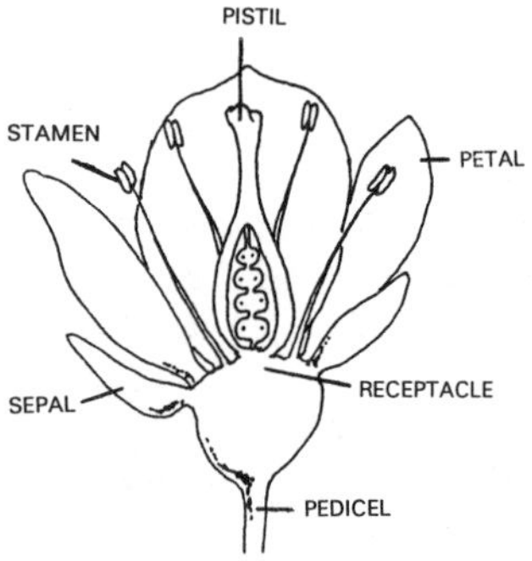

Figure 1546

Nectar gland. See **nectary**.

Nectary. A tissue or organ which produces nectar. Figure 1548.

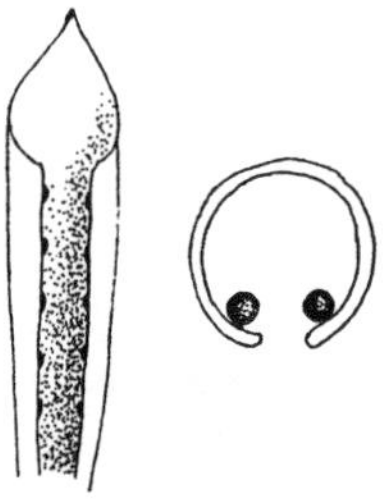

Figure 1547

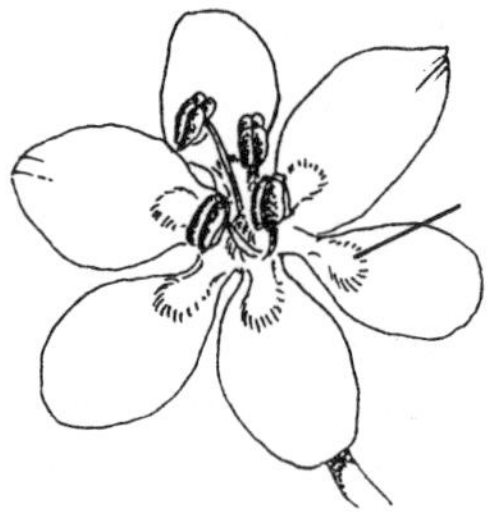

Figure 1548

Pedicel. The stalk of a single flower in an inflorescence. Figure 1546.

Peduncle. The stalk of a solitary flower or of an inflorescence. Figure 1549.

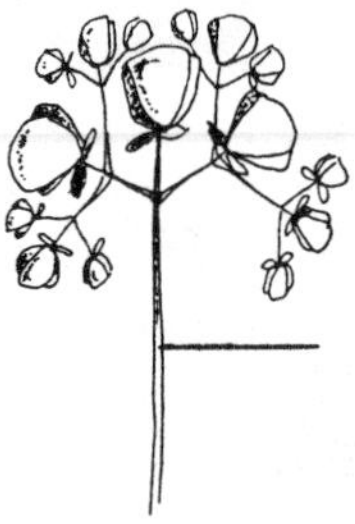

Figure 1549

Perianth. The calyx and corolla of a flower, collectively, especially when they are similar in appearance. Figure 1550.

Petal. An individual segment or member of the corolla, usually colored or white. Figure 1546.

Pistil. The female reproductive organ of a flower, typically consisting of a stigma, style, and ovary. Figure 1546. (compare **gynoecium**)

Receptacle. The portion of the pedicel upon which the flower parts are borne; in the Compositae (Asteraceae), the part of the peduncle where the flowers of the head are borne. Figures 1546 and 1551.

Sepal. A segment of the calyx. Figure 1546.

Stamen (pl. **stamens, stamina**). The male reproductive organ of a flower, consisting of an anther and filament; the angiosperm microsporophyll. Figure 1546.

Tepal. A segment of a perianth which is not differentiated into calyx and corolla; a sepal or petal. Figure 1550.

Torus (pl. **tori**). See **receptacle**.

Whorl. A ringlike arrangement of similar parts arising from a common point or node; a verticil. Figure 1552.

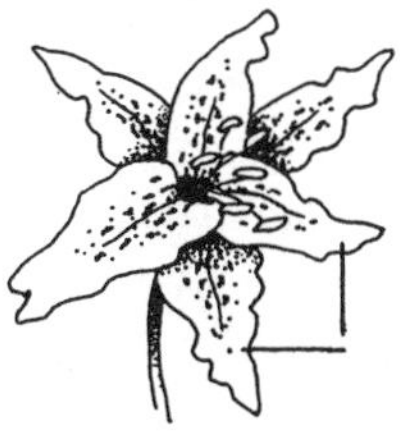

Figure 1550

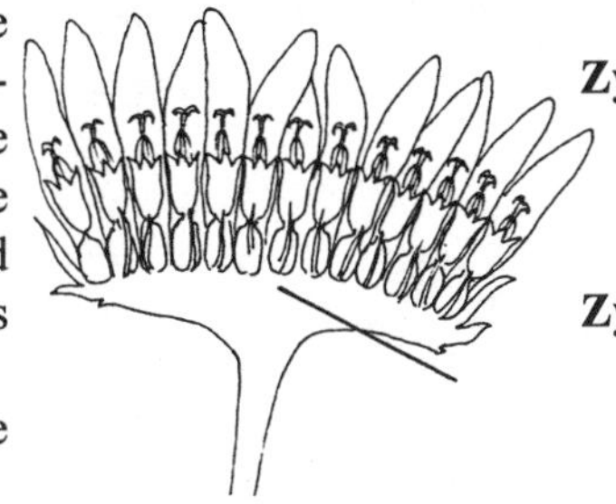

Figure 1551

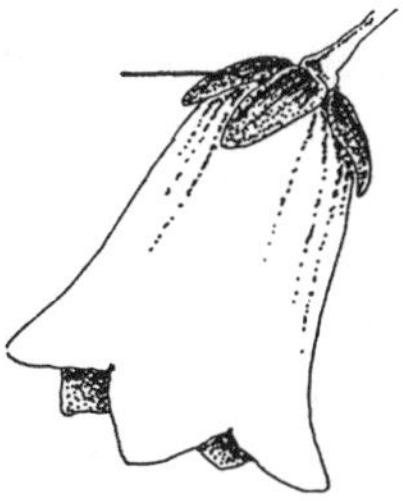

Figure 1552

FLOWER SYMMETRY

Actinomorphic. Radially symmetrical, so that a line drawn through the middle of the structure along any plane will produce a mirror image on either side. Figure 1553.

Actinomorphous. See **actinomorphic**.

Irregular. Bilaterally symmetrical; said of a flower in which all parts are not similar in size and arrangement on the receptacle. Figure 1554.

Monosymmetrical. Bilaterally symmetrical; zygomorphic. Figure 1554.

Peloria. Radial symmetry in flowers normally bilaterally symmetrical.

Regular. Radially symmetrical; said of a flower in which all parts are similar in size and arrangement on the receptacle. Figure 1553.

Stereomorphic. Radially symmetrical, so that a line drawn through the middle of the structure along any plane will produce a mirror image on either side; essentially the same as **actinomorphic**. Figure 1553.

Zygomorphic. Bilaterally symmetrical, so that a line drawn through the middle of the structure along only one plane will produce a mirror image on either side. Figure 1554.

Zygomorphous. See **zygomorphic**.

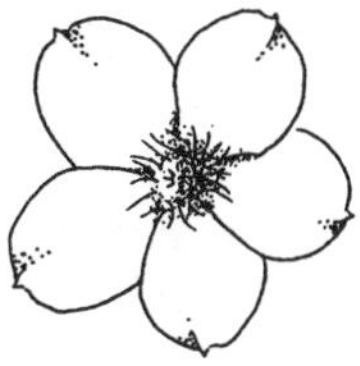

Figure 1553

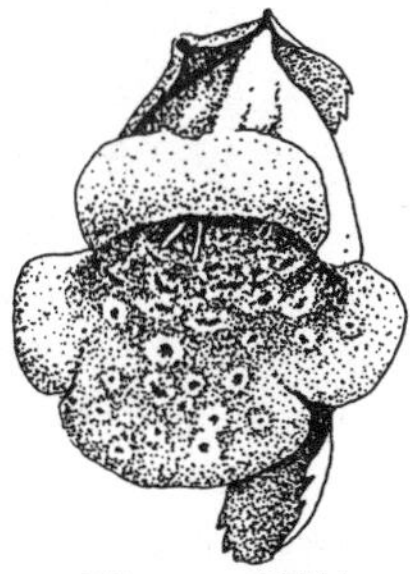

Figure 1554

INSERTION OF FLORAL STRUCTURES

Epigynous. With stamens, petals, and sepals attached to the top of the ovary, the ovary inferior to the other floral parts. Figure 1555.

Hypogynous. With stamens, petals, and sepals attached below the ovary, the ovary superior to the other floral parts. Figure 1556.

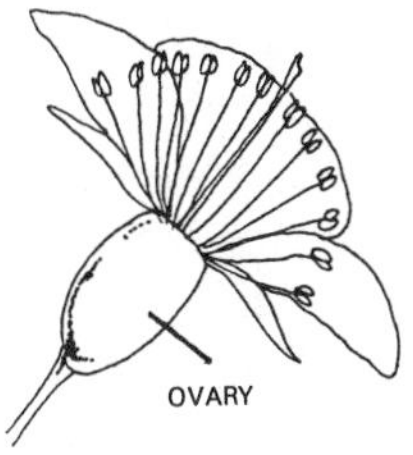

Figure 1555

Perigynous. With stamens, petals, and sepals borne on a calyx tube (hypanthium) surrounding, but not actually attached to, the superior ovary. Figure 1557.

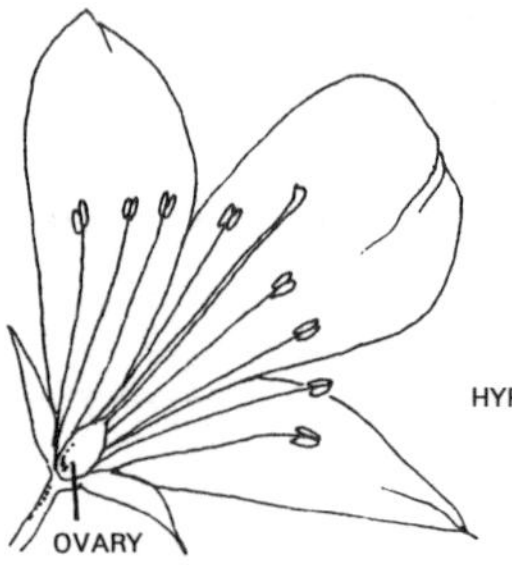

Figure 1556

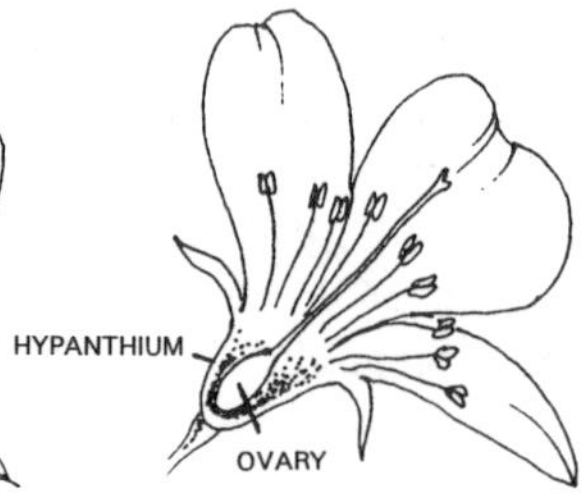

Figure 1557

POLLINATION SYSTEMS

Anemophilous. Wind pollinated; producing wind-borne pollen.

Chasmogamous. Applied to flowers which open before fertilization and are usually cross-pollinated. (compare **cleistogamous**)

Cleistogamous. Said of flowers which self-fertilize without opening. (compare **chasmogamous**)

Dichogamic. See **dichogamous**.

Dichogamous. With the pistils and stamens maturing at different times to prevent self-fertilization. (compare **homogamous**)

Entomophilous. Insect pollinated.

Geitonogamous. Pollinated by flowers of the same plant.

Homogamous. With the pistils and stamens maturing at the same time, allowing self-fertilization. (compare **dichogamous**)

Kleistogamous. See **cleistogamous**.

Metandrous. With the female flowers maturing before the male flowers; protogynous.

Ornithophilous. Pollinated by birds.

Outcrossing. Transferring pollen from the anthers of the flowers of one plant to the stigma of the flower of another plant.

Protandrous. With the anthers releasing pollen before the stigma is receptive.

Proterandrous. See **protandrous**.

Proterogynous. See **protogynous**.

Protogynous. With the stigma receptive before the anthers release pollen.

Self-pollinating. Transferring pollen from the anthers to the stigma of the same flower or to the stigma of another flower on the same plant.

FLOWER SEXUALITY

Andro-dioecious. With staminate and perfect flowers on separate plants.

Andro-monoecious. See **andro-polygamous**.

Andro-polygamous. With staminate and perfect flowers on the same plant.

Bisexual. With both male and female reproductive organs (stamens and pistils). (same as **perfect**)

Diclinous. With the stamens and pistils in separate flowers; imperfect.

Dioecious. With imperfect flowers, the staminate and pistillate flowers borne on different plants. (compare **monoecious**)

Dioicous. See **dioecious**.

Diecious. See **dioecious**.

Gyno-dioecious. With pistillate and perfect flowers on separate plants.

Hermaphroditic. With pistils and stamens in the same flower; bisexual; monoclinous; perfect.

Heterogamous. With flowers of differing sex.

Imperfect. With either stamens or pistils, but not both; unisexual.

Monecious. See **monoecious**.

Monoclinous. With pistils and stamens in the same flower; perfect.

Monoecious. Flowers imperfect, the staminate and pistillate flowers borne on the same plant. (compare **dioecious**)

Perfect. With both male and female reproductive organs (stamens and pistils); bisexual.

Pistillate. Bearing a pistil or pistils, but lacking stamens. (compare **staminate**)

Polygamo-dioecious. Mostly dioecious, but with some perfect flowers.

Polygamo-monoecious. Mostly monoecious, but with some perfect flowers.

Polygamous. With unisexual and bisexual flowers on the same plant.

Staminate. Bearing stamens but not pistils, as a male flower which does not produce fruit or seeds. (compare **pistillate**)

Triecious. See **trioecious**.

Trioecious. With male, female, and bisexual flowers on different plants.

Unisexual. With either male or female reproductive parts, but not both.

FLOWERING TIME

Diurnal. Occurring or opening in the daytime.
Equinoctial. With flowers that open regularly at a particular hour of the day.
Hibernal. Flowering in the winter.
Matutinal. Opening in the morning.
Nocturnal. Opening at night.
Nyctanthous. Night-flowering.
Nyctigamous. Opening at night.
Precocious. With the flowers developing before the leaves.
Proteranthous. With the flowers developing before the leaves.
Semperflorous. Flowering throughout the year.
Serotinous. Flowering late; with flowers developing after the leaves are fully developed.
Vernal. Flowering in the spring.
Vespertine. Opening in the evening.

NUMBERS OF FLORAL STRUCTURES

Anisomerous. With a different number of parts (usually less) than the other floral whorls, as in a flower with five sepals and petals, but only two stamens. Figure 1558.
Complete. With all of the parts typically belonging to it, as a flower with sepals, petals, stamens, and pistils. Figure 1559.

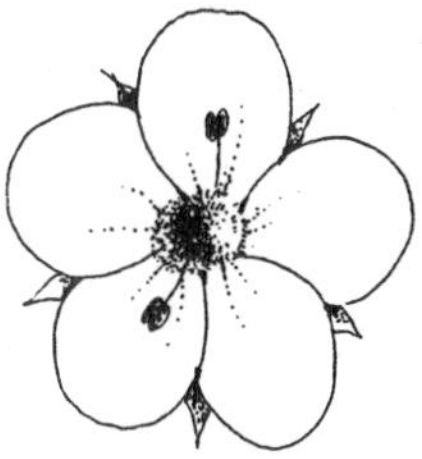

Figure 1558

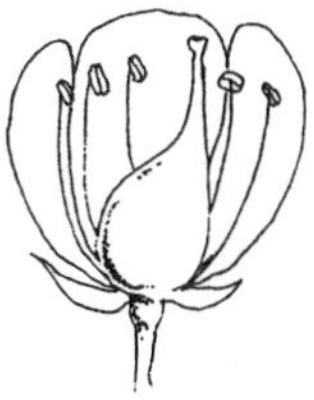

Figure 1559

Dicyclic. With two whorls.
Dimerous. With parts arranged in sets or multiples of two.
Heteromerous. With a variable number of parts, as in a flower with a different number of members in each floral whorl. Figure 1560.
Hexamerous. With parts arranged in sets or multiples of six. Figure 1561.

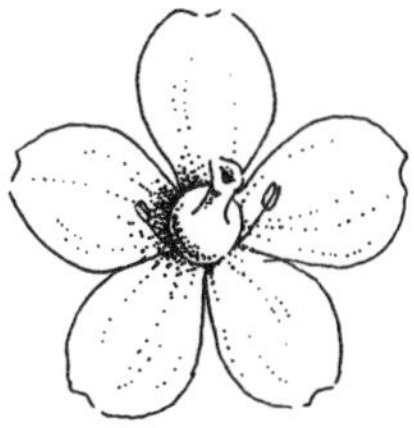

Figure 1560

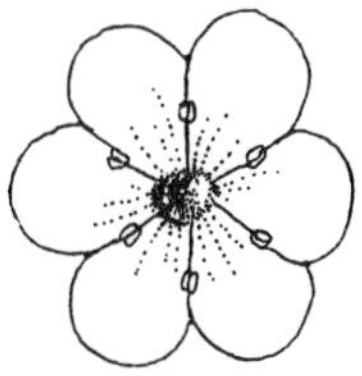

Figure 1561

Incomplete. Lacking an expected part or series of parts, as in a flower lacking one of the floral whorls (i.e. sepals, petals, stamens, or pistils).
Isomerous. With an equal number of parts, as in a flower with an equal number of members in each floral whorl.
Monocyclic. With a single whorl. Figure 1562.
Monomerous. With a single member, as in a floral whorl with only one part. Figure 1562.
Pentacyclic. With five whorls.
Pentamerous. With parts arranged in sets or multiples of five. Figure 1563.

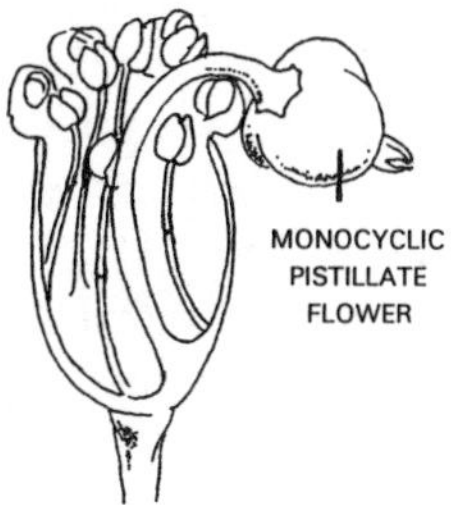

Figure 1562

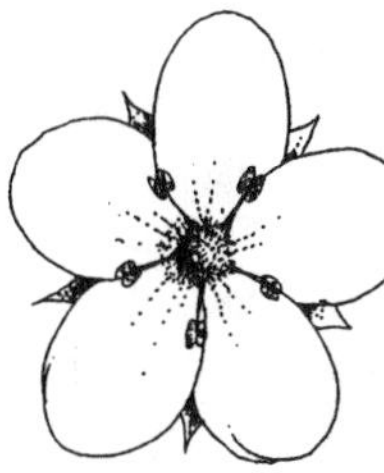

Figure 1563

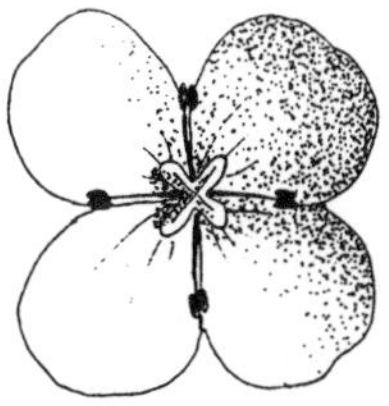

Figure 1564

Figure 1565

Polycyclic. With many whorls.
Polymerous. With many parts, as in a floral whorl with many members.
Symmetric. Said of a flower having the same number of parts in each floral whorl. Figure

1564.

Tetramerous. With parts arranged in sets or multiples of four. Figure 1564.

Trimerous. With parts arranged in sets or multiples of three. Figure 1565.

PERIANTH

Perianth Parts

Ala (pl. **alae**). One of the two lateral petals of a papilionaceous corolla. Figure 1566.

Anterior lip or lobe. The lower lip of a bilabiate corolla. Figure 1567. (compare **posterior**)

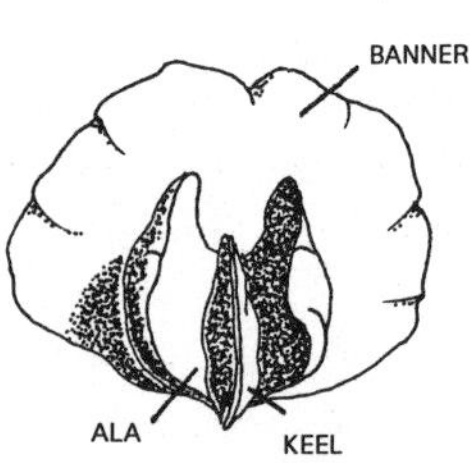

Figure 1566

Figure 1567

Banner. The upper and usually largest petal of a papilionaceous flower, as in peas and sweet peas. Figure 1566.

Blade. The broad part of a petal. Figure 1568.

Bridge. A band of tissue connecting the corolla scales, as in *Cuscuta.* Figure 1569.

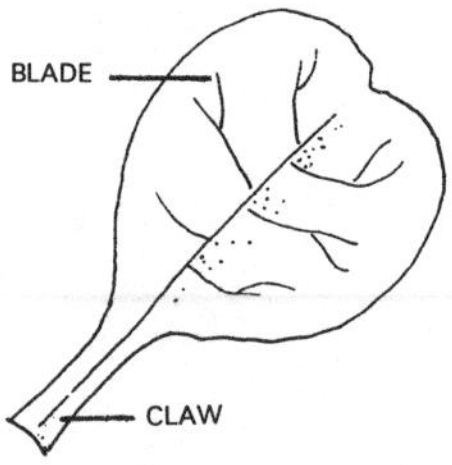

Figure 1568

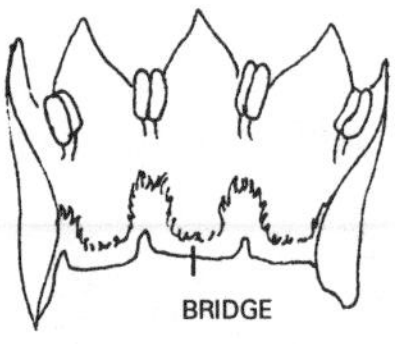

Figure 1569

Calcar. A spur or spurlike appendage. Figure 1570.

Calyx limb. See **calyx lobe.**

Calyx lobe. One of the free portions of a calyx of united sepals. Figure 1571.

Calyx tooth. See **calyx lobe.**

Calyx tube. The tube-like united portion of a calyx of united sepals. Figure 1571.

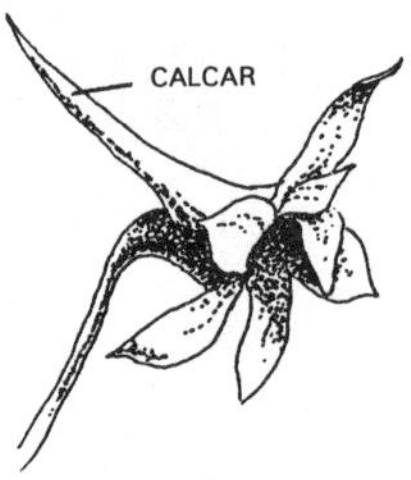

Figure 1570

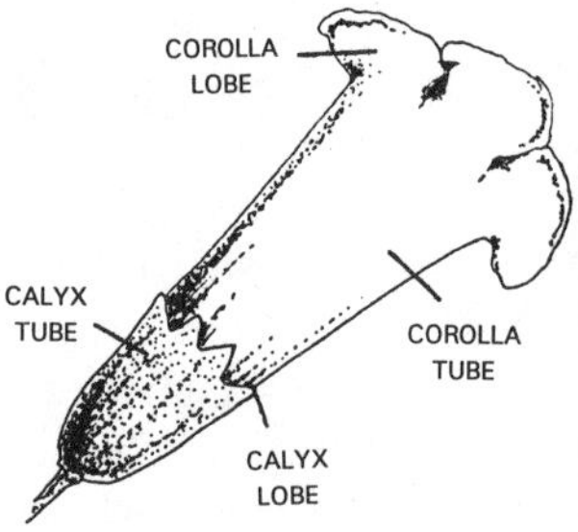

Figure 1571

Carina. A keel or ridge. Figure 1566.

Claw. The narrowed base of some petals and sepals. Figure 1568.

Corolla lobe. One of the free portions of a corolla of united petals. Figure 1571.

Corolla tube. The hollow, cylindric portion of a corolla of united petals. Figure 1571.

Corona. Petal-like or crown-like structures between the petals and stamens in some flowers; a crown. Figure 1572.

Crest. An elevated ridge or rib.

Crown. See **corona.**

Cucullus. A hood. Figure 1573.

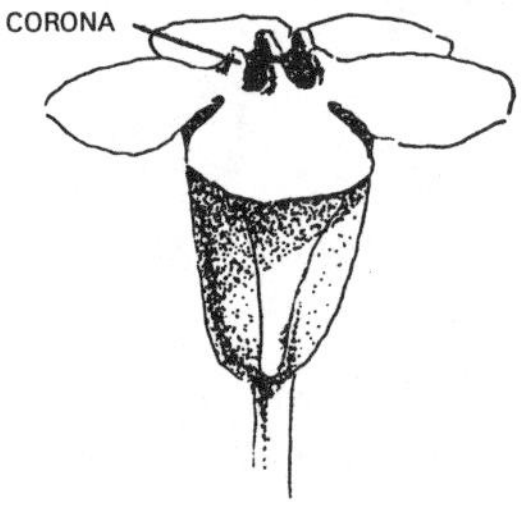

Figure 1572

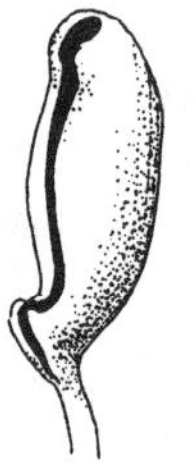

Figure 1573

Epicalyx. An involucre which resembles an outer calyx, as in *Malva.* Figure 1574.

Falls. The sepals of an *Iris.* Figure 1575.

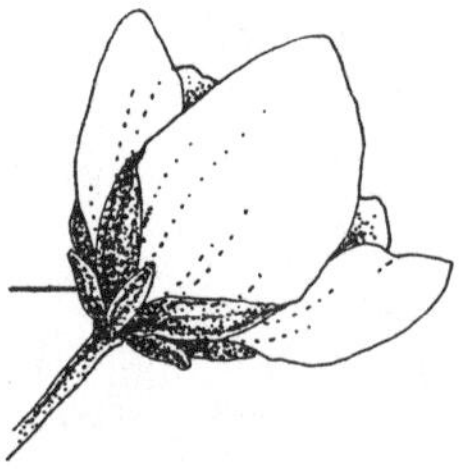

Figure 1574

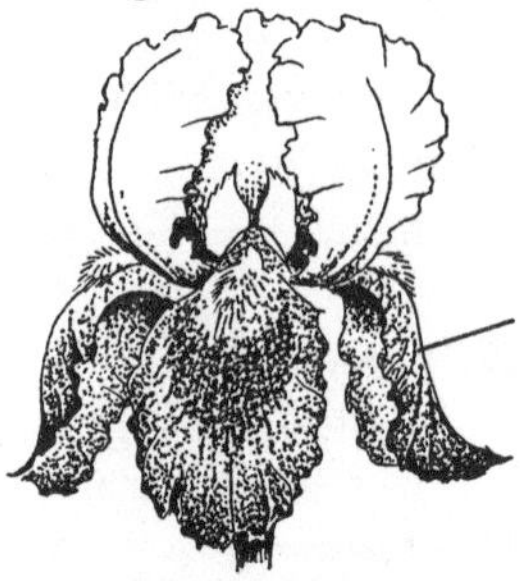

Figure 1575

Floral tube. An elongated tubular portion of a perianth. See illustration for **corolla tube**.

Fornix (pl. **fornices**). One of a set of small crests or scales in the throat of a corolla, as in many of the Boraginaceae. See illustration for **corona**.

Fringe. Hairs or bristles along the margin. Figure 1576.

Galea. The helmet-shaped or hoodlike upper lip of some two-lipped corollas. Figure 1577.

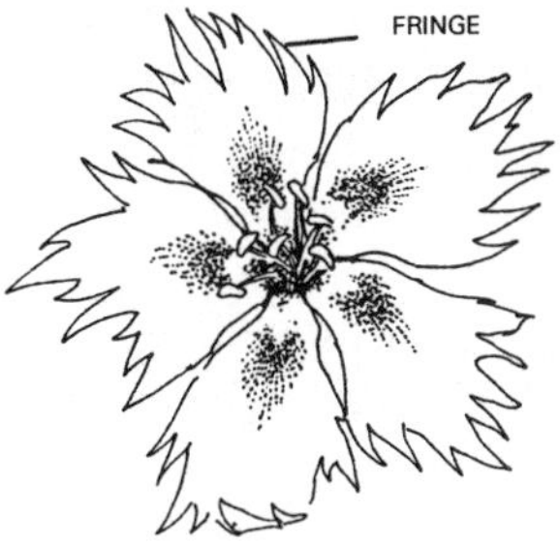

Figure 1576

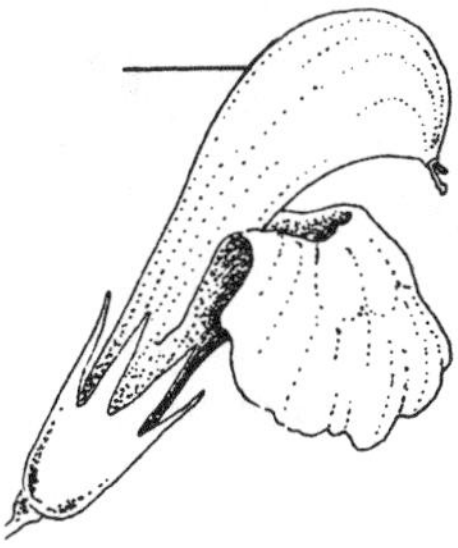

Figure 1577

Gibbosity. A swelling or protuberance; the state of being gibbous. Figure 1578.

Helmet. See **hood**.

Hood. A hollow, arched covering, as the upper petal in *Aconitum*. Figure 1579.

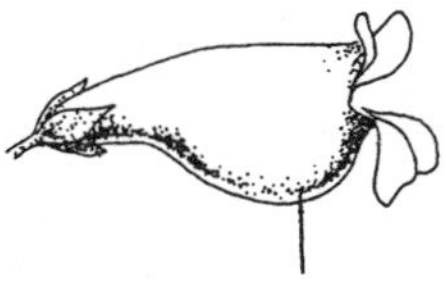

Figure 1578

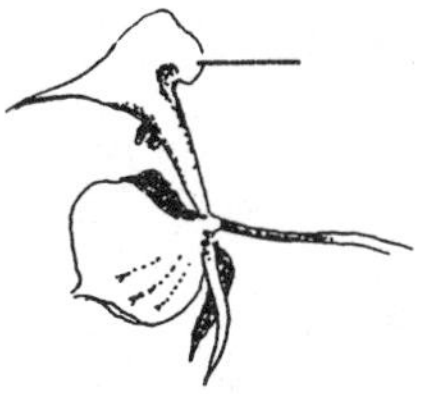

Figure 1579

Horn. A tapering projection resembling the horn of a cow. Figure 1580.

Keel. The two lower united petals of a papillonaceous flower. Figure 1566.

Labellum. Lip; the exceptional petal of an orchid blossom. Figure 1581.

Figure 1580

Labium (pl. **labia**). The lower lip of a bilabiate corolla. Figure 1582.

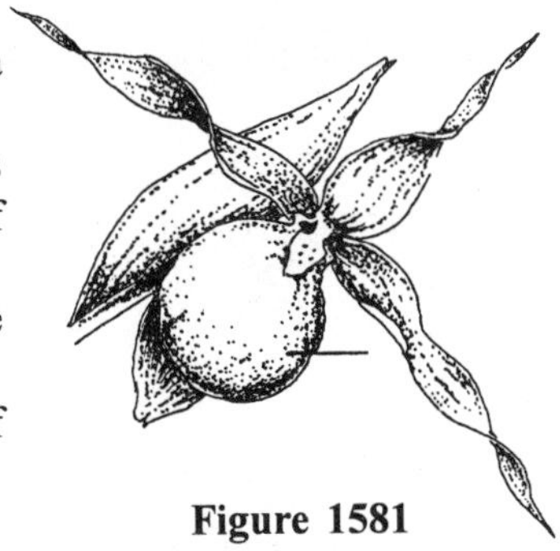

Figure 1581

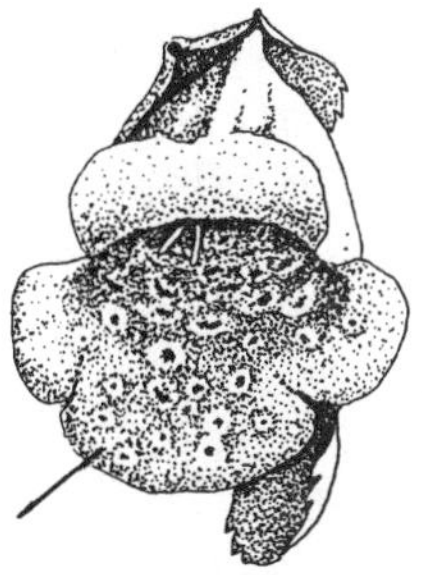

Figure 1582

Lamella (pl. **lamellae**). An erect scale inserted on the petal in some corollas and forming part of the corona. See illustration for **corona**.

Lamina. The expanded portion, or blade, of a petal. Figure 1568.

Lepanthium. A petal with a nectary.

Ligula. See **ligule**.

Ligule. The flattened part of the ray corolla in the Compositae (Asteraceae). Figure 1583.

Limb. The expanded part, or blade, of a petal. Figure 1568; the expanded part of a sympetalous corolla. Figure 1584.

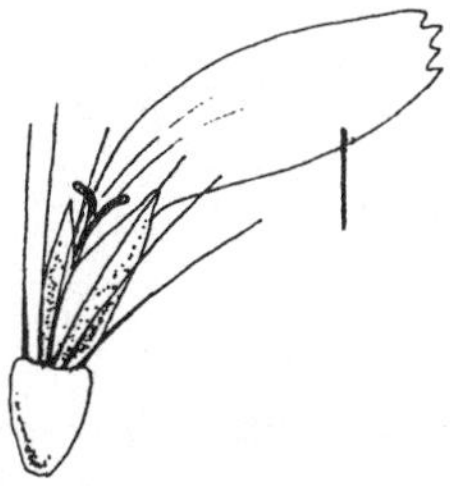

Figure 1583

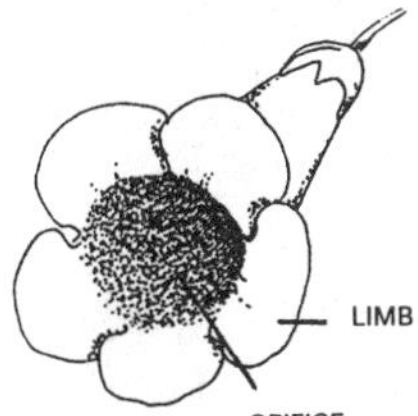

Figure 1584

Lip. One of the two projections or segments of an irregular, two-lipped corolla or calyx; a labium. Figure 1582; the exceptional petal of an orchid blossom. Figure 1581.

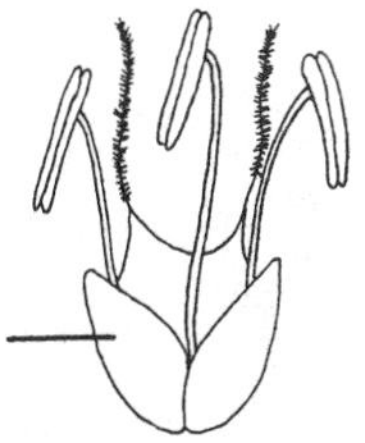

Figure 1585

Lodicule. Paired, rudimentary scales at the base of the ovary in grass flowers. Figure 1585.

Orifice. An opening or mouth, as the mouthlike opening of a tubular corolla. Figure 1584.

Palate. A raised appendage on the lower lip of a

corolla which partially or completely closes the throat. Figure 1586.

Pappus. The modified calyx of the Compositae (Asteraceae), consisting of awns, scales, or bristles at the apex of the achene. Figure 1587.

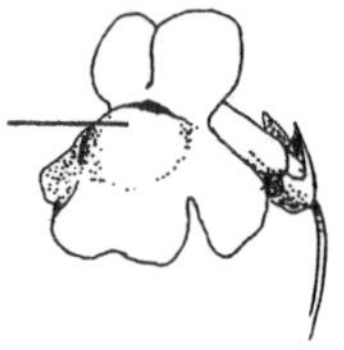

Figure 1586

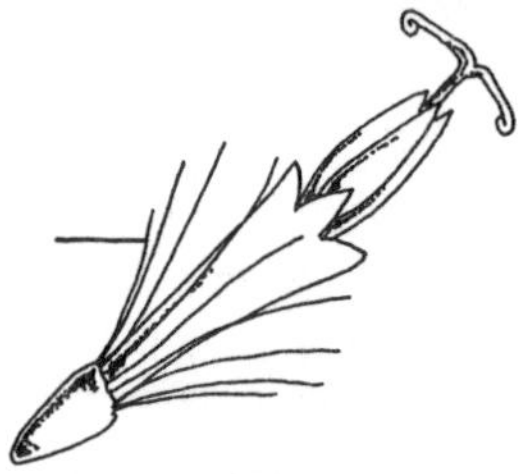

Figure 1587

Petal. An individual segment or member of the corolla, usually colored or white. Figure 1588.

Plait. A fold or pleat, as in some corollas. Figure 1589.

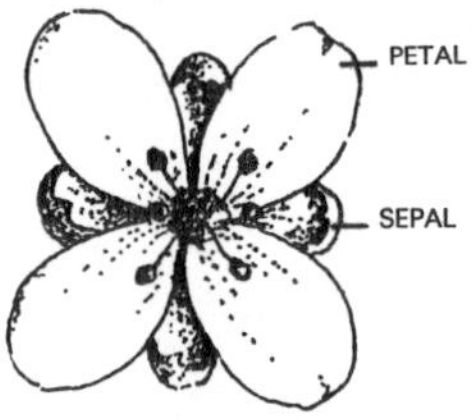

Figure 1588

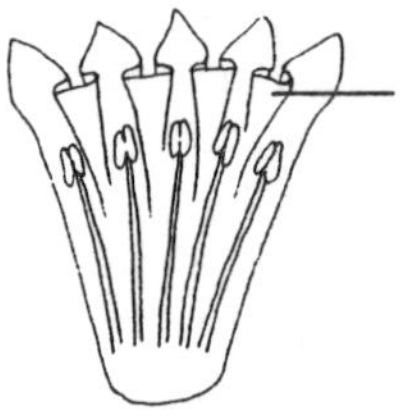

Figure 1589

Posterior. The upper lip of a bilabiate corolla. Figure 1590. (compare **anterior**)

Ray. The straplike portion of a ligulate flower (or the ligulate flower itself) in the Compositae (Asteraceae). Figure 1591.

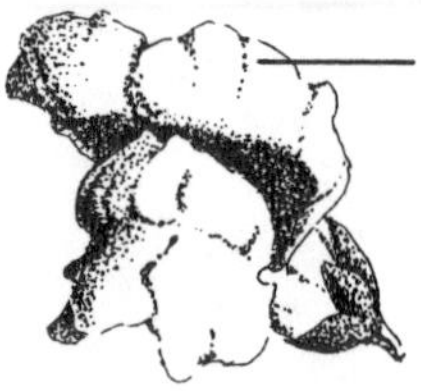

Figure 1590

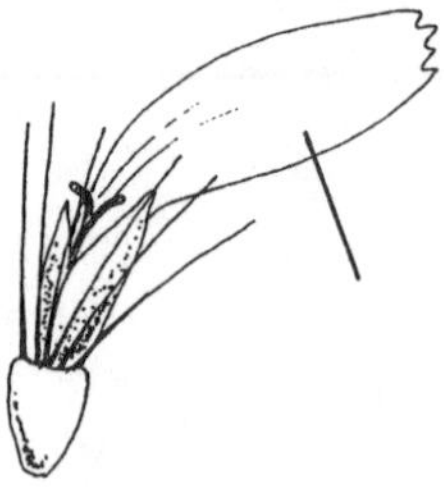

Figure 1591

Sac. A bag-shaped structure. Figure 1592.

Scale. Any thin, flat, scarious structure. Figure 1593.

Figure 1592

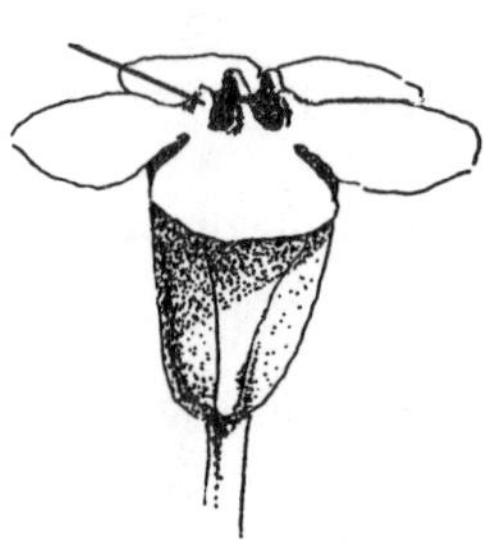

Figure 1593

Sepal. A segment of the calyx. Figure 1588.

Sinus. The cleft, depression, or recess between two lobes of a petal. Figure 1594.

Spur. A hollow, slender, saclike appendage of a petal or sepal, or of the calyx or corolla. Figure 1595.

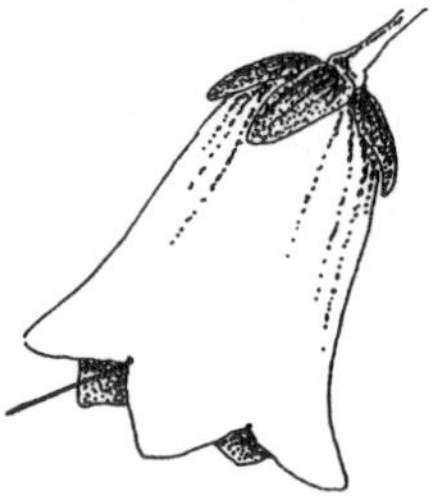

Figure 1594

Figure 1595

Squama (pl. **squamae**). A scale, as in some types of pappus in the Compositae (Asteraceae). Figure 1596.

Standard. See **banner**.

Tepal. A segment of a perianth which is not differentiated into calyx and corolla. Figure 1597.

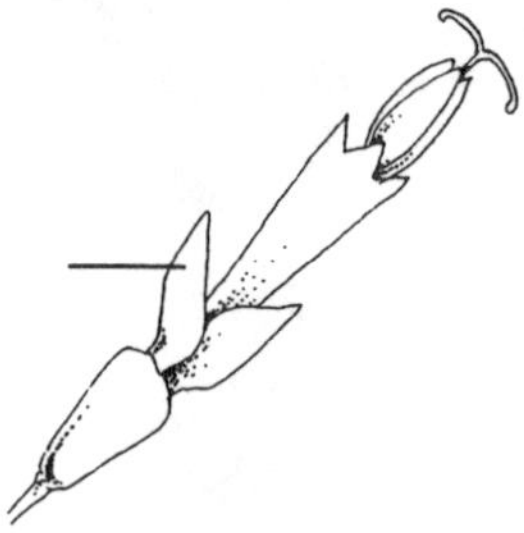

Figure 1596

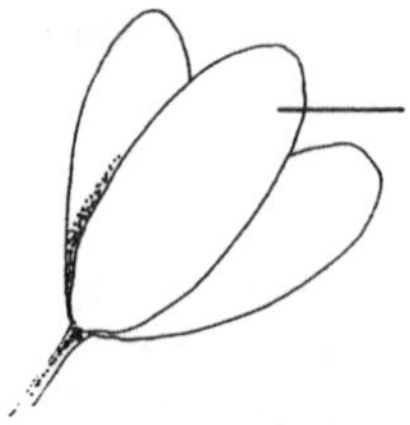

Figure 1597

Throat. The orifice of a gamopetalous corolla or gamosepalous calyx. Figure 1598; the expanded portion of the corolla between the limb and the tube. Figure 1599.

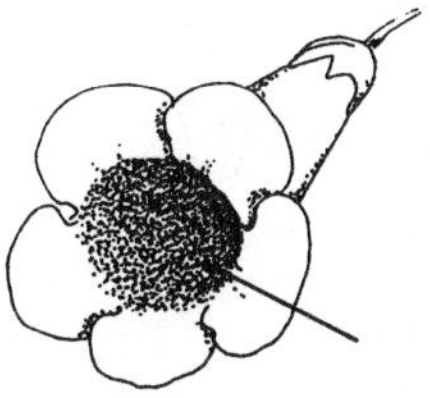

Figure 1598

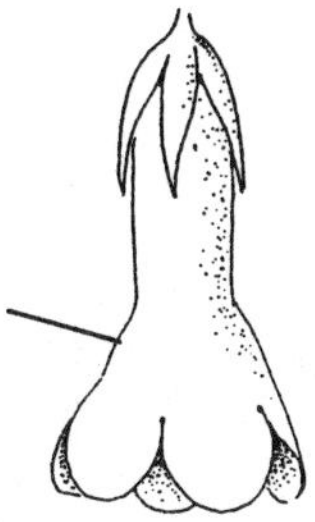

Figure 1599

Vexillum. The upper and usually largest petal of a papilionaceous flower, as in peas and sweet peas; banner. Figure 1600.

Wing. One of the two lateral petals of a papilionaceouscorolla. Figure 1600.

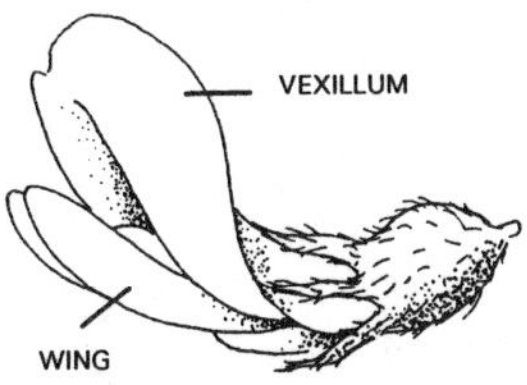

Figure 1600

Perianth Types

Achlamydeous. Lacking a perianth. Figure 1601.

Apetalous. Without petals.

Apopetalous. With separate petals. Figure 1602.

Aposepalous. With separate sepals. Figure 1602.

Figure 1601

Figure 1602

Asepalous. Without sepals.

Bipetalous. With two petals.

Calyculate. With small bracts around the calyx, as if possessing an outer calyx. Figure 1603.

Chlamydeous. With a floral whorl.

Choripetalous. See **apopetalous** or **polypetalous**.

Dichlamydeous. With two types of perianth whorls, i.e., calyx and corolla. Figure 1602.

Dipetalous. See **bipetalous**.

Double. Having a larger number of petals than usual.

Epappose. Without a pappus. Figure 1604.

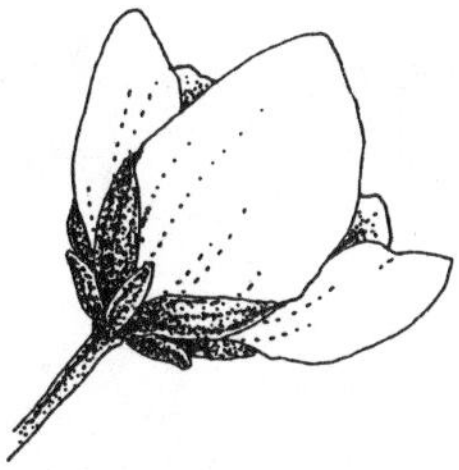

Figure 1603

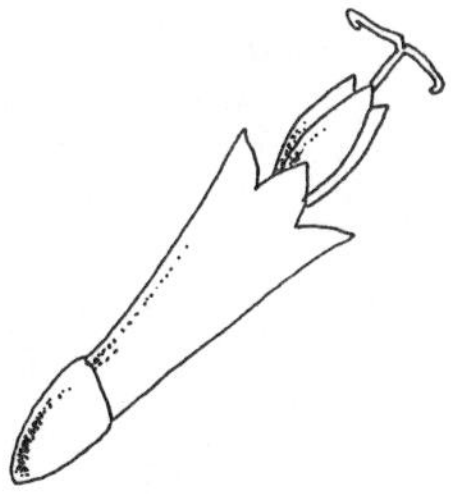

Figure 1604

Gamopetalous. With the petals united, at least partially. Figure 1605.

Gamosepalous. With the sepals united. Figure 1606.

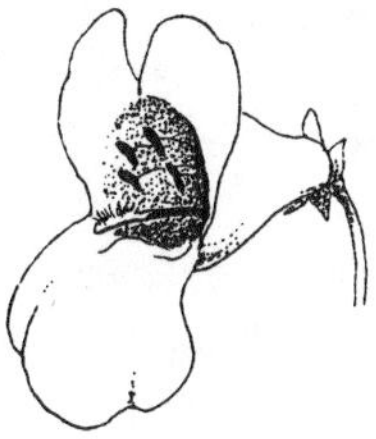

Figure 1605

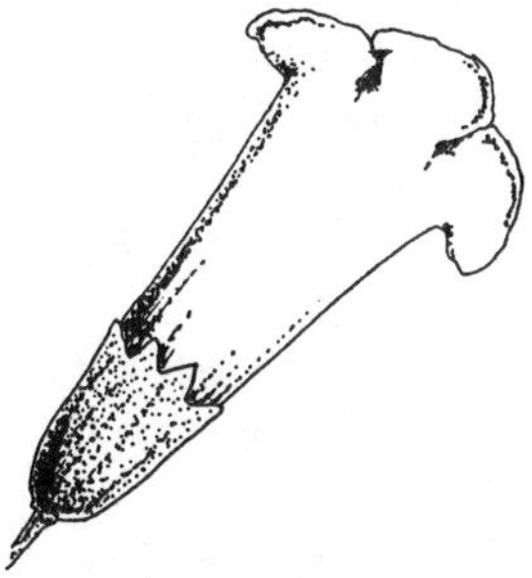

Figure 1606

Monochlamydeous. With only one type of perianth member. Figure 1607.

Monopetalous. See **sympetalous** or **gamopetalous**.

Octopetalous. With eight petals.

Octosepalous. With eight sepals.

Pappiferous. See **pappose**.

Pappose. Pappus-bearing. Figure 1608.

Petaliferous. Bearing petals.

Petalous. With petals.

Polypetalous. With a corolla of completely

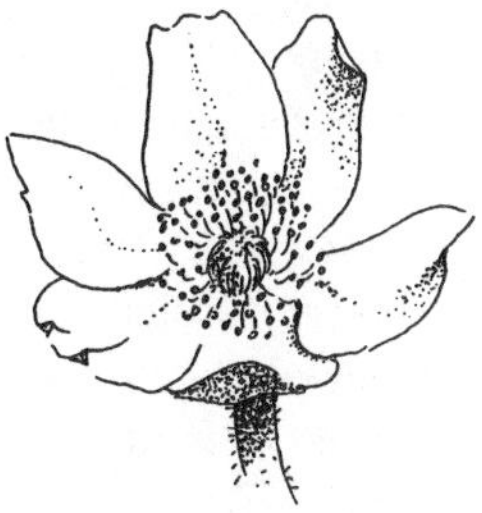

Figure 1607

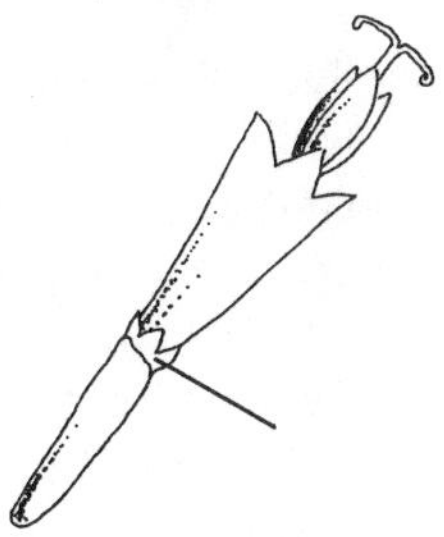

Figure 1608

separate petals. Figure 1609.

Polysepalous. With a calyx of separate sepals. Figure 1609.

Sympetalous. With the petals united, at least near the base. Figure 1610.

Synsepalous. With united sepals. Figure 1610.

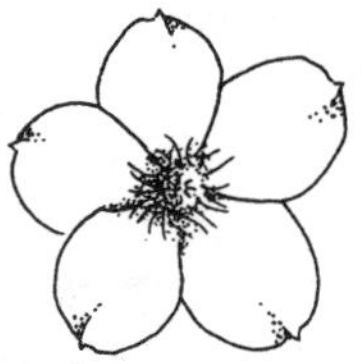

Figure 1609

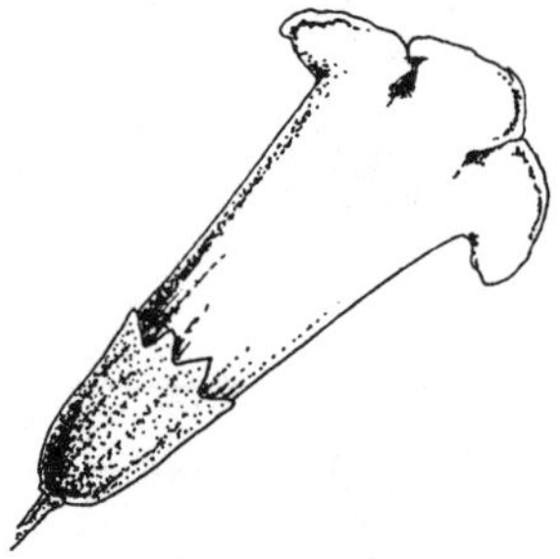

Figure 1610

Tetrapetalous. With four petals.

Tripetalous. With three petals.

Unipetalous. With only a single petal.

Perianth Forms (Figure 1611.)

Accrescent. Becoming larger with age, as a calyx which continues to enlarge after anthesis.

Alate. Winged.

Ampliate. Enlarged or expanded.

Bilabiate. Two-lipped, as in many irregular flowers.

Calcarate. With a calcar; spurred.

Calceolate. Shoe-shaped or slipper-shaped, as the labellum of some orchids.

Campanulate. Bell-shaped.

Carinate. Keeled.

Corniculate. With small hornlike protuberances.

Cornute. Horned. See illustration for **corniculate**.

Coroniform. Crown-shaped.

Cruciate. See **cruciform**.

Cruciform. Cross-shaped.

Cucullate. Hooded or hood-shaped.

Explanate. Spread out flat.

Funnelform. Gradually widening from base to apex; funnel-shaped.

Galeate. Helmet-shaped; with a galea.

Galeiform. See **galeate**.

Gibbous. Swollen or enlarged on one side; ventricose.

Inflated. Swollen or expanded; bladdery.

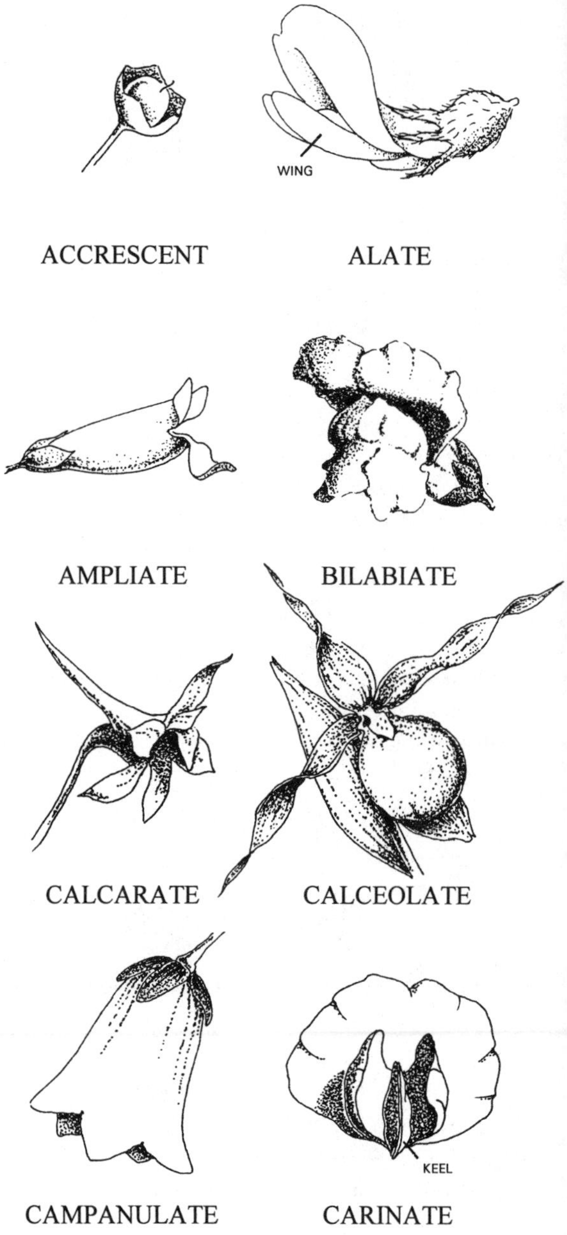

Figure 1611a.

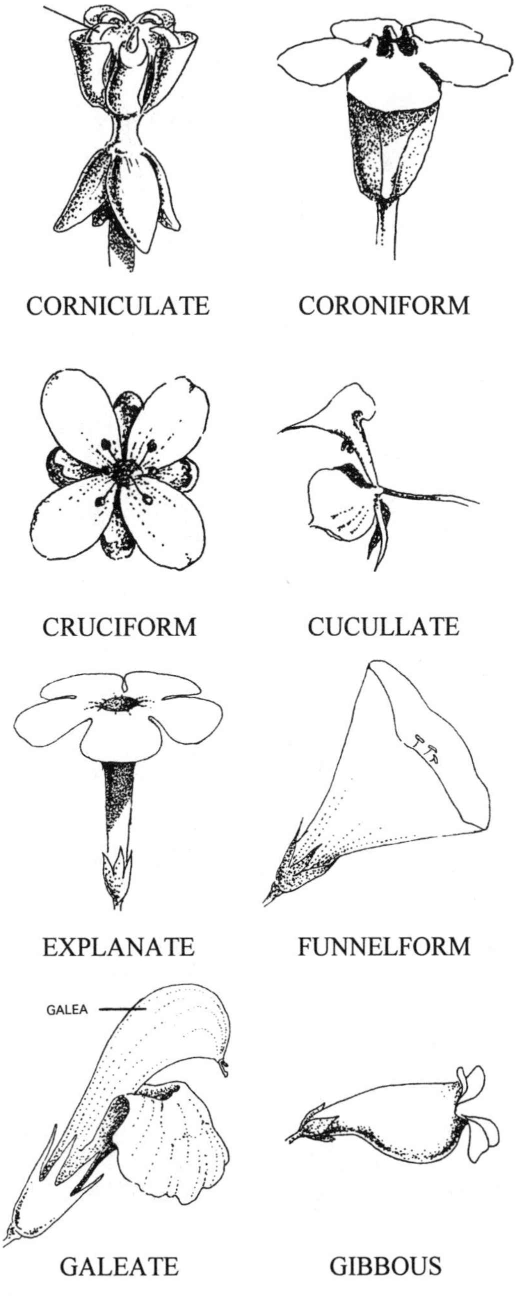

Figure 1611b.

Infundibuliform. Funnel-shaped. See illustration for **funnelform**.

Keeled. Ridged, like the keel of a boat.

Labiate. Lipped; with parts which are arranged like lips or shaped like lips.

Ligulate. Strap-shaped.

Liguliform. See **ligulate**.

Paleolate. With a lodicule.

Papilionaceous. Butterflylike, as the irregular corolla of a pea, with a banner petal, two wing petals, and a keel petal.

Personate. Two-lipped, with the throat closed by a prominent projection (palate).

Plicate. Plaited or folded, as a folding fan.

Porrect. Extended forward; resembling a parrot beak.

Reflexed. Bent backward or downward.

Ringent. Gaping; with widely spreading lips, as in some corollas.

Rotate. Disc-shaped; flat and circular, as a sympetalous corolla with widely spreading lobes and little or no tube.

Saccate. With a sac, or in the shape of a sac.

Salverform. With a slender tube and an abruptly spreading, flattened limb.

Spurred. Bearing a spur or spurs.

Strap-shaped. Elongated and flat.

Tubular. With the form of a tube or cylinder.

Urceolate. Pitcherlike; hollow and contracted near the mouth like a pitcher or urn.

Urn-shaped. See **urceolate**.

Ventricose. Inflated or swollen on one side only, as in some corollas, especially in the genus *Penstemon*.

Winged. Possessing wings.

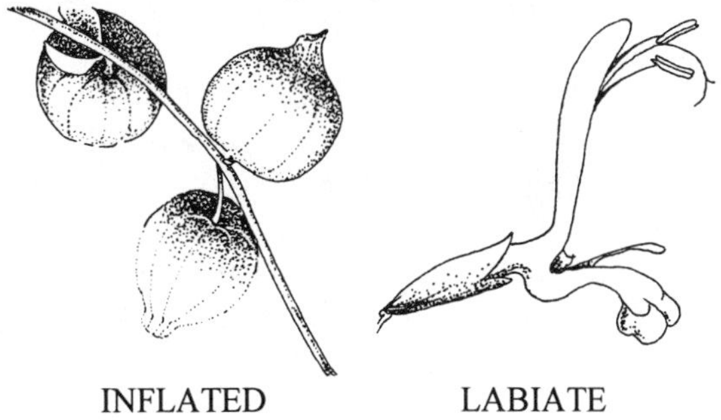

Figure 1611c.

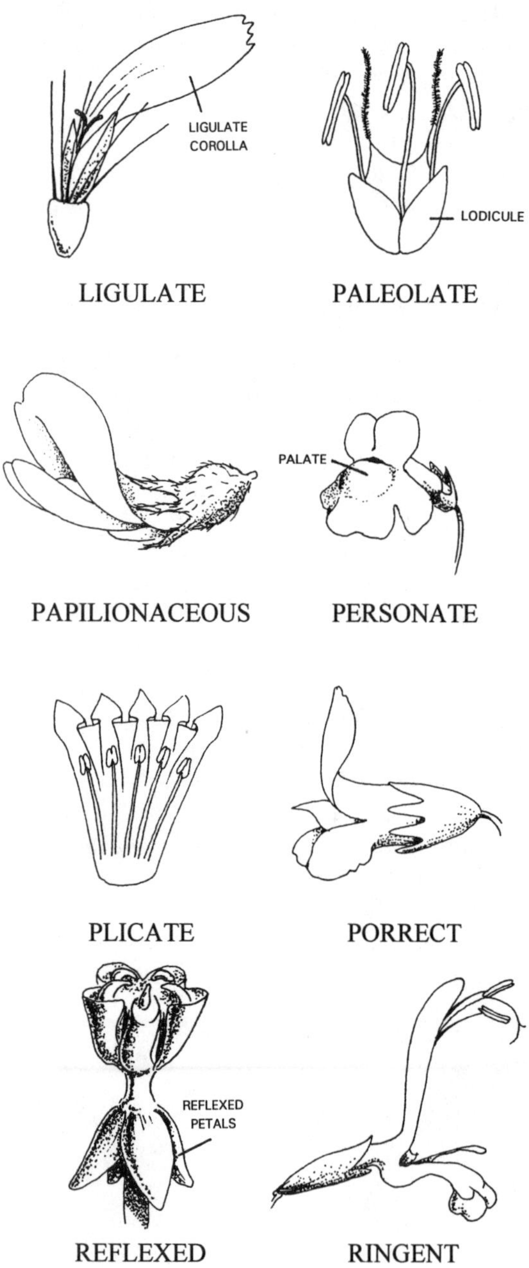

Figure 1611d.

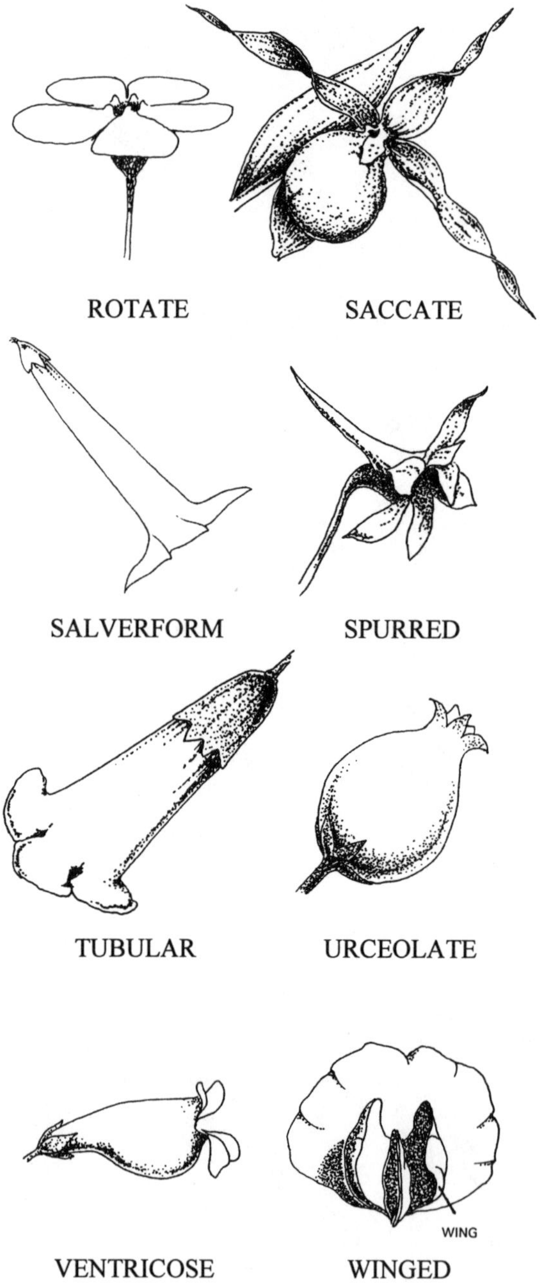

Figure 1611e.

ANDROECIUM

The male reproductive parts of a flower.

Androecium Parts

Androgynophore. Stalk supporting the androecium and gynoecium in some flowers. Figure 1612.

Androphore. Stalk supporting a group of stamens.

Anther. The expanded, apical, pollen bearing portion of the stamen. Figure 1613.

Figure 1612

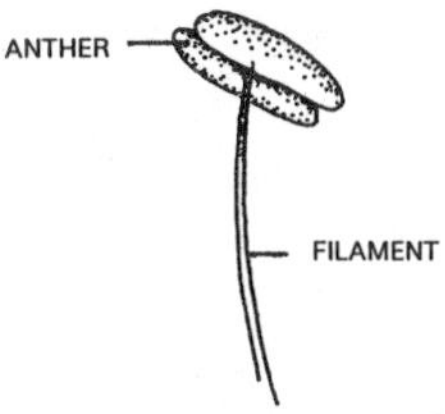

Figure 1613

Anther sac. One of the pollen bearing chambers of the anther. Figure 1614.

Cell. A hollow cavity or compartment within a structure, as the cavity of the anther containing pollen. Figure 1614.

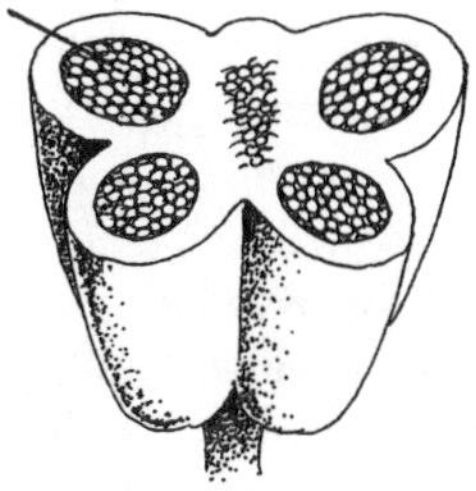

Figure 1614

Column. A structure formed by the union of staminal filaments. Figure 1615; the united filaments and style in the Orchidaceae. Figure 1616.

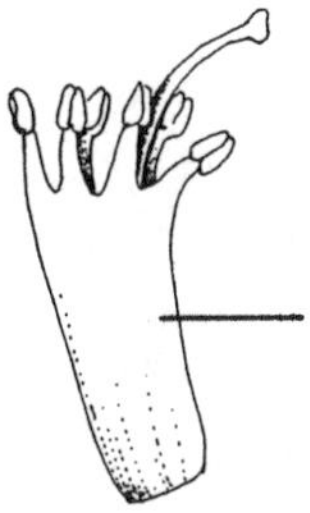

Figure 1615

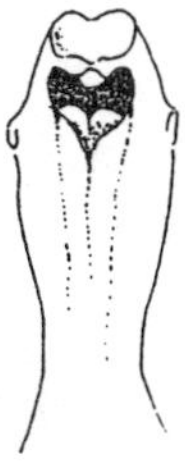

Figure 1616

Connective. The portion of the stamen connecting the two pollen sacs of an anther. Figure 1617.

Filament. The stalk of the stamen which supports the anther. Figure 1613.

Gynandrium. A column bearing stamens and pistils. Figures 1616 and 1618.

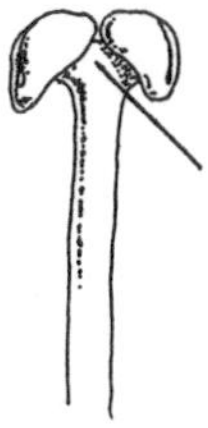

Figure 1617

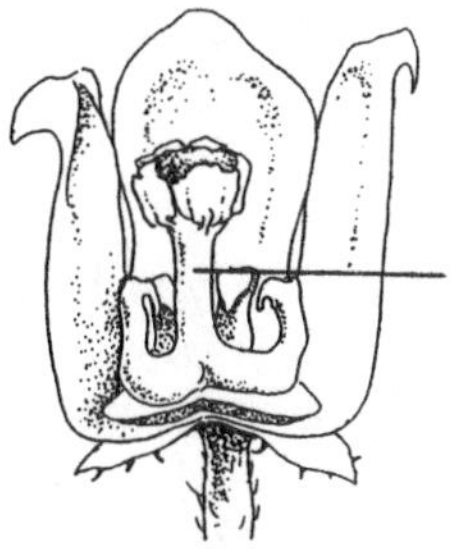

Figure 1618

Phalange. Two or more stamens joined by their filaments. Figure 1619.

Phalanx. See **Phalange**.

Pollen. The mature microspores or developing male gametophytes of a seed plant, produced in the microsporangium of a gymnosperm or in the anther of an angiosperm. Figure 1620.

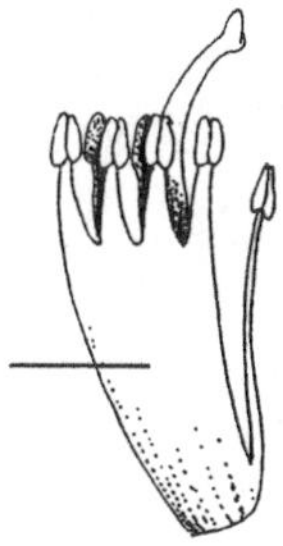

Figure 1619

Pollinium (pl. **pollinia**). A mass of waxy pollen grains transported as a unit in many members of the Orchidaceae and Asclepiadaceae. Figure 1621.

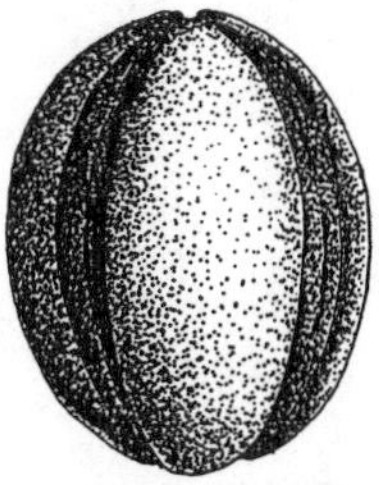

Figure 1620

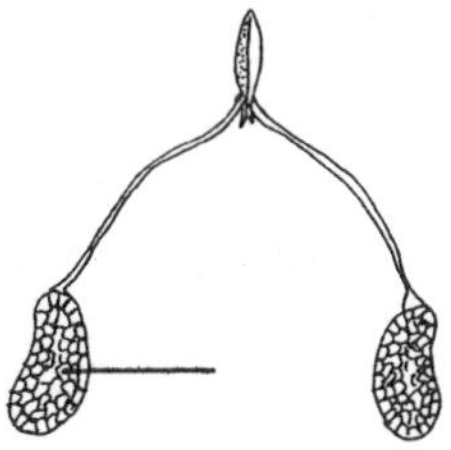

Figure 1621

Sac. A bag-shaped compartment, as the cavity of an anther. See **anther sac**.

Stalk. See **filament**.

Stamen (pl. **stamens**, **stamina**). The male reproductive organ of a flower, consisting of an anther and filament. Figure 1613.

Staminode (pl. **staminodia**). A modified stamen which is sterile, producing no pollen. Figure 1622.

Staminodium. See **staminode**.

Suture. The line of dehiscence of an anther. Figure 1623.

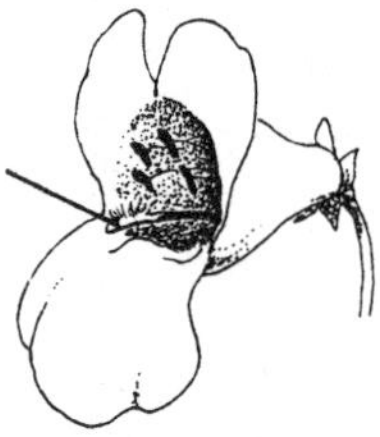

Figure 1622

Theca (pl. **thecae**). A pollen sac or cell of the anther. See **anther sac**.

Translator. The connecting structure between the pollinia of adjacent anthers in the Asclepiadaceae. Figure 1624.

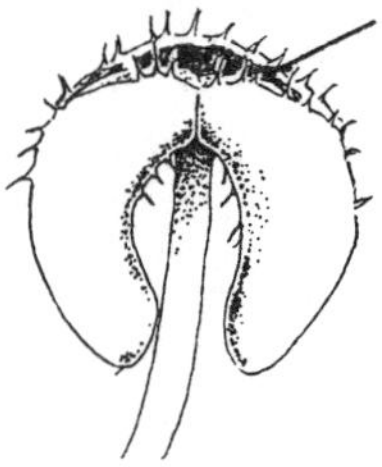

Figure 1623

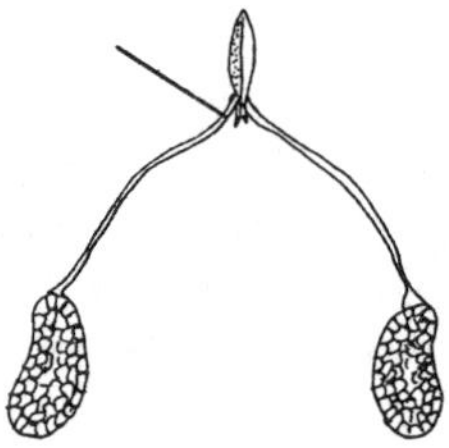

Figure 1624

Stamen Types

Abortive. Not fully or properly developed; rudimentary. Figure 1625.

Fertile. Capable of bearing seeds; capable of bearing pollen.

Filantherous. Of a stamen with a distinct anther and filament. Figure 1626.

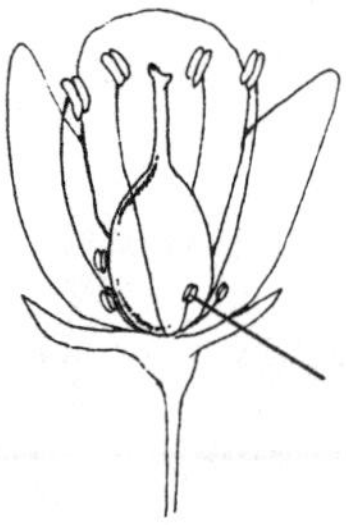

Figure 1625

Infertile. Sterile or inviable. See **abortive**.

Petalantherous. Of a stamen with a petaloid filament. See **petaloid**.

Petaloid. Petal-like in appearance. Figure 1627.

Polliniferous. Bearing pollen.

Rudimentary. Imperfectly developed; vestigial. See **abortive**.

Sterile. Infertile, as a stamen that does not bear pollen. See **abortive**.

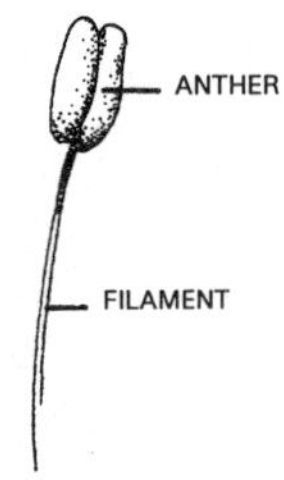

Figure 1626

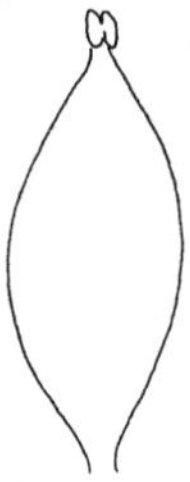

Figure 1627

Stamen Number

Anandrous. Without stamens; lacking an androecium. Figure 1628.

Astemonous. Without stamens. Figure 1628.

Diandrous. With two stamens.

Haplostemonous. With as many stamens as petals. Figure 1629.

Figure 1628

Monandrous. With a single stamen.

Octandrous. With eight stamens.

Octostemonous. With eight stamens.

Oligandrous. With few stamens.

Pentandrous. With five stamens.

Polyandrous. With many stamens (usually more than ten). Figure 1630.

Polystemonous. With many stamens. Figure 1630.

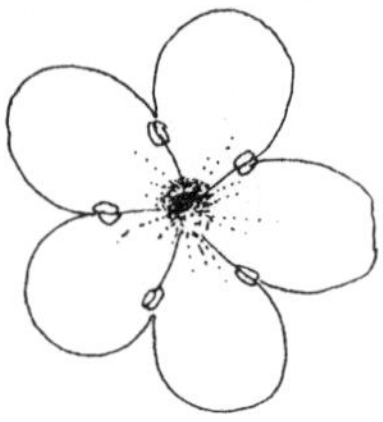

Figure 1629

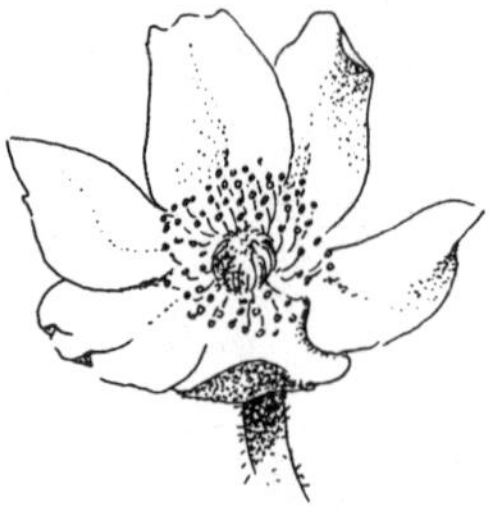

Figure 1630

Stamen Arrangement

Alternate. Stamens borne between the petals. Figure 1631. (compare **opposite**)

Antepetalous. Directly in front of (opposite) the petals. Figure 1632.

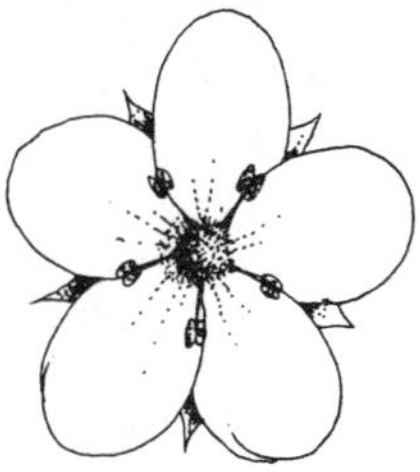

Figure 1631

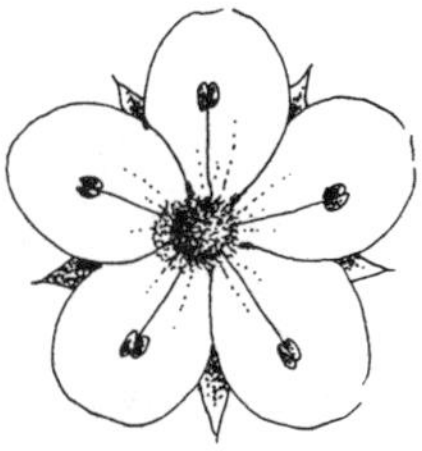

Figure 1632

Antesepalous. Directly in front of (opposite) the sepals. Figure 1631.

Antipetalous. See **Antepetalous.**

Antisepalous. See **Antesepalous.**

Didynamous. With two pairs of stamens of unequal length; occurring in pairs. Figure 1633.

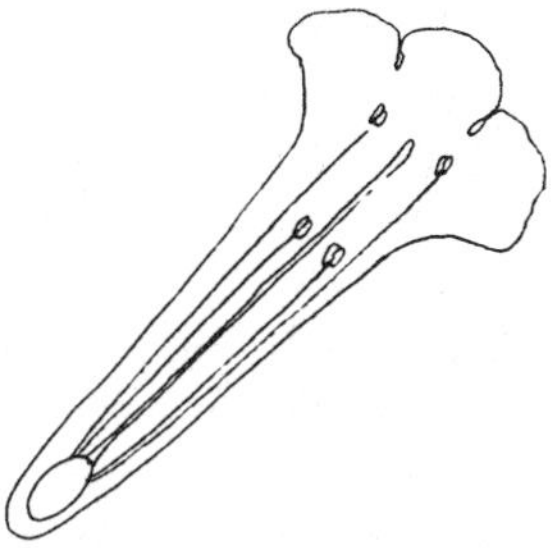

Figure 1633

Diplostemonous. With two series of stamens, the outer series opposite the sepals and the inner series opposite the petals; with twice as many stamens as petals. Figure 1634.

Excurved. Curving outward, away from the axis. Figure 1635.

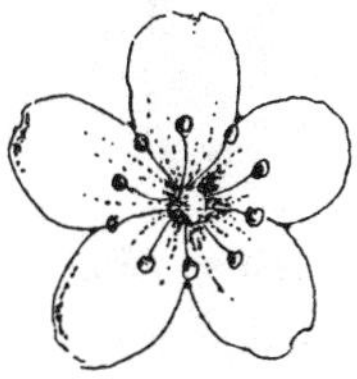

Figure 1634

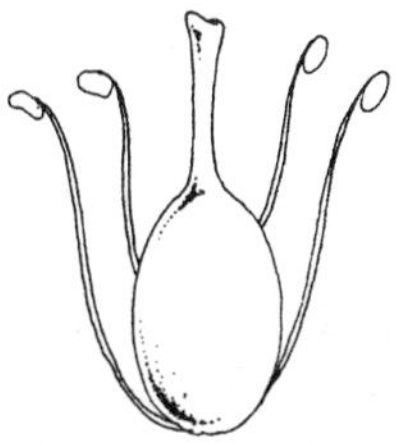

Figure 1635

Exserted. Stamens protruding from the corolla. Figure 1636.

Extrorse. Turned outward, away from the axis. Figure 1635. (compare **introrse**)

Haplostemonous. With one series of stamens. Figure 1637.

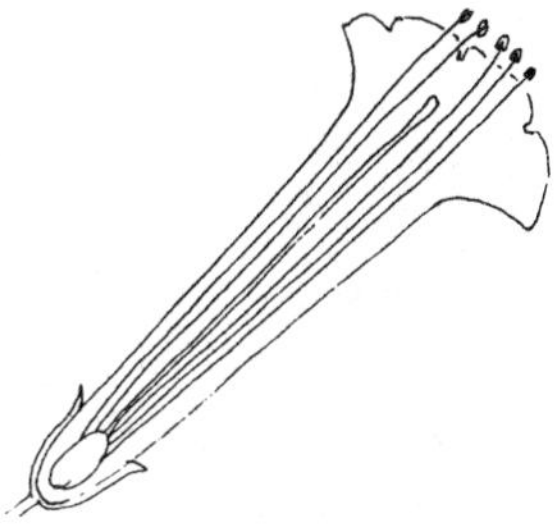

Figure 1636

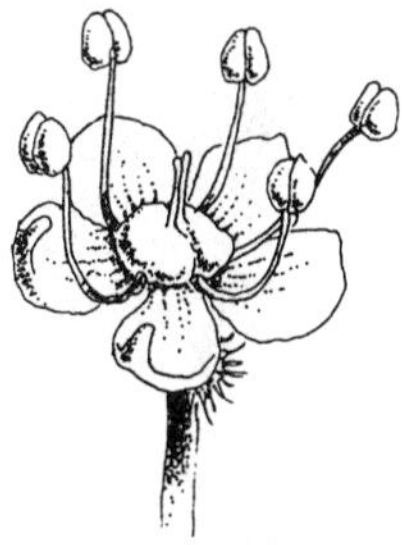

Figure 1637

Included. Stamens not projecting beyond the corolla. Figure 1638.

Incurved. Curving inward, toward the axis. Figure 1639.

Introrse. Turned inward, toward the axis. Figure 1639. (compare **extrorse**)

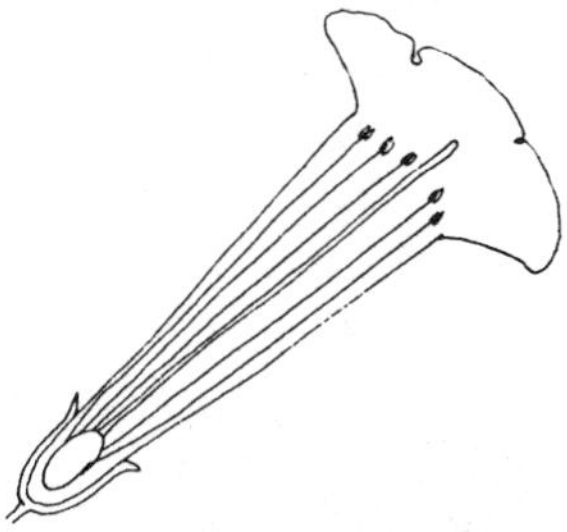

Figure 1638

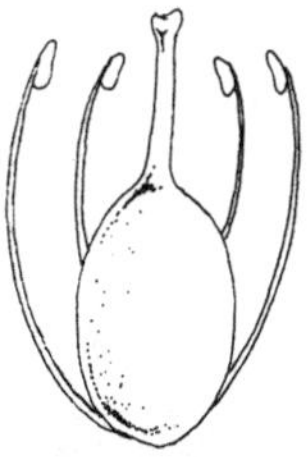

Figure 1639

Obdiplostemonous. Having two whorls of stamens, the outer whorl opposite the petals and the inner whorl opposite the sepals. Figure 1640.

Opposite. Stamens borne on the same radius as the petals. Figure 1632. (compare **alternate**)

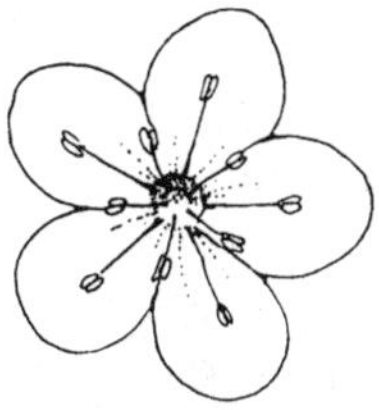

Figure 1640

Phaenantherous. With stamens exserted from the corolla. Figure 1636.

Polyadelphous. Borne in several distinct groups. Figure 1641. (compare **monadelphous** and **diadelphous**)

Tetradynamous. Having four long and two short stamens, as in most of the Cruciferae (Brassicaceae). Figure 1642.

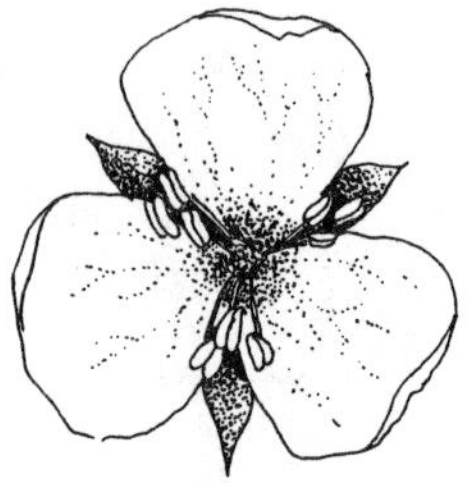

Figure 1641 **Figure 1642**

Tridynamous. With stamens arranged in two groups of three.

Stamen Fusion

Adherent. Sticking together of unlike parts, as the anthers to the style. The attachment is not as firm or solid as **adnate**.

Adnate. Fusion of unlike parts, as the stamens to the corolla. Figure 1643. (compare **connate**)

Apostemonous. With separate stamens. Figure 1644.

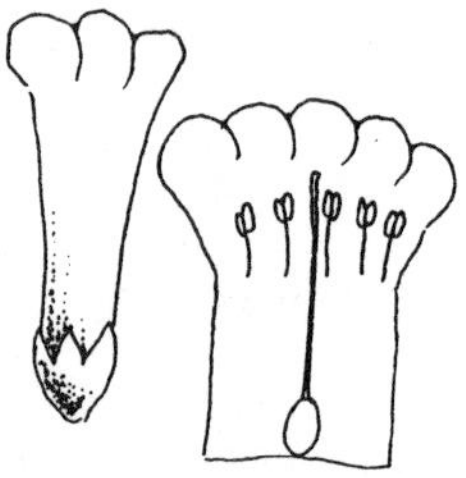

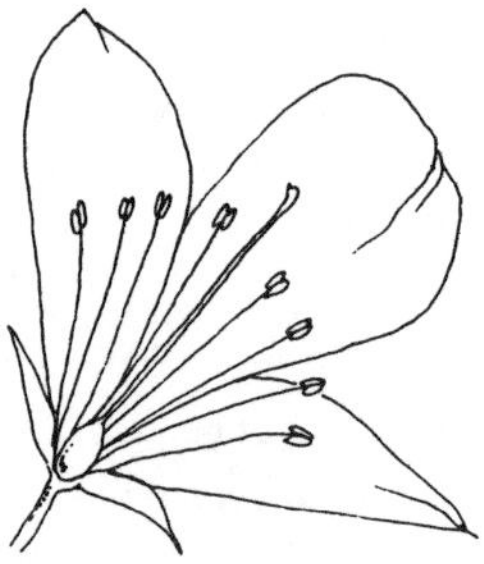

Figure 1643 **Figure 1644**

Appressed. Pressed close or flat against another organ.

Approximate. Borne close together, but not fused.

Coalescent. United together to form a single unit. Figure 1645.

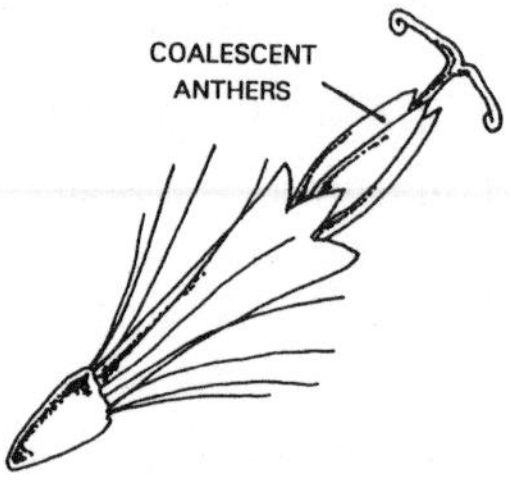

Figure 1645

Coherent. Sticking together of like parts. The attachment is not as firm or solid as **connate**.

Connate. Fusion of like parts, as the fusion of staminal filaments into a tube. Figure 1646. (compare **adnate**)

Connivent. Converging, but not actually fused or united.

Diadelphous. Stamens united into two, often unequal, sets by their filaments. Figure 1647.

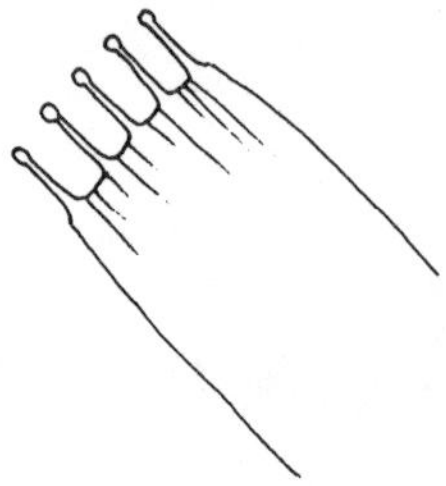

Figure 1646 **Figure 1647**

Distinct. Separate; not attached to like parts. Figure 1644. (compare **connate**)

Epipetalous. Attached to the petals. Figure 1648.

Free. Not attached to other organs. Figure 1644.

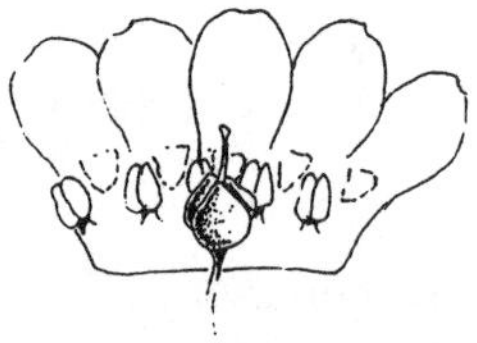

Figure 1648

Gynandrous. With the stamens adnate to the pistil. Figure 1649.

Gynostemial. See **gynandrous**.

Monadelphous. Stamens united by the filaments and forming a tube around the gynoecium. Figure 1650.

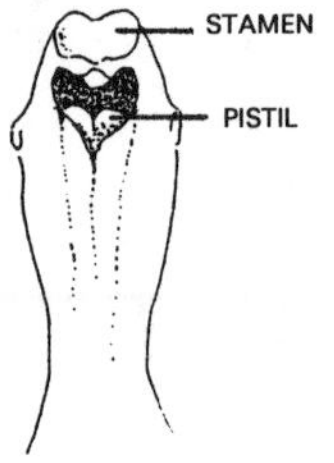

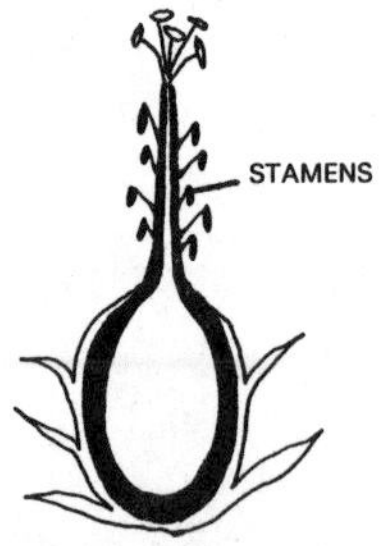

Figure 1649 **Figure 1650**

Petalostemonous. With the staminal filaments fused to the corolla and the anthers free. Figure 1651.

Polyadelphous. Borne in several distinct groups, as the stamens of some flowers. Figure 1652.

Synandrous. With united anthers. Figure 1645.

Syngenesious. With stamens united by their

anthers. Figure 1645.

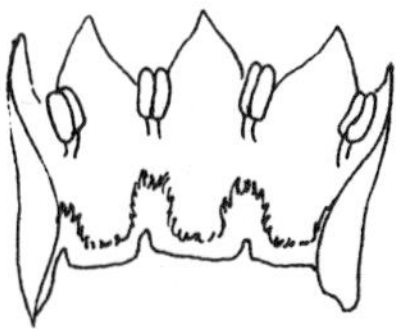

Figure 1651

Figure 1652

Anther Attachment

Basifixed. With the anther attached by the base. Figure 1653.
Dorsifixed. With the anther attached at the back. Figure 1654.
Versatile. With the anther attached near the middle rather that at one end. Figure 1655.

Figure 1653

Figure 1654

Figure 1655

Anther Dehiscence

Longitudinal. Dehiscing along the long axis of the anther. Figure 1656.
Poricidal. Dehiscing through a pore. Figure 1657.
Transverse. Dehiscing at a right angle to the long axis of the anther. Figure 1658.
Valvular. Dehiscing through flap-covered pores. Figure 1659.

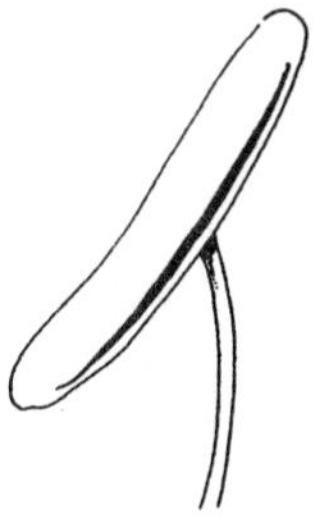

Figure 1656

Figure 1657

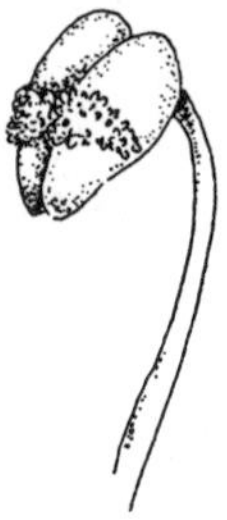

Figure 1658

Figure 1659

GYNOECIUM

The female reproductive parts of a flower.

Gynoecium Parts

Androgynophore. Stalk supporting the androecium and gynoecium in some flowers. Figure 1660.
Carpel. A simple pistil formed from one modified leaf, or that part of a compound pistil formed from one modified leaf. Figure 1661.

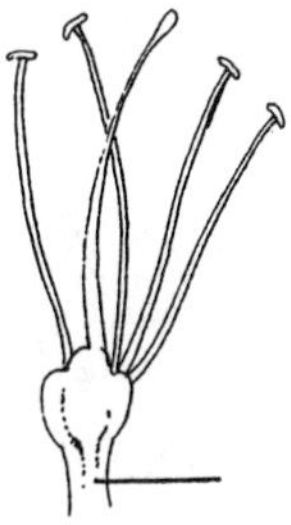

Figure 1660

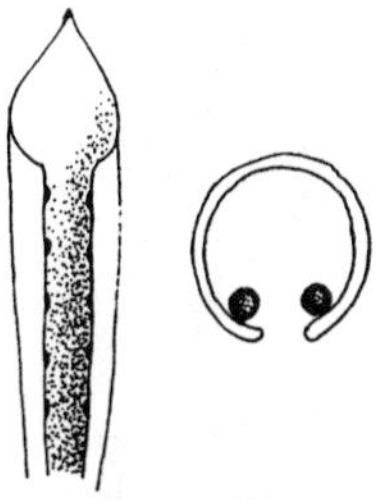

Figure 1661

Carpopodium. A stipe supporting an ovary. Figure 1662.
Cell. A hollow cavity or compartment within an ovary; a locule. Figure 1663.

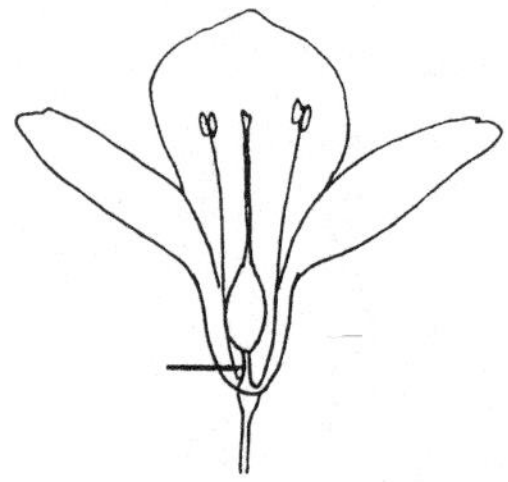

Figure 1662

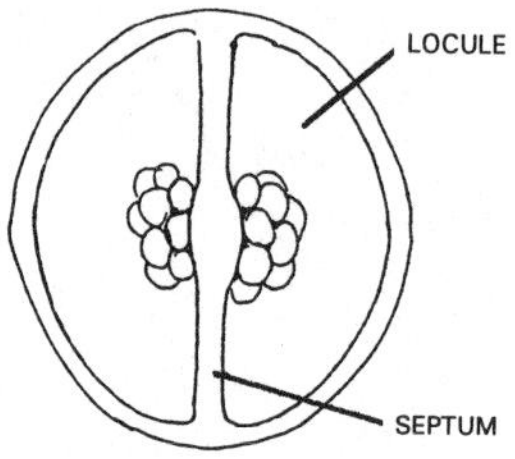

Figure 1663

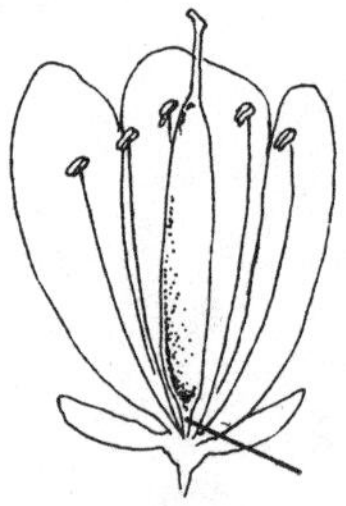

Figure 1667

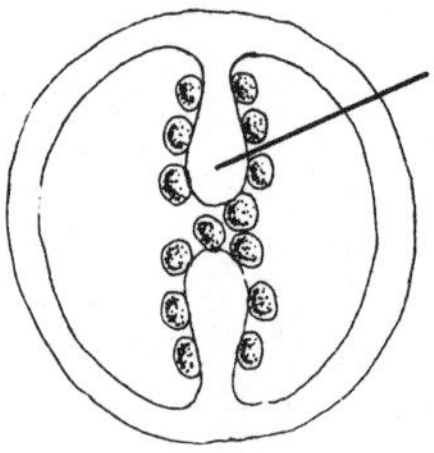

Figure 1668

Chalaza. The part of an ovule or seed where the integuments are connected to the nucellus, at the opposite end from the micropyle. Figure 1664.

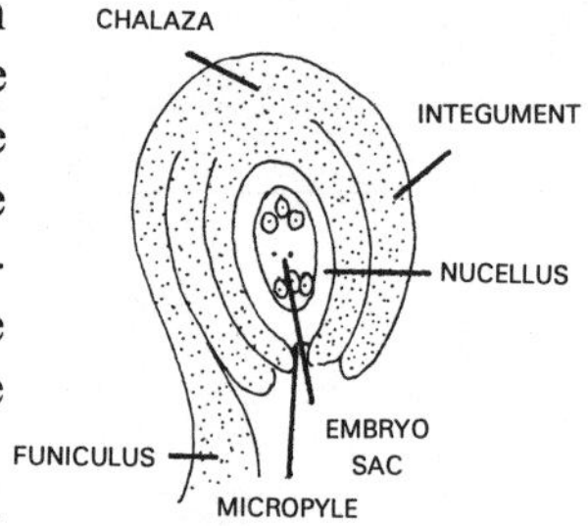

Figure 1664

Dissepiment. See **septum**.

Embryo sac. The megagametophyte within the ovule of a flowering plant. Figure 1664.

Funicle. See **funiculus**.

Funiculus (pl. **funiculi**). The stalk connecting the ovule to the placenta; the stalk of a seed. Figure 1664.

Gynandrium. A column bearing stamens and pistils. Figure 1665.

Gynobase. An elongation or enlargement of the receptacle, as in the flowers of the Boraginaceae. Figure 1666.

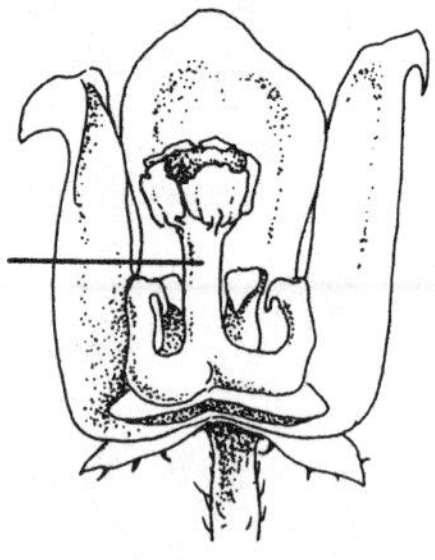

Figure 1665

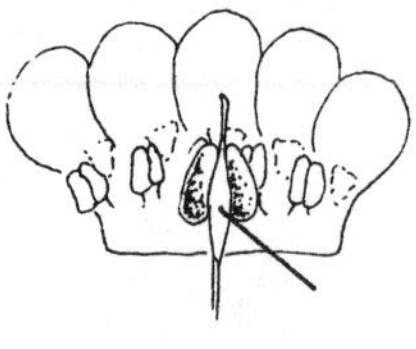

Figure 1666

Gynophore. An elongated stalk bearing the pistil in some flowers. Figure 1667.

Integument. The covering of the ovule which will become the seed coat. Figure 1664.

Intrusion. Protrusion into, as placentae into the cell of an ovary. Figure 1668.

Kernel. See **nucellus**.

Locule. The chamber or cavity of an ovary containing the ovules. Figure 1663.

Loculus (pl. **loculi**). See **locule**.

Micropyle. The opening in the integuments of the ovule. Figure 1664.

Nucellus. The part of the ovule just beneath the integuments and surrounding the female gametophyte. Figure 1664.

Ovary. The expanded basal portion of the pistil that contains the ovules. Figure 1669.

Ovule. An immature seed; the megasporangium and surrounding integuments of a seed plant. Figure 1664.

Pistil. The female reproductive organ of a flower, typically consisting of a stigma, style, and ovary. Figure 1669.

Placenta (pl. **placentae**). The portion of the ovary bearing ovules. Figure 1670.

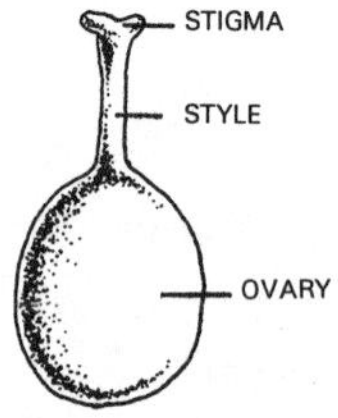

Figure 1669

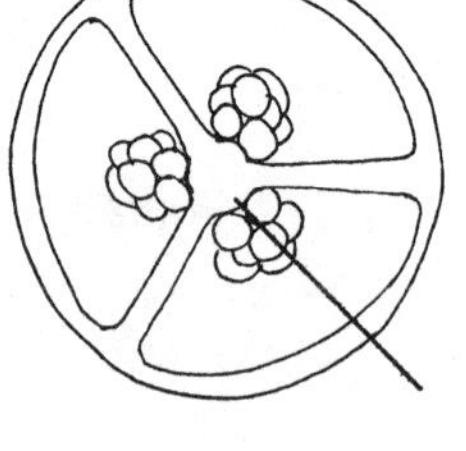

Figure 1670

Podogyne. See **carpopodium**.

Rostellum. A small beak; an extension from the upper edge of the stigma in orchids. Figure 1671.

Septum (pl. **septa**). A partition, as the partitions separating the locules of an ovary. Figure 1663.

Stigma. The portion of the pistil which is receptive to pollen. Figure 1669.

Stipe. A stalk supporting a structure, as the stalk attaching the ovary to the receptacle in some flowers. Figure 1672.

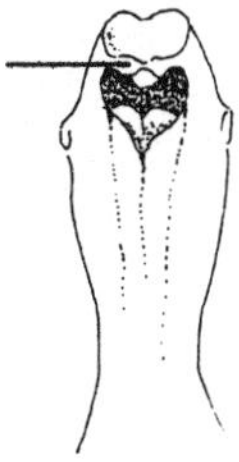

Figure 1671

Figure 1672

Style. The usually narrowed portion of the pistil connecting the stigma to the ovary. Figure 1669.

Stylopodium. A disk-like expansion or enlargement at the base of the style in the Umbelliferae (Apiaceae). Figure 1673.

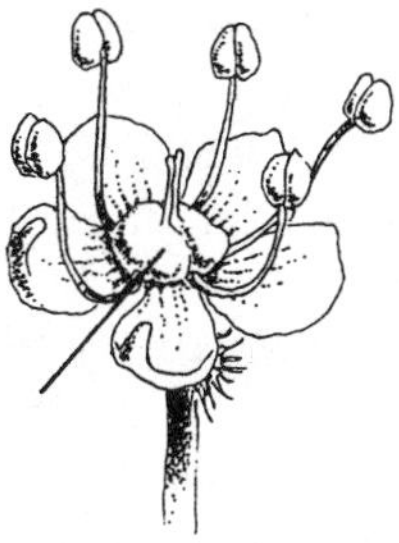

Figure 1673

Carpel Types

Astylocarpellous. Lacking a style and a stipe. Figure 1674.

Astylocarpepodic. Without a style, but with a stipe. Figure 1675.

Figure 1674

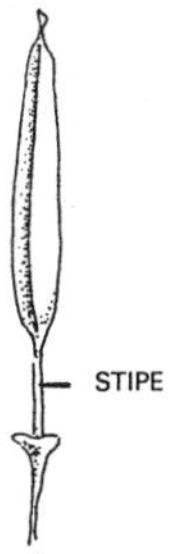

Figure 1675

Astylous. Without a style. Figure 1676.

Eccentric. Off-center; not positioned directly on the central axis. Figure 1677.

Stipitate. Borne on a stipe or stalk. Figure 1678.

Stylocarpellous. With a style, but without a stipe. Figure 1679.

Stylocarpepodic. With a style and a stipe. Figure 1678.

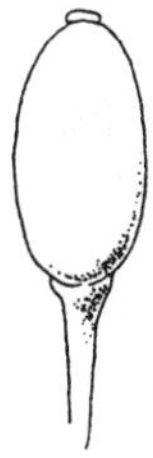

Figure 1676

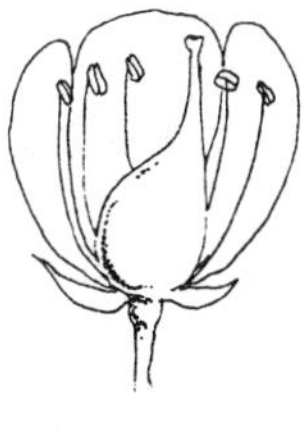

Figure 1677

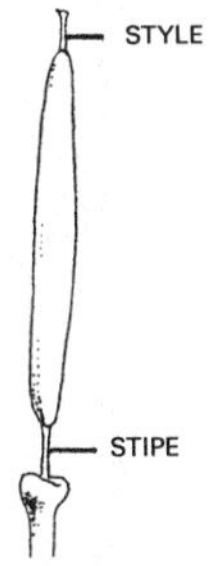

Figure 1678

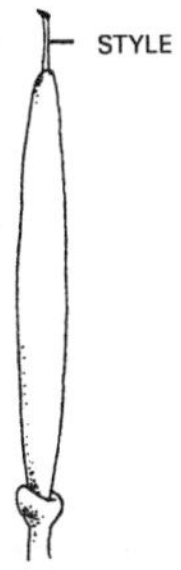

Figure 1679

Carpel Number

Acarpous. Without carpels. Figure 1680.

Bicarpellate. With two carpels. Figure 1681.

Figure 1680

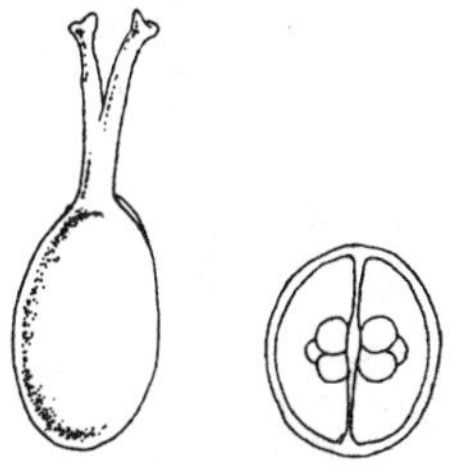

Figure 1681

Dicarpellate. See **bicarpellate**.

Digynous. With two pistils. Figure 1682.

Monocarpous. With one carpel. Figure 1683.

Monogynous. See **monocarpous**.

Octogynous. With eight pistils or styles.

Figure 1682

Figure 1684.

Figure 1683

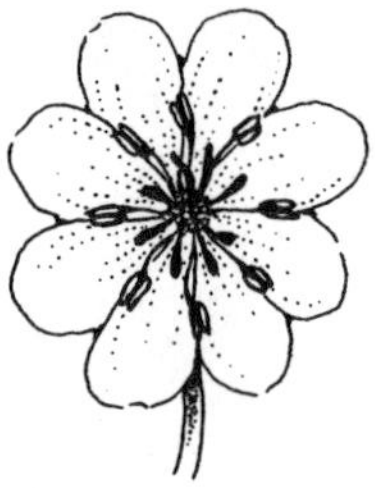

Figure 1684

Polycarpous. With many carpels. Figure 1685.

Polygynous. With many pistils or styles. Figure 1685.

Pseudomonomerous. A structure which appears to be simple, though actually derived from the fusion of separate structures, as a pistil which appears to be composed of a single carpel, though actually composed of two or more carpels.

Figure 1685

Stylodious. See **unicarpellous**.

Tricarpellary. With three carpels. Figure 1686.

Unicarpellous. With a single, free carpel. Figure 1687.

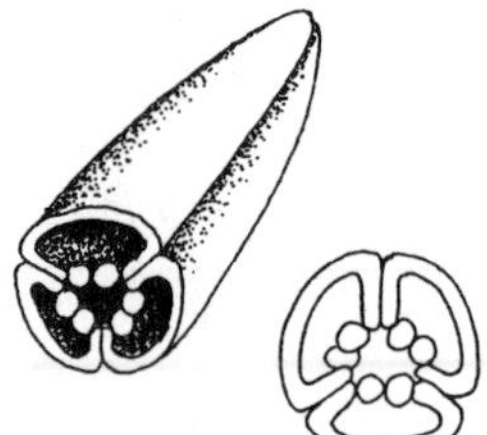

Figure 1686

Figure 1687

Carpel Fusion

Apocarpous. A flower with carpels forming separate pistils, as in a buttercup. Figure 1688. (compare **syncarpous**)

Compound ovary. An ovary of two or more carpels. Figure 1686.

Free. Not attached to other organs. Figure 1689.

Gynandrial. See **gynandrous**.

Figure 1688

Figure 1689

Gynandrous. With the stamens adnate to the pistil. Figure 1690.

Gynostemial. See **gynandrous**.

Semicarpous. With ovaries of carpels partly fused, the styles and stigmas separate. Figure 1691.

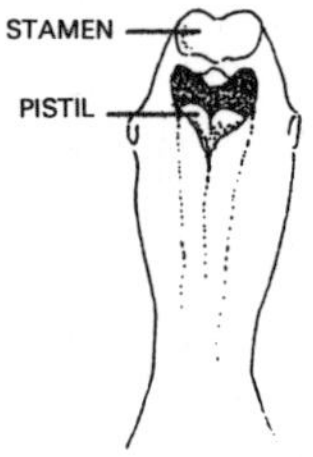

Figure 1690

Simple ovary. An ovary composed of only one carpel. Figure 1687.

Syncarpous. With united carpels. Figure 1692. (compare **apocarpous**)

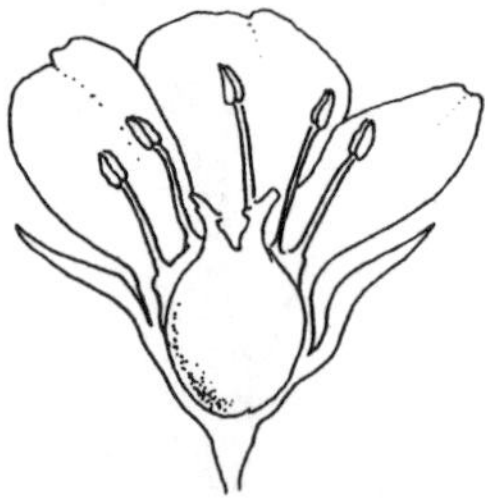

Figure 1691

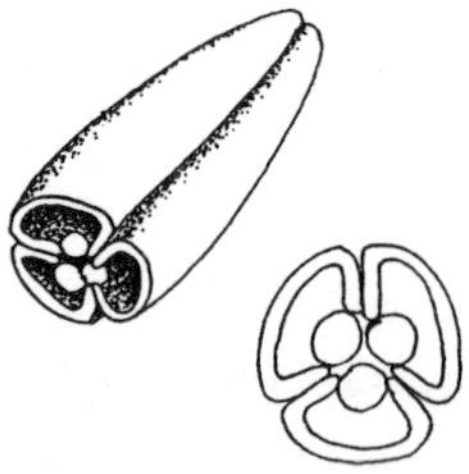

Figure 1692

Ovary Position

Half-inferior. Attached below the lower half, as a flower with a hypanthium that is fused to the lower half of the ovary, giving the appearance that the other floral whorls are arising from about the middle of the ovary. Figure 1693.

Inferior. Attached beneath, as an ovary that is attached beneath the point of attachment of the other floral whorls which appear, therefore, to arise from the top of the ovary. Figure 1694.

Superior. Attached above, as an ovary that is attached above the point of attachment of the other floral whorls. Figure 1695.

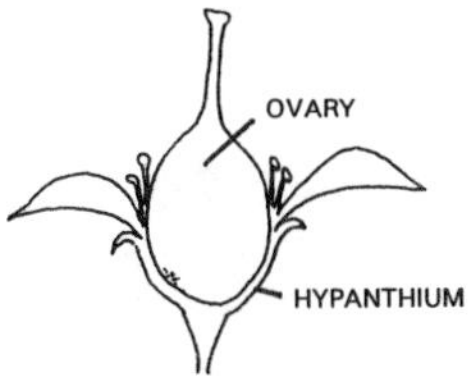

Figure 1693

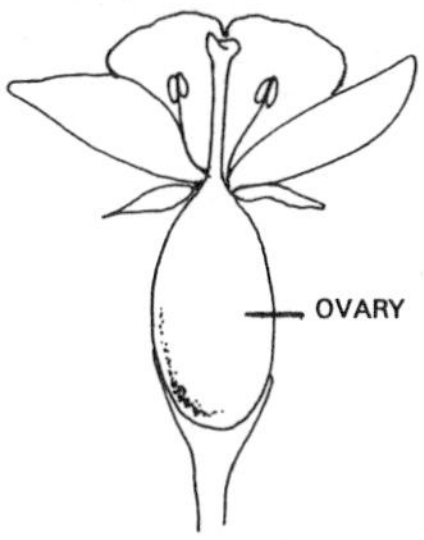

Figure 1694

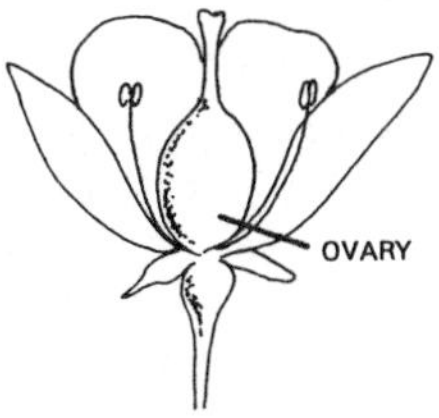

Figure 1695

Style Forms

Astylous. Without a style. Figure 1696.

Bifid. Deeply two-cleft or two-lobed, usually from the tip. Figure 1697.

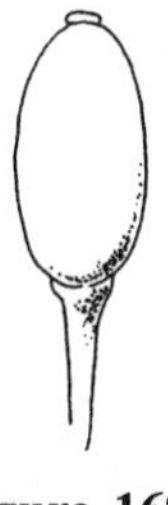

Figure 1696

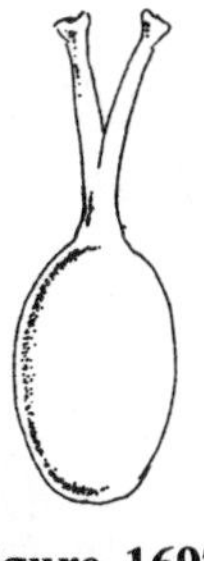

Figure 1697

Eccentric. Off-center; not positioned directly on the central axis. Figure 1698.

Gynobasic style. A style which is attached to the gynobase as well as to the carpels. Figure 1699.

Heterostylic. With styles of different lengths in flowers of the same species. Figures 1700 and 1701.

Heterostylous. See **heterostylic**.

Homostylic. With styles of more or less constant length in flowers of the same species.

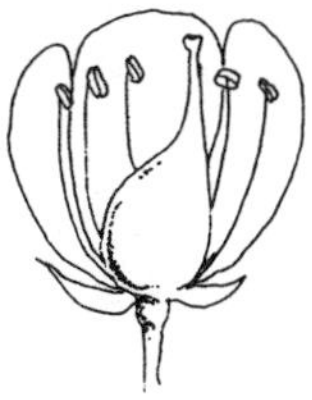

Figure 1698

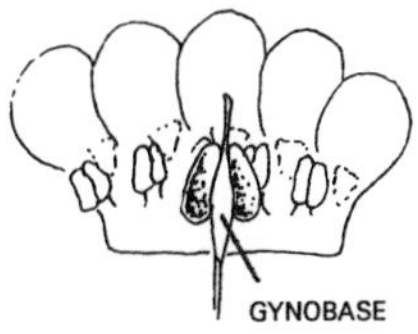

Figure 1699

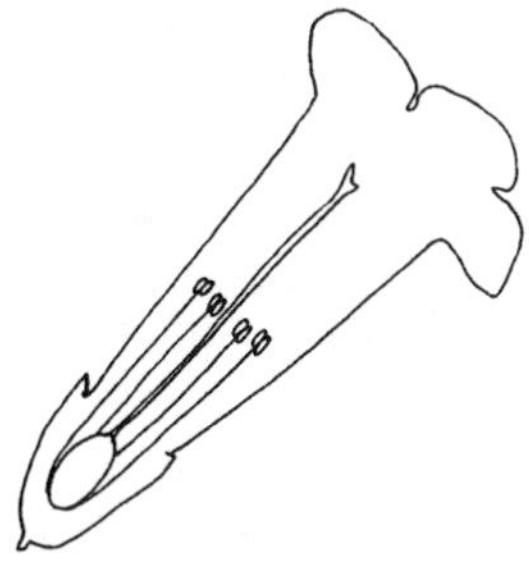

Figure 1700

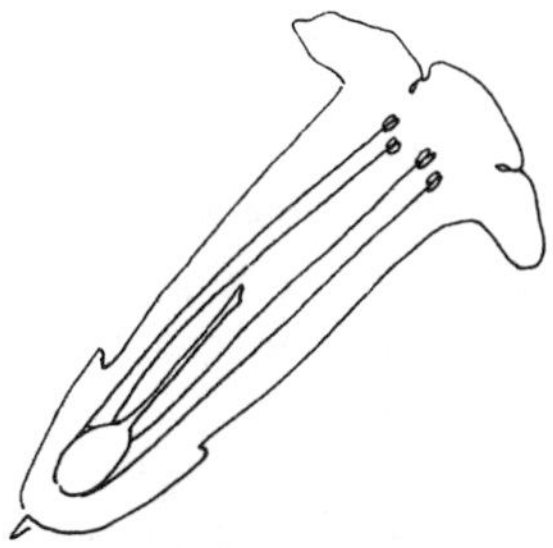

Figure 1701

Homostylous. See **homostylic**.

Macrostylous. With a long style. Figure 1700.

Monostylous. With a single style. Figure 1700.

Stylopodic. With a stylopodium. Figure 1702.

Tristylous. With three styles. Figure 1703.

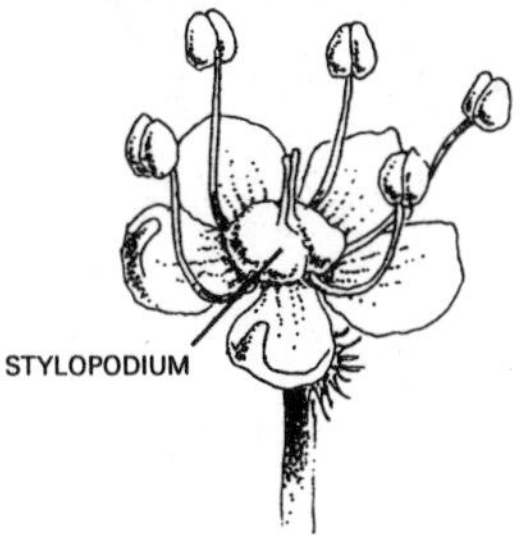

Figure 1702

Figure 1703

Placentation

Axile placentation. Ovules attached to the central axis of an ovary with two or more locules. Figure 1704.

Basal placentation. Ovules positioned at the base of a single-

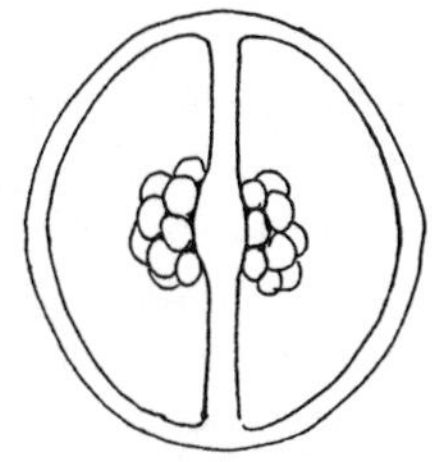

Figure 1704

loculed ovary. Figure 1705.

Free-central placentation. Ovules attached to a free-standing column in the center of a unilocular ovary. Figure 1706.

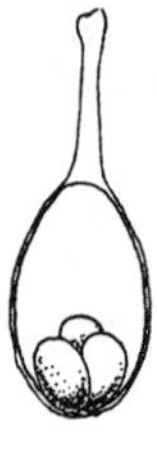

Figure 1705

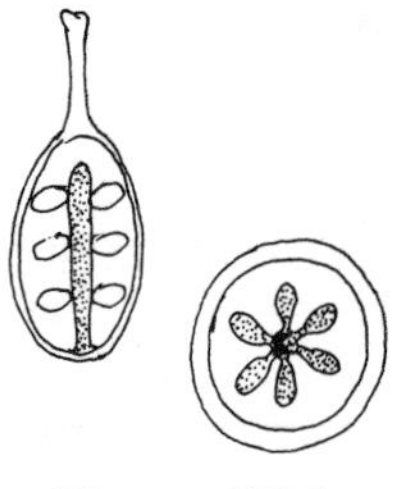

Figure 1706

Marginal placentation. Ovules attached to the juxtaposed margins of a simple pistil. Figure 1707.

Parietal placentation. Ovules attached to the walls of the ovary. Figure 1708.

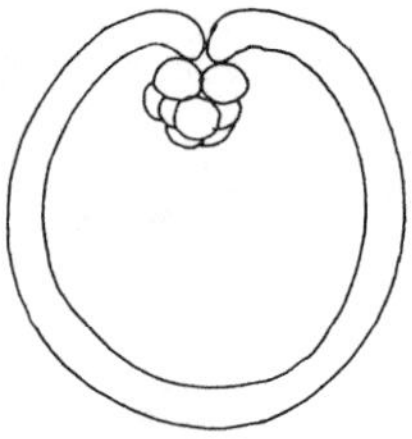

Figure 1707

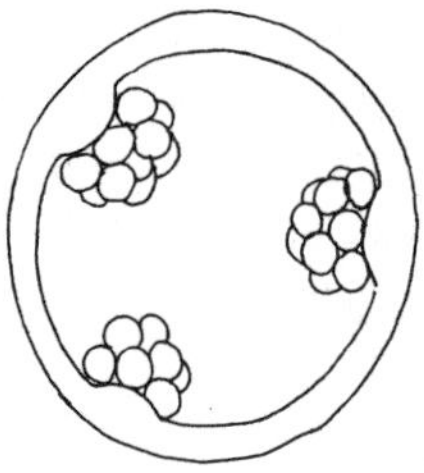

Figure 1708

Ovule Types

Amphitropous ovule. An ovule which is half-inverted and straight, with the hilum lateral. Figure 1709.

Anatropous ovule. An ovule which is inverted and straight with the micropyle situated next to the funiculus. Figure 1710.

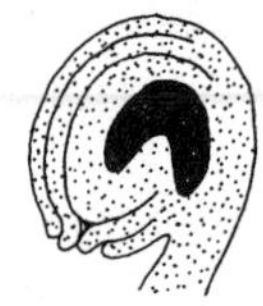

Figure 1709

Campylotropous ovule. An ovule which is curved so that the micropyle is positioned near the funiculus and the chalaza. Figure 1711.

Hemianatropous ovule. See **hemitropous ovule**.

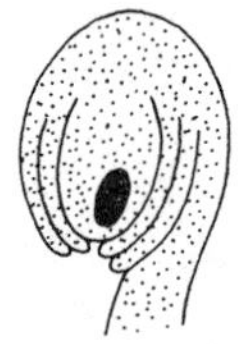

Figure 1710

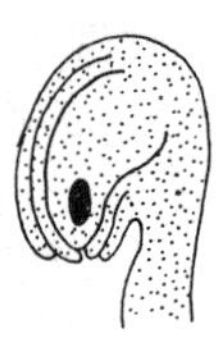

Figure 1711

Hemitropous ovule. An ovule which is half-inverted so that the funiculus is attached near the middle with the micropyle at a right angle. Figure 1712.

Orthotropous ovule. An ovule which is straight and erect. Figure 1713.

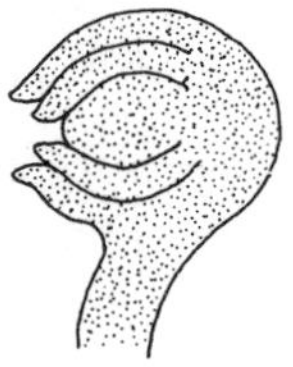

Figure 1712

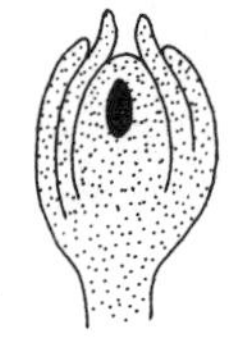

Figure 1713

FRUITS

A ripened ovary and any other structures which are attached and ripen with it.

FRUIT PARTS

Article. Section of a fruit separated from others by a constricted joint. Figure 1714.

Carpophore. A slender prolongation of the receptacle between the carpels as a central axis, as in the fruits of some members of the Umbelliferae (Apiaceae) and the Geraniaceae. Figure 1715.

Cell. A hollow cavity or compartment within an ovary containing

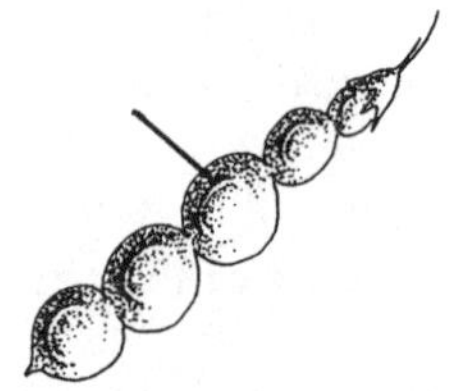

Figure 1714

ovules; a locule. Figure 1716.

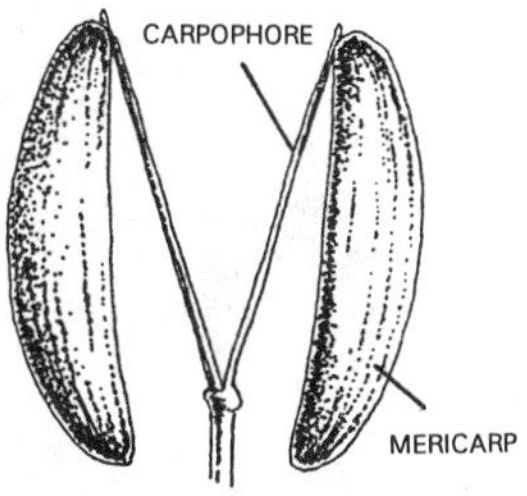

Figure 1715

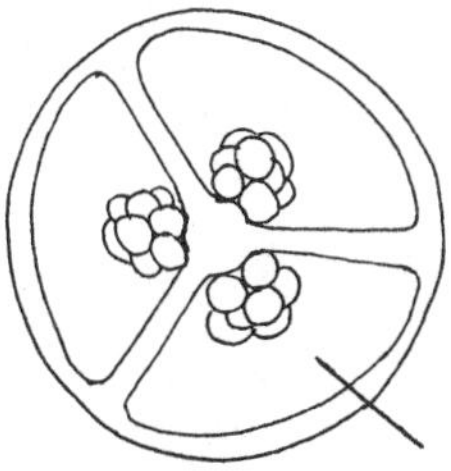

Figure 1716

Commissure. The face by which two carpels join one another, as in the Umbelliferae (Apiaceae). Figure 1717.

Cupule. A cup-shaped involucre, as in an acorn. Figure 1718.

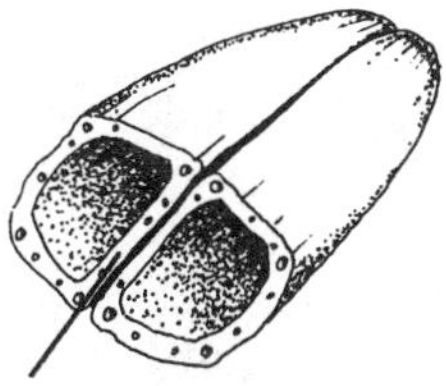

Figure 1717

Figure 1718

Dissepiment. Same as **septum**.

Endocarp. The inner layer of the pericarp of a fruit. Figure 1719. (compare **mesocarp** and **exocarp**)

Epicarp. Same as **exocarp**.

Exocarp. The outer layer of the pericarp of a fruit. Figure 1719. (compare **mesocarp** and **endocarp**)

Funiculus (pl. **funiculi**). The stalk of a seed. Figure 1720.

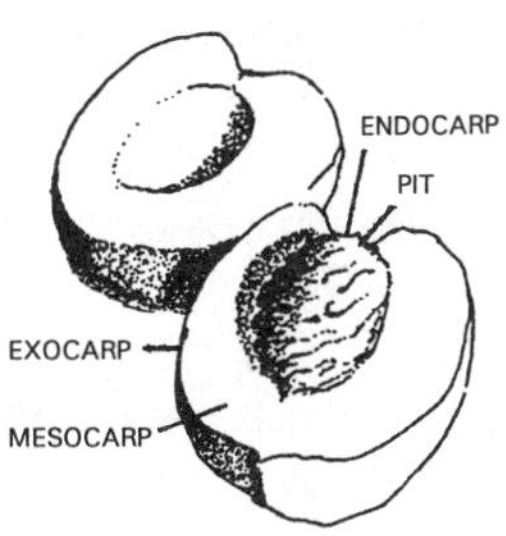

Figure 1719

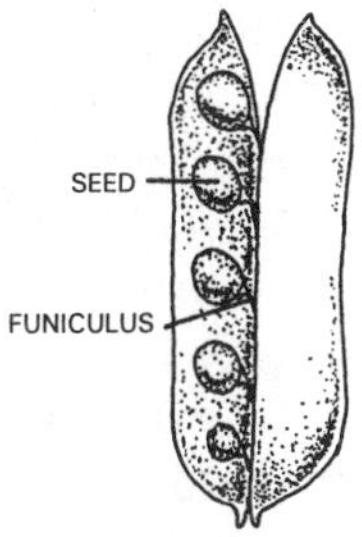

Figure 1720

Gynophore. See **stipe**.

Hypanthium. A cup-shaped extension of the floral axis usually formed from the union of the basal parts of the calyx, corolla, and androecium, commonly surrounding or enclosing the pistils. Figure 1721.

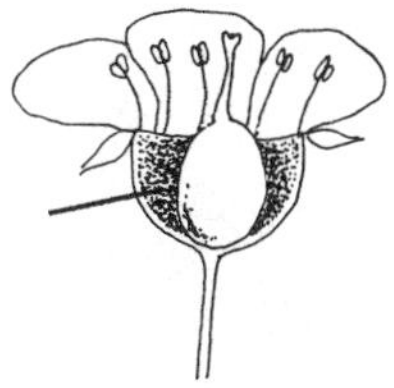

Figure 1721

Locule. The chamber or cavity of an ovary containing the seed. Figure 1716.

Loculus (pl. **loculi**). See **locule**.

Mericarp. A section of a schizocarp; one of the two halves of the fruit in the Umbelliferae (Apiaceae). Figure 1715.

Mesocarp. The middle layer of the pericarp of a fruit. Figure 1719. (compare **endocarp** and **exocarp**)

Operculum. A small lid, such as the deciduous cap of a circumscissile capsule. Figure 1722.

Ovary. The expanded basal portion of the pistil that contains the ovules; the immature fruit. Figure 1723.

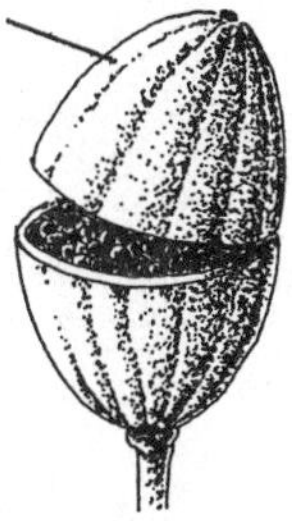

Figure 1722

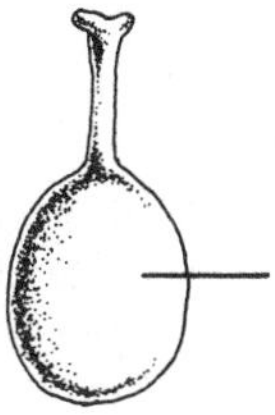

Figure 1723

Pericarp. The wall of the fruit. Figure 1724.

Pit. The stony endocarp of a drupe, as in a peach or cherry. Figure 1719.

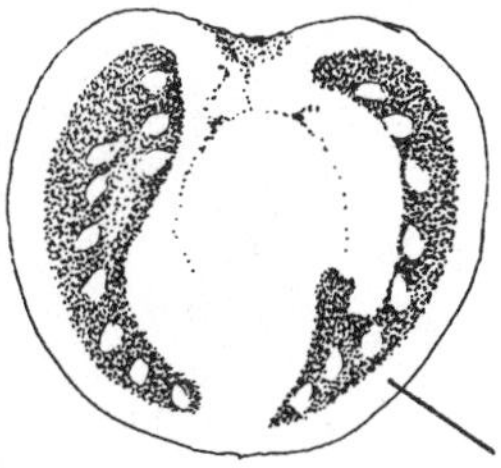

Figure 1724

Replum. Partition or septum between the two valves or compartments of silicles or siliques in the Cruciferae (Brassicaceae). Figure 1725.

Seed. A ripened ovule. Figure 1720.

Segment. A section of a fruit. Figure 1726.

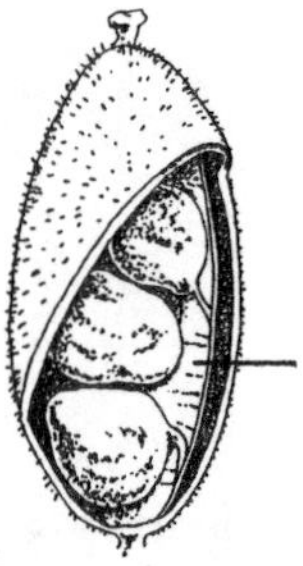

Figure 1725

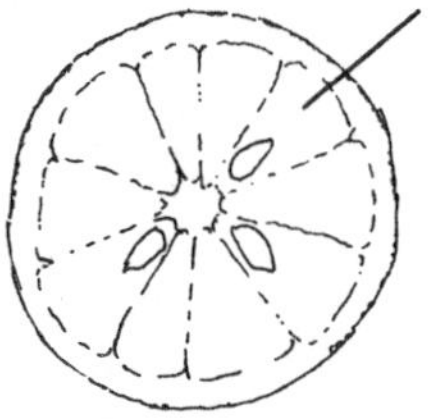

Figure 1726

Septum (pl. **septa**). A partition, as the partitions separating the locules of an ovary. Figure 1716.

Stipe. A stalk attaching the fruit to the receptacle. Figure 1727.

Stone. The hard, woody endocarp enclosing the seed of a drupe. Figure 1728.

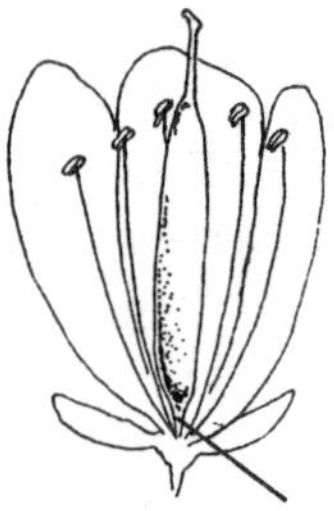

Figure 1727

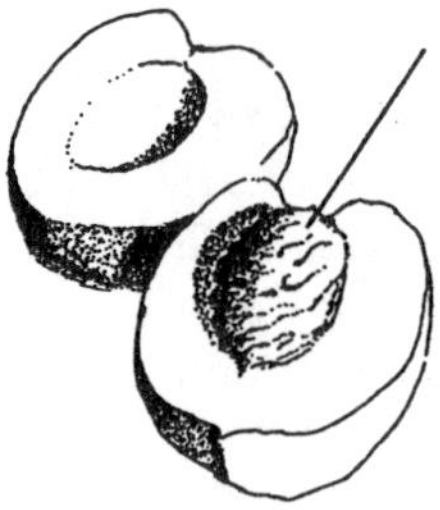

Figure 1728

Stylopodium. A disklike expansion or enlargement at the base of the style in the Umbelliferae (Apiaceae). Figure 1729.

Suture. A line of fusion; the line of dehiscence of a fruit. Figure 1730.

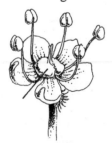

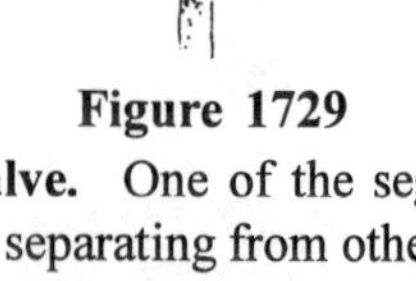

Figure 1729

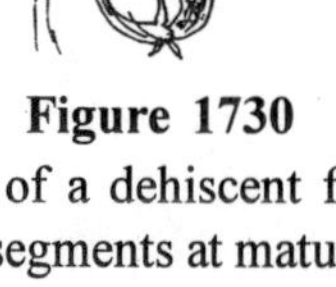

Figure 1730

Valve. One of the segments of a dehiscent fruit, separating from other such segments at maturity. Figure 1731.

Vitta (pl. **vitae**). An oil tube in the carpel walls of the fruits of the Umbelliferae (Apiaceae). Figure 1732.

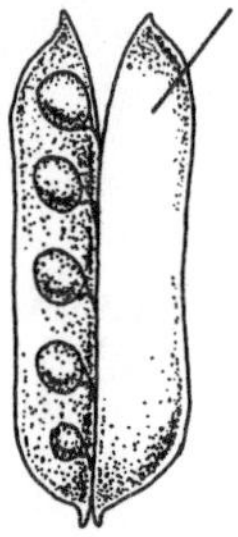

Figure 1731

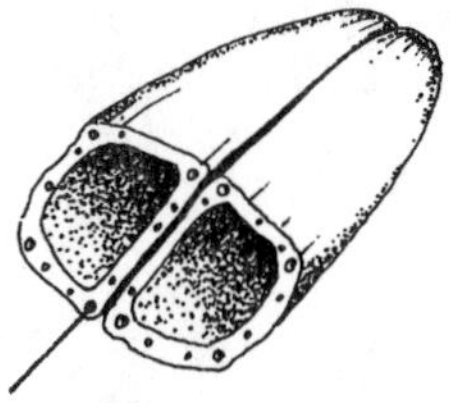

Figure 1732

FRUIT TYPES (Figure 1733.)

Accessory fruit. A fleshy fruit developing from a succulent receptacle rather than the pistil. The ripened ovaries are small achenes on the surface of the receptacle, as in the strawberry.

Achene. A small, dry, indehiscent fruit with a single locule and a single seed, and with the seed attached to the ovary wall at a single point, as in the sunflower.

Acorn. The hard, dry, indehiscent fruit of oaks, with a single, large seed and a cuplike base.

Aggregate fruit. Usually applied to a cluster or group of small fleshy fruits originating from a number of separate pistils in a single flower, as in the clustered drupelets of the raspberry.

Akene. See **achene**.

Anthocarp. A fruit with some portion of the flower besides the pericarp persisting.

Aril. A fleshy thickening of the seed coat which resembles a true fruit, as in *Taxus*.

Berry. A fleshy fruit developing from a single pistil, with several or many seeds, as the tomato. Sometimes applied to any fruit which is fleshy or pulpy throughout, i.e. lacking a pit or core.

Bur. A fruit armed with often hooked or barbed spines or appendages.

Capsule. A dry, dehiscent fruit composed of more than one carpel.

Cariopsis. See **caryopsis**.

Caryopsis. A dry, one-seeded, indehiscent fruit with the seed coat fused to the pericarp, as in the fruits of the grass family; a grain.

Circumscissile capsule. A capsule dehiscing along a transverse circular line, so that the top

Key to Common Fruit Types

1 Fruit formed from more than one flower.
  2 Fruit consisting primarily of receptacle tissue, the ripened ovaries borne inside of the hollow, inverted receptacle. . . . . **Syconium**
  2 Fruit consisting of many tightly clustered ripened ovaries. . . . . **Multiple**
1 Fruit formed from a single flower.
  3 Fruit of more than one ovary.
    4 Carpels enclosed, borne on the wall of a globose hypanthium. . . . . **Hip**
    4 Carpels not enclosed, not borne on the wall of a hypanthium.
      5 Pistils developing into fleshy drupelets on a non-fleshy receptacle. . . . . **Aggregate**
      5 Pistils developing into achenes on a fleshy receptacle. . . . . **Accessory**
  3 Fruit of a single ovary.
    6 Fruit dry at maturity.
      7 Fruit dehiscent at maturity.
        8 Fruit composed of more than one carpel.
          9 Carpels two, separated by a persistent, translucent septum.
            10 Fruit less than twice longer than wide. . . . . **Silicle**
            10 Fruit more than twice longer than wide. . . . . **Silique**
          9 Carpels two or more, not separated by a persistent, translucent septum. **(Capsule)**
            11 Capsule opening along a transverse circular line, the top separating like a lid. . . . . **Circumscissile capsule**
            11 Capsule opening along longitudinal lines or by pores.
              12 Capsule opening by pores. . . . . **Poricidal capsule**
              12 Capsule opening along longitudinal lines.
                13 Capsule dehiscing through the locules. . . . . **Loculicidal capsule**
                13 Capsule dehiscing through the septae. . . . . **Septicidal capsule**
        8 Fruit composed of a single carpel.
          14 Fruit opening along a single line of dehiscence. . . . . **Follicle**
          14 Fruit opening along two lines of dehiscence.
            15 Fruit not obviously constricted between the seeds. . . . . **Legume**
            15 Fruit obviously constricted between the seeds. . . . . **Loment**
      7 Fruit indehiscent at maturity.
        16 Fruit splitting at maturity, but carpels not dehiscing to release seeds. . . . . **Schizocarp**
        16 Fruit not splitting at maturity.
          17 Fruit winged. . . . . **Samara**
          17 Fruit not winged.
            18 Seed inseparably fused to the ovary wall. . . . . **Caryopsis, Grain**
            18 Seed not inseparably fused to the ovary wall.
              19 Fruit wall bladdery-inflated. . . . . **Utricle**
              19 Fruit wall not bladdery-inflated.
                20 Fruit wall hard and tough.
                  21 Fruit very small. . . . . **Nutlet**
                  21 Fruit larger. . . . . **Nut**
                20 Fruit wall not particularly hard and tough. . . . . **Achene**

6 Fruit fleshy at maturity.
  22 Seed one.
    23 Fruit very small. . . . . **Drupelet**
    23 Fruit larger. . . . . **Drupe**
  22 Seeds more than one.
    24 Fruit surrounded by the fleshy receptacle. . . . . **Pome**
    24 Fruit not surrounded by the receptacle.
      25 Fruit with a tough rind. . . . . **Pepo**
      25 Fruit lacking a tough rind. . . . . **Berry**

separates like a lid.

Cremocarp. See **schizocarp**.

Drupe. A fleshy, indehiscent fruit with a stony endocarp surrounding a usually single seed, as in a peach or cherry.

Drupelet. A small drupe, as in the individual segments of a raspberry fruit.

Follicle. A dry, dehiscent fruit composed of a single carpel and opening along a single side, as a milkweed pod.

Grain. A seedlike structure, as on the fruit of some *Rumex* species; a caryopsis.

Hesperidium. A fleshy berrylike fruit with a tough rind, as a lemon or orange.

Hip. A berrylike structure composed of an enlarged hypanthium surrounding numerous achenes.

Legume. A dry, dehiscent fruit derived from a single carpel and usually opening along two lines of dehiscence, as a pea pod.

Loculicidal capsule. A capsule dehiscing through the locules of a fruit rather than through the septa. (compare **septacidal** and **poricidal**)

Loment. A legume which is constricted between the seeds.

Lomentum (pl. **lomenta**). See **loment**.

Multiple fruit. A fruit formed from several separate flowers crowded on a single axis, as a mulberry or pineapple.

Nut. A hard, dry, indehiscent fruit, usually with a single seed.

Nutlet. A small nut; one of the lobes or sections of the mature ovary of some members of the Boraginaceae, Verbenaceae, and Labiatae (Lamiaceae).

Pepo. A fleshy, indehiscent, many-seeded fruit with a tough rind, as a melon or a cucumber.

Pod. Any dry, dehiscent fruit, especially a legume or follicle.

Pome. A fleshy, indehiscent fruit derived from an inferior, compound ovary, consisting of a modified floral tube surrounding a core, as in an apple.

Poricidal capsule. A capsule opening by pores, as in a poppy.

Pseudocarp. A fruit which develops from the receptacle rather than the ovary, as in a pome.

Pyxidium. See **pyxis**.

Pyxis. A circumscissile capsule, the top coming off as a lid.

Samara. A dry, indehiscent, winged fruit.

Schizocarp. A dry, indehiscent fruit which splits into separate one-seeded segments (carpels) at maturity.

Septicidal capsule. A capsule dehiscing through the septa and between the locules. (compare **loculicidal** and **poricidal**)

Silicle. A dry, dehiscent fruit of the Cruciferae (Brassicaceae), typically less than twice as long as wide, with two valves separating from the persistent placentae and septum (replum).

Silique. A dry, dehiscent fruit of the Cruciferae (Brassicaceae), typically more than twice as long as wide, with two valves separating from the persistent placentae and septum (replum).

Syconium. The fruit of a fig, consisting of an entire ripened inflorescence with a hollow, inverted receptacle bearing flowers internally.

Syncarp. A multiple fruit.

Utricle. A small, thin-walled, one-seeded, more or less bladdery-inflated fruit.

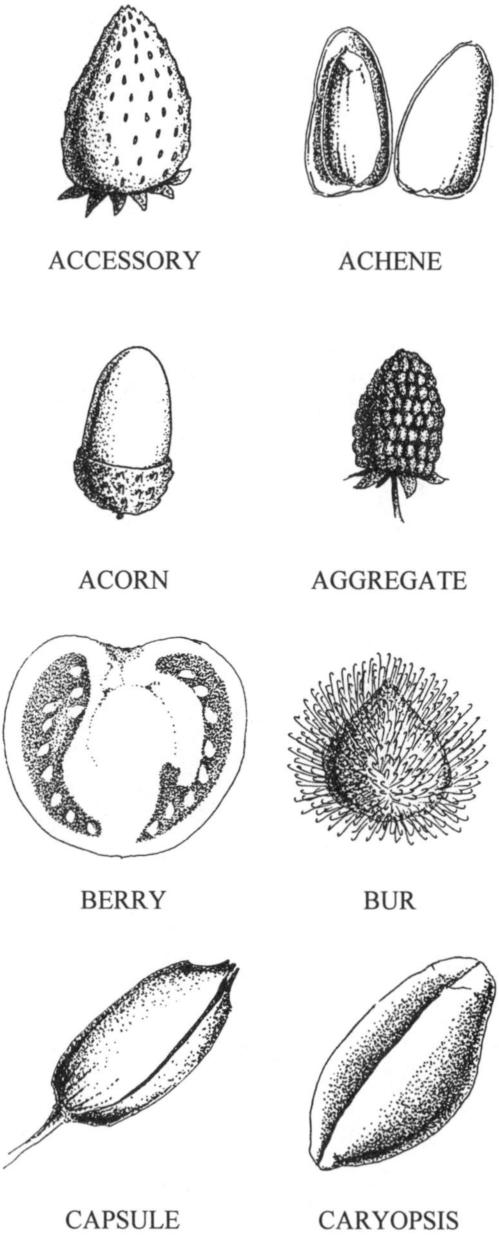

Figure 1733a.

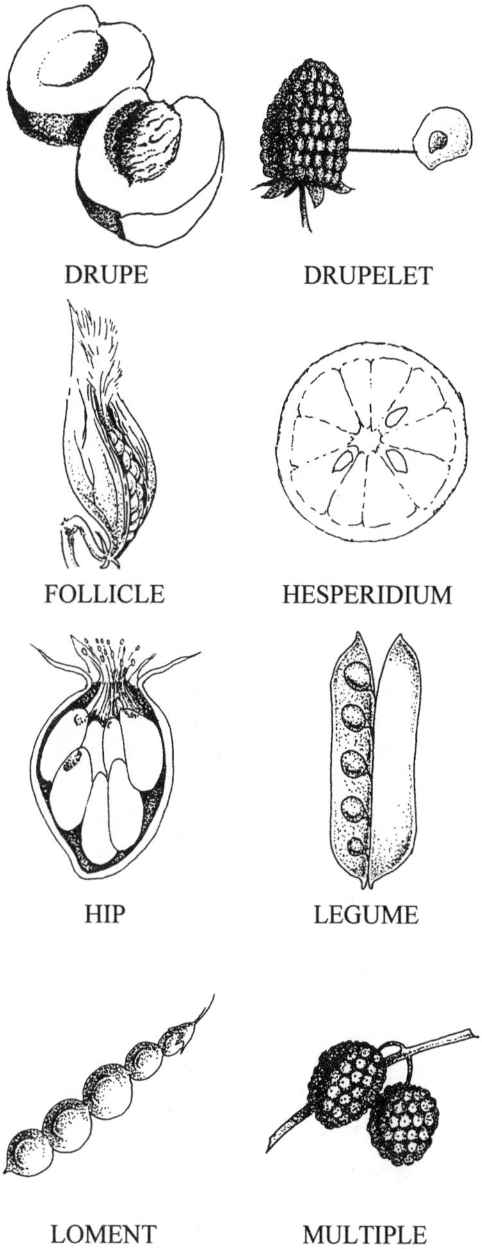

Figure 1733b.

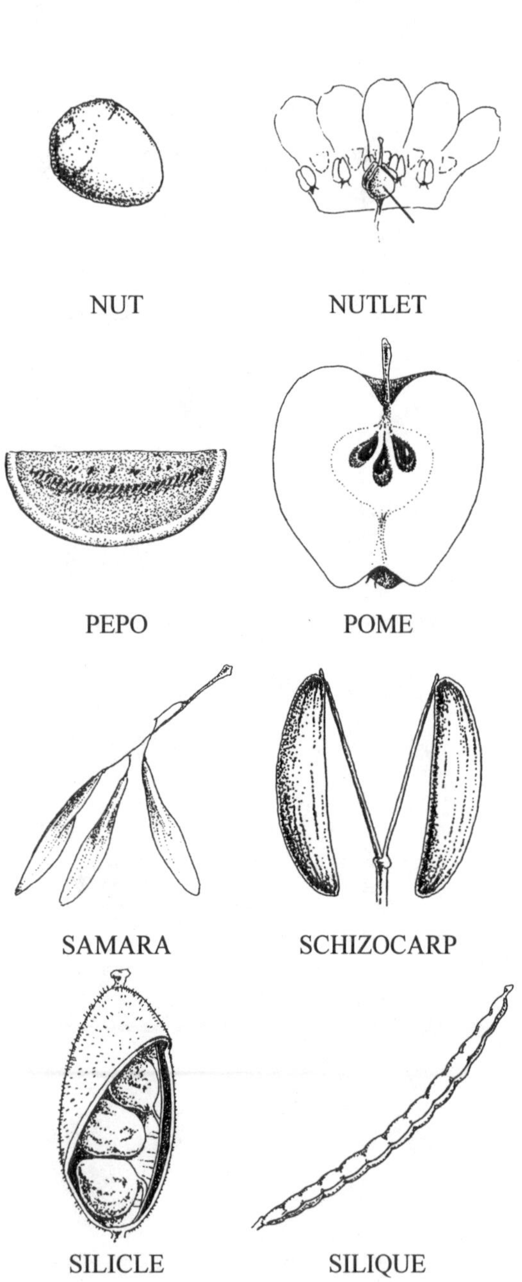

Figure 1733c.

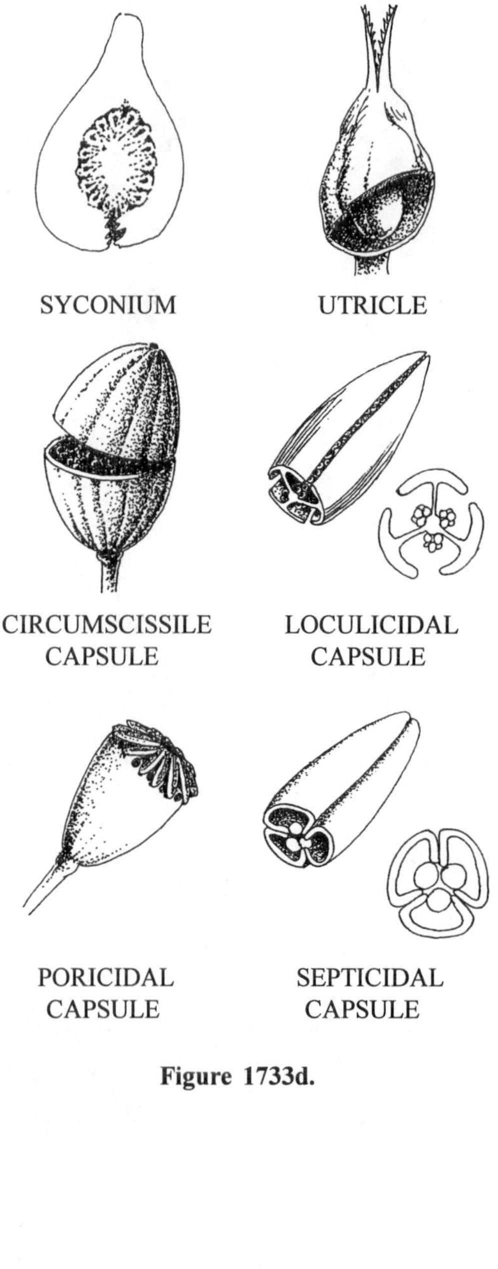

Figure 1733d.

PLANT IDENTIFICATION TERMINOLOGY

An Illustrated Glossary

Spring Lake Publishing
P.O. Box 266
Payson UT 84651